Practical Business Analytics Using R and Python

Solve Business Problems Using a Data-driven Approach

Second Edition

Umesh R. Hodeghatta, Ph.D
Umesha Nayak

Apress®

Practical Business Analytics Using R and Python: Solve Business Problems Using a Data-driven Approach

Umesh R. Hodeghatta, Ph.D
South Portland, ME, USA

Umesha Nayak
Bangalore, Karnataka, India

ISBN-13 (pbk): 978-1-4842-8753-8
ISBN-13 (electronic): 978-1-4842-8754-5
https://doi.org/10.1007/978-1-4842-8754-5

Managing Director, Apress Media LLC: Welmoed Spahr
Acquisitions Editor: Celestin Suresh John
Development Editor: Laura Berendson
Coordinating Editor: Mark Powers
Copy Editor: Kim Wimpsett

Cover designed by eStudioCalamar

Cover image by Maxim Berg on Unsplash (www.unsplash.com)

Distributed to the book trade worldwide by Apress Media, LLC, 1 New York Plaza, New York, NY 10004, U.S.A. Phone 1-800-SPRINGER, fax (201) 348-4505, e-mail orders-ny@springer-sbm.com, or visit www.springeronline.com. Apress Media, LLC is a California LLC and the sole member (owner) is Springer Science + Business Media Finance Inc (SSBM Finance Inc). SSBM Finance Inc is a **Delaware** corporation.

For information on translations, please e-mail booktranslations@springernature.com; for reprint, paperback, or audio rights, please e-mail bookpermissions@springernature.com.

Apress titles may be purchased in bulk for academic, corporate, or promotional use. eBook versions and licenses are also available for most titles. For more information, reference our Print and eBook Bulk Sales web page at www.apress.com/bulk-sales.

Any source code or other supplementary material referenced by the author in this book is available to readers on GitHub (https://github.com/Apress). For more detailed information, please visit www.apress.com/source-code.

Printed on acid-free paper

Table of Contents

About the Authors

Dr. Umesh R Hodeghatta is an engineer, scientist, and an educator. He is currently a faculty member at Northeastern University, specializing in data analytics, AI, machine learning, deep learning, natural language processing (NLP), and cybersecurity. He has more than 25 years of work experience in technical and senior management positions at AT&T Bell Laboratories, Cisco Systems, McAfee, and Wipro. He was also a faculty member at Kent State University in Kent, Ohio, and Xavier Institute of Management in Bhubaneswar, India. He earned a master's degree in electrical and computer engineering (ECE) from Oklahoma State University and a doctorate degree from the Indian Institute of Technology (IIT). His research interest is applying AI/machine learning to strengthen an organization's information security based on his expertise in information security and machine learning. As a chief data scientist, he is helping business leaders to make informed decisions and recommendations linked to the organization's strategy and financial goals, reflecting an awareness of external dynamics based on a data-driven approach.

He has published many journal articles in international journals and conference proceedings. In addition, he has authored books titled *Business Analytics Using R: A Practical Approach* and *The InfoSec Handbook: An Introduction to Information Security*, published by Springer Apress. Furthermore, Dr. Hodeghatta has contributed his services to many professional organizations and regulatory bodies. He was an executive committee member of the IEEE Computer Society (India); academic advisory member for the Information and Security Audit Association (ISACA); IT advisor for the government of India; technical advisory member of the International Neural Network Society (INNS) India; and advisory member of the Task Force on Business Intelligence & Knowledge Management. He was listed in "Who's Who in the World" for the years 2012, 2013, 2014, 2015, and 2016. He is also a senior member of the IEEE (USA).

Umesha Nayak is a director and principal consultant of MUSA Software Engineering Pvt. Ltd., which focuses on systems/process/management consulting. He is also the chief executive officer of N-U Sigma U-Square Analytics Lab, which specializes in high-end consulting on artificial intelligence and machine learning. He has 41 years' experience, of which 19 years are in providing consulting to IT/manufacturing and other organizations from across the globe. He has a master's degree in software systems and a master's degree in economics; he is certified as a CAIIB, Certified Information Systems Auditor (CISA), and Certified Risk and Information Systems Control (CRISC) professional from ISACA, PGDFM, certified lead auditor for many of the ISO standards, and certified coach, among others. He has worked extensively in banking, software development, product design and development, project management, program management, information technology audits, information application audits, quality assurance, coaching, product reliability, human resource management and culture development, and management consultancy, including consultancy in artificial intelligence and machine learning. He was a vice president and corporate executive council member at Polaris Software Lab, Chennai, prior to his current assignment. He has also held various roles such as head of quality, head of SEPG, and head of Strategic Practice Unit – Risks & Treasury at Polaris Software Lab. He started his journey with computers in 1981 with ICL mainframes and continued with minis and PCs. He was one of the founding members of information systems auditing in the banking industry in India. He has effectively guided many organizations through successful ISO 9001/ISO 27001/CMMI and other certifications and process/product improvements and solved problems through artificial intelligence and machine learning. He coauthored the book *The InfoSec Handbook: An Introduction to Information Security*, published by Apress.

Preface

Business analytics, data science, artificial intelligence (AI), and *machine learning* (ML) are hot words right now in the business community. Artificial intelligence and machine learning systems are enabling organizations to make informed decisions by optimizing processes, understanding customer behavior, maximizing customer satisfaction, and thus accelerating overall top-line growth. AI and machine learning help organizations by performing tasks efficiently and consistently, thus improving overall customer satisfaction level.

In financial services, AI models are designed to help manage customers' loans, retirement plans, investment strategies, and other financial decisions. In the automotive industry, AI models can help in vehicle design, sales and marketing decisions, customer safety features based on driving patterns of the customer, recommended vehicle type for the customer, etc. This has helped automotive companies to predict future manufacturing resources needed to build, for example, electric and driverless vehicles. AI models also help them in making better advertisement decisions.

AI can play a big role in customer relationship management (CRM), too. Machine learning models can predict consumer behavior, start a virtual agent conversation, and forecast trend analysis that can improve efficiency and response time.

Recommendation systems (AI systems) can learn users' content preferences and can select customers' choice of music, book, game, or any items the customer is planning to buy online. Recommendation systems can reduce return rates and help create better targeted content management.

Sentiment analysis using machine learning techniques can predict the opinions and feelings of users of content. This helps companies to improve their products and services by analyzing the customers' reviews and feedback.

These are a few sample business applications, but this can be extended to any business problem provided you have data; for example, an AI system can be developed for HR functions, manufacturing, process engineering, IT infrastructure and security, software development life cycle, and more.

There are several industries that have begun to adopt AI into their business decision process. Investment in analytics, machine learning, and artificial intelligence is predicted to triple in 2023, and by 2025, it is predicted to become a $47 billion market (per International Data Corp.). According to a recent research survey in the United States, nearly 34 percent of businesses are currently implementing or plan to implement AI solutions in their business decisions.

Machine learning refers to the learning algorithm of AI systems to make decisions based on past data (historical data). Some of the commonly used machine learning methods include neural networks, decision trees, k-nearest neighbors, logistic regression, cluster analysis, association rules, deep neural networks, hidden Markov models, and natural language processing. Availability and abundance of data, lower storage and processing costs, and efficient algorithms have made machine learning and AI a reality in many organizations.

AI will be the biggest disruptor to the industry in the next five years. This will no doubt have a significant impact on the workforce. Though many say AI can replace a significant number of jobs, it can actually enhance productivity and improve the efficiency of workers. AI systems can help executives make better business decisions and allow businesses to work on resources and investments to beat the competition. When decision-makers and business executives make decisions based on reliable data and recommendations arrived at through AI systems, they can make better choices for their business, investments, and employees thus enabling their business to stand out from competition.

There are currently thousands of jobs posted on job portals in machine learning, data science, and AI, and it is one of the fastest-growing technology areas, according to the Kiplinger report of 2017. Many of these jobs are going unfilled because of a shortage of qualified engineers. Apple, IBM, Google, Facebook, Microsoft, Walmart, and Amazon are some of the top companies hiring data scientists in addition to other companies such as Uber, Flipkart, Citibank, Fidelity Investments, GE, and many others including manufacturing, healthcare, agriculture, and transportation companies. Many open job positions are in San Jose, Boston, New York, London, Hong Kong, and many other cities. If you have the right skills, then you can be a data scientist in one of these companies tomorrow!

A data scientist/machine learning engineer may acquire the following skills:

- Communication skills to understand and interpret business requirements and present the final outcome
- Statistics, machine learning, and data mining skills
- SQL, NoSQL, and other database knowledge

- Knowledge of accessing XML data, connecting to databases, writing SQL queries, reading JSON, reading unstructured web data, accessing big data files such as HDFS, NoSQL MongoDB, Cassandra, Redis, Riak, CouchDB, and Neo4j
- Coding skills: Python, R or Java, C++
- Tools: Microsoft Azure, IBM Watson, SAS

This book aims to cover the skills required to become a data scientist. This book enables you to gain sufficient knowledge and skills to process data and to develop machine learning models. We have made an attempt to cover the most commonly used learning algorithms and developing models by using open-source tools such as R and Python.

Practical Business Analytics Using R and Python is organized into five parts. The first part covers the fundamental principles required to perform analytics. It starts by defining the sometimes confusing terminologies that exist in analytics, job skills, tools, and technologies required for an analytical engineer, before describing the process necessary to execute AI and analytics projects. The second and subsequent chapters cover the basics of math, probability theory, and statistics required for analytics, before delving into SQL, the business analytics process, exploring data using graphical methods, and an in-depth discussion of how to evaluate analytics model performance.

In Part II, we introduce supervised machine learning models. We start with regression analysis and then introduce different classification algorithms, including naïve Bayes, decision trees, logistic regression, and neural networks.

Part III discusses time-series models. We cover the most commonly used models including ARIMA.

Part IV covers unsupervised learning and text mining. In unsupervised learning, we discuss clustering analysis and association mining. We end the section by briefly introducing big data analytics.

In the final part, we discuss the open-source tools, R and Python, and using them in programming for analytics. The focus here is on developing sufficient programing skills to perform analytics.

Source Code

All the source code used in this book can be downloaded from `https://github.com/apress/practical-business-analytics-r-python`.

Foreword

We live in an era where mainstream media, business literature, and boardrooms are awash in breathless hype about the *data economy, AI revolution, Industry/Business 4.0,* and similar terms that are used to describe a meaningful social and business inflection point. Indeed, today's discourse describes what seems to be an almost overnight act of creation that has generated a new data-centric paradigm of societal and business order that, according to the current narrative, is without precedent and requires the creation of an entirely new set of practitioners and best practices out of thin air. In this brave new world, goes the narrative, everyone has already been reinvented as a "data-something" (data scientist, data analyst, data storyteller, data citizen, etc.) with the associated understanding of what that means for themselves and for business. These newly minted "data-somethings" are already overhauling current practice by operationalizing data-driven decision-making practices throughout their organizations. They just don't know it yet.

Look under the covers of most operating organizations, however, and a different picture appears. There is a yawning chasm of skillsets, common language, and run-rate activity between those operating in data-centric and data-adjacent roles and those in many operating functions. As such, in many organizations, data-centric activity is centered in only a few specialized areas or is performed only when made available as a feature in the context of commercial, off-the-shelf software (COTS). Similarly, while there is certainly more data volume and more variation in the said data and all of that varied data is arriving at a much higher velocity than in past, organizations often do not possess either the infrastructure sophistication or the proliferation of skillsets to validate whether the data is any good (veracity of data) and actually use it for anything of value. And with the exception of a few organizations for which data is their only or primary business, most companies use data and associated analytic activities as a means to accomplish their primary business objective more efficiently, not as an end in and of itself.

A similar discontinuity occurred a few decades ago, with the creation of the Internet. In the years that followed, entire job categories were invented (who needed a "webmaster" or "e-commerce developer" in 1990?), and yet it took almost a decade and a half before there was a meaningful common language and understanding between practitioners and operational business people at many organizations. The resulting proliferation of business literature tended to focus on "making business people more technical" so they could understand this strange new breed of practitioners who held the keys to the new world. And indeed, that helped to a degree. But the accompanying reality is that the practitioners also needed to be able to understand and communicate in the language of businesses and contextualize their work as an enabler of business rather than the point of business.

Academia and industry both struggled to negotiate the balance during the Internet age and once again struggle today in the nascent data economy. Academia, too often erred on the side of theoretical courses of study that teach technical skills (mathematics, statistics, computer science, data science) without contextualizing the applications to business, while on the other hand, industry rushes to apply techniques they lack the technical skill to properly understand to business problems. In both groups, technologists become overly wedded to a given platform, language, or technique at the expense of leveraging the right tool for the job. In all cases, the chasm of skills and common language among stakeholders often leads to either or both of incorrect conclusions or under-utilized analytics. Neither is a good outcome.

There is space both for practitioners to be trained in the practical application of techniques to business context and for business people to not only understand more about the "black box" of data-centric activity but be able to perform more of that activity in a self-service manner. Indeed, the democratization of access to both high-powered computer and low-code analytical software environments makes it possible for a broader array of people to become practitioners, which is part of what the hype is all about.

Enter this book, which provides readers with a stepwise walk-through of the mathematical underpinnings of business analytics (important to understand the proper use of various techniques) while placing those techniques in the context of specific, real-world business problems (important to understand the appropriate application of those techniques). The authors (both of whom have longstanding industry experience together, and one of whom is now bringing that experience to the classroom in a professionally oriented academic program) take an evenhanded approach to technology choices by ensuring that currently fashionable platforms such as R and Python are represented primarily as alternatives that can accomplish equivalent tasks, rather than endpoints in and of themselves. The stack-agnostic approach also helps readers prepare as to how they might incorporate the next generation of available technology, whatever that may be in the future.

As would-be practitioners in business, I urge you to read this book with the associated business context in mind. Just as with the dawn of the Internet, the true value of the data economy will only begin to be realized when *all* the "data-somethings" we work with act as appropriately contextualized practitioners who use data in the service of the business of their organizations.

Dan Koloski

Professor of the Practice and Head of Learning Programs

Roux Institute at Northeastern University

October 2022

Dan Koloski is a professor of the practice in the analytics program and director of professional studies at the Roux Institute at Northeastern University.

Professor Koloski joined Northeastern after spending more than 20 years in the IT and software industry, working in both technical and business management roles in companies large and small. This included application development, product management and partnerships, and helping lead a spin-out and sale from a venture-backed company to Oracle. Most recently, Professor Koloski was vice president of product management and business development at Oracle, where he was responsible for worldwide direct and channel go-to-market activities, partner integrations, product management, marketing/branding, and mergers and acquisitions for more than $2 billion in product and cloud-services business. Before Oracle, he was CTO and director of strategy of the web business unit at Empirix, a role that included product management, marketing, alliances, mergers and acquisitions, and analyst relations. He also worked as a freelance consultant and Allaire-certified instructor, developing and deploying database-driven web applications.

Professor Koloski earned a bachelor's degree from Yale University and earned his MBA from Harvard Business School in 2002.

PART I

Introduction to Analytics

CHAPTER 1

An Overview of Business Analytics

1.1 Introduction

Today's world is data-driven and knowledge-based. In the past, knowledge was gained mostly through observation now, knowledge is secured not only through observation but also by analyzing data that is available in abundance. In the 21st century, knowledge is acquired and applied by analyzing data available through various applications, social media sites, blogs, and much more. The advancement of computer systems complements knowledge of statistics, mathematics, algorithms, and programming. Enormous storage and exten computing capabilities have ensured that knowledge can be quickly derived from huge amounts of data and be used for many other purposes. The following examples demonstrate how seemingly obscure or unimportant data can be used to make better business decisions:

- A hotel in Switzerland welcomes you with your favorite drink and dish; you are so delighted!
- You are offered a stay at a significantly discounted rate at your favorite hotel on your birthday or marriage anniversary when traveling to your destination.
- Based on your daily activities and/or food habits, you are warned about the high probability of becoming a diabetic so you can take the right steps to avoid it.

U. R. Hodeghatta and U. Nayak, *Practical Business Analytics Using R and Python*,
https://doi.org/10.1007/978-1-4842-8754-5_1

- You enter a grocery store and find that your regular monthly purchases are already selected and set aside for you. The only decision you have to make is whether you require all of them or want to remove some from the list. How happy you are!

There are many such scenarios that are made possible by analyzing data about you and your activities that is collected through various means—including mobile phones, your Google searches, visits to various websites, your comments on social media sites, your activities using various computer applications, and more. The use of data analytics in these scenarios has focused on your individual perspective. Now, let's look at scenarios from a business perspective.

- As a hotel business owner, you are able to provide competitive yet profitable rates to your prospective customers. At the same time, you can ensure that your hotel is completely occupied all the time by providing additional benefits, including discounts on local travel and local sightseeing offers tied to other local vendors.
- As a taxi business owner, you are able to repeatedly attract the same customers based on their travel history and preferences of taxi type and driver.
- As a fast-food business owner you are able to offer discounted rates to attract customers on slow days. These discounts enable you to ensure full occupancy on those days also.
- You are in the human resources (HR) department of an organization and are bogged down by high attrition. But now you are able to understand the types of people you should focus on recruiting based on the characteristics of those who perform well and who are more loyal and committed to the organization.
- You are in the business of designing, manufacturing, and selling medical equipment used by hospitals. You are able to understand the possibility of equipment failure well before the equipment actually fails, by carrying out analysis of the errors or warnings captured in the equipment logs.

All these scenarios are possible by analyzing data that the businesses and others collect from various sources. There are many such possible scenarios. The application of data analytics to the field of business is called *business analytics*.

You have most likely observed the following scenarios:

- You've been searching, for the past few days, on Google for adventurous places to visit. You've also tried to find various travel packages that might be available. You suddenly find that when you are on Facebook, Twitter, or other websites, they show a specific advertisement of what you are looking for, usually at a discounted rate.
- You've been searching for a specific item to purchase on Amazon (or any other site). Suddenly, on other sites you visit, you find advertisements related to what you are looking for or find customized mail landing in your mailbox, offering discounts along with other items you might be interested in.
- You've also seen recommendations that Netflix, Amazon, Walmart.com, etc., make based on your searches, your wish list, or previous purchases or movies you have watched. Many times you've also likely observed these sites offering you discounts or promoting new products based on the huge amount of customer data these companies have collected.

All of these possibilities are now a reality because of data analytics specifically used by businesses.

1.2 Objectives of This Book

Many professionals are interested in learning analytics. But not all of them have rich statistical or mathematical backgrounds. This book is the right place for techies as well as those who are not so technical to get started with business analytics. You'll learn about data analytics processes, tools, and techniques, as well as different types of analytics, including predictive modeling and big data. This is an introductory book in the field of business analytics.

The following are some of the advantages of this book:

- It offers the right mix of theory and hands-on labs. The concepts are explained using business scenarios or case studies where required.
- It is written by industry professionals who are currently working in the field of analytics on real-life problems for paying customers.

This book provides the following:

- Practical insights into the use of data that has been collected, collated, purchased, or available for free from government sources or others. These insights are attained via computer programming, statistical and mathematical knowledge, and expertise in relevant fields that enable you to understand the data and arrive at predictive capabilities.
- Information on the effective use of various techniques related to business analytics.
- Explanations of how to effectively use the programming platform R or Python for business analytics.
- Practical cases and examples that enable you to apply what you learn from this book.
- The dos and don'ts of business analytics.
- The book does *not* do the following:
    - Deliberate on the definitions of various terms related to analytics, which can be confusing.
    - Elaborate on the fundamentals behind any statistical or mathematical technique or particular algorithm beyond certain limits.
    - Provide a repository of all the techniques or algorithms used in the field of business analytics (but does explore many of them).
    - Advanced concepts in the field of neural networks and deep learning.
    - Not a guide to programming R or Python.

1.3 Confusing Terminology

Many terms are used in discussions of this topic—for example, *data analytics, business analytics, big data analytics*, and *data science*. Most of these are, in a sense, the same. However, the purpose of the analytics, the extent of the data that's available for analysis, and the difficulty of the data analysis may vary from one to the other. Finally, regardless of the differences in terminology, we need to know how to use the data effectively for our businesses. These differences in terminology should not get in the way of applying techniques to the data (especially in analyzing it and using it for various purposes, including understanding it, deriving models from it, and then using these models for predictive purposes).

In layperson's terms, let's look at some of this terminology:

- *Data analytics* is the analysis of data, whether huge or small, in order to understand it and decide how to use the knowledge hidden within it. An example is the analysis of data related to various classes of travelers.
- *Business analytics* is the application of data analytics to business. An example of how business analytics might be applied is offering specific discounts to different classes of travelers based on the amount of business they offer or have the potential to offer.
- *Data science* is an interdisciplinary field (including disciplines such as statistics, mathematics, and computer programming) that derives knowledge from data and applies it for predictive or other purposes. Expertise about underlying processes, systems, and algorithms is used. An example is the application of t-values and p-values from statistics in identifying significant model parameters in a regression equation.
- *Big data analytics* is the analysis of huge amounts of data (for example, trillions of records) or the analysis of difficult-to-crack problems. Usually, this requires an enormous amount of storage and/or computing power, including massive amounts of memory to hold the data and a huge number of high-speed processors to crunch the data and get its essence. An example is the analysis of geospatial data captured by satellites to identify weather patterns and make related predictions.

- *Data mining and machine learning*

  Data mining refers to the process of mining data to find meaningful patterns hidden within the data. The term data mining brings to mind mining earth to find useful minerals. Just as important minerals are often buried deep within earth and rock, useful information is often hidden in data, neither easily visible nor understandable. Data mining uses techniques such as supervised machine learning and unsupervised machine learning to perform mining tasks. Since advanced analytics requires data mining, we will cover these in later chapters.

1.4 Drivers for Business Analytics

Though many data techniques, algorithms, and tools have been around for a very long time, recent developments in technology, computation, and algorithms have led to tremendous popularity of analytics. We believe the following are the growth drivers for business analytics in recent years:

- Increasing numbers of relevant computer packages and applications. One example is the Python and R programming environment with its various data sets, documentation on its packages, and open-source libraries with built-in algorithms.
- Integration of data from various sources and of various types, including both structured and unstructured data, such as data from flat files, data from relational databases, data from log files, data from Twitter messages, and more. An example is the consolidation of information from data files in a Microsoft SQL Server database with data from a Twitter message stream.
- Growth of seemingly infinite storage and computing capabilities by clustering multiple computers and extending these capabilities via the cloud. An example is the use of Apache Hadoop clusters to distribute and analyze huge amounts of data.

- Emergence of many new techniques and algorithms to effectively use statistical and mathematical concepts for business analysis. One example is the recent development in natural language processing (NLP) and neural networks.
- Business complexity arising from globalization. An economic or political situation in a particular country can affect the sales in that country. The need for business survival and growth in a highly competitive world requires each company and business to deep dive into data to understand customer behavior patterns and take advantage of them.
- Availability of many easy-to-use programming tools, platforms, and frameworks.

A note of caution here Not all business problems require complicated analytics solutions. Some may be easy to understand and to solve by using techniques such as the visual depiction of data.

Now let's discuss each of these drivers for business analytics in more detail.

1.4.1 Growth of Computer Packages and Applications

Computer packages and applications have completely flooded modern life. This is true at both an individual and business level. This is especially true with our extensive use of smartphones, which enable the following:

- Communication with others through email packages
- Activities in social media and blogs
- Business communications through email, instant messaging, and other tools
- Day-to-day searches for information and news through search engines
- Recording of individual and business financial transactions through accounting packages
- Recording of our travel details via online ticket-booking websites or apps

- Recording of our various purchases in e-commerce websites
- Recording our daily exercise routines, calories burned, and diets through various applications

We are surrounded by many computer packages and applications that collect a lot of data about us. This data is used by businesses to make them more competitive, attract more business, and retain and grow their customer base. With thousands of apps on platforms such as Android, iOS, and Windows, capturing data encompasses nearly all the activities carried out by individuals across the globe (who are the consumers for most of the products and services). This has been enabled further by the reach of hardware devices such as computers, laptops, mobile phones, and smartphones even to remote places.

1.4.2 Feasibility to Consolidate Data from Various Sources

Technology has grown by leaps and bounds over the last few years. The growth of technology coupled with almost unlimited storage capability has enabled us to consolidate related or relevant data from various sources—right from flat files to database data to data in various formats. This ability to consolidate data from various sources has provided a great deal of momentum to effective business analysis.

1.4.3 Growth of Infinite Storage and Computing Capability

The growth of server technologies, processing storage, and memories have changed the computation power today. This advancement in technology has made processing huge data by AI algorithms simpler and faster. Similarly, the memory and storage capacity of individual computers has increased drastically, whereas external storage devices have provided a significant increase in storage capacity. This has been augmented by cloud-based storage services that can provide a virtually unlimited amount of storage. The growth of cloud platforms has also contributed to virtually unlimited computing capability. Now you can rent the processing power of multiple CPUs and graphical processing units (GPUs) coupled with huge memory and large storage to carry out any analysis—however big the data is. This has reduced the need to rely on a sampling of data for analysis. Instead, you can take the entire population of data available with you and analyze it by using the power of cloud storage and computing capabilities that was not possible 20 years back.

1.4.4 Survival and Growth in the Highly Competitive World

Without the Internet, life can come to a standstill. Everyone is connected on social media, and information is constantly flowing across the world, both good and bad, reliable and unreliable. Internet users are also potential customers who are constantly discussing about their experiences of products and services on social media. Businesses have become highly competitive. Businesses want to take advantage of such information by analyzing such data. Each business is targeting the same customer and that customer's spending capability. Using social media, businesses are fiercely competing with each other. To survive, businesses have to find the best ways to target other businesses that require their products and services as well as the end consumers who require their products and services. Data or business analytics has enabled this effectively. Analytics provides various techniques to find hidden patterns in the data and provides enough knowledge for businesses to make better business decisions.

1.4.5 Business Complexity Growing Out of Globalization

Economic globalization that cuts across the boundaries of the countries where businesses produce goods or provide services has drastically increased the complexities of business. Businesses now have the challenge of catering to cultures that may have been previously unknown to them. With the large amount of data now possible to acquire (or already at their disposal), businesses can easily gauge differences between local and regional cultures, demands, and practices including spending trends and preferences.

1.4.6 Easy-to-Use Programming Tools and Platforms

In addition to commercially available data analytics tools, many open-source tools or platforms such as Python, R, Spark, PyTorch, TensorFlow, and Hadoop are available for easy development of analytics applications and developing models. These powerful tools are easy to use and well documented. They usually require an understanding of basic programming concepts with much knowledge of complex programming skills. Hadoop is useful in the effective and efficient analysis of big data.

1.5 Applications of Business Analytics

Business analytics has been applied effectively to many fields, including retail, e-commerce, travel (including the airline business), hospitality, logistics, and manufacturing. It also finds many applications in marketing and sales, human resources, finance, manufacturing, product design, service design, and customer service and support. Furthermore, business analytics has been applied to a whole range of other businesses, including energy, production, pharmaceutical, and other industries, for finding patterns and predicting failures of machines, equipment, and processes.

In this section, we discuss some of the areas in which data/business analytics is used effectively to benefit organizations. These examples are only illustrative and not exhaustive.

1.5.1 Marketing and Sales

Marketing and sales teams are the ones that have heavily used business analytics to identify appropriate approaches to marketing to reach a maximum number of potential customers at an optimized or reduced effort. These teams use business analytics to identify which marketing channel would be most effective (for example, emails, websites, or direct telephone contacts). They also use business analytics to determine which offers make sense to which types of customers (in terms of geographical regions, for instance) and to specifically tune their offers.

A marketing and sales team in the retail business would like to enable retail outlets (physical or online) to promote new products along with other products, as a bundled offer, based on the purchasing pattern of consumers. Logistics companies always want to optimize delivery time and commitment to customers by sticking to delivery commitments. This is always an important factor for businesses to tie up for their services. Similarly, an airline company would like to promote exciting offers based on a customer's travel history, thus encouraging customers to travel again and again via the same airline, thereby creating a loyal customer over a period of time. A travel agency would like to determine whether people like adventurous vacations, spiritual vacations, or historical vacations based on past customer data, and analytics can provide the inputs needed to focus marketing and sales efforts according to those specific interests of the people—thus optimizing the time spent by the marketing and sales team.

1.5.2 Human Resources

Retention is the biggest problem faced by an HR department in any industry, especially in the service industry. An HR department can identify which employees have high potential for retention by processing past employee data. Similarly, an HR department can also analyze which competence (qualification, knowledge, skill, or training) has the most influence on the organization's or team's capability to deliver quality output within committed timelines.

1.5.3 Product Design

Product design is not easy and often involves complicated processes. Risks factored in during product design, subsequent issues faced during manufacturing, and any resultant issues faced by customers or field staff can be a rich source of data that can help you understand potential issues with a future design. This analysis may reveal issues with materials, issues with the processes employed, issues with the design process itself, issues with the manufacturing, or issues with the handling of the equipment installation or later servicing. The results of such an analysis can substantially improve the quality of future designs by any company. Another interesting aspect is that data can help indicate which design aspects (color, sleekness, finish, weight, size, or material) customers like and which ones customers do not like.

1.5.4 Service Design

Like products, services are also carefully designed and priced by organizations. Identifying components of the service (and what are not) also depends on product design and cost factors compared to pricing. The length of the warranty, coverage during warranty, and pricing for various services can also be determined based on data from earlier experiences and from target market characteristics. Some customer regions may more easily accept "use and throw" products, whereas other regions may prefer "repair and use" kinds of products. Hence, the types of services need to be designed according to the preferences of regions. Again, different service levels (responsiveness) may have different price tags and may be targeted toward a specific segment of customers (for example, big corporations, small businesses, or individuals).

1.5.5 Customer Service and Support Areas

After-sales service and customer service are important aspects that no business can ignore. A lack of effective customer service can lead to negative publicity, impacting future sales of new versions of the product or of new products from the same company. Hence, customer service is an important area in which data analysis is applied significantly. Customer comments on the Web or on social media (for example, Twitter) provide a significant source of understanding about the customer pulse as well as the reasons behind the issues faced by customers. A service strategy can be accordingly drawn up, or necessary changes to the support structure may be carried out, based on the analysis of the data available to the industry.

1.6 Skills Required for an Analytics Job

Till now we have discussed various terminologies of analytics, reasons for the exponential growth in analytics, and illustrative examples of some of the applications of business analytics. Let's now discuss the skills required to perform data analytics and create models. Typically, this task requires substantial knowledge of the following topics:

1. Communications skills to understand the problem and the requirements.

Having a clear understanding of the problem/data task is one of the most important requirements. If the person analyzing the data does not understand the underlying problem or the specific characteristics of the task, then the analysis performed by the data analytics person can lead to the wrong conclusions or lead the business in the wrong direction. Also, if an individual does not know the specific domain in which the problem is being solved, then one should consult the domain expert to perform the analysis. Not understanding requirements, and just having only programming skills along with statistical or mathematical knowledge, can sometimes lead to proposing impractical (or even dangerous) suggestions for the business. These suggestions also waste the time of core business personnel.

2. Tools, techniques, and algorithms that can be applied to the data.

Understanding data and data types, preprocessing data, exploring data, choosing the right techniques, selecting supervised learning or unsupervised learning algorithms, and creating models are an essential part of an analytics job. It is equally important to apply the proper analysis techniques and algorithms to suitable situations or analyses. The depth of this knowledge may vary from the job titles and experience. For example, linear regression or multiple linear regression (supervised method) may be suitable if you know (based on business characteristics) that there exists a strong relationship between a response variable and various predictors. Clustering (unsupervised method) can allow you to cluster data into various segments. Using and applying business analytics effectively can be difficult without understanding these techniques and algorithms.

Having knowledge of tools is important. Though it is not possible to learn all the tools that are available, knowing as many tools helps fetch job interviews. Computer knowledge is required for a capable data analytics person as well so that there is no dependency on other programmers who don't understand the statistics or mathematics behind the techniques or algorithms. Platforms such as R, Python, and Hadoop have reduced the pain of learning programming, even though at times we may have to use other complementary programming languages.

3. Data structures and data storage or data warehousing techniques, including how to query the data effectively.

Knowledge of data structures and data storage/data warehousing eliminates dependence on database administrators and database programmers. This enables you to consolidate data from varied sources (including databases and flat files), arrange them into a proper structure, and store them appropriately in a data repository required for the analysis. The capability to query such a data repository is another additional competence of value to any data analyst.

4. Statistical and mathematical concepts (probability theory, linear algebra, matrix algebra, calculus, and cost-optimization algorithms such as gradient descent or ascent algorithms).

Data analytics and data mining techniques use many statistical and mathematical concepts on which various algorithms, measures, and computations are based. Good knowledge of statistical and mathematical concepts is essential to properly use the concepts to depict, analyze, and present the data and the results of the analysis. Otherwise, the wrong interpretations, wrong models, and wrong theories can lead others in the wrong direction by misinterpreting the results because the application of the technique or interpretation of the result itself was wrong.

Statistics contribute to a significant aspect of effective data analysis. Similarly, the knowledge discovery enablers such as machine learning have contributed significantly to the application of business analytics. Another area that has given impetus to business analytics is the growth of database systems, from SQL-oriented ones to NoSQL ones. All these combined, along with easy data visualization and reporting capabilities, have led to a clear understanding of what the data tells us and what we understand from the data. This has led to the vast application of business and data analytics to solve problems faced by organizations and to drive a competitive edge in business through the application of this understanding.

There are umpteen tools available to support each piece of the business analytics framework. Figure 1-1 presents some of these tools, along with details of the typical analytics framework.

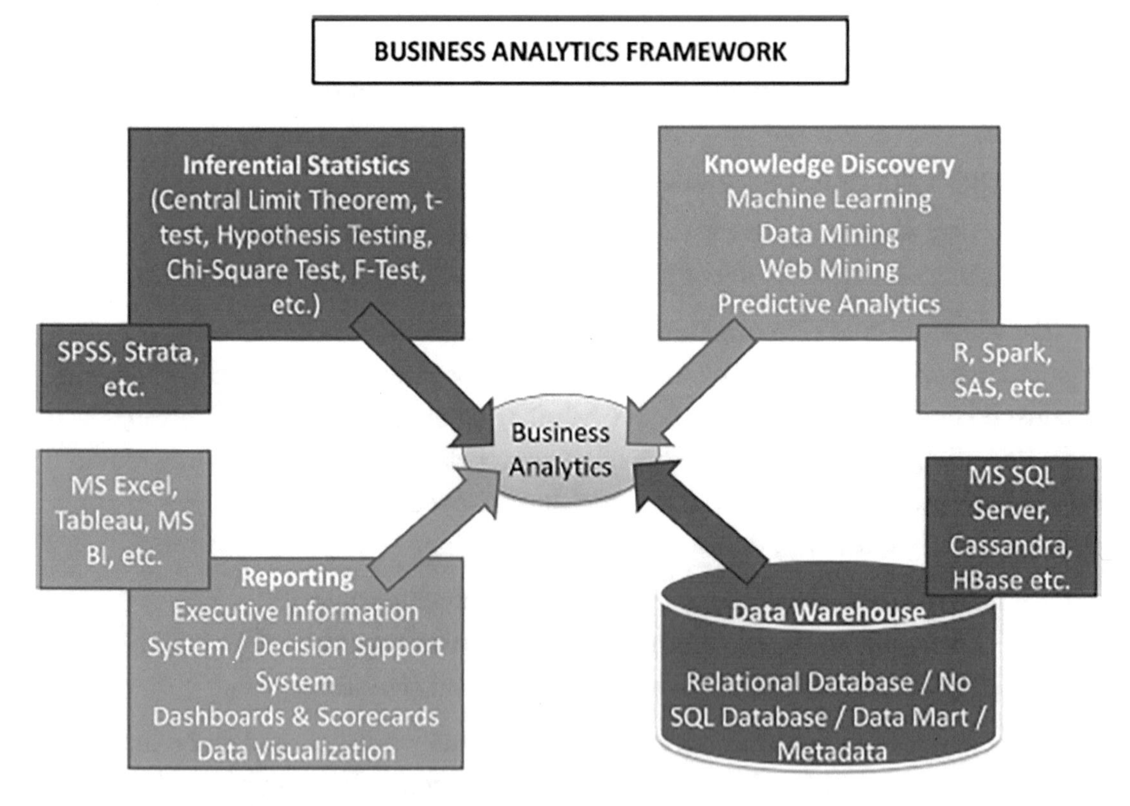

Figure 1-1. *Business analytics framework*

1.7 Process of an Analytics Project

Earlier we discussed the various data analytics principles, tools, algorithms, and skills required to perform data and business analytics. In this section, we briefly touch upon the typical process of business analytics and data mining projects, which is shown in Figure 1-2.

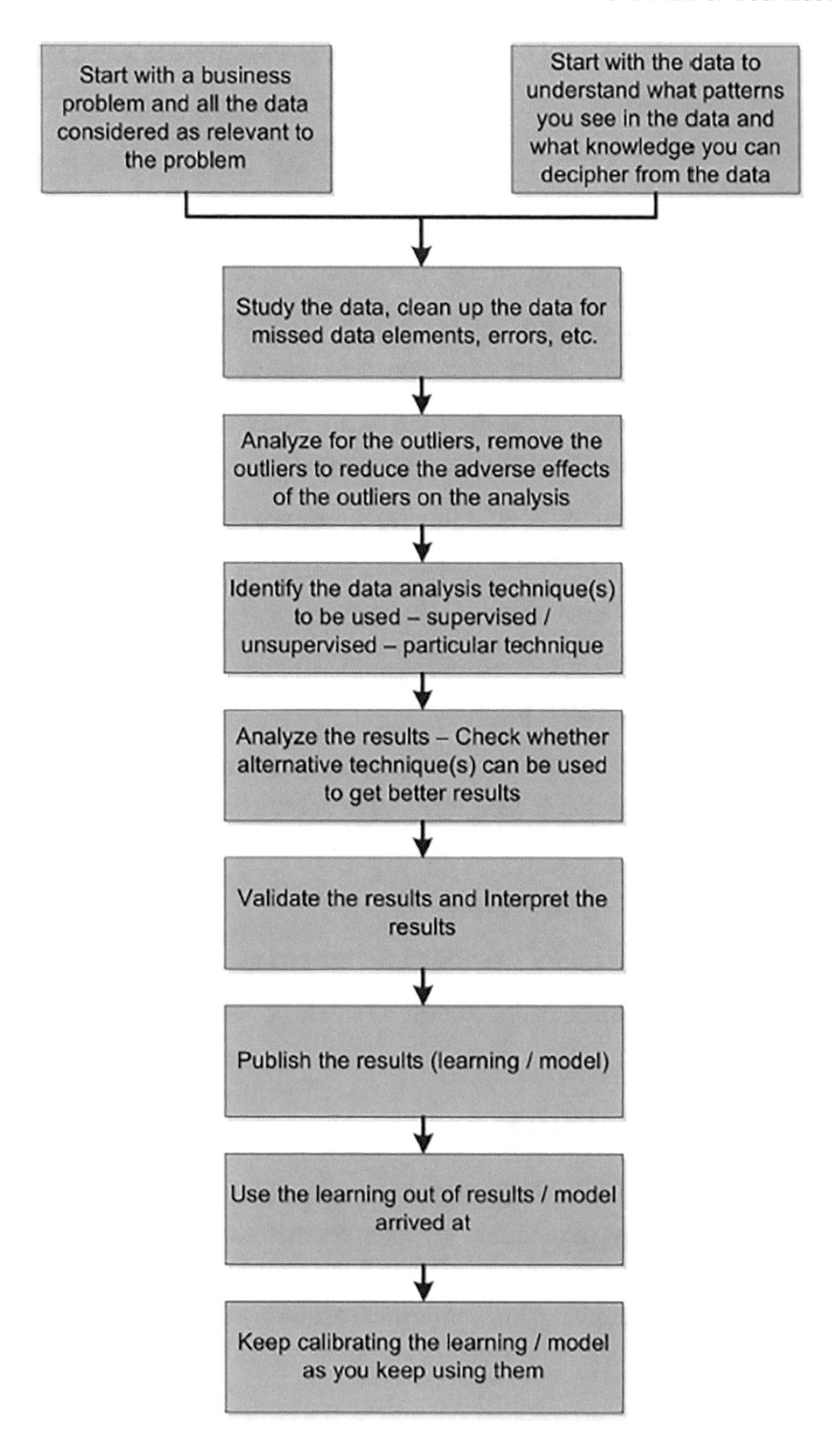

Figure 1-2. *Life cycle of a business analytics project*

The steps of the typical AI and business analytics project process are as follows:

1. Start with a business problem and all the data considered as relevant to the problem.

   or

   Start with the data to understand what patterns you see in the data and what knowledge you can decipher from the data.

2. Study the data, data types, and preprocess data; clean up the data for missing values; and any other data elements or errors.

3. Check for the outliers in the data and remove them from the data set to reduce their adverse impact on the analysis.

4. Identify the data analysis technique(s) to be used (for example, supervised or unsupervised).

5. Apply identified data analytics techniques using appropriate tools.

6. Analyze the results and check whether alternative technique(s) can be used to get better results.

7. Validate the results, understand the results, and interpret the results. Check whether it is performing as per the business requirements.

8. Publish the results (learning/model).

9. Use the learning from the results/model arrived at.

10. Keep calibrating the learning/model as you keep using it.

1.8 Chapter Summary

In this chapter, you saw how knowledge has evolved. You also looked at many scenarios in which data analytics helps individuals. The chapter included many examples of business analytics helping businesses to grow and compete effectively. You were also provided with examples of how business analytics results are used by businesses effectively.

You briefly went through the skills required for a business analyst. In particular, you understood the importance of the following: understanding the business and business problems, data analysis techniques and algorithms, computer programming, data structures and data storage/warehousing techniques, and statistical and mathematical concepts required for data analytics.

Finally, we briefly explained the process for executing an analytics project.

CHAPTER 2

The Foundations of Business Analytics

Uncertainty and randomness are bound to exist in most business decisions. Probability quantifies the uncertainty that we encounter every day. This chapter discusses the fundamentals of statistics, such as mean, variance, standard deviation, probability theory basics, types of probability distributions, and the difference between population and sample, which are essential for any analytics modeling. We will provide demonstrations using both Python and R.

2.1 Introduction

We all have studied statistics at some point of time in our education. However, we may never have gained a true appreciation of why applying some of that statistical knowledge is important. In the context of data and business analytics, knowledge of statistics can provide insight into characteristics of a data set you have to analyze that will help you determine the right techniques and methods to be employed for further analysis. There are many terms in statistics such as *mean, variance, median, mode*, and *standard deviation*, among others. We will try to provide a context for these terms with a simple example from our daily lives before explaining the terms from a business angle. Further, we will cover the basics of probability theory and different probability distributions and why they are necessary for business data analytics.

Imagine you are traveling and have just reached the bank of a muddy river, but there are no bridges or boats or anyone to help you to cross the river. Unfortunately, you do not know to swim. When you look around in this confused situation where there is no help available to you, you notice a sign, as shown in Figure 2-1.

U. R. Hodeghatta and U. Nayak, *Practical Business Analytics Using R and Python*,
https://doi.org/10.1007/978-1-4842-8754-5_2

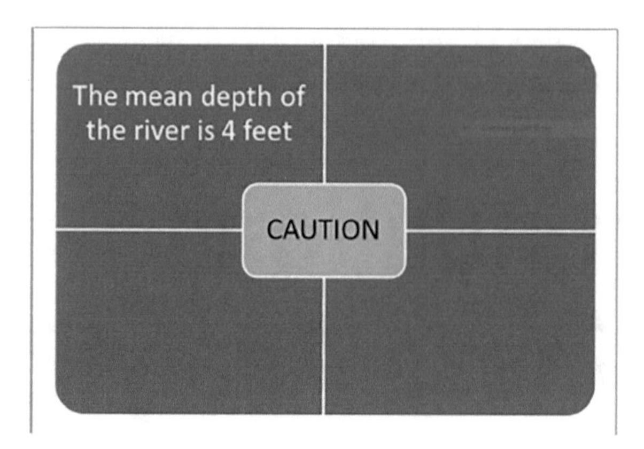

Figure 2-1. *A sign on the bank of a river*

The sign says, "The mean depth of the river is 4 feet." Say this value of mean is calculated by averaging the depth of the river at each square-foot area of the river. This leads us to the following question: "What is average or mean?" *Average* or *mean* is the quantity arrived at by summing up the depth at each square foot and dividing this sum by the number of measurements (i.e., number of square feet measured).

Your height is 6 feet. Does Figure 2-1 provide enough information for you to attempt to cross the river by walking? If you say "yes," definitely I appreciate your guts. I would not dare to cross the river because I do not know whether there is any point where the depth is more than my height. If there are points with depths like 7 feet, 8 feet, 10 feet, or 12 feet, then I will not dare to cross as I do not know where these points are, and at these points I am likely to drown.

Suppose the sign also says "Maximum depth is 12ft and minimum depth is 1ft" (see Figure 2-2). I am sure this additional information will scare you since you now know that there are points where you can get drowned. Maximum depth is the measure at one or more points that are the largest of all the values measured. Again, with this information you may not be sure that the depth of 12 feet is at one point or at multiple points. Minimum sounds encouraging (this is the lowest of the values observed) for you to cross the river, but again you do not know whether it is at one point or multiple points.

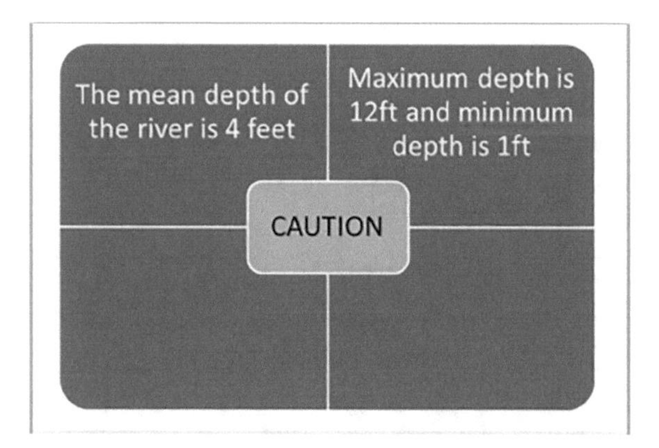

Figure 2-2. *The sign indicating mean, maximum, and minimum depths*

Suppose, in addition to the previous information, that the sign (shown in Figure 2-3) also says "Median of the depth is 4.5ft." Median is the middle point of all the measured depths if all the measured depths are arranged in ascending order. This means 50 percent of the depths measured are less than this, and also 50 percent of the depths measured are above this. You may not still dare to cross the river as 50 percent of the values are above 4.5 feet and the maximum depth is 12 feet.

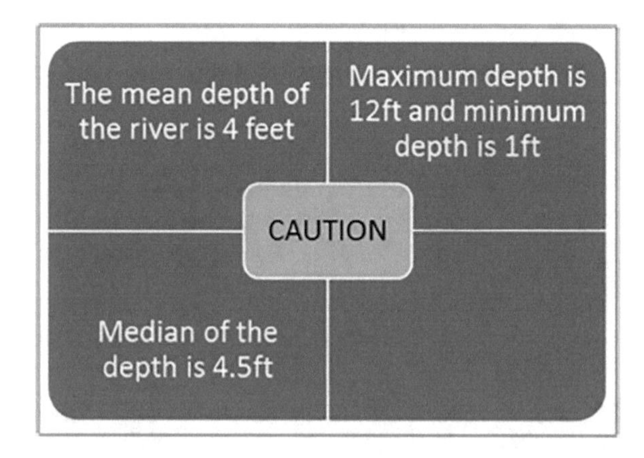

Figure 2-3. *The sign indicating mean, maximum, minimum, and median depths*

Suppose, in addition to the previous information, that the sign (shown in Figure 2-4) also says "Quartile 3 is 4.75ft." Quartile 3 is the point below which 75 percent of the measured values fall when the measured values are arranged in ascending order. This also means there are 25 percent of the measured values that have greater depth than this. You may not be still comfortable crossing the river as you know the maximum depth is 12 feet and there are 25 percent of the points above 4.75 feet.

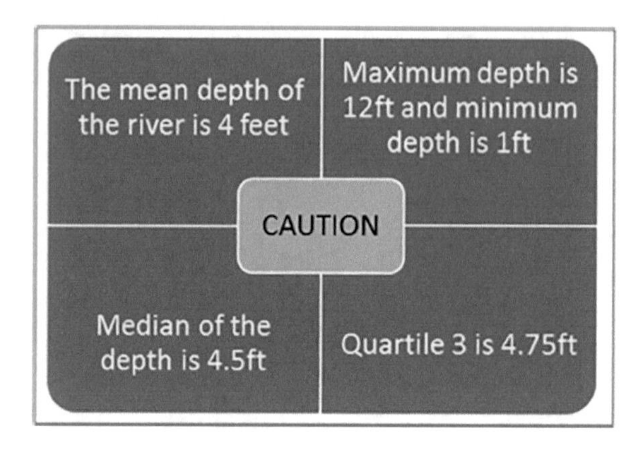

Figure 2-4. *The sign adds a quartile measurement*

Suppose, in addition to the previous information, that the sign (shown in Figure 2-5) also says "Percentile 90 is 4.9ft and percentile 95 is 5ft." Suppose this is the maximum information available. You now know that only 5 percent of the measured points are of depth more than 5 feet. You may now want to take a risk if you do not have any other means other than crossing the river by walking or wading through as now you know that there are only 5 percent of the points with depth more than 5 feet. Your height is 6 feet. You may hope that 98 or 99 percentile may be still 5.5 feet. You may now believe that the maximum points may be rare and you can, by having faith in God, cross the river safely.

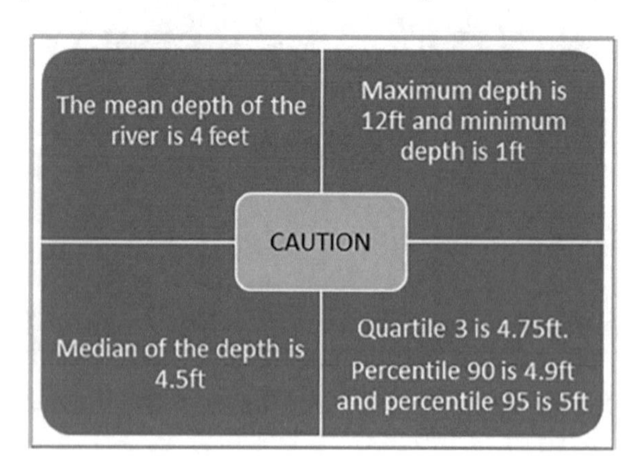

Figure 2-5. *The sign completes its percentile measurements*

In spite of the previous cautious calculations, you may still drown if you reach rare points of depth of more than 6 feet (like the maximum point of depth). But, with the foregoing information, you know that your risk is substantially less compared to your risk at the initial time when you had only limited information (that the mean depth of the river is 4 feet).

This is the point we wanted to make through the river analogy: with one single parameter of measurement, you may not be able to describe the situation clearly and may require more parameters to elaborate the situation. Each additional parameter calculated may increase the clarity required to make decisions or to understand the phenomenon clearly. Again, another note of caution: there are many other parameters than the ones discussed earlier that are of interest in making decisions or understanding any situation or context.

Statistical parameters such as mean or average, median, quartile, maximum, minimum, range, variance, and standard deviation describe the data very clearly. As shown in the example discussed earlier, one aspect of the data may not provide all the clarity necessary, but many related parameters provide better clarity with regard to data or situation, or the context

Later in this chapter, we will discuss how to calculate all these parameters using R as well as Python. Before that, we need to understand the important aspect—the meaning of population and sample.

2.2 Population and Sample

2.2.1 Population

In simple terms, *population* means the complete set of data. In the case of the river example, it means that measurements have been taken across the river at each and every square foot without leaving out any area. This also means that all the possible values are taken into consideration. When we consider the entire possible set of values, we say that we are considering the "population." The following are examples of population: the population of all the employees of the entire information technology (IT) industry, population of all the employees of a company, population of all the transaction data of an application, population of all the people in a country, population of all the people in a state, population of all the Internet users, and population of all the users of e-commerce sites. The list of examples is unlimited.

However, when we have to analyze the data, it is difficult to get the entire population, especially when the data size is enormous. This is because:

- It is not always possible to gather the data of the entire population. For example, in the previous example, how can we measure the volume of the entire river by taking every square inch of the river flow? It is practically not possible. Similarly, in many business situations, we may not be able to acquire the data of the entire population.
- It takes substantial time to process the data, and the time taken to analyze may be prohibitively high in terms of the requirements related to the application of the data. For example, if the entire transaction data related to all the purchases of all the users has to be analyzed before you recommend a particular product to a user, the amount of processing time taken may be so huge that you may miss the opportunity to suggest the product to the user who has to be provided the suggestions quickly when he is in session on the Internet.
- It takes substantial processing power (i.e., memory or CPU power) to hold the data and process it; not everyone has the infrastructure to deploy such a huge processing capability.
- Though processing the entire population is possible with “Big Data” tools and technologies, not every organization is equipped with such technology power.

2.2.2 Sample

In simple terms, *sample* means a section or subset of the population selected for analysis. Examples of samples are the following: randomly selected 100, 000 employees from the entire IT industry or randomly selected 1, 000 employees of a company or randomly selected 1,000,000 transactions of an application or randomly selected 10,000,000 Internet users or randomly selected 5,000 users each from each ecommerce site, and so on. Sample can also be selected using stratification (i.e., based on some rules of interest). For example, all the employees of the IT industry whose income is greater than $100,000 or all the employees of a company whose salary is greater than $50,000 or the top 100,000 transactions by amount per transaction (e.g., minimum $1,000 per transaction or all Internet users who spend more than two hours per day, etc.).

Several sampling techniques are available to ensure that the data integrity is maintained and the conclusions and hypothesis can be applied to the population. Several sampling techniques have been in practice since the old days of statistics and research design. We will briefly mention them without going into the details as they are covered extensively in many statistics books. The popular sampling methods used in data analytics are as follows:

- **Random sampling:** This is the most common probability sampling technique as every single sample is selected randomly from the population data set. This gives an opportunity for each record in the data set an equal chance (probability) to be chosen to be a part of the sample. For example, the HR department wants to conduct a social event. Therefore, it wants to select 50 people out of 300. To provide an equal opportunity to everyone, HR picks the names randomly from a jar containing all the names of employees.
- **Systematic sampling:** The systematic sampling method chooses the sample at regular intervals. For example, your analysis requires transactional data. This can be collected over 30-day intervals with a size of 1,000 out of a 10,000 population daily. Over a period of 30 days, the total sample collected is around 30,000.
- **Cluster sampling:** The entire population data is divided into different clusters, and samples are collected randomly from each cluster. For example, you are creating a predictive model for your organization. The organization is spread across the globe in 30 different countries. Your sample data should consist of data from all these different (30) clusters for your model to provide an accurate prediction. If the data is taken from only one geographical location, then the model is biased and does not perform well for other geographical locations.

The following are the benefits of sampling:

- Data is now substantially less, not losing characteristics of "population" data, and is focused on the application. Hence, data can be processed quickly, and the information from this analysis can be applied quickly.
- The data can be processed easily with a lesser requirement for computing power.

However, of late, we have higher computing power at our hands because of cloud technologies and the possibility to cluster computers for better computing power. Though such large computing power allows us, in some cases, to use the entire population for analysis, sampling definitely helps carry out the analysis relatively easily and faster in many cases. However, sampling has a weakness: if the samples are not selected properly, then the analysis results may be wrong. For example, for analyzing the data for the entire year, only this month's data is taken. This sample selection may not give the required information as to how the changes have happened over the months.

2.3 Statistical Parameters of Interest

In this section, we will dive into the world of statistical parameters and derive a more in-depth understanding. At the same time, we will use R to calculate these statistical parameters. The same can be calculated using Python or any other programming language. The purpose of this section is to provide an overview of the concepts and a practical understanding of the statistical terms used in any data analytics tasks.

2.3.1 Mean

Mean is also known as *average* in general terms. If we have to summarize any data set quickly, then the common measure used is mean. Some examples of the usage of the mean are the following:

- For a business, mean profitability over the last five years may be one good way to represent the organization's profitability in order to judge its success.
- For an organization, mean employee retention over last five years may be a good way to represent employee retention in order to judge the success of the human resources (HR) policies.
- For a country, mean gross domestic product (GDP) over the last five years may be a good way to represent the health of the economy.
- For a business, mean growth in sales or revenue over a period of the last five years may be a good way to represent growth.

- For a business, mean reduction in cost of operations over a period of the last five years may be a good way to understand operational efficiency improvement.

Normally, a mean or average figure gives a sense of what the figure is likely to be for the next year based on the performance for the last number of years. However, there are limitations of using or relying only on this parameter.

Let's look at a few examples to understand more about using mean or average:

- Good Luck Co. Pvt. Ltd earned a profit of $1,000,000; $750,000; $600,000; $500,000; and $500,000 over the last five years. Mean or average profit over the last five years is calculated as sum (all the profits over the last 5 years)/No. of years; i.e., ($1,000,000 + $750,000 + $600,000 + $500,000 + $500,000) / 5 = $670,000.

  This calculation is depicted both in R and Python in Figures 2-6, 2-7, 2-8, and 2-9. We have demonstrated the different ways to represent data and calculate mean. In the first method, Figures 2-6 and 2-7, we have assigned numbers to individual variables, and in the second method, Figures 2-8 and 2-9, data is in an array.

```
> #GoodLuck Co. Pvt. Ltd - Profit figures for last 5 years
> Year1Prof<-1000000
> Year2Prof<-750000
> Year3Prof<-600000
> Year4Prof<-500000
> Year5Prof<-500000
> #To calculate the mean or average profit, you require to
> # sum all the 5 years profit and divide bu the number of years
> SumYrsProfs<-Year1Prof+Year2Prof+Year3Prof+Year4Prof+Year5Prof
> MeanProf<- SumYrsProfs/5
> MeanProf
[1] 670000


>
```

Figure 2-6. *How to calculate mean in R*

```
Year1Profit = 1000000

Year2Profit =  750000

Year3Profit = 600000

Year4Profit = 500000

Year5Profit = 500000

# Total Profit is summation of 5 yeras profit

TotalProfit = Year1Profit + Year2Profit + Year3Profit + Year4
Profit + Year5Profit


# Mean or Average Profit is Total/n

MeanProfit = TotalProfit/5


print(MeanProfit)
```

Figure 2-7. *How to calculate mean in Python*

```
> #Alternate method to calculate Mean
> #GoodLuck Co. Pvt. Ltd for the last 5 years
> # Creating a list of profits
> Prof5Yrs<-c(1000000,750000,600000,500000,500000)
> MeanProf<-mean(Prof5Yrs)
> MeanProf
[1] 670000
```

Figure 2-8. *Alternative and simple way for calculating mean in R*

```
#Create an array with all the years profit
ProfYears = [1000000, 750000, 600000, 500000, 500000]
# Add all array (all the elements of array) and divide by the toa
MeanProf = (sum(ProfYears)/len(ProfYears))
print(MeanProf)

670000.0
```

Figure 2-9. *Alternative and simple way for calculating mean in Python*

There could be many different ways in any programming language to perform a task. The mean for the previous example is calculated using R and Python in a couple ways in Figures 2-6, 2-7 2-8, and 2-9.

Similarly, for the other examples we can work out the mean or average value if we know the individual figures for the years.

The problem with the mean or average as a single parameter is as follows:

- Any extreme high or low figure in one or more of the years can skew the mean, and thus the mean may not appropriately represent the likely figure next year. For example, consider that there was very high profit in one of the years because of a volatile international economy that led to severe devaluation of the local currency. Profits for five years of a company were, respectively, €6,000,000; €4,000,000; €4,500,000; €4,750,000; and €4,250,000. The first-year profit of €6,000,000 was on account of steep devaluation of the euro in the international market. If the effective value of profit without taking into consideration devaluation during the first year is €4,000,000, then the average or mean profit on account of increased profit would be €400,000, as shown in Figure 2-10.

```
> ##Duck&Duck LLP
> # Data for 5 years profit - first year profit significantly
increased on account
> # of increased foreign exchange rate due to devaluation of
local currency
> Prof5Yrs<- c(6000000,4000000,4500000,4750000,4250000)
> MeanProf <- mean(Prof5Yrs)
> MeanProf
[1] 4700000
> Prof5Yrs<-c(4000000,4000000,4500000,4750000,4250000)
> MeanProf<-mean(Prof5Yrs)
> MeanProf
[1] 4300000
```

***Figure 2-10.** Actual mean profit and effective mean profit example*

- Using mean or average alone will not show the volatility in the figure over the years effectively. Also, mean or average does not depict the trend as to whether it is decreasing or increasing. Let's take an example. Suppose the revenue of a company over the last five years is, respectively, $22,000,000; $15,000,000; $32,000,000; $18,000,000; and $10,000,000. The average revenue of the last five years is $19,400,000. If you notice the figures, the revenue is quite volatile; that is, compared to first year, it decreased significantly in the second year, jumped up by a huge number during the third year, then decreased significantly during the fourth year, and continued to decrease further significantly during the fifth year. The average or mean figure does not depict either this volatility in revenue or trending downwardness in revenue. Figure 2-11 shows this downside of mean as a measure.

```
> #Sam&George LLP
> # Data for 5 years revenu - large variations over the years
> Rev5Yrs <- c(22000000,15000000,32000000,18000000,10000000)
> MeanRev<-mean(Rev5Yrs)
> MeanRev
[1] 19400000
```

***Figure 2-11.** Downside of mean as a statistical parameter*

2.3.2 Median

Median is the middle value by ordering the values in either ascending order or descending order. In many circumstances, median may be more representative than mean. It clearly divides the data set at the middle into two equal partitions; that is, 50 percent of the values will be below the median, and 50 percent of the values will be above the median. Examples are as follows:

- Age of workers in an organization to know the vitality of the organization
- Productivity of the employees in an organization
- Salaries of the employees in an organization pertaining to a particular skill set

Let us consider the age of 20 workers in an organization as 18, 20, 50, 55, 56, 57, 58, 47, 36, 57, 56, 55, 54, 37, 58, 49, 51, 54, 22, and 57. From a simple examination of these figures, you can make out that the organization has more aged workers than youngsters and there may be an issue of knowledge drain in a few years if the organizational retirement age is 60. Let us also compare mean and median for this data set. The following figure shows that 50 percent of the workers are above 54 years of age and are likely to retire early (i.e., if we take 60 years as retirement age, they have only 6 years to retirement), which may depict the possibility of significant knowledge drain. However, if we use the average figure of 47.35, it shows a better situation (i.e., about 12.65 years to retirement). But, it is not so if we look at the raw data: 13 of the 20 employees are already at the age of 50 or older, which is of concern to the organization. Figure 2-12 shows a worked-out example of median using R.

```
> #Sam&George LLP
> # Data of employees age of 20 workers
> WorkAge<-c(18,20,50,56,57,58,47,36,57,56,55,54,37,58,49,51,54,22,57)
> #Workers age if arranged in ascending order
> # 18,20,22,36,37,47,49,50,51, 54,54,55,55,56,56,57,57.57.58.58
> #Middle number will be 10th and 11th,i.e, 54, 54
> #Hence, median age of worker is 54 years
> MedWorkAge<-median(WorkAge)
> MedWorkAge
[1] 54
> MeanWorkAge<-mean(WorkAge)
> MeanWorkAge
[1] 47.35
```

Figure 2-12. *How to calculate median using R*

We repeat the same in Python, but we use the `statistics()` library to calculate the median, as shown in Figure 2-13.

```
In [39]:  #SamiGeorge LLP
          # Data of employees age of 20 workers
          WorkersAge = [18, 20, 50, 55, 56, 57, 58, 47, 36, 57, 56, 55, 54, 37, 58, 49, 51, 54, 22, 57]

          #Median of WorkersAge
          # Using statistics() python Library
          import statistics
          statistics.median(WorkersAge)

Out[39]:  54.0

In [40]:  # Finding mean using statistucs() python Library
          statistics.mean(WorkersAge)

Out[40]:  47.35

In [ ]:
```

Figure 2-13. *How to calculate median using Python*

However, if the 10th value had been 54 and the 11th value had been 55, respectively, then the median would have been (54+55)/2, i.e., 54.5.

Let us take another example of a productivity of a company. Let the productivity per day in terms of items produced per worker be 20, 50, 55, 60, 21, 22, 65, 55, 23, 21, 20, 35, 56, 59, 22, 23, 25, 30, 35, 41, 22, 24, 25, 24, and 25, respectively. The median productivity is 25 items per day, which means that there are 50 percent of the workers in the organization who produce less than 25 items per day, and there are 50 percent of the employees who produce more than 25 items per day. Mean productivity is 34.32 items per day because some of the workers have significantly higher productivity than the median worker, which is evident from the productivity of some of the workers; that is, 65 items per day, 60 items per day, 59 items per day, 56 items per day, 56 items per day, 55 items per day, etc. The analysis from R in Figure 2-14 clearly shows the difference between mean and median.

```
> ProdWorkDay<- c(20, 50, 55, 60, 21, 22, 65, 55, 23, 21, 20, 35,
56, 59, 22, 23, 25, 30, 35, 41, 22, 24, 25, 24,25)
> MedProd<-median(ProdWorkDay)
> MedProd
[1] 25
> MeanProd<-mean(ProdWorkDay)
> MeanProd
[1] 34.32


>
```

Figure 2-14. *Difference between mean and median highlighted*

If you have to work out median through manual calculations, you have to arrange the data points in ascending or descending order and then select the value of the middle term if there are an odd number of values. If there are an even number of values, then you have to sum up the middle two terms and then divide the sum by 2 as mentioned in the previous discussions.

If you notice from the previous discussion, instead of only mean or median alone, looking at both mean and median gives a better idea of the data.

2.3.3 Mode

Mode is the data point in the data set that occurs the most. For example, in our data set related to the age of workers, 57 occurs the maximum number of times (i.e., three times). Hence, 57 is the mode of the workers' age data set. This shows the pattern of repetition in the data.

There is no built-in function in R to compute mode. Hence, we have written a function and have computed the mode as shown in Figure 2-15. We have used the same data set we used earlier (i.e., `WorkAge`).

```
> ##MODE
> WorkAge
 [1] 18 20 50 56 57 58 47 36 57 56 55 54 37 58 49 51 54 22
[19] 57
> # We are creating a function by name CalMode to calculate mode
> # This function is used to compute highest number of occurances
> # of the same term
> CalMode<- function(dataset)
+ {
+   UniDataSet <-unique(dataset)
+   UniDataSet[which.max(tabulate(match(dataset,UniDataSet)))]
+ }
> #Using CalMode function on WorkAge data
> CalMode(WorkAge)
[1] 57

>
```

Figure 2-15. *Calculation of mode using a function created in R*

In the previous function, `unique()` creates a set of unique numbers from the data set. In the case of the `WorkAge` example, the unique numbers are 18, 20, 50, 55, 56, 57, 58, 47, 36, 54, 37, 49, 51, and 22. The `match()` function matches the numbers between the ones in the data set and the unique numbers set we got and provides the position of each unique number in the original data set. The function `tabulate()` returns the number of times each unique number is occurring in the data set. The function `which.max()` returns the position of the maximum times repeating number in the unique numbers set.

In Python, the `statistics()` library provides the `mode()` function to calculate the mode of the data set given. We can also use the `max()` function with the *key argument*, as shown in Figure 2-16. Some of the functions of R are not applicable directly in Python. The purpose of the previous description is to demonstrate the concepts.

```
In [8]:  statistics.mode(WorkersAge)
 Out[8]: 57

In [11]:  max(WorkersAge, key=WorkersAge.count)
 Out[11]: 57
```

Figure 2-16. *The* `max()` *function with a key argument*

2.3.4 Range

The *range* is a simple but essential statistical parameter. It depicts the distance between the end points of the data set arranged in ascending or descending order (i.e., between the maximum value in the data set and the minimum value in the data set). This provides the measure of overall dispersion of the data set.

The R command `range(dataset)` provides the minimum and maximum values (see Figure 2-17) on the same data set used earlier (i.e., `WorkAge`).

```
> ##Range
> RangeWorkAge<-range(WorkAge)
> RangeWorkAge
[1] 18 58

>
```

Figure 2-17. *How to calculate range using R*

The difference between maximum and minimum values is the range.

In Python, we write our own function, `myRange()`, which accepts an array of numbers, calculates the maximum value and the minimum value, and returns the two values to calculate the range of an array of numbers, as shown in Figure 2-18. The `max()` function finds the maximum value of the list, and the `min()` function finds the minimum value in the list.

```
In [6]:  #SamiGeorge LLP
         # Data of employees age of 20 workers
         WorkersAge = [18, 20, 50, 55, 56, 57, 58, 47, 36, 57, 56, 55, 54, 37, 58, 49, 51, 54, 22, 57]

In [14]:  def myRange(alist):
              maxval = max(alist)
              minval = min(alist)
              return (minval, maxval)

In [15]:  myRange(WorkersAge)

Out[15]: (18, 58)
```

Figure 2-18. *How to calculate range using Python*

2.3.5 Quantiles

Quantiles are also known as *percentiles*. Quantiles divide the data set arranged in ascending or descending order into equal partitions. The *median* is nothing but the data point dividing the data arranged in ascending or descending order into two sets of equal number of elements. Hence, it is also known as the 50th percentile. On the other hand, *quartiles* divide the data set arranged in ascending order into four sets of equal number of data elements. The first quartile (also known as Q1 or as the 25th percentile) will have 25 percent of the data elements below it and 75 percent of the data elements above it. The second quartile (also known as Q2 or the 50th percentile or median) will have 50 percent of the data elements below it and 50 percent of the data elements above it. The third quartile (also known as Q3 or the 75th percentile) has 75 percent of the data elements below it and 25 percent of the data elements above it. *Quantile* is a generic word, whereas *quartile* is specific to a particular percentile. For example, Q1 is the 25th percentile. Quartile 4 is nothing but the 100th percentile.

Quantiles, quartiles, or percentiles provide us with the information that the mean is not able to provide us. In other words, quantiles, quartiles, or percentiles provide us additional information about the data set in addition to mean.

Let us take the same two data sets as given in the section "Median" and work out the quartiles. Figures 2-19A and 2-19B show the working of the quartiles. In the following code, we have the data in an array called `WorkAge`. We call the built-in function `quantile()` in R to find out the different quantiles. In this example, we want the "quartile" and hence set the value to 0.25. By setting the value of the `probs` parameter, you can decide how you want the data ratios to be split.

```
> ##Quartiles
> # Sam&George LLP
> # Data of Employee Age of 20 workers
> WorkAge
 [1] 18 20 50 56 57 58 47 36 57 56 55 54 37 58
[15] 49 51 54 22 57
> # Let us calculate the quartiles.
> # However, there is no function in R like quartile()
> # Instead we use qunatiles() only
> QuartWorkAge<-quantile(WorkAge, probs = seq(0,1,0.25))
> QuartWorkAge
  0%  25%  50%  75% 100%
18.0 42.0 54.0 56.5 58.0
> #Now let us calculate median using the median()
function as we used earlier
> MedWorkAge<-median(WorkAge)
> MedWorkAge
[1] 54
> #Now let us calculate the median of the qork age
using qunatile() function
> MediWorkAge<-quantile(WorkAge, probs=0.50)
> MediWorkAge
50%
 54


>
```

Figure 2-19A. *Calculating quantiles or percentiles using R*

```
>
> #If we want to divide the data set into 5 partitions
> #Or we want to find out the 20th, 40th, 60th, 80th and 100th percentile
> #We will do the same as follows
> PercentWorkAge <- quantile(WorkAge, probs = seq(0, 1, 0.20))
> PercentWorkAge
  0%  20%  40%  60%  80% 100%
18.0 36.8 50.6 55.0 57.0 58.0
>
```

Figure 2-19B. *Calculating quantiles or percentiles using R*

Similarly, you can divide the data set into 20 sets of equal number of data elements by using the quantile function with `probs = seq(0, 1, 0.05)`, as shown in Figure 2-20.

```
> ##Sam&George LLP
> ##Divide the data into 20 sets of equal numbers
> TwentySplit<-quantile(WorkAge, probs=seq(0,1,0.05))
> TwentySplit
  0%    5%  10%  15%  20%  25%  30%  35%  40%  45%
18.0 19.8 21.6 31.8 36.6 42.0 47.8 49.3 50.2 51.3
 50%  55%  60%  65%  70%  75%  80%  85%  90%  95%
54.0 54.0 54.8 55.7 56.0 56.5 57.0 57.0 57.2 58.0
100%
58.0

>
```

***Figure 2-20.** Partitioning the data into a set of 20 sets of equal number of data elements*

As you can observe from Figure 2-20, the minimum value of the data set is seen at the 0 percentile, and the maximum value of the data set is seen at the 100 percentile. As you can observe, typically between each 5 percentiles you can see one data element.

We use the `statistics()` function in Python to calculate the quartiles. The `statistics.quantiles()` function in Python returns the quantiles of the data that correspond to the numbers ***n*** set in the function. The function returns the corresponding n-1 quantiles. For example, if the n is set as 10 for deciles, the `statistics.quantiles()` method will return 10-1=9 cut points of equal intervals, as shown in the Figure 2-21.

```
In [19]:  statistics.quantiles(WorkersAge)
 Out[19]: [39.5, 54.0, 56.75]

In [20]:  statistics.quantiles(WorkersAge,n=10)
 Out[20]: [20.2, 36.2, 47.6, 50.4, 54.0, 55.0, 56.0, 57.0, 57.9]

In [21]:  statistics.quantiles(WorkersAge,n=5)
 Out[21]: [36.2, 50.4, 55.0, 57.0]

In [ ]:
```

***Figure 2-21.** Partitioning the data into equal number of data elements*

As evident from this discussion, quartiles and various quantiles provide additional information about the data distribution in addition to the information provided by mean or median (even though median is nothing but the second quartile).

2.3.6 Standard Deviation

The measures mean and median depict the center of the data set, or *distribution*. On the other hand, *standard deviation* specifies the spread of the data set or data values.

The standard deviation is manually calculated as follows:

1. First the mean of the data set or distribution is calculated.
2. Then the distance of each value from the mean is calculated (this is known as the *deviation*).
3. Then the distance as calculated previously is squared.
4. Then the squared distances are summed up.
5. Then the sum of the squared distances arrived at earlier is divided by the number of values minus 1 to adjust for the degrees of freedom.

The squaring in step 3 is required to understand the real spread of the data as the negatives and positives in the data set compensate for each other or cancel out the effect of each other when we calculate or arrive at the mean.

Let us take the age of the workers example shown in Figure 2-22 to calculate the standard deviation.

Worker	Age	Deviation from the Mean	Square of the Deviation
1	18	=47.35-18=29.35	861.4225
2	20	=47.35-20=27.35	748.0225
3	50	=47.35-50=02.65	7.0225
4	55	=47.35-55=-7.65	58.5225
5	56	=47.35-56=-8.65	74.8225
6	57	=47.35-57=-9.65	93.1225
7	58	=47.35-58=-10.65	113.4225
8	47	=47.35-47=-0.35	0.1225
9	36	=47.35-36=11.35	128.8225
10	57	=47.35-57=-9.65	93.1225
11	56	=47.35-56=-8.65	74.8225
12	55	=47.35-55=-7.65	58.5225
13	54	=47.35-54=-6.65	44.2225
14	37	=47.35-37=10.35	107.1225
15	58	=47.35-58=-10.65	113.4225
16	49	=47.35-49=-1.65	2.7225
17	51	=47.35-51=-3.65	13.3225
18	54	=47.35-54=-6.65	44.2225
19	22	=47.35-22=25.35	642.6225
20	57	=47.35-57=-9.65	93.1225
		Total	3372.55
		Standard Deviation	=squre root(3372.55/(20-1)) =13.32301

Figure 2-22. *Manual calculation of standard deviation using the WorkAge data set*

In R, this calculation can be done easily through the simple command `sd(dataset)`. Figure 2-23 shows the example of finding standard deviation using R.

```
> ####Standard Deviation
> WorkAge1<-c(18,20,50,55,56,57,58,47,36,57,56,55,54,
37,58,49,51,54,22,57)
> #Standard Deviation in R is calculated using sd command
> StdDevWorkAge1<-sd(WorkAge1)
> StdDevWorkAge1
[1] 13.32301

>
```

Figure 2-23. *Calculating standard deviation using R*

Similarly the standard deviation can be calculated using the `stdev()` function of the `statistics()` library, as shown in Figure 2-24.

```
In [23]:  #Calculating standard deviation of WorkersAge
          statistics.stdev(WorkersAge)

Out[23]:  13.323011355506207

In [3]:   #Calculating standard deviation of WorkersAge
          round(statistics.stdev(WorkersAge), 2)

Out[3]:   13.32
```

Figure 2-24. *Calculating standard deviation using Python*

Normally, as per the rules of the normal curve (a data set that consists of a large number of items is generally said to have a normal distribution or normal curve):

- +/- 1 standard deviation denotes that 68 percent of the data falls within it.
- +/- 2 standard deviation denotes that 95 percent of the data falls within it.
- +/- 3 standard deviation denote that 99.7 percent of the data falls within it.

In total, around 99.7 percent of the data will be within +/- 3 standard deviations.

As you can see from Figure 2-25, in the case of a normally distributed data (where the number of data points is typically greater than 30 (i.e., more the better), it is observed that about 68 percent of the data falls within +/- one standard deviation from the center of the distribution (i.e., mean). Similarly, about 95 percent (or around 95.2 percent, as shown in Figure 4-17B) of the data values fall within +/- two standard deviations from the center. About 99.7 percent of the data values fall within +/- three standard deviations from the center. A curve shown here is known as typically a *bell curve* or *normal distribution curve*. For example, profit or loss of all the companies in a country is normally distributed around the center value (i.e., mean of the profit or loss).

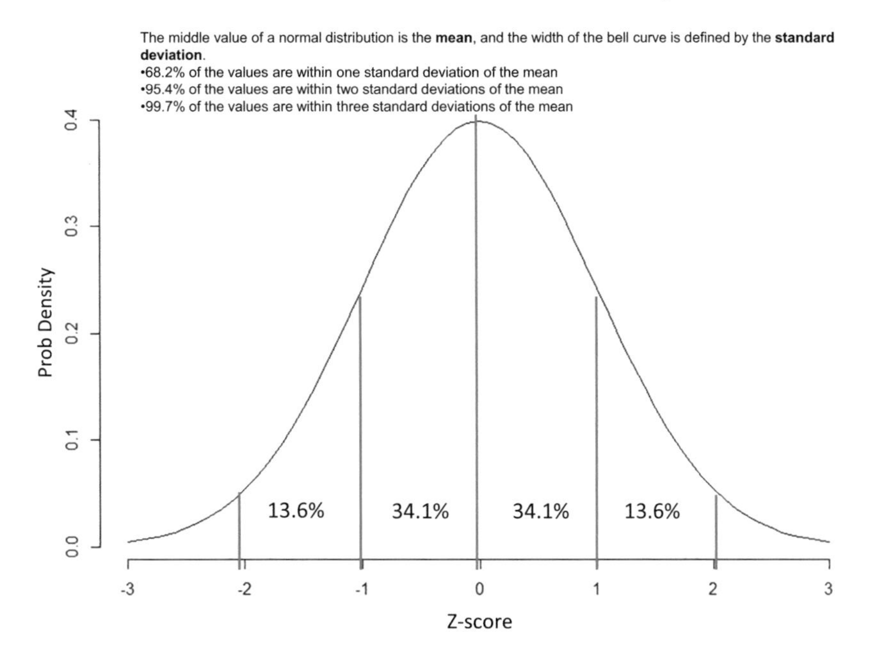

Figure 2-25. *Bell curve showing data coverage within various standard deviations*

The higher the standard deviation, the higher is the spread from the mean—i.e., it indicates that the data points vary from each other significantly and shows the heterogeneity of the data. The lower the standard deviation, the lower the spread from the mean—i.e., it indicates that the data points vary less from each other and shows the homogeneity of the data.

However, standard deviation along with other factors such as mean, median, quartiles, and percentiles give us substantial information about the data or explain the data more effectively.

2.3.7 Variance

Variance is another way of depicting the spread. In simple terms, it is the square of the standard deviation, as shown in Figure 2-26. Variance provides the spread of squared deviation from the mean value. It is another way of representing the spread compared to the standard deviation. Mathematically, as mentioned earlier, the variance is the square of the standard deviation. We are continuing to use the `WorkAge` data set we used earlier in this chapter.

```
> ##Variance
> WorkAge1
 [1] 18 20 50 55 56 57 58 47 36 57 56 55 54 37 58
[16] 49 51 54 22 57
> WorkAgeStdDev<-sd(WorkAge1)
> WorkAgeStdDev
[1] 13.32301
> WorkAgeVar<-var(WorkAge1)
> WorkAgeVar
[1] 177.5026
> WorkAgeStdDev*WorkAgeStdDev
[1] 177.5026


>
```

Figure 2-26. *Calculating variance using R*

```
In [25]:  #Calculating variance
          statistics.variance(WorkersAge)

Out[25]: 177.50263157894736
```

Figure 2-27. *Calculating variance using Python*

2.3.8 Summary Command in R

The command `summary(dataset))` provides the following information on the data set, which covers most of the statistical parameters discussed. This command gives us the output such as minimum value, first quartile, median (i.e., the second quartile), mean, third quartile, and maximum value. This is an easy way of getting the summary information through a single command (see Figure 2-28, which has a screenshot from R).

```
> ##Summary() command in R provides you the 6
> #important aspects of data i.e, Min Value, 1st
> #quartile, Median (2nd quartile), Mean, 3rd
> #quartile and Max value
> summary(WorkAge1)
   Min. 1st Qu.  Median    Mean 3rd Qu.    Max.
  18.00   44.50   54.00   47.35   56.25   58.00
```

***Figure 2-28.** Finding out major statistical parameters in R using summary() command*

If you use the `summary(dataset)` command, then if required you can use additional commands, like `sd(dataset)`, `var(dataset)`, etc., to obtain the additional parameters of interest related to the data.

2.4 Probability

The concepts of probability and related distributions are as important to business analytics as to the field of pure statistics. Some of the important concepts used in business analytics such as Bayesian theory and decision trees, etc., are based on the concepts of probability.

As you are aware, *probability* in simple terms is the chance of an event happening. In some cases, we may have some prior information related to the event; in other cases, the event may be random—that is, we may not have prior knowledge of the outcome. A popular way to describe the probability is with the example of tossing a coin or tossing a dice. A coin has two sides, and when it is tossed, the probability of either the head or the tail coming up is 1/2 because in any throw either it can be the head or the tail that comes up. You can validate this by tossing up the coin many times and observing that the probability of either the head or the tail coming up is around 50 percent (i.e., 1/2). Similarly, the probability of any one of the numbers being rolled using the dice is 1/6, which can be again validated by tossing the dice many times.

If an event is not likely to happen, the probability of the same is 0. However, if an event is sure to happen, the probability of this is 1. However, the probability of an event is always between 0 and 1 and depends upon the chance of it happening or the uncertainty associated with its happening.

Mathematically, the probability of any event "e" is the ratio of the number of outcomes of an event to the total number of possible outcomes. It is denoted as P (e).

P(e) = n/N, where "n" is the outcome of an event and "N" is the total number of outcomes.

Any given two or more events can happen independent of each other. Similarly, any two or more events can happen exclusive of each other.

Example 1: Can you travel at the same time to two destinations in opposite directions? If you travel toward the west direction, you can't travel toward the east direction at the same time.

Example 2: If we are making profit in one of the client accounts, we cannot make loss in the same account.

Examples 1 and 2 are types of events that exclude the happening of a particular event when the other event happens; they are known as *mutually exclusive events.*

Example 3: A person "tossing a coin" and "raining" can happen at the same, but neither impacts the outcome of the other.

Example 4: A company may make profit and at the same time have legal issues. One event (of making profit) does not have an impact on the other event (of having legal issues).

Examples 3 and 4 are types of events that do not impact the outcome of each other; they are known as mutually independent events. These are also the examples of mutually nonexclusive events as both outcomes can happen at the same time.

2.4.1 Rules of Probability

Probability defines four fundamental rules. They are explained below.

2.4.1.1 Probability of Mutually Exclusive Events

The probability of two mutually exclusive events (say X and Y) happening at the same time is 0 as by definition both do not happen at the same time: P(X and Y) = $P(X \cap Y) = 0$.

The probability of n mutually exclusive events (say A1, A2...AN) happening at the same time is 0 as by definition all these events do not happen at the same time: P(A1 and A2 and...AN) = $P(A1 \cap A2 \cap ... \cap AN) = 0$.

However, the probability of one of the two mutually exclusive events (say X or Y) happening is the sum of the probability of each of these events happening:

P(X or Y) = $P(X \cup Y) = P(X) + P(Y)$.

Example: What is the probability of a dice showing 1 or 3?

P(A) = 1/ 6 and P(3) = 1/6

P(1 or 3) = P(1) + P(3) = 1/6 + 1/6 = 2/6 = 1/3

2.4.1.2 Probability of Mutually Nonexclusive Events

Mutually nonexclusive events are the ones that are not mutually exclusive; two events A and B are mutually exclusive if the events A and B have at least one common outcome.

The probability of two mutually nonexclusive events (i.e., the sum of the probability of each of these events happening minus the probability of both of these events happening at the same time, or in other words together) is P(X or Y) = P(X U Y) = P(X) + P(Y) – P(X ∩ Y).

The probability of n mutually nonexclusive events is P(A1 or A2 or …or AN) = P(A1 U A2 U…U AN) = P(A1) + P(A2) +…+ P(AN) – P(A1 ∩ A2 ∩ A3 ∩ … ∩ AN).

Two sets are nonmutually exclusive if they share common elements. Let us consider two sets of all numbers from 1 to 10 and the set of all odd numbers from 1 to 16:

Set A = {1,2,3,4,5,6,7,8,9,10}

Set B = {1,3,5,7,9,11,13,15}

Two sets are nonmutually exclusive since they share the common elements of {1,3,5,7,9}.

Example: An urn contains two red, four green, five blue, and three yellow marbles. If a single random marble is chosen from the box, what is the probability that it is a red or green marble?

P(red or gree) = P(red) + p(green) – P(red)*P(green)

= 2/14 + 4/14 – 2/14*4/14 = 0.14 + 0.28 – 0.04 = 0.38

2.4.1.3 Probability of Mutually Independent Events

The probability of two mutually independent events happening at the same time is P(X and Y) = P(X) * P(Y).

2.4.1.4 The Probability of the Complement

The probability of event A is P(A) and the probability of its complement, an event not occurring, is P (Ac) = 1 – P (A).

2.4.2 Probability Distributions

Random variables are important in analysis. Probability distributions depict the distribution of the values of a random variable. The distributions can help in selecting the right algorithms, and hence plotting the distribution of the data is an important part of the analytical process; this is performed as a part of exploratory data analysis (EDA). The following are some important probability distributions:

- Normal distribution
- Binomial distribution
- Poisson distribution
- Uniform distribution
- Chi-squared distribution
- Exponential distribution

We will not be discussing all of these. There are many more types of distributions possible including F-distribution, hypergeometric distribution, joint and marginal probability distributions, and conditional distributions. We will discuss only normal distribution, binomial distribution, and Poisson distribution in this chapter. We will discuss more about these distributions and other relevant distributions in later chapters.

2.4.2.1 Normal Distribution

A huge amount of data is considered to be normally distributed if the distribution is normally centered around the mean, as shown in Figure 2-29. Normal distribution is observed in real life in many situations. On account of the bell shape of the distribution, the normal distribution is also called *bell curve*. The properties of normal distribution typically having 68 percent of the values within +/- 1 standard deviation, 95% of the values within +/- 2 standard deviation, and 99.7 percent of the values within +/- 3 standard deviation are the ones heavily used in most of the analytical techniques and so are also the properties of standard normal curve. The standard normal curve has a mean of 0 and a standard deviation of 1. Z-score, used to normalize the values of the features in a regression, is based on the concept of standard normal distribution. The normal distribution is a bell-shaped curve and is generated using the `pnorm()` function in R.

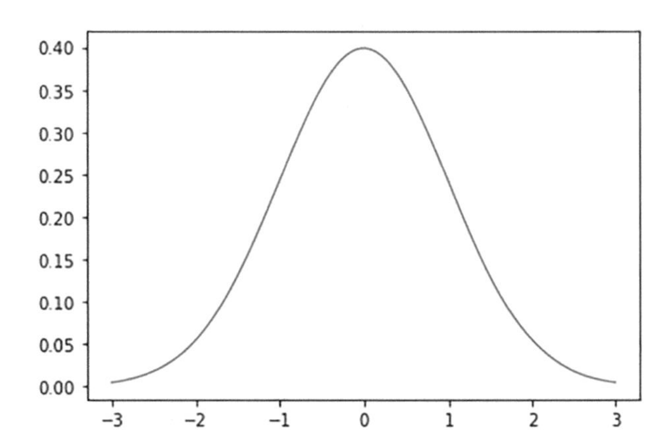

Figure 2-29. *Typical normal distribution function*

One of the important examples of the application of the normal distribution is a statistical process control that is applied in most of the industries to understand the process performance and control it. Another important example is of the distribution of the performance of the employees to identify a few employees as high performers and some as low performers, whereas most of the employees perform around the average.

Consider that an organization has 150 employees; the average grade of the employees is 3, the minimum grade is 1, and the maximum is 5. The standard deviation is 1.5. The percentage of employees getting a grade of 4 and above is calculated using R, as shown in Figure 2-30. We use the `pnorm()` function to calculate the probability.

```
> ##Normal distribution
> pnorm(4, mean=3,sd=1.5,lower.tail=FALSE)
[1] 0.2524925
```

Figure 2-30. *Example of a normal distribution problem solved in R*

Please note that we are interested in the upper tail as we want to know the percentage of employees who have received a grade of 4 or 5. The answer here is 25.25 percent.

2.4.2.2 Binomial Distribution

Binomial distribution normally follows where success or failure is measured, as shown in Figure 2-31. In a cricket match, tossing a coin is an important event at the beginning of the match to decide which side bats (or fields) first. Tossing a coin and calling for "head" wins you the toss if "head" is the outcome. Otherwise, if the "tail" is the outcome, you lose the toss.

```
> ## Binomial Distribution
> pbinom(4, size=15, prob=0.2)
[1] 0.8357663
```

Figure 2-31. *Example of a binomial distribution problem solved in R*

As the sales head of an organization, you have submitted responses to 15 tenders. There are five contenders in each tender. You may be successful in some or all or may be successful in none. As there are five contenders in each tender and each tender can be won by only one company; the probability of winning the tender is 0.2 (i.e., 1/5). You want to win more than four of the fifteen tenders. You can find out the probability of winning four or fewer tenders employing binomial distribution using R (see Figure 2-31). We use the `pbinorm()` function to find out the probability of this example given the binomial distribution of the data.

Please note that the `pbinom()` function uses the cumulative probability distribution function for binomial distribution as we are interested in knowing the chance of winning four or more tenders. As you can see, the probability of winning four or fewer tenders is 83.58 percent. The probability of winning more than four tenders is (100 – 83.58) = 16.42 percent.

Typical binomial distribution looks like Figure 2-32.

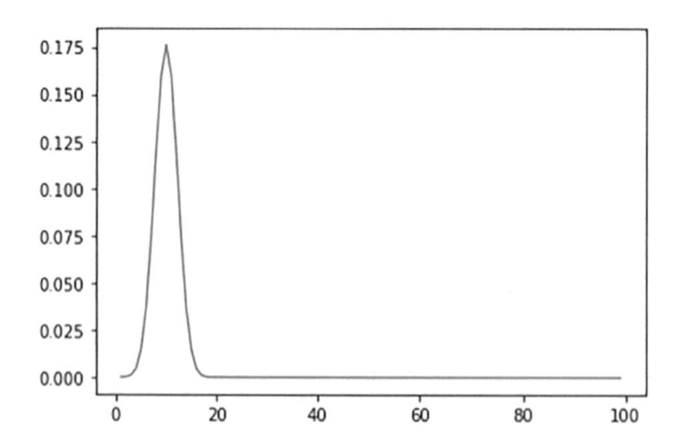

Figure 2-32. *Typical binomial distribution function*

2.4.2.3 Poisson Distribution

Poisson distribution represents the independent events happening in a time interval. The arrival of calls at a call center or the arrival of customers in a banking hall or the arrival of passengers at an airport/bus terminus follow Poisson distribution.

Let us take an example of the number of customers arriving at a particular bank's specific branch office. Suppose an average of 20 customers are arriving per hour. We can find out the probability of 26 or more customers arriving at the bank's branch per hour using R and Poisson distribution (see Figure 2-33). We use the `ppois()` function to find out the outcome.

```
> ##Poisson distribution
> ppois(25, lambda=20, lower=FALSE)
[1] 0.112185
```

***Figure 2-33.** Example of a Poisson distribution problem solved in R*

Please note that we have used `lower = FALSE` as we are interested in the upper tail and we want to know the probability of 26 or more customers arriving at the bank's branch per hour. The answer here is 11.22 percent.

A typical Poisson distribution function may look like in Figure 2-34.

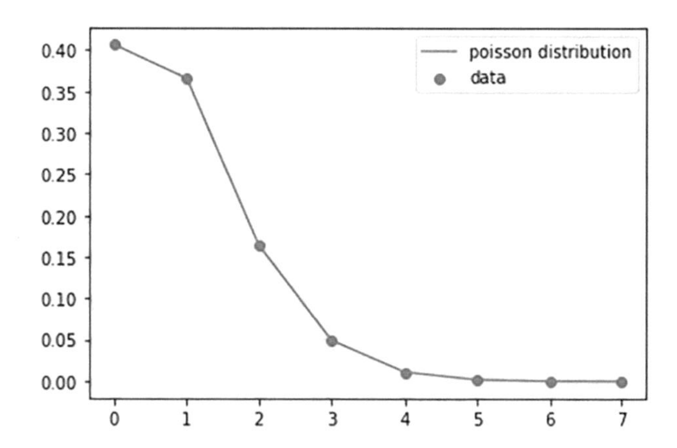

***Figure 2-34.** Typical Poisson distribution function*

2.4.3 Conditional Probability

The conditional probability is the probability of event A occurring given that event B has already occurred. It is the probability of one event occurring with some relationship with the other events. This conditional probability is written as P(A|B), meaning the probability of A given B has occurred.

If A and B are both independent events (where event A occurring has no impact on event B occurring), then the conditional probability of (A | B) is the probability of event A, P(A).

P(A}B) = P(A)

If events A and B are not independent,

P(A|B) = P(A AND B) / P(B)

Conditional probability has many areas of application and is also used quite frequently in analytics algorithms. For example, the enrollment of students to a specific university depends on university program ranking as well as tuition fees. Similarly, the weather reports of rain in your area reported by the news channel or radio channel depend on the many conditional things such as the following:

- The wind direction
- The wind speed
- The cold front pushing in your area

Here are some more practical examples of conditional probability:

- Imagine that you are an electronics salesperson selling TVs. The probability of a new customer buying TV on any day is 20 percent. However, if there is any sports events happening, for example Wimbledon or PGA Tournament or a major league basketball game, then the sales are higher with a probability of 48 percent. We can represent the conditional probability of selling a TV if a sports event happening in that month as P(Selling TV | Sports event month); the symbol "|" represents "given that." In this case, the probability of selling TV depends on the sports event.
- Four guests, A, B, C, and D, are expected to attend an event in my new home. Each guest has an equal chance of attending, that is, 25 percent, provided the weather condition is favourable. However, if the weather forecast says there is a 50percent chance of rain, the probability will change: P(Guests Attending | Weather condition).

Example 1

In a group of 100 electronics shopping customers, 40 bought TV (event A), 30 purchased sound systems (event B), and 20 purchased a TV system and a sound system. If a customer chosen at random bought a TV system, what is the probability they also bought sound system?

Step 1: P(A) is given in the question as 40 percent, or 0.4.

Step 2: P(B) is given that is 0.3.

Step 3: $P(A \cap B)$, this is the intersection of A and B, purchasing both A and B is 0.2.

Step 3: Now use the formula to calculate the conditional probability.

$P(B|A) = P(A \cap B) / P(A) = 0.2 / 0.4 = 0.5$.

The probability that a customer bought a sound system, given that they purchased a TV system, is 50 percent.

2.5 Computations on Data Frames

All our discussions in the previous paragraphs were focused on single-dimensional data—for example, a vector with a single feature of the workers, such as the age of workers.

But most of the data we need to analyze is multidimensional and requires thorough knowledge of how to do this. Often you will encounter two-dimensional data with rows of data pertaining to various features represented through columns.

The data set depicted in Figure 2-35 is a data frame. The data frame is nothing but a table structure in R where each column represents the values of a variable and each row represents data related to a case or an instance. In this data frame, we have data related to the name, age, and salary of 10 employees. The data of each employee is depicted through a row, the features or aspects of the employee are depicted through the labels of the columns, and this type of data is captured in the corresponding columns.

```
> ##
> ##Reading three dimensional data
> # Employee name, Employee Age and Employee Salary
> # reading data to a R data frame
> EmpData = read.csv("empdata.csv")
> #printing the contents of the dataframe
> EmpData
   ID EmpName EmpAge EmpSal
1   1    John     18  18000
2   2   Craig     28  28000
3   3    Bill     32  32000
4   4    Nick     42  42000
5   5   Umesh     50  50000
6   6    Rama     55  55000
7   7     Ken     57  57000
8   8     Zen     58  58000
9   9 Roberts     59  59000
10 10    Andy     59  59000
> # Displaying summary statistics of the data frame
> summary(EmpData)
       ID            EmpName       EmpAge           EmpSal
 Min.   : 1.00   Andy    :1   Min.   :18.00   Min.   :18000
 1st Qu.: 3.25   Bill    :1   1st Qu.:34.50   1st Qu.:34500
 Median : 5.50   Craig   :1   Median :52.50   Median :52500
 Mean   : 5.50   John    :1   Mean   :45.80   Mean   :45800
 3rd Qu.: 7.75   Ken     :1   3rd Qu.:57.75   3rd Qu.:57750
 Max.   :10.00   Nick    :1   Max.   :59.00   Max.   :59000
                 (Other):4


>
```

Figure 2-35. *Dataframe in R and its statsitcs*

As you can see in the figure, the command `summary(dataset)` can be used here also to obtain the summary information pertaining to each feature (i.e., the data in each column).

You can now compute additional information required if any (as shown in Figure 2-36).

```
> #Computing standard deviation EMployee Age
> StdDevEmpAge <- sd(EmpData$EmpAge)
> StdDevEmpAge
[1] 14.97999
> #Computing standard deviation of Employee salary
> StdDevEmpSal<-sd(EmpData$EmpSal)
> StdDevEmpSal
[1] 14979.99


>
```

***Figure 2-36.** Computation of standard deviation in R on EmpData data frame features*

As shown, any column from the data set can be accessed using data set name followed by `$column_name`.

In Python, we use the Pandas data frame to perform all the data analysis and to create various models. The same operations described can be performed in Python, as shown in Figure 2-37 and Figure 2-38.

```
import pandas as pd

df = pd.read_csv("empdata.csv", sep=",")

df
```

***Figure 2-37.** Python/Pandas data frame*

Output:

Out[22]:

	ID	EmpName	EmpAge	EmpSal
0	1	John	18	18000
1	2	Craig	28	28000
2	3	Bill	32	32000
3	4	Nick	42	42000
4	5	Umesh	50	50000
5	6	Rama	55	55000
6	7	Ken	57	57000
7	8	Zen	58	58000
8	9	Roberts	59	59000
9	10	Andy	59	59000

Figure 2-38. *Python/Pandas data frame output*

The statistics of the data frame can be explained using the `describe()` function, as shown in Figure 2-39 and Figure 2-40.

```
df.describe()
```

Figure 2-39. *Pandas data frame statistics function describe()*

Out[24]:

	ID	EmpAge	EmpSal
count	10.00000	10.000000	10.000000
mean	5.50000	45.800000	45800.000000
std	3.02765	14.979987	14979.986649
min	1.00000	18.000000	18000.000000
25%	3.25000	34.500000	34500.000000
50%	5.50000	52.500000	52500.000000
75%	7.75000	57.750000	57750.000000
max	10.00000	59.000000	59000.000000

Figure 2-40. *Pandas data frame describe() output*

2.6 Scatter Plot

Scatter plots are an important kind of plot in the analysis of data. These plots depict the relationship between two variables. Scatter plots are normally used to show cause and effect relationships, but any relationship seen in the scatter plots need not always be a cause and effect relationship. Figure 2-41 shows how to create a scatter plot in R, and Figure 2-42 shows the actual scatter plot generated. The underlying concept of correlation will be explained in detail in subsequent chapters about regression.

```
> plot(EmpData$EmpSal ~ EmpData$EmpAge, type="b")

>
```

Figure 2-41. *Code for creating a scatter plot in R*

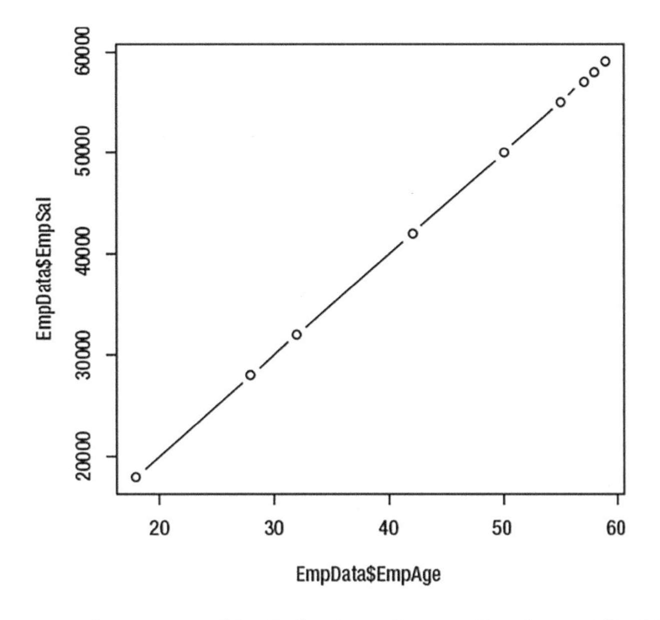

Figure 2-42. *Scatter plot created in R (using the method specified in Figure 4-31A)*

As you can see from this example, there is a direct relationship between employee age and employee salary. The salary of the employees grows in direct proportion to their age. This may not be true in a real scenario. Figure 2-42 shows that the salary of the employee increases proportionate to their age and increases linearly. Such a relationship is known as a *linear relationship*. Please note that `type = "b"` along with the `plot(dataset)` command has created both point and line graph.

Let us now consider another data frame named `EmpData1` with one more additional feature (also known as a *column* or *field*) and with different data in it. In Figure 2-43 you can see the data and summary of the data in this data frame. As you can see in Figure 2-44, one more feature has been added, namely, `EmpPerGrade`, and also have changed the values of salary from the earlier data frame, that is `EmpData`. `EmpData1` has the following data now.

```
>
> EmpData1 <- data.frame(EmpName1, EmpAge1, EmpSal, EmpPerGrade)
> EmpData1
   EmpName1 EmpAge1 EmpSal EmpPerGrade
1      John      28  28000           0
2    George      32  34000           5
3    Jaison      36  40000           5
4   Roberts      38  42000           4
5    Ronnie      40  44000           4
6    Rajesh      44  46000           4
7    Raghav      48  47000           3
8    Sherry      52  48000           3
9      Bill      56  48500           2
10  William      58  49000           1
> summary(EmpData1)
    EmpName1     EmpAge1            EmpSal          EmpPerGrade
 Bill    :1   Min.   :28.0   Min.   :28000   Min.   :0.00
 George  :1   1st Qu.:36.5   1st Qu.:40500   1st Qu.:2.25
 Jaison  :1   Median :42.0   Median :45000   Median :3.50
 John    :1   Mean   :43.2   Mean   :42650   Mean   :3.10
 Raghav  :1   3rd Qu.:51.0   3rd Qu.:47750   3rd Qu.:4.00
 Rajesh  :1   Max.   :58.0   Max.   :49000   Max.   :5.00
 (Other):4
> plot(EmpData1$EmpSal ~ EmpData1$EmpAge1, type = "b")
```

***Figure 2-43.** Data from data frame EmpData1*

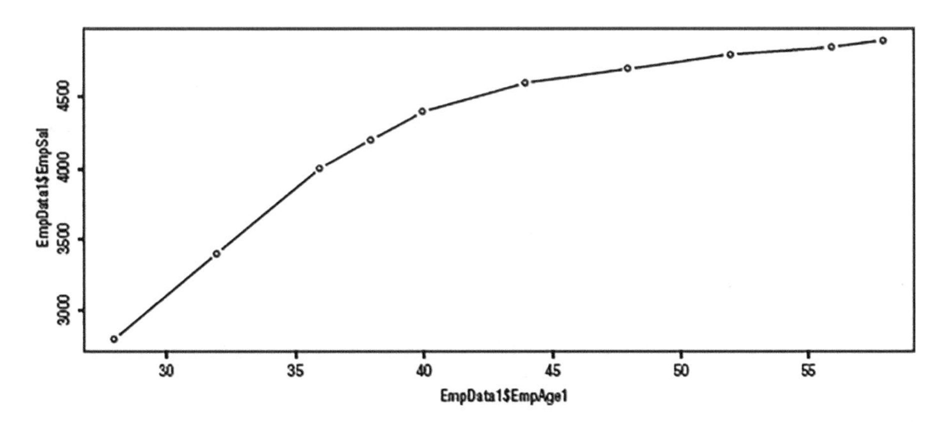

Figure 2-44. *Scatter plot from R showing the changed relationship between two features of data frame EmpData1*

Now, as you can see from Figure 2-45, the relationship between the employee age and the employee salary has changed; as you can observe, as the age grows, the increase in employee salary is not proportional but tapers down. This is normally known as a *quadratic relationship.*

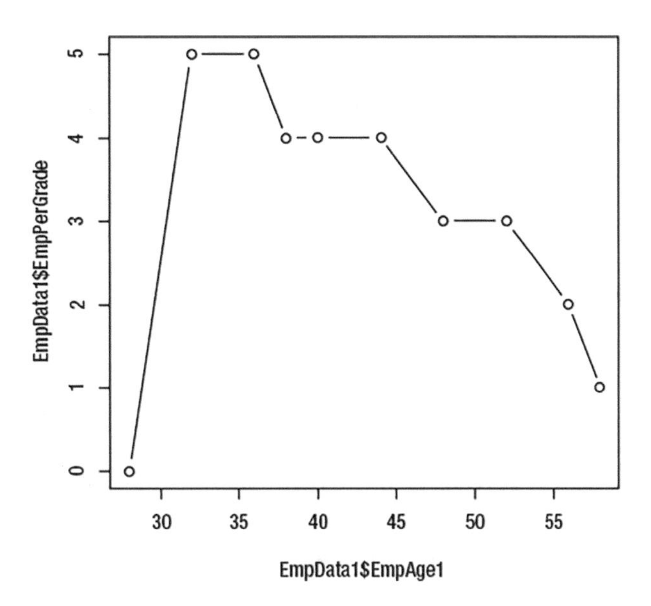

Figure 2-45. *Scatter plot from R showing the changed relationship between two features of data frame EmpData1*

In Figure 2-45, you can see the relationship plotted between employee age and employee performance grade. Ignore the first data point as it was for a new employee joined recently and he was not graded. Hence, the data related to performance grade is 0. Otherwise, as you can observe, as the age progresses (as per the previous data), the performance has come down. In this case, there is an inverse relationship between employee age and employee performance (i.e., as the age progresses, performance is degrading). This is again not a true data and is given only for illustration.

The same can be plotted using Python Pandas, as shown in Figure 2-46.

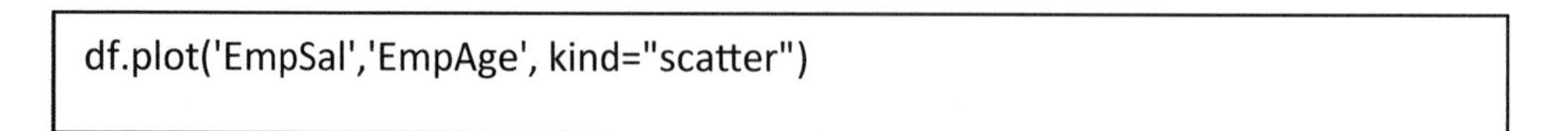

```
df.plot('EmpSal','EmpAge', kind="scatter")
```

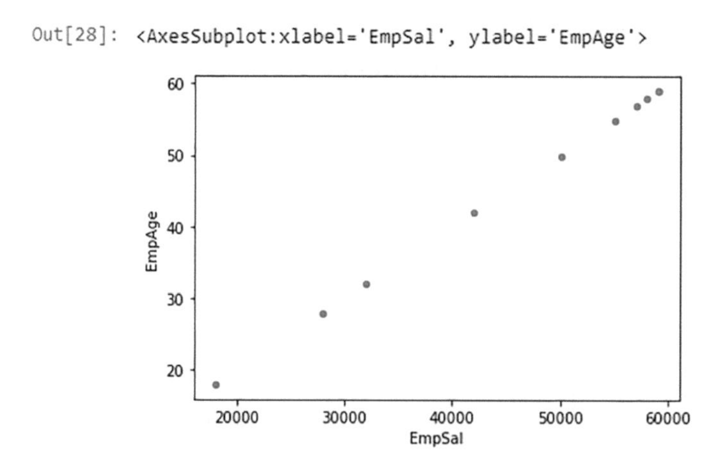

***Figure 2-46.** Scatter plot usingScatter plotsPython Pandas the Pandas plot() function*

2.7 Chapter Summary

- In this chapter, you learned about various statistical parameters of interest in descriptive analytics (mean, median, quantiles, quartiles, percentiles, standard deviation, variance, and mode). You saw how to compute these using R and Python. You also learned how most of these parameters can be gotten through a simple command like `summary(dataset)` using R and Python.

- You explored another important aspect of business analytics known as probability, which has a bearing on many of the techniques used in business analytics (Bayesian techniques, decision trees, etc.).
- You learned the important probability distributions and what they are. You also looked at a few examples of probability distributions and also how to solve business problems employing some popular distributions like normal distribution, binomial distribution, and Poisson distribution, using R and Python.
- You learned one of the important data structures of R and Python: `pandas()` data frames. You learned how to get the summary data from the data contained in these data frames.
- You explored how scatter plots can show the relationship between various features of the data frame and hence enable us to better understand these relationships graphically and easily.

CHAPTER 3

Structured Query Language Analytics

3.1 Introduction

Structured Query Language (SQL) is a popular programming language created to define, populate, manipulate, and query databases. It is not a general-purpose programming language, as it typically works only with databases and cannot be used for creating desktop applications, web applications, or mobile applications. There may be some variations in SQL syntax when it comes to different database engines. By going through the documentation of the database engine you are going to use, you can easily ascertain the differences yourself. SQL is a powerful tool for business and data analysts, and hence we are going to cover it in detail.

SQL is typically pronounced as "sequel" or by its individual letters.

To demonstrate the usage of SQL, in this chapter we have used SQL version of a popular database engine: PostgreSQL.

Typically, SQL can be categorized into the following three broad categories:

- **SQL Data Definition Language (DDL)**: DDL statements are basically used to define or modify the data structure of the database. The commands used are `CREATE TABLE`, `ALTER TABLE`, `DROP TABLE`, `RENAME TABLE`, and `TRUNCATE TABLE`. DDL can also be used to define and modify database objects other than the table, using, e.g., `VIEW`, `INDEX`, `FUNCTION`, and `STORED PROCEDURE`. In essence, SQL DDL is used to create the database schema.

U. R. Hodeghatta and U. Nayak, *Practical Business Analytics Using R and Python*,
https://doi.org/10.1007/978-1-4842-8754-5_3

- **SQL Data Manipulation Language (DML):** These statements are basically used to query the data and add or modify the data. The commands used are `SELECT`, `INSERT`, `UPDATE`, and `DELETE`. Some call the query portion of the DML the Data Query Language (DQL). The `SELECT` command falls into this category.
- **SQL Data Control Language (DCL):** These statements are basically used to provide or remove rights of access on the database objects. The commands used are `GRANT`, `REVOKE`, and `DENY`.

In this chapter, we will not delve deep into definitions and theory, as there are hundreds of books already available on this topic. Typical data types used in a database are `CHAR` to store string data of fixed length; `VARCHAR` to store variable-size string; `TEXT`, etc., to store a string; `SMALLINT/MEDIUMINT/BIGINT/NUMERIC/REAL/DOUBLE PRECISION`, etc., to store numeric data; `MONEY` to store monetary values; `DATE/TIMESTAMP/INTERVAL` to store date, date and time, and time interval values; `BOOLEAN` to store logical data like `TRUE/FALSE`, `YES/NO`, `ON/OFF`; etc. There may be many other data types available for a particular database engine. These data types also vary from one database engine to the other. We are not going to discuss the data types and the limitations related to them here because it is beyond the scope of this chapter. To get a clear understanding of these, please refer to the corresponding documentation from the database engine you use.

This book is meant to provide you with an understanding of the practical use of SQL, particularly in the context of business analytics. Please note that the intention of this book is not to provide a comprehensive chapter covering all the aspects of SQL but to demonstrate those aspects that are most useful to a data engineer or a data scientist.

3.2 Data Used by Us

To familiarize you with SQL, we will use some manufacturing data as many factories are becoming smart by using automation and sensors and have started using artificial intelligence and machine learning to better manage their factories. To start with, we will create three tables: `machine_details`, `machine_status`, and `machine_issues`.

The first table, `machine_details`, consists of the machine details such as machine ID, machine name, and manufacturing line ID. The second table, `machine_status`, consists of the individual machine status as the factory works by getting the details from various sensors. It captures the details at various timestamps of the individual machine data from the corresponding sensors built into the machines. The third table, machine_issues, captures issues with the machines on various dates. While the first table is a master table that is created, the second table gets populated automatically during the running of the factory, and the third table is updated manually by the machine maintenance department.

Let's create the first table using the psql shell interface. The code used is provided here:

```
postgres=# CREATE TABLE machine_details (
postgres(#      machine_id        INTEGER     PRIMARY KEY      NOT NULL,
postgres(#      machine_name      CHARACTER(10)                NOT NULL,
postgres(#      mfg_line          INTEGER                      NOT NULL
postgres(# );
```

Please note that we have represented the key words from SQL in uppercase letters and the details provided by us in lowercase letters. However, technically it is not necessary to type the key words in capital letters. Please note you should end every SQL statement with a semicolon (;). INTEGER can be specified as INT, and CHARACTER can be specified as CHAR.

Caution Do not use hyphens (-) in the names of table, columns, and so on, as they are not recognized. You can use an underscore (_).

To check if the table was created, you can use the \d command on the psql (short form for PostgreSQL) shell interface.

Let's now create the second table, machine_status. The code used for this purpose is provided here:

```
postgres=# CREATE TABLE machine_status (
postgres(#      machine_id          INTEGER         NOT NULL,
postgres(#      pressure_sensor_reading             NUMERIC(7,2),
postgres(#      temp_sensor_reading                 NUMERIC(6,2),
postgres(#      date_and_time                       TIMESTAMP,
postgres(#      FOREIGN KEY (machine_id)
postgres(#      REFERENCES machine_details (machine_id)
postgres(# );
```

Let's now create the third table, machine_issues. The code used for this purpose is provided here:

```
postgres=# CREATE TABLE machine_issues (
postgres(#      date              DATE            NOT NULL,
postgres(#      machine_id        INTEGER         NOT NULL,
postgres(#      issue_descrip     VARCHAR(50),
postgres(#      FOREIGN KEY (machine_id)
postgres(#      REFERENCES machine_details (machine_id)
postgres(# );
```

Now, let's check the creation of all the three tables using the \d command on the psql interface. The code used (i.e., input) is provided here:

```
postgres=# \d
```

The output obtained is provided here:

```
             List of relations
 Schema     | Name             | Type    |  Owner
-----------+-----------------+--------+-------------
 public     | machine_details  | table   | postgres
 public     | machine_issues   | table   | postgres
 public     | machine_status   | table   | postgres
(3 rows)
```

We used SQL DDL to create the previous table structures in our database. In the previous scripts, we used constraints like PRIMARY KEY, NOT NULL, and FOREIGN KEY. A *primary key* represents the column that identifies each row uniquely, which means that a primary key field/column cannot have a NULL value or a duplicate value. A *foreign key* is a column or a combination of columns referencing the primary key of some other table. The NOT NULL constraint ensures that the column must be filled and cannot hold a NULL/blank value. This means that such a column cannot have a missing value. But it does not mean that such a field cannot have a wrong or inaccurate value. In addition, we can use other constraints like CHECK and UNIQUE as relevant.

Currently, all the three tables created by us do not have any data populated in them. Now, we will insert some data into them. First let's populate the table machine_details. We use the INSERT command for this purpose, as follows:

```
postgres=# INSERT INTO machine_details (machine_id, machine_name, mfg_line)
postgres-# VALUES
postgres-#      (1, 'Machine001', 01),
postgres-#      (2, 'Machine002', 01),
postgres-#      (3, 'Machine003', 01),
postgres-#      (4, 'Machine004', 01),
postgres-#      (5, 'Machine005', 01),
postgres-#      (7, 'Machine007', 02),
postgres-#      (8, 'Machine008', 02),
postgres-#      (9, 'Machine009', 02),
postgres-#      (10, 'Machine010', 02),
postgres-#      (11, 'Machine011', 02);
```

Even though the table machine_status is intended to be updated automatically during the manufacturing using the factory interface, for the sake of your understanding of the exercises in this chapter, we will add a few records to it manually using the INSERT command, as follows:

```
postgres=# INSERT INTO machine_status (machine_id, pressure_sensor_reading,
temp_sensor_reading, date_and_time)
postgres-# VALUES
postgres-#      (1, 25.25, 125.26, '2022-04-04 09:10:10'),
postgres-#      (7, 25.50, 125.55, '2022-04-04 09:10:10'),
postgres-#      (2, 55.50, 250.25, '2022-04-04 09:10:10'),
postgres-#      (8, 55.75, 250.55, '2022-04-04 09:10:10'),
postgres-#      (3, 44.40, 220.50, '2022-04-04 09:10:10'),
postgres-#      (9, 44.25, 220.25, '2022-04-04 09:10:10'),
postgres-#      (4, 20.15, 190.25, '2022-04-04 09:10:10'),
postgres-#      (10,20.00, 190.00, '2022-04-04 09:10:10'),
postgres-#      (5,100.35, 500.55, '2022-04-04 09:10:10'),
postgres-#      (11,100.55,500.45, '2022-04-04 09:10:10'),
postgres-#      (1, 25.50, 125.40, '2022-04-05 09:10:10'),
postgres-#      (7, 25.55, 125.75, '2022-04-05 09:10:10'),
postgres-#      (2, 55.25, 250.00, '2022-04-05 09:10:10'),
postgres-#      (8, 55.50, 250.75, '2022-04-05 09:10:10'),
postgres-#      (3, 44.25, 220.25, '2022-04-05 09:10:10'),
```

```
postgres-#       (9, 44.50, 220.40, '2022-04-05 09:10:10'),
postgres-#       (4, 20.25, 190.50, '2022-04-05 09:10:10'),
postgres-#       (10,20.10, 190.25, '2022-04-05 09:10:10'),
postgres-#       (5,100.50, 500.00, '2022-04-05 09:10:10'),
postgres-#       (11,100.75,500.25, '2022-04-05 09:10:10');
```

Similarly, let's insert a few records into the machine_issues table manually on behalf of the maintenance department. The code is provided here:

```
postgres=# INSERT INTO machine_issues (date, machine_id, issue_descrip)
postgres-# VALUES
postgres-#       ('2021-12-31', 1, 'Taken off for preventive maintenance'),
postgres-#       ('2021-12-31', 2, 'Taken off for preventive maintenance'),
postgres-#       ('2021-12-31', 3, 'Taken off for preventive maintenance'),
postgres-#       ('2021-12-31', 4, 'Taken off for preventive maintenance'),
postgres-#       ('2021-12-31', 5, 'Taken off for preventive maintenance'),
postgres-#       ('2021-12-31', 7, 'Taken off for preventive maintenance'),
postgres-#       ('2021-12-31', 8, 'Taken off for preventive maintenance'),
postgres-#       ('2021-12-31', 9, 'Taken off for preventive maintenance'),
postgres-#       ('2021-12-31',10, 'Taken off for preventive maintenance'),
postgres-#       ('2021-12-31',11, 'Taken off for preventive maintenance'),
postgres-#       ('2022-02-02', 1, 'Break down bearing issue'),
postgres-#       ('2022-02-28', 7, 'Break down bearing issue'),
postgres-#       ('2022-03-05', 5, 'Break down leakage issue'),
postgres-#       ('2022-03-31',10, 'Break down valve issue');
```

Any SQL statement will throw an error, pointing out the location of the error, if there is anything wrong with the syntax or data type against the defined data type. Otherwise, in this case, the INSERT statement will show, upon execution, the number of records added.

You can check on the status of the insertion of the records using the DML/DQL statement SELECT. The SELECT statement to query the machine_details table is provided here:

```
postgres=# SELECT * FROM machine_details;
```

The output returned by the previous SELECT statement is provided here:

```
machine_id  | machine_name | mfg_line
------------+--------------+-------------
          1 | Machine001   |          1
          2 | Machine002   |          1
          3 | Machine003   |          1
          4 | Machine004   |          1
          5 | Machine005   |          1
          7 | Machine007   |          2
          8 | Machine008   |          2
          9 | Machine009   |          2
         10 | Machine010   |          2
         11 | Machine011   |          2
(10 rows)
```

The SELECT statement to query on the machine_status table is provided here:

```
postgres=# SELECT * FROM machine_status;
```

The output returned by the previous SELECT statement is provided next.

Note You need to know that SQL automatically wraps the result/output horizontally if there is not enough space to display the results. We have tried to neatly format the wrapped-up output manually, wherever possible, throughout this chapter for better readability. However, at some places the output is too large to format it properly. In such cases, the output has been kept as it is.

```
machine_id | pressure_sensor_reading | temp_sensor_reading |       date_and_time
-----------+-------------------------+---------------------+--------------------
         1 |                   25.25 |              125.26 | 2022-04-04 09:10:10
         7 |                   25.50 |              125.55 | 2022-04-04 09:10:10
         2 |                   55.50 |              250.25 | 2022-04-04 09:10:10
         8 |                   55.75 |              250.55 | 2022-04-04 09:10:10
         3 |                   44.40 |              220.50 | 2022-04-04 09:10:10
         9 |                   44.25 |              220.25 | 2022-04-04 09:10:10
         4 |                   20.15 |              190.25 | 2022-04-04 09:10:10
```

```
        10 |                20.00 |          190.00 | 2022-04-04 09:10:10
         5 |               100.35 |          500.55 | 2022-04-04 09:10:10
        11 |               100.55 |          500.45 | 2022-04-04 09:10:10
         1 |                25.50 |          125.40 | 2022-04-05 09:10:10
         7 |                25.55 |          125.75 | 2022-04-05 09:10:10
         2 |                55.25 |          250.00 | 2022-04-05 09:10:10
         8 |                55.50 |          250.75 | 2022-04-05 09:10:10
         3 |                44.25 |          220.25 | 2022-04-05 09:10:10
         9 |                44.50 |          220.40 | 2022-04-05 09:10:10
         4 |                20.25 |          190.50 | 2022-04-05 09:10:10
        10 |                20.10 |          190.25 | 2022-04-05 09:10:10
         5 |               100.50 |          500.00 | 2022-04-05 09:10:10
        11 |               100.75 |          500.25 | 2022-04-05 09:10:10
(20 rows)
```

The SELECT statement to query the machine_issues table is provided here:

```
postgres=# SELECT * FROM machine_issues;
```

The output returned by the previous SELECT statement is provided here:

```
    date    | machine_id |             issue_descrip
------------+------------+----------------------------------------
 2021-12-31 |          1 | Taken off for preventive maintenance
 2021-12-31 |          2 | Taken off for preventive maintenance
 2021-12-31 |          3 | Taken off for preventive maintenance
 2021-12-31 |          4 | Taken off for preventive maintenance
 2021-12-31 |          5 | Taken off for preventive maintenance
 2021-12-31 |          7 | Taken off for preventive maintenance
 2021-12-31 |          8 | Taken off for preventive maintenance
 2021-12-31 |          9 | Taken off for preventive maintenance
 2021-12-31 |         10 | Taken off for preventive maintenance
 2021-12-31 |         11 | Taken off for preventive maintenance
 2022-02-02 |          1 | Break down bearing issue
 2022-02-28 |          7 | Break down bearing issue
 2022-03-05 |          5 | Break down leakage issue
 2022-03-31 |         10 | Break down valve issue
(14 rows)
```

Another way to confirm the proper insertion of the records is to add RETURNING *; at the end of the INSERT statement.

We can look for DISTINCT or UNIQUE values in a field/column of a table using DISTINCT before the field name in the SELECT statement. For example, we can find out the distinct/unique issues with the machines from the machine_issues table that are provided. The code used in this regard is provided here:

```
postgres=# SELECT DISTINCT issue_descrip
postgres-# FROM machine_issues;
```

The output returned by the previous SELECT statement is provided here:

```
           issue_descrip
--------------------------------------
 Break down leakage issue
 Taken off for preventive maintenance
 Break down valve issue
 Break down bearing issue
(4 rows)
```

We have discussed the creation of the table and the insertion of the records into the table because sometimes as a business analyst or data engineer you may be required to create and populate your own data tables.

3.3 Steps for Business Analytics

Business analysts or data engineers typically work as follows:

1. Initial exploration and understanding of the data.
2. Understanding wrong data, missing data, or issues with the data.
3. Cleaning up the data for further analysis.
4. Further exploration and reporting on the data. This may include shaping the data.

We will explore these steps in detail.

3.3.1 Initial Exploration and Understanding of the Data

Once you have the data, it is important as a business analyst to explore the data and understand it thoroughly. In this regard, a SELECT statement along with its various clauses like WHERE, HAVING, ORDER BY, GROUP BY, TOP, IN, EXISTS, etc., help you explore and understand the data by slicing and dicing it. Further, as your various tables have relationships with each other, the various joins with other tables like INNER JOIN, LEFT JOIN, CROSS JOIN, FULL JOIN, and OUTER JOIN along with set operations like UNION, INTERSECT, and EXCEPT help you explore the data.

Based on the problem/challenge at hand, business analysts must ask themselves questions and then explore the answers using the data. From the machine-related data that we have, let's try to formulate questions related to a few of the problems.

- Which machines had breakdown issues and when?
- Which machines were taken off for preventive maintenance and when?
- If the tolerance range for the pressure for machine 2 are between 50 kPa to 60 kPa, check if the pressure readings sent by the corresponding sensor are within these tolerance limits.
- If the tolerance range for the pressure for machine 2 and machine 8 are between 50 kPa to 60 kPa, check if the pressure readings sent by the corresponding sensor are within these tolerance limits.
- If the tolerance range for the pressure for machine 3 and machine 9 are between 43kPa to 45kPa and the tolerance range for temperature are between 210 degrees Celsius to 230 degrees Celsius, check if both these parameters are within these tolerance limits.

To address these queries, the SELECT statement helps us. Let's answer the first question, i.e., which machines have breakdown issues. If we only require the machine ID and the corresponding breakdown issue, then we can use the simple SELECT statement as follows:

```
postgres=# SELECT machine_id, date, issue_descrip FROM machine_issues WHERE
issue_descrip LIKE '%Break%';
```

The output returned by the previous SELECT statement looks like this:

```
machine_id |    date    |        issue_descrip
------------+------------+-----------------------------------
          1 | 2022-02-02 | Break down bearing issue
          7 | 2022-02-28 | Break down bearing issue
          5 | 2022-03-05 | Break down leakage issue
         10 | 2022-03-31 | Break down valve issue
(4 rows)
```

The LIKE operator matches the pattern specified within the single quotes with the data in the field and throws out those rows that have the matches to the pattern specified. Here, the pattern selected says that in the field issue_descrip we are looking for the word *Break* wherever it is within this field.

Additional Tip In the previous SELECT query, instead of LIKE '%Break%', you can use LIKE '%Break_down%'. The underscore between the two words stands for one character including the space. However, you should note that the words or phrases you are looking for in the LIKE statement are case sensitive.

Similarly, you can check for machine IDs that underwent the preventive maintenance using the following SELECT statement on the machine_issues table:

```
postgres=# SELECT machine_id, date, issue_descrip FROM machine_issues WHERE
issue_descrip LIKE '%preventive%';
```

The output returned by the previous SELECT statement is given here:

```
machine_id |    date    |              issue_descrip
------------+------------+-------------------------------------------
          1 | 2021-12-31 | Taken off for preventive maintenance
          2 | 2021-12-31 | Taken off for preventive maintenance
          3 | 2021-12-31 | Taken off for preventive maintenance
          4 | 2021-12-31 | Taken off for preventive maintenance
          5 | 2021-12-31 | Taken off for preventive maintenance
```

```
          7 | 2021-12-31 | Taken off for preventive maintenance
          8 | 2021-12-31 | Taken off for preventive maintenance
          9 | 2021-12-31 | Taken off for preventive maintenance
         10 | 2021-12-31 | Taken off for preventive maintenance
         11 | 2021-12-31 | Taken off for preventive maintenance
(10 rows)
```

Suppose in the earlier queries we also want the machine name; then the `machine_name` is available in another table, i.e., in the `machine_details` table and not in the table we queried earlier i.e., in the `machine_issues` table. Hence, in order to get the desired result, we have to use the query joining `machine_issues` table with `machine_details` table, as follows:

```
postgres=# SELECT m.machine_id, k.machine_name, m.date, m.issue_descrip
postgres-# FROM machine_issues m, machine_details k WHERE
postgres-# (m.machine_id = k.machine_id AND m.issue_descrip LIKE
'%preventive%');
```

The output returned by the previous query is provided here:

```
machine_id | machine_name |    date    |            issue_descrip
-----------+--------------+------------+-------------------------------------
         1 | Machine001   | 2021-12-31 | Taken off for preventive maintenance
         2 | Machine002   | 2021-12-31 | Taken off for preventive maintenance
         3 | Machine003   | 2021-12-31 | Taken off for preventive maintenance
         4 | Machine004   | 2021-12-31 | Taken off for preventive maintenance
         5 | Machine005   | 2021-12-31 | Taken off for preventive maintenance
         7 | Machine007   | 2021-12-31 | Taken off for preventive maintenance
         8 | Machine008   | 2021-12-31 | Taken off for preventive maintenance
         9 | Machine009   | 2021-12-31 | Taken off for preventive maintenance
        10 | Machine010   | 2021-12-31 | Taken off for preventive maintenance
        11 | Machine011   | 2021-12-31 | Taken off for preventive maintenance
(10 rows)
```

In the previous query, `m` and `k` are the aliases used for the tables `machine_issues` and `machine_details`, respectively. This is used to reduce the need for the repetition of the entire name of the associated table before the column name concerned where multiple tables are joined.

If you want to count the number of preventive maintenance per machine, then you can use the aggregate function COUNT in the SELECT statement on the issue_descrip field and use the GROUP BY function to get the count, as follows:

```
postgres=# SELECT m.machine_id, k.machine_name, m.date, count
(m.issue_descrip)
postgres-# FROM machine_issues m, machine_details k WHERE
postgres-# (m.machine_id = k.machine_id AND m.issue_descrip LIKE
'%preventive%')
postgres-# GROUP BY (m.machine_id, k.machine_name, m.date);
```

The output returned by the previous query is provided here:

```
machine_id | machine_name      |    date    | count
-----------+-------------------+------------+---------
         1 |     Machine001    | 2021-12-31 |     1
         2 |     Machine002    | 2021-12-31 |     1
         3 |     Machine003    | 2021-12-31 |     1
         4 |     Machine004    | 2021-12-31 |     1
         5 |     Machine005    | 2021-12-31 |     1
         7 |     Machine007    | 2021-12-31 |     1
         8 |     Machine008    | 2021-12-31 |     1
         9 |     Machine009    | 2021-12-31 |     1
        10 |     Machine010    | 2021-12-31 |     1
        11 |     Machine011    | 2021-12-31 |     1
(10 rows)
```

Caution You should ensure that the GROUP BY clause includes all the fields in the SELECT statement other than the aggregate function. Otherwise, you will get an error. Alternatively, the field not used in the GROUP BY clause needs to have the AGGREGATE function on it in the SELECT statement.

Additional Tip There are various aggregate functions like SUM, COUNT, MAX, MIN, and AVG. You may use them to aggregate the data depending upon the context.

Let's take the next question; i.e., if the tolerance range for the pressure for machine 2 are between 50 kPa to 60 kPa, check if the pressure readings sent by the corresponding sensor are within these tolerance limits.

We have all the required data related to this query in the `machine_status` table. Let's query this table for the answer, as follows:

```
postgres=# SELECT machine_id, pressure_sensor_reading FROM machine_status
postgres-# WHERE (machine_id = 2 AND pressure_sensor_reading BETWEEN 50
and 60);
```

The output of the previous query is provided here:

```
 machine_id | pressure_sensor_reading
------------+--------------------------------
          2 |                  55.50
          2 |                  55.25
(2 rows)
```

This query will return only those records that have the `pressure_sensor_reading` between 50 kPa and 60 kPa. It does not specify if there are any records pertaining to machine 2 that have the `pressure_sensor_reading` outside the range of 50kPa to 60kPa. Hence, the answer to our question addressed by this query is partial. To get the other part of the answer, we need to query the `machine_status` table additionally as follows:

```
postgres=# SELECT machine_id, pressure_sensor_reading FROM machine_status
postgres-# WHERE (machine_id = 2 AND pressure_sensor_reading NOT BETWEEN 50
and 60);
```

The output returned by the previous query is provided here:

```
 machine_id | pressure_sensor_reading
------------+--------------------------------
(0 rows)
```

The previous output clearly shows that all the values of the pressure_sensor_reading for machine 2 are within the tolerance limits specified.

If we want to understand the MAX and MIN values of the pressure_sensor_reading instead of the individual reading, thereby getting the maximum and minimum values within the tolerance limits, we can use the following query:

```
postgres=# SELECT machine_id, max(pressure_sensor_reading), min
(pressure_sensor_reading)
postgres-# FROM machine_status
postgres-# WHERE (machine_id = 2 AND pressure_sensor_reading BETWEEN
50 AND 60)
postgres-# GROUP BY machine_id;
```

The output of the previous query is provided here:

```
 machine_id |  max  |  min
------------+-------+-------
          2 | 55.50 | 55.25
(1 row)
```

Caution The WHERE clause cannot have AGGREGATE functions. If you include AGGREGATE functions in the WHERE clause, you will get an error. If you want to include AGGREGATE functions to filter the data, then you need to use the HAVING clause instead.

Let's take up the next question: if the tolerance range for the pressure for machine 2 and machine 8 are between 50 kPa to 60 kPa, check if the pressure readings sent by the corresponding sensor are within these tolerance limits. For this again, we need to query only one table, i.e., machine_status, as follows:

```
postgres=# SELECT machine_id, pressure_sensor_reading FROM machine_status
postgres-# WHERE ((machine_id = 2 OR machine_id =8) AND
(pressure_sensor_reading
postgres(# NOT BETWEEN 50 AND 60));
```

The output returned by the previous query is provided here:

```
machine_id | pressure_sensor_reading
-----------+---------------------------------
(0 rows)
```

You can clearly validate the above result using the following query where you can see that all the pressure readings are within the tolerance limits specified:

```
postgres=# SELECT machine_id, pressure_sensor_reading FROM machine_status
postgres-# WHERE ((machine_id = 2 OR machine_id = 8) AND
(pressure_sensor_reading
postgres(# BETWEEN 50 AND 60));
```

The output returned by the previous query is provided here:

```
machine_id | pressure_sensor_reading
-----------+----------------------------------
         2 |                 55.50
         8 |                 55.75
         2 |                 55.25
         8 |                 55.50
(4 rows)
```

Additional Tip In the previous query, you need to use OR between the two `machine_ids`. Otherwise, zero records will be thrown out as the output as it is impossible to have `machine_id` = 2 the same as `machine_id` = 8. OR looks for applicability to one of the aspects. Even if one aspect turns out to be true, it will be included. AND looks for applicability to all of the aspects. Only if all the aspects turn out to be true will the query return the related output.

Let's take up our next question, which is more complicated than the ones we dealt with previously: if the tolerance range for the pressure for machine 3 and machine 9 are between 43kPa to 45kPa and the tolerance range for temperature are between 210 degrees Celsius to 230 degrees Celsius, check if both these parameters are within these tolerance limits. The query (code) related to the same is provided here:

```
postgres=# SELECT machine_id, pressure_sensor_reading, temp_sensor_reading
postgres-# FROM machine_status
postgres-# WHERE ((machine_id = 3 OR machine_id = 9) AND
(pressure_sensor_reading
postgres(# BETWEEN 43 AND 45) AND (temp_sensor_reading BETWEEN 210 AND 230));
```

The output of the previous query is provided here:

```
 machine_id | pressure_sensor_reading | temp_sensor_reading
------------+-------------------------+---------------------------
          3 |                   44.40 |              220.50
          9 |                   44.25 |              220.25
          3 |                   44.25 |              220.25
          9 |                   44.50 |              220.40
(4 rows)
```

We can add ORDER BY machine id ASC to get the details in the ascending order of the machine_id to allow better readability of the data when there are more records in the table. Here is the query and the result:

```
postgres=# SELECT machine_id, pressure_sensor_reading, temp_sensor_reading
postgres-# FROM machine_status
postgres-# WHERE ((machine_id = 3 OR machine_id = 9) AND
(pressure_sensor_reading
postgres(# BETWEEN 43 AND 45) AND (temp_sensor_reading BETWEEN 210 AND 230))
postgres-# ORDER BY machine_id ASC;
```

The output returned by the previous query is provided here:

```
 machine_id | pressure_sensor_reading | temp_sensor_reading
------------+-------------------------+---------------------------
          3 |                   44.40 |              220.50
          3 |                   44.25 |              220.25
          9 |                   44.25 |              220.25
          9 |                   44.50 |              220.40
(4 rows)
```

Additional Tip Use of ASC is optional. When it is not specified explicitly, it is assumed to be so, and the records will be thrown out in the ascending order only. Using DESC instead of ASC will give you the data in the descending order of the machine_id, i.e., in this case basically starting with 9.

To understand very clearly and be doubly sure that there are no records pertaining to machine 3 or 9, which are not within the tolerance limits specified in the question, let's use the following query:

```
postgres=# SELECT machine_id, pressure_sensor_reading, temp_sensor_reading
postgres-# FROM machine_status
postgres-# WHERE ((machine_id = 3 OR machine_id = 9) AND (pressure_
sensor_reading
postgres(# NOT BETWEEN 43 AND 45) AND (temp_sensor_reading NOT BETWEEN 210
AND 230));
```

The output returned by the previous code/query is provided here:

```
machine_id | pressure_sensor_reading | temp_sensor_reading
-----------+-------------------------+-----------------------------
(0 rows)
```

3.3.2 Understanding Incorrect and Missing Data, and Correcting Such Data

All the data available in the world may not be clean, i.e., correct and complete. It may be easy for a domain expert to check if the data is complete, but it may be difficult for a data engineer or a data scientist to understand if the data available is correct. In cases where they are not domain experts, it is advisable for them to have the discussion with the domain experts or those involved in the process related to the data to understand the characteristics and bounds of the data. This is essential most of the time to understand the accuracy and integrity of the data.

Queries in the SELECT statement filtering on IS NULL or = ' ' for the value of a field can help us understand the presence of NULL/blank or no values in our data.

In the previous examples, our data is clean as we created the data based on the records. However, to help you to understand the process, we will be adding/creating some incomplete and incorrect records to these tables. Please note that you can use the INSERT/UPDATE functions to do so. After modifying the data, the details in the previous three tables are as follows. The queries in this regard and the outputs are provided here.

Code:

```
postgres=# SELECT * from machine_details;
```

Output:

```
machine_id | machine_name      | mfg_line
-----------+-------------------+-------------
         1 |     Machine001    |       1
         2 |     Machine002    |       1
         3 |     Machine003    |       1
         4 |     Machine004    |       1
         5 |     Machine005    |       1
         7 |     Machine007    |       2
         8 |     Machine008    |       2
         9 |     Machine009    |       2
        10 |     Machine010    |       2
        11 |     Machine011    |       2
        12 |     Machine012    |       3
        13 |                   |       3
        14 |                   |       3
        15 |                   |       3
        16 |                   |       3
        17 |                   |       0
(16 rows)
```

Code:

```
postgres=# SELECT * from machine_status;
```

Output:

```
machine_id | pressure_sensor_reading | temp_sensor_reading |     date_and_time
-----------+-------------------------+---------------------+---------------------
         1 |                   25.25 |              125.26 | 2022-04-04 09:10:10
         7 |                   25.50 |              125.55 | 2022-04-04 09:10:10
         2 |                   55.50 |              250.25 | 2022-04-04 09:10:10
         8 |                   55.75 |              250.55 | 2022-04-04 09:10:10
         3 |                   44.40 |              220.50 | 2022-04-04 09:10:10
         9 |                   44.25 |              220.25 | 2022-04-04 09:10:10
         4 |                   20.15 |              190.25 | 2022-04-04 09:10:10
        10 |                   20.00 |              190.00 | 2022-04-04 09:10:10
         5 |                  100.35 |              500.55 | 2022-04-04 09:10:10
        11 |                  100.55 |              500.45 | 2022-04-04 09:10:10
         1 |                   25.50 |              125.40 | 2022-04-05 09:10:10
         7 |                   25.55 |              125.75 | 2022-04-05 09:10:10
         2 |                   55.25 |              250.00 | 2022-04-05 09:10:10
         8 |                   55.50 |              250.75 | 2022-04-05 09:10:10
         3 |                   44.25 |              220.25 | 2022-04-05 09:10:10
         9 |                   44.50 |              220.40 | 2022-04-05 09:10:10
         4 |                   20.25 |              190.50 | 2022-04-05 09:10:10
        10 |                   20.10 |              190.25 | 2022-04-05 09:10:10
         5 |                  100.50 |              500.00 | 2022-04-05 09:10:10
        11 |                  100.75 |              500.25 | 2022-04-05 09:10:10
        12 |                         |                     | 2022-04-09 09:10:10
         3 |                         |                     | 2022-04-09 09:10:10
        14 |                         |                     | 2022-04-09 09:10:10
        15 |                         |                     | 2022-04-09 09:10:10
        16 |                         |                     | 2022-04-09 09:10:10
(25 rows)
```

Code:

```
postgres=# SELECT * from machine_issues;
```

Output:

```
    date    | machine_id |             issue_descrip
------------+------------+------------------------------------
 2021-12-31 |          1 | Taken off for preventive maintenance
 2021-12-31 |          2 | Taken off for preventive maintenance
 2021-12-31 |          3 | Taken off for preventive maintenance
 2021-12-31 |          4 | Taken off for preventive maintenance
 2021-12-31 |          5 | Taken off for preventive maintenance
 2021-12-31 |          7 | Taken off for preventive maintenance
 2021-12-31 |          8 | Taken off for preventive maintenance
 2021-12-31 |          9 | Taken off for preventive maintenance
 2021-12-31 |         10 | Taken off for preventive maintenance
 2021-12-31 |         11 | Taken off for preventive maintenance
 2022-02-02 |          1 | Break down bearing issue
 2022-02-28 |          7 | Break down bearing issue
 2022-03-05 |          5 | Break down leakage issue
 2022-03-31 |         10 | Break down valve issue
 2022-04-09 |         13 | Taken off for preventive maintenance
 2022-04-09 |         14 | Taken off for preventive maintenance
 2022-04-09 |         16 | Taken off for preventive maintenance
(17 rows)
```

You can now see some columns with no/blank details in two of these tables. Some of the table columns have been defined with the NOT NULL constraint and hence will not allow any NULL/no/blank value. However, other columns without this constraint may have such NULL/no/blank value. We can find out such rows and columns in each of the tables as follows.

Code:

```
postgres=# SELECT * FROM machine_details
postgres-# WHERE machine_name = ' ';
```

Output:

```
machine_id | machine_name | mfg_line
-----------+--------------+-------------
        13 |              |      3
        14 |              |      3
        15 |              |      3
        16 |              |      3
        17 |              |      0
(5 rows)
```

Please note that we know that machine_id and mfg_line are integer fields and cannot be NULL as defined by the constraints on those columns set while creating the table. Hence, we have included here only the machine_name column in our query.

In the previous result, you can observe mfg_line = 0 for the machine_id = 17. You may understand in discussion with the procurement team or manufacturing team that this machine is a new machine and is yet to be deployed on any manufacturing line. Hence mfg_line is currently set as 0, which means it is yet to be deployed.

As you know that the machine-naming convention is based on the machine_id, you can easily update without reference to anybody else the names of the machines in the table machine_details using the UPDATE function, as follows:

```
postgres=# UPDATE machine_details
postgres-# SET machine_name = 'Machine013'
postgres-# WHERE machine_id = 13;
```

You should repeat the previous statement with the appropriate details to set the machine_name for different machine_id.

Additional Tip Instead of repeating many times the previous query for different machine_ids, you can use the CASE statement with the UPDATE statement. The CASE statement is described later in this chapter.

After doing this, we can now observe from the following query that all the values in the table machine_details are complete and correct.

Code:

```
postgres=# SELECT * FROM machine_details;
```

Output:

```
machine_id | machine_name    | mfg_line
-----------+-----------------+------------
         1 |    Machine001   |          1
         2 |    Machine002   |          1
         3 |    Machine003   |          1
         4 |    Machine004   |          1
         5 |    Machine005   |          1
         7 |    Machine007   |          2
         8 |    Machine008   |          2
         9 |    Machine009   |          2
        10 |    Machine010   |          2
        11 |    Machine011   |          2
        12 |    Machine012   |          3
        13 |    Machine013   |          3
        14 |    Machine014   |          3
        15 |    Machine015   |          3
        16 |    Machine016   |          3
        17 |    Machine017   |          0
(16 rows)
```

Now, let's look at the details from the table machine_status.

Code:

```
postgres=# SELECT * FROM machine_status
postgres-# WHERE (pressure_sensor_reading IS NULL OR temp_sensor_reading IS NULL);
```

Output:

```
machine_id | pressure_sensor_reading | temp_sensor_reading |     date_and_time
-----------+-------------------------+---------------------+-------------------
        12 |                         |                     | 2022-04-09 09:10:10
        13 |                         |                     | 2022-04-09 09:10:10
        14 |                         |                     | 2022-04-09 09:10:10
        15 |                         |                     | 2022-04-09 09:10:10
        16 |                         |                     | 2022-04-09 09:10:10
(5 rows)
```

If you think through the previous result, you may conclude that at the specified time when the automated system tried to get the data from the sensors, possibly the sensors were unavailable, which means possibly the machines were down or taken off for preventive maintenance. Alternatively, it is possible that the sensors were faulty and returning a NULL value. From the data available with us, we can check if the machine was taken off for preventive maintenance or was on maintenance on account of the machine breakdown. Let's combine the data from the two tables, i.e., machine_status and machine_issues, to check on these aspects.

Code:

```
postgres=# SELECT m.machine_id, k.pressure_sensor_reading,
k.temp_sensor_reading, k.date_and_time, m.date, m.issue_descrip
postgres-# FROM machine_issues m, machine_status k
postgres-# WHERE (k.pressure_sensor_reading IS NULL OR
k.temp_sensor_reading IS NULL)
postgres-# AND (m.machine_id = k.machine_id);
```

Output:

```
machine_id | pressure_sensor_reading | temp_sensor_reading |
  date_and_time       |    date    |      issue_descrip
-----------+-------------------------+---------------------+
---------------------+------------+-----------------
        13 |                         |                     |
  2022-04-09 09:10:10 | 2022-04-09 | Taken off for preventive maintenance
        14 |                         |                     |
  2022-04-09 09:10:10 | 2022-04-09 | Taken off for preventive maintenance
        16 |                         |                     |
  2022-04-09 09:10:10 | 2022-04-09 | Taken off for preventive maintenance
(3 rows)
```

Additional Tip Alternatively, you can use the TIMESTAMP data type converted to the DATE data type so that it is easy to see the date values side by side. For this purpose, you may use the type cast `k.date_and_time::DATE` in the SELECT function. The CAST feature is very useful for converting the data type from one type to another. There are other ways of type casting. Please refer to the relevant documentation pertaining to the database engine you use.

As you can see, three machines with machine_ids 13, 14, and 16 were taken off for preventive maintenance on April 9, 2022. However, as you can see two other machines with machine_ids 12 and 15 from the machine_status table are not in this list, but they have the pressure_sensor_reading as NULL and temp_sensor_reading as NULL on April 9, 2022, which means that possibly the sensors were not readable by our application populating the data into the table machine_status, but the machines were working. Otherwise, their numbers would have been listed in the machine_issues table on April 9, 2022 (presuming that the machine maintenance department has not missed out to enter the issues with these machines in the machine_issues table). Now, we cannot retain this faulty data in the machine_status table as it can skew all our analysis. Hence, we have two options, i.e., either to remove the rows with null values on April 9, 2022, for machine_ids 12 and 15 from the machine_status table or to substitute some values within tolerance limits like average values or median values for the said machine from the table (if we are sure in discussion with the floor personnel that these two machines were working on that day and we have a good amount of the data from which we can derive these average or median values) for both the pressure_sensor_reading and temp_sensor_reading. The decision we must take is based on the impact of each action. If we are the business analysts, we would inform the concerned technicians to rectify the sensors and delete the blank value rows pertaining to machine_ids 12 and 15 for April 9, 2022, from the machine_status table. We will also remove the rows with blank values pertaining to the machines taken off for preventive maintenance in the machine_status table. For this purpose, we can use the DELETE function of SQL.

Code:

```
postgres=# DELETE FROM machine_status
postgres-# WHERE (pressure_sensor_reading IS NULL OR temp_sensor_reading
IS NULL)
postgres-# AND (date_and_time = '2022-04-09 09:10:10');
```

The status returned is DELETE 5. This indicates that five rows are deleted.

Alternatively, you have to substitute the missing sensor values with the average sensor value or median sensor value using the UPDATE function for two machines that were not taken off for preventive maintenance. This means that the nonexistent values of these are replaced with the average of the existing values of that column or median of the existing values of that column or the most representative value for that column, for the particular machine_id, based on the nature of the sensor. For example, for a particular machine_id we have only three rows, and for one of the columns, we have two values populated and one value is not populated. We are sure from the other data available to us that particular value is missing and needs to be filled. In this case, we can use the average of the two existing values or the most representative value for that field.

3.3.3 Further Exploration and Reporting on the Data

After the cleaning of the wrong/incorrect data by deletion/correction or the substitution of the missing values with appropriate but maybe approximate values, we need to explore the data more and come up with further insights. For this purpose, various JOINs in SQL will help.

Let's first explore the JOINs.

INNER JOIN (also known as EQUIJOIN) works like the intersection of two sets in the case of joining the two tables. The JOIN is made on a common field in both the tables. It returns a new row with all the fields included in the SELECT statement while getting the relevant data from the joined tables where the common field matches for both the tables. It ignores the rows in both the tables where the common field on which join is made does not match. In our case, we can have an inner join on both the machine_details table and the machine_status table as both tables have the machine_id column in common.

Code:

```
postgres=# SELECT m.machine_id, m.machine_name, m.mfg_line,
k.pressure_sensor_reading, k.temp_sensor_reading, k.date_and_time
postgres-# FROM machine_details m
postgres-# INNER JOIN machine_status k
postgres-# ON m.machine_id = k.machine_id
postgres-# ORDER BY m.machine_id ASC;
```

Output:

```
machine_id | machine_name | mfg_line | pressure_sensor_reading |
 temp_sensor_reading |     date_and_time
-----------+-------------+---------+-------------------------+
---------------------+-----------------
         1 | Machine001   |        1 |                   25.25 |
             125.26 | 2022-04-04 09:10:10
         1 | Machine001   |        1 |                   25.50 |
             125.40 | 2022-04-05 09:10:10
         2 | Machine002   |        1 |                   55.50 |
             250.25 | 2022-04-04 09:10:10
         2 | Machine002   |        1 |                   55.25 |
             250.00 | 2022-04-05 09:10:10
         3 | Machine003   |        1 |                   44.25 |
             220.25 | 2022-04-05 09:10:10
         3 | Machine003   |        1 |                   44.40 |
             220.50 | 2022-04-04 09:10:10
         4 | Machine004   |        1 |                   20.15 |
             190.25 | 2022-04-04 09:10:10
         4 | Machine004   |        1 |                   20.25 |
             190.50 | 2022-04-05 09:10:10
         5 | Machine005   |        1 |                  100.35 |
             500.55 | 2022-04-04 09:10:10
         5 | Machine005   |        1 |                  100.50 |
             500.00 | 2022-04-05 09:10:10
         7 | Machine007   |        2 |                   25.50 |
             125.55 | 2022-04-04 09:10:10
         7 | Machine007   |        2 |                   25.55 |
             125.75 | 2022-04-05 09:10:10
         8 | Machine008   |        2 |                   55.50 |
             250.75 | 2022-04-05 09:10:10
         8 | Machine008   |        2 |                   55.75 |
             250.55 | 2022-04-04 09:10:10
         9 | Machine009   |        2 |                   44.25 |
             220.25 | 2022-04-04 09:10:10
```

```
         9 | Machine009    |        2 |                  44.50 |
             220.40 | 2022-04-05 09:10:10
        10 | Machine010   |        2 |                  20.00 |
             190.00 | 2022-04-04 09:10:10
        10 | Machine010   |        2 |                  20.10 |
             190.25 | 2022-04-05 09:10:10
        11 | Machine011   |        2 |                 100.75 |
             500.25 | 2022-04-05 09:10:10
        11 | Machine011   |        2 |                 100.55 |
             500.45 | 2022-04-04 09:10:10
(20 rows)
```

Please note that we have used the ORDER BY clause to order the data returned by the query in the order of machine_id in the machine_details table. Alternatively, you can use ORDER BY on date_and_time from the machine_status table. Also, you can use ORDER BY both together, i.e., date_and_time from the machine_status table and machine_id from the machine_details table.

Code:

```
postgres=# SELECT m.machine_id, m.machine_name, m.mfg_line,
k.pressure_sensor_reading, k.temp_sensor_reading, k.date_and_time
postgres-# FROM machine_details m
postgres-# INNER JOIN machine_status k
postgres-# ON m.machine_id = k.machine_id
postgres-# ORDER BY k.date_and_time ASC, m.machine_id ASC;
```

Output:

```
 machine_id | machine_name | mfg_line | pressure_sensor_reading |
  temp_sensor_reading |     date_and_time
------------+--------------+----------+-------------------------+
---------------------+---------------------
          1 | Machine001    |        1 |                  25.25 |
             125.26 | 2022-04-04 09:10:10
          2 | Machine002    |        1 |                  55.50 |
             250.25 | 2022-04-04 09:10:10
```

```
         3 | Machine003    |          1 |                   44.40 |
           220.50 | 2022-04-04 09:10:10
         4 | Machine004    |          1 |                   20.15 |
           190.25 | 2022-04-04 09:10:10
         5 | Machine005    |          1 |                  100.35 |
           500.55 | 2022-04-04 09:10:10
         7 | Machine007    |          2 |                   25.50 |
           125.55 | 2022-04-04 09:10:10
         8 | Machine008    |          2 |                   55.75 |
           250.55 | 2022-04-04 09:10:10
         9 | Machine009    |          2 |                   44.25 |
           220.25 | 2022-04-04 09:10:10
        10 | Machine010    |          2 |                   20.00 |
           190.00 | 2022-04-04 09:10:10
        11 | Machine011    |          2 |                  100.55 |
           500.45 | 2022-04-04 09:10:10
         1 | Machine001    |          1 |                   25.50 |
           125.40 | 2022-04-05 09:10:10
         2 | Machine002    |          1 |                   55.25 |
           250.00 | 2022-04-05 09:10:10
         3 | Machine003    |          1 |                   44.25 |
           220.25 | 2022-04-05 09:10:10
         4 | Machine004    |          1 |                   20.25 |
           190.50 | 2022-04-05 09:10:10
         5 | Machine005    |          1 |                  100.50 |
           500.00 | 2022-04-05 09:10:10
         7 | Machine007    |          2 |                   25.55 |
           125.75 | 2022-04-05 09:10:10
         8 | Machine008    |          2 |                   55.50 |
           250.75 | 2022-04-05 09:10:10
         9 | Machine009    |          2 |                   44.50 |
           220.40 | 2022-04-05 09:10:10
        10 | Machine010    |          2 |                   20.10 |
           190.25 | 2022-04-05 09:10:10
        11 | Machine011    |          2 |                  100.75 |
           500.25 | 2022-04-05 09:10:10
(20 rows)
```

Based on the purpose for which you use the report, you can choose one of the suitable methods.

Additional Tip In the SELECT clause you should mention the fields in the order you want to have them in your report.

Here is another example of INNER JOIN. Here we are joining the tables machine_details and machine_issues and ordering the fields on the date from the machine_issues table and within that machine_id from the machine_details table.

Code:

```
postgres=# SELECT l.date, m.machine_id, m.machine_name, m.mfg_line,
l.issue_descrip
postgres-# FROM machine_details m
postgres-# INNER JOIN machine_issues l
postgres-# ON m.machine_id = l.machine_id
postgres-# ORDER BY l.date ASC, m.machine_id ASC;
```

Output:

```
    date    | machine_id | machine_name | mfg_line  |         issue_descrip
------------+------------+--------------+-----------+-----------------------
 2021-12-31 |          1 | Machine001   |         1 | Taken off for preventive
                                                      maintenance
 2021-12-31 |          2 | Machine002   |         1 | Taken off for preventive
                                                      maintenance
 2021-12-31 |          3 | Machine003   |         1 | Taken off for preventive
                                                      maintenance
 2021-12-31 |          4 | Machine004   |         1 | Taken off for preventive
                                                      maintenance
 2021-12-31 |          5 | Machine005   |         1 | Taken off for preventive
                                                      maintenance
 2021-12-31 |          7 | Machine007   |         2 | Taken off for preventive
                                                      maintenance
```

```
 2021-12-31 |          8 | Machine008    |          2 | Taken off for preventive
                                                       maintenance
 2021-12-31 |          9 | Machine009    |          2 | Taken off for preventive
                                                       maintenance
 2021-12-31 |         10 | Machine010    |          2 | Taken off for preventive
                                                       maintenance
 2021-12-31 |         11 | Machine011    |          2 | Taken off for preventive
                                                       maintenance
 2022-02-02 |          1 | Machine001    |          1 | Break down bearing issue
 2022-02-28 |          7 | Machine007    |          2 | Break down bearing issue
 2022-03-05 |          5 | Machine005    |          1 | Break down leakage issue
 2022-03-31 |         10 | Machine010    |          2 | Break down valve issue
 2022-04-09 |         13 | Machine013    |          3 | Taken off for preventive
                                                       maintenance
 2022-04-09 |         14 | Machine014    |          3 | Taken off for preventive
                                                       maintenance
 2022-04-09 |         16 | Machine016    |          3 | Taken off for preventive
                                                       maintenance
(17 rows)
```

You can also combine more than two tables using the inner join as long as you have a relationship between the first table to the second, second to the third, and so on. Let's explore the inner join on all three tables, i.e., machine_status, machine_issues, machine_details. The following query demonstrates the corresponding results.

Code:

```
Postgres-# SELECT m.machine_id, m.machine_name, m.mfg_line, k.pressure_
sensor_reading, k.temp_sensor_reading, k.date_and_time, l.issue_descrip
postgres-# FROM machine_status k
postgres-# INNER JOIN machine_issues l
postgres-#      ON l.machine_id = k.machine_id
postgres-# INNER JOIN machine_details m
postgres-#      ON l.machine_id = m.machine_id
postgres-# ORDER BY m.machine_id ASC;
```

Output:

```
machine_id | machine_name | mfg_line | pressure_sensor_reading | temp_sensor_reading |
   date_and_time     |    date    |              issue_descrip
------------+--------------+----------+-------------------------+---------------------+
---------------------+------------+--------------------------------------
         1 | Machine001   |        1 |                   25.25 |              125.26 |
2022-04-04 09:10:10 | 2021-12-31 | Taken off for preventive maintenance
         1 | Machine001   |        1 |                   25.50 |              125.40 |
2022-04-05 09:10:10 | 2021-12-31 | Taken off for preventive maintenance
         1 | Machine001   |        1 |                   25.25 |              125.26 |
2022-04-04 09:10:10 | 2022-02-02 | Break down bearing issue
         1 | Machine001   |        1 |                   25.50 |              125.40 |
2022-04-05 09:10:10 | 2022-02-02 | Break down bearing issue
         2 | Machine002   |        1 |                   55.50 |              250.25 |
2022-04-04 09:10:10 | 2021-12-31 | Taken off for preventive maintenance
         2 | Machine002   |        1 |                   55.25 |              250.00 |
2022-04-05 09:10:10 | 2021-12-31 | Taken off for preventive maintenance
         3 | Machine003   |        1 |                   44.25 |              220.25 |
2022-04-05 09:10:10 | 2021-12-31 | Taken off for preventive maintenance
         3 | Machine003   |        1 |                   44.40 |              220.50 |
2022-04-04 09:10:10 | 2021-12-31 | Taken off for preventive maintenance
         4 | Machine004   |        1 |                   20.15 |              190.25 |
2022-04-04 09:10:10 | 2021-12-31 | Taken off for preventive maintenance
         4 | Machine004   |        1 |                   20.25 |              190.50 |
2022-04-05 09:10:10 | 2021-12-31 | Taken off for preventive maintenance
         5 | Machine005   |        1 |                  100.35 |              500.55 |
2022-04-04 09:10:10 | 2021-12-31 | Taken off for preventive maintenance
         5 | Machine005   |        1 |                  100.50 |              500.00 |
2022-04-05 09:10:10 | 2021-12-31 | Taken off for preventive maintenance
         5 | Machine005   |        1 |                  100.35 |              500.55 |
2022-04-04 09:10:10 | 2022-03-05 | Break down leakage issue
         5 | Machine005   |        1 |                  100.50 |              500.00 |
2022-04-05 09:10:10 | 2022-03-05 | Break down leakage issue
         7 | Machine007   |        2 |                   25.50 |              125.55 |
2022-04-04 09:10:10 | 2021-12-31 | Taken off for preventive maintenance
```

```
          7 | Machine007    |       2 |                25.55 |          125.75 |
 2022-04-05 09:10:10 | 2021-12-31 | Taken off for preventive maintenance
          7 | Machine007    |       2 |                25.50 |          125.55 |
 2022-04-04 09:10:10 | 2022-02-28 | Break down bearing issue
          7 | Machine007    |       2 |                25.55 |          125.75 |
 2022-04-05 09:10:10 | 2022-02-28 | Break down bearing issue
          8 | Machine008    |       2 |                55.50 |          250.75 |
 2022-04-05 09:10:10 | 2021-12-31 | Taken off for preventive maintenance
          8 | Machine008    |       2 |                55.75 |          250.55 |
 2022-04-04 09:10:10 | 2021-12-31 | Taken off for preventive maintenance
          9 | Machine009    |       2 |                44.25 |          220.25 |
 2022-04-04 09:10:10 | 2021-12-31 | Taken off for preventive maintenance
          9 | Machine009    |       2 |                44.50 |          220.40 |
 2022-04-05 09:10:10 | 2021-12-31 | Taken off for preventive maintenance
         10 | Machine010    |       2 |                20.00 |          190.00 |
 2022-04-04 09:10:10 | 2021-12-31 | Taken off for preventive maintenance
         10 | Machine010    |       2 |                20.10 |          190.25 |
 2022-04-05 09:10:10 | 2021-12-31 | Taken off for preventive maintenance
         10 | Machine010    |       2 |                20.00 |          190.00 |
 2022-04-04 09:10:10 | 2022-03-31 | Break down valve issue
         10 | Machine010    |       2 |                20.10 |          190.25 |
 2022-04-05 09:10:10 | 2022-03-31 | Break down valve issue
         11 | Machine011    |       2 |               100.75 |          500.25 |
 2022-04-05 09:10:10 | 2021-12-31 | Taken off for preventive maintenance
         11 | Machine011    |       2 |               100.55 |          500.45 |
 2022-04-04 09:10:10 | 2021-12-31 | Taken off for preventive maintenance
(28 rows)
```

However, from the previous result, you can see that using INNER JOIN has thrown out a lot of duplicated rows and the complexity of such an output makes it less usable. Hence, we suggest you use the `INNER JOIN` on more than two tables only when it really makes sense in terms of the utility of such a report/output.

Let's now explore `LEFT JOIN` (also known as `LEFT OUTER JOIN`). This gets all the details from the first table (i.e., left table) and populates the corresponding details from the next table (i.e., second table), both based on the fields specified in the `SELECT` statement when the data in the common field in both the tables match. Where the data for the common field from the first table (i.e., left table) does not match with the next table (i.e., second table), then the details from the first table are still selected, and the fields from the second table are marked as NULL in the result returned. Only those fields that are included in the SELECT statement are returned by the query. This enables us to check on the data pertaining to the rows in the first table, which is not there in the second table. For example, we can use this to find out if there was any machine in the organization (all the details of the machines in the organization are captured in the machine_details table), which either was not taken off for preventive maintenance or did not have any breakdown or any other issue (such issues are captured in our case in the machine_issues table). The LEFT JOIN on the tables machine_details (i.e., left table) and machine_issues (i.e., second table) is demonstrated with the query and the results shown next.

Code:

```
postgres=# SELECT m.machine_id, m.machine_name, m.mfg_line, l.date,
l.issue_descrip
postgres-# FROM machine_details m
postgres-# LEFT JOIN machine_issues l
postgres-#      ON l.machine_id = m.machine_id
postgres-# ORDER BY m.machine_id ASC, l.date DESC;
```

Output:

```
machine_id  | machine_name | mfg_line |    date    |          issue_descrip
------------+--------------+----------+------------+-------------------------
          1 | Machine001   |        1 | 2022-02-02 | Break down bearing issue
          1 | Machine001   |        1 | 2021-12-31 | Taken off for preventive
                                                     maintenance
          2 | Machine002   |        1 | 2021-12-31 | Taken off for preventive
                                                     maintenance
          3 | Machine003   |        1 | 2021-12-31 | Taken off for preventive
                                                     maintenance
          4 | Machine004   |        1 | 2021-12-31 | Taken off for preventive
                                                     maintenance
          5 | Machine005   |        1 | 2022-03-05 | Break down leakage issue
          5 | Machine005   |        1 | 2021-12-31 | Taken off for preventive
                                                     maintenance
          7 | Machine007   |        2 | 2022-02-28 | Break down bearing issue
```

```
         7 | Machine007   |          2 | 2021-12-31 | Taken off for preventive
                                                      maintenance
         8 | Machine008   |          2 | 2021-12-31 | Taken off for preventive
                                                      maintenance
         9 | Machine009   |          2 | 2021-12-31 | Taken off for preventive
                                                      maintenance
        10 | Machine010   |          2 | 2022-03-31 | Break down valve issue
        10 | Machine010   |          2 | 2021-12-31 | Taken off for preventive
                                                      maintenance
        11 | Machine011   |          2 | 2021-12-31 | Taken off for preventive
                                                      maintenance
        12 | Machine012   |          3 |            |
        13 | Machine013   |          3 | 2022-04-09 | Taken off for preventive
                                                      maintenance
        14 | Machine014   |          3 | 2022-04-09 | Taken off for preventive
                                                      maintenance
        15 | Machine015   |          3 |            |
        16 | Machine016   |          3 | 2022-04-09 | Taken off for preventive
                                                      maintenance
        17 | Machine017   |          0 |            |
(20 rows)
```

Here, you can see that those machines with `machine_ids` 12, 15, and 17 did not have any preventive maintenance or breakdown issues. As you know, `machine_id` 17, i.e., Machine017, is yet to be used on a manufacturing line, and hence no issues or preventive maintenance are carried out on it. However, machines 12 and 15 did not undergo any preventive maintenance or breakdown issues so far. This possibly may suggest the organization's maintenance team to take them for preventive maintenance if the period by which preventive maintenance to be carried out has been already breached from the last preventive maintenance date or the deployment date if they were new machines deployed and are being used currently.

Let's now explore the `RIGHT JOIN` (also known as the `RIGHT OUTER JOIN`). This returns all the rows in the second table (the `RIGHT` table) and the corresponding values from the first table where the common field (used for the join) in the second table matches with the common field in the first table. Where the left table does not have its common field data (used for the join) from the right table, still all the data from the second table (i.e., `RIGHT` table) is returned along with the columns from the first table populated with NULL values. Only those fields listed in the SELECT statement will be returned. This enables us to check on the data pertaining to the rows in the second table, which is not there in the first table. Let's take the machine_status table as the second table (RIGHT table) and the machine_id table as the first table. Here is the query and the output pertaining to the RIGHT JOIN.

Code:

```
postgres=# SELECT m.machine_id, m.machine_name, m.mfg_line, k.pressure_
sensor_reading, k.temp_sensor_reading, k.date_and_time
postgres-# FROM machine_status k
postgres-# RIGHT JOIN machine_details m
postgres-#       ON m.machine_id = k.machine_id
postgres-# ORDER BY m.machine_id ASC;
```

Output:

```
 machine_id | machine_name | mfg_line | pressure_sensor_reading |
 temp_sensor_reading |     date_and_time
------------+--------------+----------+-------------------------+
---------------------+---------------------
          1 | Machine001   |        1 |                   25.25 |
            125.26 | 2022-04-04 09:10:10
          1 | Machine001   |        1 |                   25.50 |
            125.40 | 2022-04-05 09:10:10
          2 | Machine002   |        1 |                   55.25 |
            250.00 | 2022-04-05 09:10:10
          2 | Machine002   |        1 |                   55.50 |
            250.25 | 2022-04-04 09:10:10
          3 | Machine003   |        1 |                   44.40 |
            220.50 | 2022-04-04 09:10:10
          3 | Machine003   |        1 |                   44.25 |
            220.25 | 2022-04-05 09:10:10
          4 | Machine004   |        1 |                   20.15 |
            190.25 | 2022-04-04 09:10:10
          4 | Machine004   |        1 |                   20.25 |
            190.50 | 2022-04-05 09:10:10
          5 | Machine005   |        1 |                  100.35 |
            500.55 | 2022-04-04 09:10:10
```

```
       5 | Machine005    |          1 |                    100.50 |
           500.00 | 2022-04-05 09:10:10
       7 | Machine007    |          2 |                     25.50 |
           125.55 | 2022-04-04 09:10:10
       7 | Machine007    |          2 |                     25.55 |
           125.75 | 2022-04-05 09:10:10
       8 | Machine008    |          2 |                     55.50 |
           250.75 | 2022-04-05 09:10:10
       8 | Machine008    |          2 |                     55.75 |
           250.55 | 2022-04-04 09:10:10
       9 | Machine009    |          2 |                     44.50 |
           220.40 | 2022-04-05 09:10:10
       9 | Machine009    |          2 |                     44.25 |
           220.25 | 2022-04-04 09:10:10
      10 | Machine010    |          2 |                     20.00 |
           190.00 | 2022-04-04 09:10:10
      10 | Machine010    |          2 |                     20.10 |
           190.25 | 2022-04-05 09:10:10
      11 | Machine011    |          2 |                    100.55 |
           500.45 | 2022-04-04 09:10:10
      11 | Machine011    |          2 |                    100.75 |
           500.25 | 2022-04-05 09:10:10
      12 | Machine012    |          3 |                           |
                  |
      13 | Machine013    |          3 |                           |
                  |
      14 | Machine014    |          3 |                           |
                  |
      15 | Machine015    |          3 |                           |
                  |
      16 | Machine016    |          3 |                           |
                  |
      17 | Machine017    |          0 |                           |
                  |
(26 rows)
```

The previous output clearly shows that sensors on the machines 12, 13, 14, 15, and 16 are possibly not yet deployed or activated or automation is yet to be carried out to obtain the feed from them. These machines are surely deployed as these machines are on manufacturing line 3, and as we have seen earlier, some of these underwent preventive maintenance too. As we know, we are yet to deploy machine with machine_id 17.

Let's now look at FULL JOIN (also known as FULL OUTER JOIN). Here, if we are joining two tables, then the results will be a combination of both LEFT JOIN and RIGHT JOIN, which means that the records from both the tables will be included in the result with the columns from the other table, which do not have any contents, populated with NULL values. Let's now carry out a FULL OUTER JOIN on two tables, i.e., machine_status and machine_issues. The result includes the matching records from both the tables as well as the records from the both the tables that do not match. The query and the results are provided next.

Code:

```
Postgres-# SELECT k.machine_id, l.date, k.date_and_time, l.issue_descrip,
k.pressure_sensor_reading, l.temp_sensor_reading
postgres-# FROM machine_issues l
postgres-# FULL JOIN machine_status k
postgres-# ON k.machine_id = l.machine_id;
```

Output:

```
 machine_id |    date    |    date_and_time    |             issue_descr              |
   pressure_sensor_reading | temp_sensor_reading
------------+------------+---------------------+--------------------------------------+
---------------------------+-------------
          1 | 2022-02-02 | 2022-04-04 09:10:10 | Break down bearing issue             |
                     25.25 |              125.26
          1 | 2021-12-31 | 2022-04-04 09:10:10 | Taken off for preventive maintenance |
                     25.25 |              125.26
          7 | 2022-02-28 | 2022-04-04 09:10:10 | Break down bearing issue             |
                     25.50 |              125.55
          7 | 2021-12-31 | 2022-04-04 09:10:10 | Taken off for preventive maintenance |
                     25.50 |              125.55
```

```
 2 | 2021-12-31 | 2022-04-04 09:10:10 | Taken off for preventive maintenance |
          55.50       |                  250.25
 8 | 2021-12-31 | 2022-04-04 09:10:10 | Taken off for preventive maintenance |
          55.75       |                  250.55
 3 | 2021-12-31 | 2022-04-04 09:10:10 | Taken off for preventive maintenance |
          44.40       |                  220.50
 9 | 2021-12-31 | 2022-04-04 09:10:10 | Taken off for preventive maintenance |
          44.25       |                  220.25
 4 | 2021-12-31 | 2022-04-04 09:10:10 | Taken off for preventive maintenance |
          20.15       |                  190.25
10 | 2022-03-31 | 2022-04-04 09:10:10 | Break down valve issue               |
          20.00       |                  190.00
10 | 2021-12-31 | 2022-04-04 09:10:10 | Taken off for preventive maintenance |
          20.00       |                  190.00
 5 | 2022-03-05 | 2022-04-04 09:10:10 | Break down leakage issue             |
         100.35       |                  500.55
 5 | 2021-12-31 | 2022-04-04 09:10:10 | Taken off for preventive maintenance |
         100.35       |                  500.55
11 | 2021-12-31 | 2022-04-04 09:10:10 | Taken off for preventive maintenance |
         100.55       |                  500.45
 1 | 2022-02-02 | 2022-04-05 09:10:10 | Break down bearing issue             |
          25.50       |                  125.40
 1 | 2021-12-31 | 2022-04-05 09:10:10 | Taken off for preventive maintenance |
          25.50       |                  125.40
 7 | 2022-02-28 | 2022-04-05 09:10:10 | Break down bearing issue             |
          25.55       |                  125.75
 7 | 2021-12-31 | 2022-04-05 09:10:10 | Taken off for preventive maintenance |
          25.55       |                  125.75
 2 | 2021-12-31 | 2022-04-05 09:10:10 | Taken off for preventive maintenance |
          55.25       |                  250.00
 8 | 2021-12-31 | 2022-04-05 09:10:10 | Taken off for preventive maintenance |
           55.50      |                  250.75
 3 | 2021-12-31 | 2022-04-05 09:10:10 | Taken off for preventive maintenance |
           44.25      |                  220.25
```

```
        9 | 2021-12-31 | 2022-04-05 09:10:10 | Taken off for preventive maintenance |
                44.50      |               220.40
        4 | 2021-12-31 | 2022-04-05 09:10:10 | Taken off for preventive maintenance |
                20.25      |               190.50
       10 | 2022-03-31 | 2022-04-05 09:10:10 | Break down valve issue               |
                20.10      |               190.25
       10 | 2021-12-31 | 2022-04-05 09:10:10 | Taken off for preventive maintenance |
                20.10      |               190.25
        5 | 2022-03-05 | 2022-04-05 09:10:10 | Break down leakage issue             |
               100.50      |               500.00
        5 | 2021-12-31 | 2022-04-05 09:10:10 | Taken off for preventive maintenance |
               100.50      |               500.00
       11 | 2021-12-31 | 2022-04-05 09:10:10 | Taken off for preventive maintenance |
               100.75      |               500.25
       13 | 2022-04-09 |                     | Taken off for preventive maintenance |
                           |
       16 | 2022-04-09 |                     | Taken off for preventive maintenance |
                           |
       14 | 2022-04-09 |                     | Taken off for preventive maintenance |
                           |
(31 rows)
```

As you can see in the previous output, wherever the left table did not have the corresponding data in the right table, then the columns pertaining to the right table have been populated with the NULL (blanks) values and have been included in the output. In the same way, if there are no rows that are matching available in the right table, the columns pertaining to the left table will be populated with the NULL (blanks) values. However, in our case there are only the left table rows, which are not available in the right table, and hence the fields pertaining to the right table are populated with NULL values where there are no matching records in the left table.

A FULL JOIN is typically used to check on the similarity or dissimilarity between the contents of the two tables. In the case of our example, this may be more confusing than really useful. However, as you saw in the earlier examples, INNER JOIN, LEFT JOIN, and RIGHT JOIN were all useful.

Additional Tip It is not necessary to include all the columns from all the tables in the SELECT statement for the JOINs. You need to include them based on the purpose of the query. However, in our examples we have included all just to make you aware of the outputs in detail. We have not explored the SELF JOIN in this chapter. A SELF JOIN is either an INNER JOIN or a LEFT JOIN or a RIGHT JOIN on the same table.

Let's now explore another important SQL condition expression, i.e., CASE. This is very useful in understanding, in exploring the data, and even for reporting. This can be used for the cleaning of the data as well. CASE statements typically use a WHEN….THEN…..ELSE structure within them. ELSE here is optional. Let's say we want to find out the number of instances of the preventive maintenance and number of instances of breakdown; the CASE statement can be used as follows.

Code:

```
postgres=# SELECT
postgres-#      SUM (CASE
postgres(#              WHEN issue_descrip LIKE '%Break_down%' THEN 1
postgres(#              ELSE 0
postgres(#           END
postgres(#      ) AS "No. of Break Down Cases",
postgres-#      SUM (CASE
postgres(#              WHEN issue_descrip LIKE '%preventive_
                        maintenance%' THEN 1
postgres(#              ELSE 0
postgres(#           END
postgres(#      ) AS "No. of Preventive Maintenance Cases"
postgres-# FROM machine_issues;
```

Output:

```
 No. of Break Down Cases | No. of Preventive Maintenance Cases
-------------------------+-------------------------------------
                       4 |                                  13
(1 row)
```

Suppose you want to look at the machine_id, number of breakdown cases and number of preventive maintenance cases; then you use the query using CASE as follows.

Code:

```
postgres=# SELECT machine_id,
postgres-#      SUM (CASE
postgres(#              WHEN issue_descrip LIKE '%Break_down%' THEN 1
postgres(#              ELSE 0
postgres(#           END
postgres(#      ) AS "No. of Break Down Cases",
postgres-#      SUM (CASE
postgres(#              WHEN issue_descrip LIKE '%preventive_
                        maintenance%' THEN 1
postgres(#              ELSE 0
postgres(#           END
postgres(#      ) AS "No. of Preventive Maintenance Cases"
postgres-# FROM machine_issues
postgres-# GROUP BY machine_id
postgres-# ORDER BY machine_id;
```

Output:

```
 machine_id | No. of Break Down Cases | No. of Preventive Maintenance Cases
------------+-------------------------+-------------------------------------
          1 |                       1 |                                   1
          2 |                       0 |                                   1
          3 |                       0 |                                   1
          4 |                       0 |                                   1
          5 |                       1 |                                   1
          7 |                       1 |                                   1
          8 |                       0 |                                   1
          9 |                       0 |                                   1
         10 |                       1 |                                   1
         11 |                       0 |                                   1
         13 |                       0 |                                   1
         14 |                       0 |                                   1
         16 |                       0 |                                   1
(13 rows)
```

Please note that when you have multiple CASE statements with WHEN and THEN, the execution starts with the first WHEN and THEN. If that WHEN statement is evaluated to be TRUE, then the THEN statement following it will be executed, and the next WHEN statements will not be evaluated. However, if the first WHEN statement is evaluated to be FALSE, then the next WHEN statement following this statement will be evaluated. If all the WHEN statements evaluate to be false, then the ELSE statement will be executed.

Let's look at an example of this. The query and the result are shown next.

Code:

```
postgres=# SELECT machine_id, pressure_sensor_reading, temp_sensor_reading,
postgres-#      CASE WHEN (pressure_sensor_reading IS NULL OR temp_sensor_
readin IS NULL) THEN 'machine_not_used or faulty'
postgres-#           ELSE 'machine_is_working'
postgres-#      END machine_health
postgres-# FROM machine_status
postgres-# ORDER BY machine_id;
```

Output:

```
 machine_id | pressure_sensor_reading | temp_sensor_reading |   machine_health
------------+-------------------------+---------------------+-------------------
          1 |                   25.25 |              125.26 | machine_is_working
          1 |                   25.50 |              125.40 | machine_is_working
          2 |                   55.50 |              250.25 | machine_is_working
          2 |                   55.25 |              250.00 | machine_is_working
          3 |                   44.25 |              220.25 | machine_is_working
          3 |                   44.40 |              220.50 | machine_is_working
          4 |                   20.15 |              190.25 | machine_is_working
          4 |                   20.25 |              190.50 | machine_is_working
          5 |                  100.35 |              500.55 | machine_is_working
          5 |                  100.50 |              500.00 | machine_is_working
          7 |                   25.50 |              125.55 | machine_is_working
          7 |                   25.55 |              125.75 | machine_is_working
          8 |                   55.50 |              250.75 | machine_is_working
          8 |                   55.75 |              250.55 | machine_is_working
          9 |                   44.25 |              220.25 | machine_is_working
          9 |                   44.50 |              220.40 | machine_is_working
```

```
         10 |                 20.00 |              190.00 | machine_is_working
         10 |                 20.10 |              190.25 | machine_is_working
         11 |                100.75 |              500.25 | machine_is_working
         11 |                100.55 |              500.45 | machine_is_working
(20 rows)
```

At this point of time, the machine_status table has all the records with no NULL values for pressure_sensor_reading and temp_sensor_reading. Hence only the ELSE statement got executed and the result for all the records is, i.e., health_status is machine_is_working.

Now, let's add a record/row with NULL values for pressure_sensor_reading and temp_sensor_reading, date_and_time, and check what will be returned by the previous query.

Code:

```
postgres=# INSERT INTO machine_status (machine_id, pressure_sensor_reading,
temp_sensor_reading, date_and_time)
postgres-# VALUES (16, NULL, NULL, NULL);
Returned result: INSERT 0 1
postgres=# select * from machine_status;
```

Output:

```
 machine_id | pressure_sensor_reading | temp_sensor_reading |    date_and_time
------------+-------------------------+---------------------+---------------------
          1 |                   25.25 |              125.26 | 2022-04-04 09:10:10
          7 |                   25.50 |              125.55 | 2022-04-04 09:10:10
          2 |                   55.50 |              250.25 | 2022-04-04 09:10:10
          8 |                   55.75 |              250.55 | 2022-04-04 09:10:10
          3 |                   44.40 |              220.50 | 2022-04-04 09:10:10
          9 |                   44.25 |              220.25 | 2022-04-04 09:10:10
          4 |                   20.15 |              190.25 | 2022-04-04 09:10:10
         10 |                   20.00 |              190.00 | 2022-04-04 09:10:10
          5 |                  100.35 |              500.55 | 2022-04-04 09:10:10
         11 |                  100.55 |              500.45 | 2022-04-04 09:10:10
          1 |                   25.50 |              125.40 | 2022-04-05 09:10:10
          7 |                   25.55 |              125.75 | 2022-04-05 09:10:10
```

```
         2 |                  55.25 |              250.00 | 2022-04-05 09:10:10
         8 |                  55.50 |              250.75 | 2022-04-05 09:10:10
         3 |                  44.25 |              220.25 | 2022-04-05 09:10:10
         9 |                  44.50 |              220.40 | 2022-04-05 09:10:10
         4 |                  20.25 |              190.50 | 2022-04-05 09:10:10
        10 |                  20.10 |              190.25 | 2022-04-05 09:10:10
         5 |                 100.50 |              500.00 | 2022-04-05 09:10:10
        11 |                 100.75 |              500.25 | 2022-04-05 09:10:10
        16 |                        |                     |
(21 rows)
```

As you can see, the machine_status table now has a row/record with a NULL value for the pressure_sensor_reading, temp_sensor_reading, and date_and_time columns.

Let's now run the CASE query used before this INSERT and look at what happens.

Code:

```
postgres=# SELECT machine_id, pressure_sensor_reading, temp_sensor_reading,
postgres-#      CASE WHEN (pressure_sensor_reading IS NULL OR temp_sensor_
reading IS NULL) THEN 'machine_not_used or faulty'
postgres-#           ELSE 'machine_is_working'
postgres-#      END machine_health
postgres-# FROM machine_status
postgres-# ORDER BY machine_id;
```

Output:

```
machine_id | pressure_sensor_reading | temp_sensor_reading |       machine_health
-----------+-------------------------+---------------------+--------------------
         1 |                   25.25 |              125.26 | machine_is_working
         1 |                   25.50 |              125.40 | machine_is_working
         2 |                   55.50 |              250.25 | machine_is_working
         2 |                   55.25 |              250.00 | machine_is_working
         3 |                   44.25 |              220.25 | machine_is_working
         3 |                   44.40 |              220.50 | machine_is_working
         4 |                   20.15 |              190.25 | machine_is_working
         4 |                   20.25 |              190.50 | machine_is_working
```

```
         5 |              100.35 |               500.55 | machine_is_working
         5 |              100.50 |               500.00 | machine_is_working
         7 |               25.50 |               125.55 | machine_is_working
         7 |               25.55 |               125.75 | machine_is_working
         8 |               55.50 |               250.75 | machine_is_working
         8 |               55.75 |               250.55 | machine_is_working
         9 |               44.25 |               220.25 | machine_is_working
         9 |               44.50 |               220.40 | machine_is_working
        10 |               20.00 |               190.00 | machine_is_working
        10 |               20.10 |               190.25 | machine_is_working
        11 |              100.75 |               500.25 | machine_is_working
        11 |              100.55 |               500.45 | machine_is_working
        16 |                     |                      | machine_not_used
                                                          or faulty
(21 rows)
```

3.3.3.1 Additional Examples of the Useful SELECT Statements

Here, we will explore some additional examples of useful SELECT statements. First let's explore the SELECT statement with the HAVING clause.

The HAVING clause follows the GROUP BY clause. The HAVING clause provides filtering or searching on AGGREGATE functions like MAX, MIN, AVG, SUM, and COUNT on the results returned by the GROUP BY clause.

Let's say we want to understand the average (AVG) of the pressure_sensor_reading and temp_sensor_reading and HAVING pressure_sensor_reading > 20 and temp_sensor_rating > 120 to check if the sensors are working properly or not. The following SELECT statement with the HAVING clause can be used.

Code:

```
postgres=# SELECT machine_id, AVG(pressure_sensor_reading) avg_pres_sens_
rdg, AVG(temp_sensor_reading) avg_temp_sens_rdg
postgres-# FROM machine_status
postgres-# GROUP BY machine_id
postgres-# HAVING AVG(pressure_sensor_reading) > 20 AND AVG(temp_sensor_
reading) > 120
postgres-# ORDER BY machine_id;
```

Output:

```
machine_id |  avg_pres_sens_rdg    |  avg_temp_sens_rdg
-----------+----------------------+---------------------
         1 |  25.3750000000000000 | 125.3300000000000000
         2 |  55.3750000000000000 | 250.1250000000000000
         3 |  44.3250000000000000 | 220.3750000000000000
         4 |  20.2000000000000000 | 190.3750000000000000
         5 | 100.4250000000000000 | 500.2750000000000000
         7 |  25.5250000000000000 | 125.6500000000000000
         8 |  55.6250000000000000 | 250.6500000000000000
         9 |  44.3750000000000000 | 220.3250000000000000
        10 |  20.0500000000000000 | 190.1250000000000000
        11 | 100.6500000000000000 | 500.3500000000000000
(10 rows)
```

Additional Tip You can use the ROUND function on the AVG function to round the AVG value to the required decimal places. For example, use ROUND(AVG(pressure_sensor_reading, 2) to round the returned value to two decimals.

As you can note, the details of the machine_id 16 are not shown in the previous result as the fields of pressure_sensor_reading and temp_sensor_reading are NULL.

Additional Tip We have given an alias to the calculated AVG fields to provide a meaningful heading to such Average columns.

Let's now see how we can use ANY and ALL with the SELECT statement. These allow the comparison of any or all values with the result returned by a subquery. These are always used with comparison operators like = (i.e., equal to), != (i.e., not equal to), <= (i.e., less than or equal to), < (i.e., less than), >= (i.e., greater than or equal to), > (i.e., greater than).

Code:

```
postgres=# SELECT machine_id, pressure_sensor_reading
postgres-# FROM machine_status
postgres-# WHERE pressure_sensor_reading < ANY
postgres-#      (SELECT AVG(pressure_sensor_reading)
postgres(#       FROM machine_status
postgres(#       GROUP BY machine_id)
postgres-# ORDER BY pressure_sensor_reading;
```

Output:

```
 machine_id | pressure_sensor_reading
------------+-----------------------
         10 |                  20.00
         10 |                  20.10
          4 |                  20.15
          4 |                  20.25
          1 |                  25.25
          7 |                  25.50
          1 |                  25.50
          7 |                  25.55
          3 |                  44.25
          9 |                  44.25
          3 |                  44.40
          9 |                  44.50
          2 |                  55.25
          2 |                  55.50
          8 |                  55.50
          8 |                  55.75
          5 |                 100.35
          5 |                 100.50
         11 |                 100.55
(19 rows)
```

As the highest value of the average of any machine is 100.65 (pertaining to machine_id 11), the < ANY will throw up all the values of pressure_sensor_reading less than this value and any value of pressure_sensor_reading above this value will not be selected.

Hence, as you can see the only value not selected here is 100.75 (pertaining to machine_id 11), which is higher than the previous average of 100.65.

Let's now see what happens if we use > ANY. The results are as follows.

Code:

```
postgres=# SELECT machine_id, pressure_sensor_reading
postgres-# FROM machine_status
postgres-# WHERE pressure_sensor_reading > ANY
postgres-#      (SELECT AVG(pressure_sensor_reading)
postgres(#       FROM machine_status
postgres(#       GROUP BY machine_id)
postgres-# ORDER BY pressure_sensor_reading;
```

Output:

```
 machine_id | pressure_sensor_reading
------------+------------------------
         10 |                   20.10
          4 |                   20.15
          4 |                   20.25
          1 |                   25.25
          1 |                   25.50
          7 |                   25.50
          7 |                   25.55
          9 |                   44.25
          3 |                   44.25
          3 |                   44.40
          9 |                   44.50
          2 |                   55.25
          8 |                   55.50
          2 |                   55.50
          8 |                   55.75
          5 |                  100.35
          5 |                  100.50
         11 |                  100.55
         11 |                  100.75
(19 rows)
```

As you can see, all the values are selected except the value lower than the lowest average value of 20.05 (pertaining to machine_id 10). This lowest value pertains to machine 10.

The previous queries may not be of much use in the context of our data but may be useful in a different context with a different data set.

Additional Tip Please note that the null value is not considered while calculating the AVG. This is applicable in our case with respect to machine_id 16. There is no average calculated for machine_id 16 used previously.

Similar to the previous example, you can use ALL before the subquery where the comparison will be carried out with the value returned by the subquery. As NULL values in your table can create issues with such queries, you need to be careful. Where it is possible, rows with the NULL values may be deleted in the copy of the data set used by you for your ease of analysis.

Similarly, UNION and INTERSECT may also be useful if we have to combine the results of two or more queries. UNION is used when we want to combine the results of two or more queries. This ensures the inclusion of the data from both the queries into the final result set. If there is a common record in both the queries, UNION ensures that only one of them is retained in the result set. INTERSECT is used when we want to select the common rows from two or more queries. For example, if you want to find out which machines have the issues reported, you can use INTERSECT on the machine_ids from the machine_details (i.e., master file of machines) table with the machine_ids from the machine_issues (which captures the issues with the machines) table.

Code:

```
postgres=# SELECT machine_id
postgres-# FROM machine_details
postgres-# INTERSECT
postgres-# SELECT machine_id
postgres-# FROM machine_issues
postgres-# ORDER BY machine_id;
```

Output:

```
machine_id
------------
          1
          2
          3
          4
          5
          7
          8
          9
         10
         11
         13
         14
         16
(13 rows)
```

Caution For the UNION and INTERSECT operators to work, each query that takes part in the SELECT statement should have same number of columns with the same order and compatible data types.

Please note that typically in a SELECT statement, the clauses are executed in the following order: FROM > ON > OUTER > WHERE > GROUP BY > HAVING > SELECT > ORDER BY, according to www.designcise.com.

Let's now explore EXISTS, NOT EXISTS, IN, and NOT IN.

Code:

```
postgres=# SELECT * FROM machine_details
postgres-# WHERE EXISTS
postgres-#      (SELECT 1
postgres(#       FROM machine_issues
postgres(#       WHERE machine_issues.machine_id = machine_details.machine_id)
postgres-# ORDER BY machine_details.machine_id;
```

Output:

```
machine_id | machine_name | mfg_line
-----------+--------------+---------
         1 | Machine001   |        1
         2 | Machine002   |        1
         3 | Machine003   |        1
         4 | Machine004   |        1
         5 | Machine005   |        1
         7 | Machine007   |        2
         8 | Machine008   |        2
         9 | Machine009   |        2
        10 | Machine010   |        2
        11 | Machine011   |        2
        13 | Machine013   |        3
        14 | Machine014   |        3
        16 | Machine016   |        3
(13 rows)
```

In this query, for every record in the outer table (i.e., in the first FROM statement), the subquery within the brackets is evaluated, and if the match is made, then the output from the first table as per the first SELECT statement is returned. Then for the next record in the outer table (i.e., in the first FROM statement), the subquery within the brackets is evaluated, and if the match is made, then the output from the first table as per the first SELECT statement is returned. This way the result of the subquery check is made for every record in the first table. Hence, this type of query is very inefficient in terms of the resource utilization.

The same result can be obtained using IN as follows.

Code:

```
postgres=# SELECT * FROM machine_details
postgres-# WHERE machine_id IN (SELECT machine_id FROM machine_issues);
```

Output:

```
machine_id | machine_name | mfg_line
-----------+------------- +---------
         1 | Machine001   |        1
         2 | Machine002   |        1
         3 | Machine003   |        1
         4 | Machine004   |        1
         5 | Machine005   |        1
         7 | Machine007   |        2
         8 | Machine008   |        2
         9 | Machine009   |        2
        10 | Machine010   |        2
        11 | Machine011   |        2
        13 | Machine013   |        3
        14 | Machine014   |        3
        16 | Machine016   |        3
(13 rows)
```

NOT EXISTS is exactly the opposite of EXISTS. NOT EXISTS evaluates TRUE only when there is no match of the rows from the first table (part of the first FROM clause) with that of the row in the table of the subquery and returns the corresponding result of the first SELECT statement. Let's change the previously executed query with NOT EXISTS and check what happens.

Code:

```
postgres=# SELECT * FROM machine_details
postgres-# WHERE NOT EXISTS
postgres-#      (SELECT 1
postgres(#       FROM machine_issues
postgres(#       WHERE machine_issues.machine_id = machine_details.
machine_id)
postgres-# ORDER BY machine_details.machine_id;
```

Output:

```
machine_id | machine_name | mfg_line
-----------+--------------+---------
        12 | Machine012   |        3
        15 | Machine015   |        3
        17 | Machine017   |        0
(3 rows)
```

The same result can be obtained by using NOT IN.

Code:

```
postgres=# SELECT * FROM machine_details
postgres-# WHERE machine_id NOT IN (SELECT machine_id FROM machine_issues);
```

Output:

```
machine_id | machine_name | mfg_line
-----------+--------------+---------
        12 | Machine012   |        3
        15 | Machine015   |        3
        17 | Machine017   |        0
(3 rows)
```

EXISTS OR NOT EXISTS gets executed when there is at least one record that is evaluated TRUE by the subquery.

LIMIT in the SELECT statement will help you sample the data or limit the output of the query in terms of number of records. This will help in case of huge data or in case you want really TOP few records or check if the query works decently and hence want to limit the returned records. The following is an example.

Code:

```
postgres=# SELECT machine_id, date, issue_descrip
postgres-# FROM machine_issues
postgres-# WHERE issue_descrip LIKE '%preventive_maintenance'
postgres-# LIMIT 5;
```

Output:

```
machine_id |    date    |                 issue_descrip
-----------+------------+-----------------------------------
         1 | 2021-12-31 | Taken off for preventive maintenance
         2 | 2021-12-31 | Taken off for preventive maintenance
         3 | 2021-12-31 | Taken off for preventive maintenance
         4 | 2021-12-31 | Taken off for preventive maintenance
         5 | 2021-12-31 | Taken off for preventive maintenance
(5 rows)
```

As you can see, only the first five records returned by the query are output as result.

`OFFSET N ROWS FETCH NEXT N ROWS ONLY;` or `FETCH FIRST N ROWS ONLY;` can be used as an alternative if you want to inspect different parts of the output before you execute the query fully in case of huge data being returned. Examples are provided here.

Code:

```
postgres=# SELECT machine_id, date, issue_descrip
postgres-# FROM machine_issues
postgres-# WHERE issue_descrip LIKE '%preventive_maintenance'
postgres-# OFFSET 5 ROWS
postgres-# FETCH NEXT 5 ROWS ONLY;
```

Output:

```
machine_id |    date    |                 issue_descrip
-----------+------------+-----------------------------------
         7 | 2021-12-31 | Taken off for preventive maintenance
         8 | 2021-12-31 | Taken off for preventive maintenance
         9 | 2021-12-31 | Taken off for preventive maintenance
        10 | 2021-12-31 | Taken off for preventive maintenance
        11 | 2021-12-31 | Taken off for preventive maintenance
(5 rows)
```

Code:

```
postgres=# SELECT machine_id, date, issue_descrip
postgres-# FROM machine_issues
postgres-# WHERE issue_descrip LIKE '%preventive_maintenance'
postgres-# FETCH FIRST 5 ROWS ONLY;
```

Output:

```
machine_id |    date    |                    issue_descrip
-----------+------------+------------------------------------
         1 | 2021-12-31 | Taken off for preventive maintenance
         2 | 2021-12-31 | Taken off for preventive maintenance
         3 | 2021-12-31 | Taken off for preventive maintenance
         4 | 2021-12-31 | Taken off for preventive maintenance
         5 | 2021-12-31 | Taken off for preventive maintenance
(5 rows)
```

3.4 Chapter Summary

- In this chapter, you saw how SQL can act as an excellent utility for data analytics.
- You learned how to create the tables in the Postgres database using PostgreSQL.
- You learned how to insert the data into the tables, if required.
- You learned how to update these tables.
- You also learned how to query on the tables and get the results presented to you.
- You also learned how these queries can be used to provide you the insights related to your business problems and throw up the intelligence hidden in the data.
- You also learned how you can handle the missing values.
- Overall, you learned how SQL in general and PostgreSQL in particular helps you in carrying out business analytics effectively.

CHAPTER 4

Business Analytics Process

This chapter covers the process and life cycle of business analytics projects. We discuss various steps in the analytic process, from understanding requirements to deploying a model in production. We also discuss the challenges at each stage of the process and how to overcome those challenges.

4.1 Business Analytics Life Cycle

The purpose of business analytics is to derive information from data in order to make appropriate business decisions. Data can be real time or historical. Typically, you build a business model from historical data, and once you are satisfied with the accuracy of the model, you can deploy it in production for real-time data analysis. Though there are many standards referenced by various industry experts and tool manufacturers, the process is more or less the same. In this section, we present the basic business analytics process with the help of a flow chart. The same process is followed in all case studies that are discussed in this book.

The life cycle of a business analytics project consists of eight phases. Figure 4-1 shows the sequence of phases. Depending on the business problem and data, this sequence may sometimes change. The outcome of each phase determines the next phase and the tasks to be performed in that phase. The whole process does not end after the model is deployed. A new set of data that was not seen before in the model can trigger a new set of problems that is addressed as a new model. The process is repeated from the first step once again, with additional new data inputs. Subsequent processes will benefit from the experiences of the previous ones.

U. R. Hodeghatta and U. Nayak, *Practical Business Analytics Using R and Python*,
https://doi.org/10.1007/978-1-4842-8754-5_4

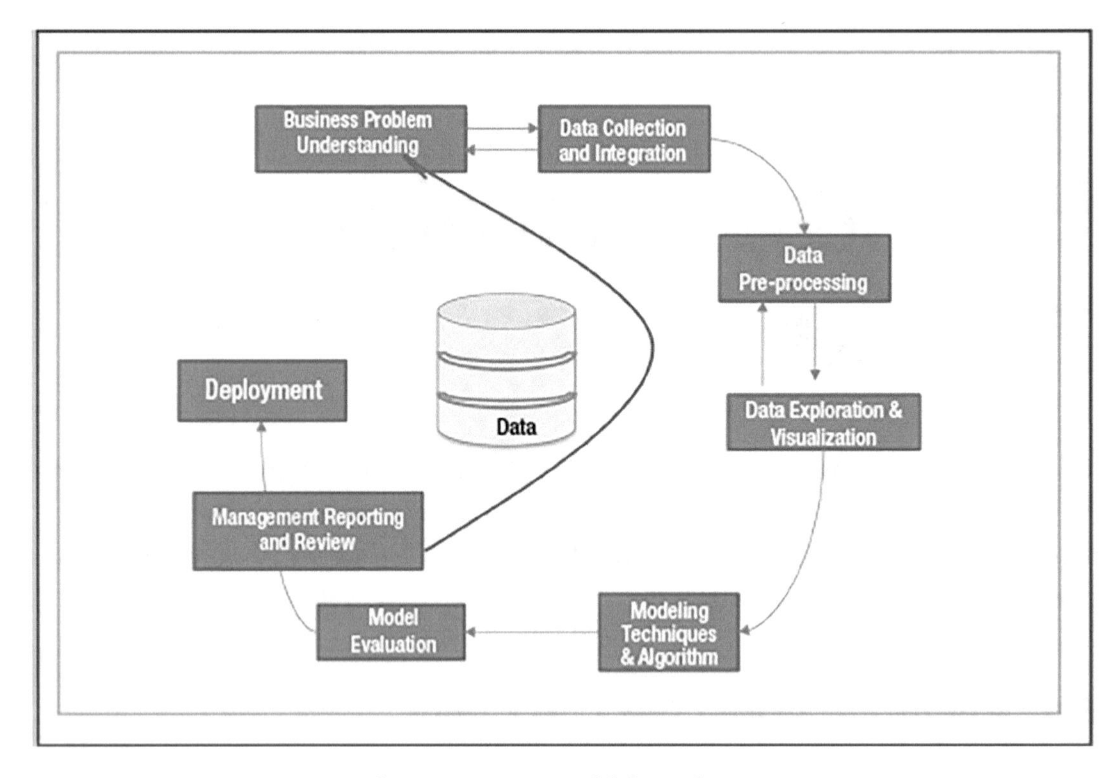

Figure 4-1. *Business analytics process and life cycle*

The typical process of business analytics and data mining projects is as follows:

1. Start with a business problem and all the data considered as relevant to the problem.

   or

   Start with the data to understand what patterns you see in the data and what knowledge you can decipher from the data.

2. Study the data and data types, preprocess data, clean up the data for missing values, and fix any other data elements or errors.

3. Check for the outliers in the data and remove them from the data set to reduce their adverse impact on the analysis.

4. Identify the data analysis techniques to be used (for example, supervised or unsupervised).

5. Apply the identified data analytics techniques using the appropriate tools.

6. Analyze the results and check whether alternative techniques can be used to get better results.

7. Validate the results, understand the results, and interpret the results. Check whether the techniques are performing as per the business requirements.

8. Publish the results (learning/model).

9. Use the learning from the results/model arrived at.

10. Keep calibrating the learning/model as you keep using it.

Let's discuss each phase in more detail.

4.1.1 Phase 1: Understand the Business Problem

This is the first phase of the project. The focus here is to understand the problem, objectives, and requirements of the stakeholders and business. It is crucial to understand from the stakeholders what problem you are expected to solve, how to obtain data, the expected model performance, and the deployment environment. Once this is clear, convert this into a data analytics problem with the aim of solving it by using appropriate methods to achieve the objective.

The key purpose of analytics is to solve a business problem. Before you solve the problem, you need to thoroughly understand the problem from a business perspective so you know what the client really wants you to solve. For example, is the client trying to find a pattern in the data, predict the price of a product to be launched, predict a customer segment to target, figure out who is defaulting on a loan, optimize a software development cycle, or improve the quality of operations? This should be clarified at the beginning of the process and should be documented. Also, it is important to determine the availability of the data, whether you have access to the data source or internal data warehouse, the data format, the data quality, the amount of data available, and the data stream for the final model deployment.

Documentation is an equally important task, as our memories are not very good at recalling every detail. As part of this task, you need to document the business objective, data source, risks, and limitations of the project. Define the timeline, the infrastructure needed to support the model, and the expertise required to execute the project.

4.1.2 Phase 2: Data Collection

In this phase, data is collected from various sources. Getting quality data is the most important factor determining the accuracy of the results. Data can be from either a primary source or a secondary source. A *primary source* is your own source—data from an internal database or from your own research. You have full control of the data collected. A *secondary source* is data from other, secondary places that sell data, or data that already exists apart from the business database. Some examples of secondary sources include official government reports, newspaper articles, and census data. Depending on the type of business problem being solved, you might use data from a primary source or secondary source or from both. The advantage of using primary data is that you are designing the type of information required for solving the problem at hand, so the data is of good quality, and less time is required to prepare the data for further analysis. An abundance of data may exist inside the organization in various databases. For example:

Operational database: This includes data related to day-to-day business transactions such as vendor management, supply chains, customer complaints, and customer feedback.

Manufacturing and production database: This includes data related to the manufacturing process, production details, supply-chain management, production schedule, and repair and maintenance of machinery.

HR and finance database: This includes data related to HR and finance functions such as employee leave, personnel data, job skills, attrition, and salaries.

IT database: This includes data related to information systems such as IT assets inventory, licensing details, software, and various logs.

Data warehouse: A data warehouse is an integrated database created from multiple databases within the organization. Data warehouse technology typically cleans, normalizes, and preprocesses data before it is stored. This helps in analysis, reporting, and the decision-making process. Data warehouse technology also supports online analytical processing (OLAP) and other functionalities such as data summarization and multidimensional analysis (slice and dice).

NoSQL database: NoSQL databases are developed to overcome the limitations of relational databases and to meet the challenges of the enormous amounts of data being generated on the Web. A NoSQL database can store both structured and unstructured data. Also, NoSQL databases need not have a schema before storing the data.

Metadata: This is data about the data. It is descriptive information about a particular data set, object, or resource, including how it is formatted and when and by whom it was collected (or example, web metadata, XML, and JSON format).

Most organizations have data spread across various databases. Pulling data from multiple sources is a required part of solving business analytics tasks. Sometimes, data may be stored in databases for different purposes than the objective you are trying to solve. Thus, the data has to be prepared to ensure it addresses the business problem prior to any analytics process. This process is sometimes referred to as *data munging* or *data wrangling*, which is covered later in the chapter.

4.1.2.1 Sampling

Many times, unless you have a big data infrastructure, only a sample of the population is used to build analytical modeling. A sample is "a smaller collection of units from a population used to determine truths about that population" (Field, 2005). The sample should be representative of the population. Choosing a sampling technique depends on the type of business problem.

For example, you might want to study the annual gross domestic product (GDP) per capita for several countries over a period of time and the periodic behavior of such series in connection with business cycles. Monthly housing sales over a period of 6–10 years show cyclic behavior, but for 6–12 months, the sales data may show seasonal behavior. Stock market data over a period of 10–15 years may show a different trend than over a 100-day period. Similarly, forecasting sales based on previous data over a period of time, or analyzing Twitter sentiments and trends over a period of time, is cyclic data. If the fluctuations are not of a fixed period, they are cyclic. If the changes are in a specific period of the calendar, the pattern is seasonal. Time-series data is data obtained through repeated measurements over a particular time period.

For time-series data, the sample should contain the time period (date or time or both) and only a sample of measurement records for that particular day or time instead of the complete data collected. For example, the Dow Jones volume is traded over 18 months. The data is collected for every millisecond, so the volume of this data for a day is huge. Over a 10-month period, this data can be in terabytes.

The other type of data is not time dependent. It can be continuous or discrete data, but time has no significance in such data sets. For example, you might look at the income or job skills of individuals in a company, the number of credit transactions in a retail store, or age and gender information. There is no relationship between any two data records.

Unless you have big data infrastructure, you can just take a sample of records for any analysis. Use a randomization technique and take steps to ensure that all the members of a population have an equal chance of being selected. This method is called *probability sampling*. There are several variations on this type of sampling.

> *Random sampling*: A sample is picked randomly, and every member has an equal opportunity to be selected. An example is a random sample of people (say 10,000) of New York City from the total population of the city. Each individual picked from the population has an equal probability of selection.
>
> *Stratified sampling*: The population is divided into groups, and data is selected randomly from a group, or *strata*. For example, the population of New York City may be divided into different income levels, say, high, medium, and low. Data for your analysis is selected randomly from each group equally.
>
> *Systematic sampling*: You select members systematically—say, every 10th member—in that particular time or event. For example, you are analyzing social media behavior and collecting social media tweets every 2 hours systematically for the next 30 days.

The details of calculating sample sizes are beyond the scope of this book and are covered extensively in statistics and other research methodology books. However, to enhance your understanding, here is a simple formula for calculating a sample:

If the population standard deviation is known, then

$$n = (z \times sigma / E)^{\wedge 2}$$

If standard deviation is unknown,

$$n = (p)(1 - p) * (z / E)^{\wedge 2}$$

where n is the sample size,
sigma is the standard deviation,
p is the proportion of the population (estimate)
e is the marginal error
Z value can be found from the Z-table.

4.1.3 Phase 3: Data Preprocessing and Preparation

The raw data received or collected may not be applied directly to build the model. The raw data may contain errors such as formatting conflicts, wrong data types, and missing values. As described in the previous phase, data can be collected from various sources, including an internal database or data warehouse. Data in a database may be susceptible to noise and inconsistencies because its data is collected from various end-user applications developed and handled by different people over a period of time. Multiple databases may have to be queried to get data for analytics purpose. During this process, data may have missing values or junk characters, be of different data types, and represent different types of observations or the same observations multiple times. Thus, before performing any data analysis, it is essential to understand the data in terms of its types, variables and characteristics of variables, and data tables.

In this step, data is cleaned and normalized. You fix any missing values in the data, and the data is then prepared for further analysis. This process may be repeated several times, depending on the data quality and the model's accuracy.

4.1.3.1 Data Types

Data can be either qualitative or quantitative. *Qualitative data* is not numerical—for example, type of car, favorite color, or favorite food. *Quantitative data* is numeric. Additionally, quantitative data can be divided into categories of *discrete* or *continuous* data (described in more detail later in this section).

Quantitative data is often referred to as *measurable data*. This type of data allows statisticians to perform various arithmetic operations, such as addition and multiplication, and to find population parameters, such as mean or variance. The observations represent counts or measurements, and thus all values are numerical. Each observation represents a characteristic of the individual data points in a population or a sample.

Discrete: A variable can take a specific value that is separate and distinct. Each value is not related to any other value. Some examples of discrete data types include the number of cars per family, the number of times a person drinks water during a day, or the number of defective products on a production line.

Continuous: A variable can take numeric values within a specific range or interval. Continuous data can take any possible value that the observations in a set can take. For example, with temperature readings, each reading can take on any real number value on a thermometer.

Nominal data: The order of the data is arbitrary, or no order is associated with the data. For example, race or ethnicity has the values black, brown, white, Indian, American, and so forth; no order is associated with the values.

Ordinal data: This data is in a particular defined order. Examples include Olympic medals, such as Gold, Silver, and Bronze, and Likert scale surveys, such as disagree, agree, strongly agree. With ordinal data, you cannot state, with certainty, whether the intervals between values are equal.

Interval data: This data has meaningful intervals between measurements, but there is no true starting zero. A good example is the temperature measurement in Kelvin or the height of a tsunami. Interval data is like ordinal data, except the intervals between values are equally split. The most common example is the temperature in degrees Fahrenheit. The difference between 29 and 30 degrees is the same magnitude as the difference between 58 and 59 degrees.

Ratio data: The difference between two values has the same measurement meaning. For example, two cities' heights above sea level can be expressed as a ratio. Likewise, the difference in water level of two reservoirs can be expressed as a ratio, such as twice as much as X reservoir.

Before the analysis, understand the variables you are using and prepare all of them with the right data type. Many tools support the transformation of variable types.

4.1.3.2 Data Preparation

After the preliminary data type conversions, the next step is to study the data. You need to check the values and their association with the data. You also need to find missing values, null values, empty spaces, and unknown characters so they can be removed from the data before the analysis. Otherwise, this can impact the accuracy of the model. This section describes some of the criteria and analysis that can be performed on the data.

Handling Missing Values

Sometimes a database may not contain values for some variables as the values weren't recorded. There may be several records with no values recorded in the database. When you perform analysis with the missing values, the results you obtain may not represent the truth. Missing values have to be addressed properly before the analysis. The following are some of the methods used to resolve this issue:

Ignore the values: This is not a very effective method. This is usually recommended when the label is missing and you're not able to interpret the value's meaning. Also, if there are only one or two records of many variables of the same row is missing, then you can delete this row of records and ignore the values. If the size of the sample is large and deleting one or two sample values does not make a difference, then you can delete the record.

Fill in the values with average value or mode: This is the simplest method. Determine the average value or mode value for all the records of an attribute and then use this value to fill in all the missing values. This method depends on the type of problem you are trying to solve. For example, if you have time-series data, this method is not recommended. In time-series data, let's say you are collecting moisture content of your agricultural land soil every day. Data is collection is over, say, every 24 hours for one month or may be for one year. Since moisture varies every day depending on the weather condition, it is not recommended to impute a mean or mode value for a missing value.

Fill in the values with an attribute mean belonging to the same bin: If your data set has an attribute classified into categories—for example, income group as high, low, and average—and another attribute, say, *population*, has several missing values, you can fill in the missing values with a mean value for that particular category. In this example, if the *population* of the high-income group is missing, use the mean population for the high-*income* group.

The following is another example of how binning can be used to fill in missing values:

1. Data set *income*: 100, 210, 300, 400, 900, 1000, 1100, 2000, 2100, 2500.
2. Partition the data into different bins based on the distance between two observations.

   Bin 1: 100, 210, 300, 400
   Bin 2: 900, 1000, 1100
   Bin 3: 2000, 2100, 2500
3. Take the average of the entries in each bin.

   Bin 1 Average: 252.5
   Bin 2 Average: 1,000
   Bin 3 Average: 2,200
4. Use the average bin values to fill in the missing value for a particular bin.

   Predict the values based on the most probable value: Based on the other attributes in the data set, you can fill in the value based on the most probable value it can take. You can use some of the statistical techniques such as Bayes' theorem or a decision tree to find the most probable value.

Handling Duplicates, Junk, and Null Values

Duplicate, junk, and null characters should be cleaned from the database before the analytics process. The same process discussed for handling empty values can be used. The only challenge is to identify the junk characters, which is sometimes hard and can cause a lot of issues during the analysis.

4.1.3.3 Data Transformation

After a preliminary analysis of data, sometimes you may realize that the raw data you have may not provide good results or doesn't seem to make any sense. For example, data may be skewed, data may not be normally distributed, or measurement scales may be different for different variables. In such cases, data may require transformation. Common transformation techniques include normalization, data aggregation, and smoothing. After the transformation, before presenting the analysis results, the inverse transformation should be applied.

Normalization

Certain techniques such as regression assume that the data is normally distributed and that all the variables should be treated equally. Sometimes the data we collect for various predictor variables may differ in their measurement units, which may have an impact on the overall equation. This may cause one variable to have more influence over another variable. In such cases, all the predictor variable data is normalized to one single scale. Some common normalization techniques include the following:

> *Z-score normalization (zero-mean normalization)*: The new value is created based on the mean and standard deviations. The new value A' for a record value A is normalized by computing the following:
>
> $$A' = (A - mean_A) / SD_A$$
>
> where $mean_A$ is the mean, and SD_A is the standard deviation of attribute A.
>
> This type of transformation is useful when we do not know the minimum and maximum values of an attribute or when there is an outlier dominating the results.

For example, you have a data set, say,

A = [2,3,4,5,6,7]

Mean = 4.5

SD = 1.87

New Z-Transformation value of 2 is = (2-4.5)/1.87 = -1.3363

Similarly, you can calculate for all the values as shown in Table 4-1.

Table 4-1. *Example of Z-Transformation*

Original Value	After Z-Transformation
2	-1.33630621
3	-0.801783726
4	-0.267261242
5	0.267261242
6	0.801783726
7	1.33630621

Min-max normalization: In this transformation, values are transformed within the range of values specified. Min-max normalization performs linear transformations on the original data set. The formula is as follows:

New value $A' = ((A - Min_A) / (Max_A - Min_A))(\text{Range of } A') + Min_{A'}$

Range of $A' = Max_{A'} - Min_{A'}$

Min-max transformation maps the value to a new range of values defined by the range $[Max_{A'} - Min_{A'}]$. For example, for the new set of values to be in the range of 0 and 1, the new Max = 1 and the new Min = 0; the old value is 50, with a min value of 12 and a max of 58. Then:

$A' = ((50 - 12) / (58 - 12))*1+0 = 0.82$

Data aggregation: Sometimes a new variable may be required to better understand the data. You can apply mathematical functions such as sum or average to one or more variables to create a new set of variables.

Sometimes, to conform to the normal distribution, it may be necessary to use `log()` or exponential functions or use a *box-cox* transformation.

A box-cox transformation method is used for transforming a non-normally distributed data into a nearly normally distributed data.

The basic idea is to find a data value for λ such that the transformed data is as close to normally distributed as possible. The formula for a box-cox transformation is as follows:

$$A' = (A' - 1) / \lambda$$

where λ is a value greater than 1.

4.1.4 Phase 4: Explore and Visualize the Data

The next step is to understand the characteristics of the data, such as its distribution, trends, and relationships among variables. This exploration enables you to get a better understanding of the data. You can become familiar with the data you collected, identify outliers and various data types, and discover your first insights into the data. The information you uncover in this phase can help form the basis of later hypotheses regarding hidden information in the sample. Data is visualized using various plots. This process is referred to as *exploratory data analysis* (EDA), and this is covered elaborately in Chapter 5.

4.1.5 Phase 5: Choose Modeling Techniques and Algorithms

At this point, you have the data ready to perform further analysis. Based on the problem you are trying to solve, an appropriate method has to be selected. Analytics is about explaining the past and predicting the future. It combines knowledge of statistics, machine learning, databases, and artificial intelligence.

You decide whether to use unsupervised or supervised machine-learning techniques in this phase. Is it a classification problem or a regression problem? Will you use a descriptive or predictive technique? Will you use linear regression or logistic regression methods? These choices depend on both the business requirements and the data you have. A proper training set is created in this stage, and various parameters are calibrated for optimal values and the model. The tools are also chosen in this phase. This may require programming skills or specific tool skills. Once you choose a model, the next step is to create the model and evaluate its performance. In the following sections, we will explain each of the different analytic model types (descriptive analytics, predictive analytics) and also explain the two learning models (supervised machine learning and unsupervised machine learning).

4.1.5.1 Descriptive Analytics

Descriptive analytics, also *referred* to *as EDA,* explains the patterns hidden in the data. These patterns can be the number of market segments, sales numbers based on regions, groups of products based on reviews, software bug patterns in a defect database, behavioral patterns in an online gaming user database, and more. These patterns are purely based on historical data and use basic statistics and data visualization techniques.

4.1.5.2 Predictive Analytics

Prediction consists of two methods: classification and regression analysis.

Classification is a data analysis in which data is classified into different classes. For example, a credit card can be approved or denied, flights at a particular airport are on time or delayed, and a potential employee will be hired or not. The class prediction is based on previous behaviors or patterns in the data. The task of the classification model is to determine the class of data from a new set of data that was not seen before.

Regression predicts the value of a numerical variable (continuous variable)—for example, company revenue or sales numbers. Most books refer to prediction as the prediction of a value of a continuous variable. However, classification is also prediction, as the classification model predicts the class of new data of an unknown class label.

4.1.5.3 Machine Learning

Machine learning is about making computers learn and perform tasks based on past historical data. Learning is always based on observations from the data available. The emphasis is on making computers build mathematical models based on that learning and perform tasks automatically without the intervention of humans. The system cannot always predict with 100 percent accuracy because the learning is based on past data available, and there is always a possibility of new data arising that was never learned earlier by the machine. Machines build models based on iterative learning to find hidden insights. Because there is always a possibility of new data, this iteration is important because the machines can independently adapt to new changes. Machine learning has been around for a long time, but recent developments in computing, storage, and programming; new complex algorithms; and big data architectures such as Hadoop; have helped it gain momentum. There are two types of machine learning: supervised machine learning and unsupervised machine learning.

Supervised Machine Learning

As the name suggests, in supervised machine learning, a machine builds a predictive model with the help of a training data set with a specific target label. In a *classification* problem, the output variable is typically a category such as Yes or No, Fraudulent or Not Fraudulent. In a *regression* problem, the output variable is a numerical/continuous value, such as revenue or income. Figure 4-2 shows a typical supervised machine learning model.

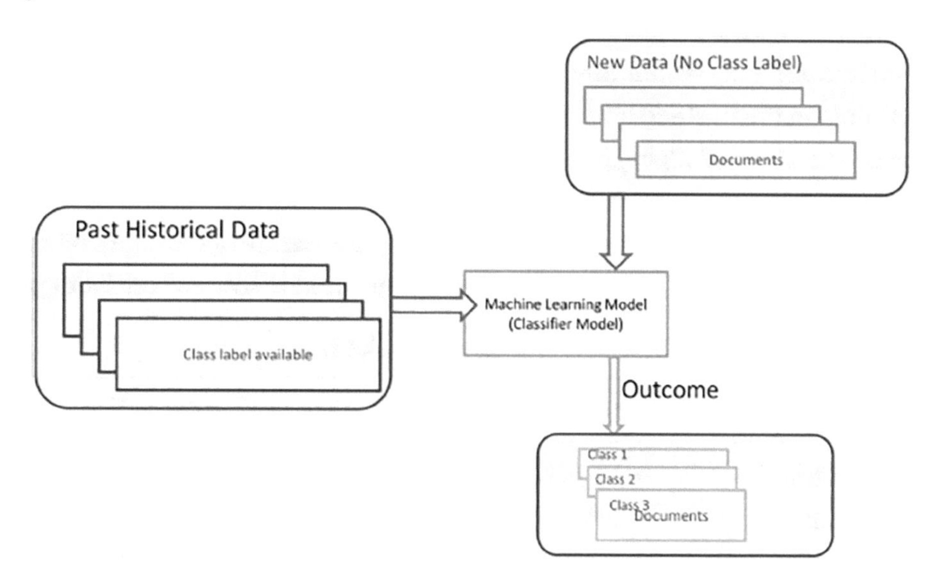

Figure 4-2. *Supervised machine learning model*

We divide any sample data into two sets. The first data set is called the training data set and this is used to train the model. The second data set, called the test data (sometimes referred to as validation set), is used to test the model performance. In this example, the data has a set of documents (data) that are already categorized and labeled into different classes. The labeling process is done by an expert who understands the different classes based on the expertise. The data set is called the *training data set*. The algorithm learns training data, which has class labels, and creates a model. Once the model is ready, it accepts the new set of documents whose labels are unknown and classifies them into proper class. Common classification supervised-learning algorithms include logistic regression, support vector machines, naïve Bayes, k-nearest neighbor, and decision trees.

Unsupervised Machine Learning

In unsupervised machine learning, the data has no class label. Thus, the entire data is fed into the algorithm to learn. Association rules and clustering are examples of unsupervised learning. In unsupervised learning, the outcome could be different clusters or finding associations patterns between two variables. Because there is no outcome class to identify by itself, further analysis is required to properly understand the results of the model. The analysis of grouping observations into clusters is called *clustering analysis*. *Association rules* find the associations among items purchased in department stores in a transactional database. This analysis is performed based on past data available in the transactional database to look for associations among various items purchased by customers.

Depending on the business problem you are trying to solve, different methods are selected, and different algorithms are used. The next few chapters discuss various classification, regression, clustering, and association techniques in detail.

The techniques and algorithms used are based on the nature of the data available. Table 4-2 summarizes the variable type and algorithms used to solve an analytics problem.

Table 4-2. *Analytics Methods Based on Data Types*

Response Variable Continuous	Response Variable Categorical	No Response Variable
• Linear regression • Neural network • K-nearest neighbor (KNN)	• Logistic regression • KNN • Neural network • Decision/classification trees • Naïve Bayes	• Cluster analysis • Principal component analysis • Association rule

Having more variables in the data set may not always provide the desired results. However, if you have more predictor variables, you need more records. For example, if you want to find out the relationship between one Y and one single predictor X, then 15 data points may give you results. But if you have 10 predictor variables, 15 data points is not enough. Then how much is enough? Statisticians and many researchers have worked on this and given a rough estimate. For example, a procedure by Delmater and Hancock (2001) indicates that you should have $6 \times m \times p$ records for any predictive models, where p is number of variables and m is the number of outcome classes. The more records you have, the better the prediction results. Hence, in *big* data processing, you can eliminate the need for sampling and try to process all the available data to get better results.

4.1.6 Phase 6: Evaluate the Model

At this point, you have a model, but how do you know that it will work for you? What is the performance of the model? In this phase, you evaluate the model by using standard methods that measure the accuracy of the model and its performance in the field. It is important to evaluate the model and to be certain that the model achieves the business objectives specified by business leaders. This requires in-depth knowledge of statistics, machine learning, algorithms, and interpreting results. In regression models we use R-squared, adjusted R-squared, and mean squared error as the performance measures. For classification models, we use F-score, precision, recall, and receiver operating curve (ROC) to measure the performance of the model. For unsupervised learning techniques like clustering, we use the cluster variance values to determine the clustering accuracy, and for relationship mining, we use lyft, support, and confidence levels as performance measures. This is explained in depth in the subsequent chapters on the performance evaluation of models.

4.1.7 Phase 7: Report to Management and Review

In this phase, you present your mathematical model to the management team and stakeholders. It is important to align your mathematical output of the model with the business objectives and be able to communicate to the management in a language nontechnical people can understand. Once management is satisfied with the results and the model, the model is ready for deployment. If any changes are made, the preceding cycle is repeated.

Typically, the following points are addressed during the presentation of the model and its use in solving the business problems.

4.1.7.1 Problem Description

First, specify the problem defined by the business and solved by the model. In this step, you are revalidating the precise problem intended to be solved and connecting the management to the objective of the data analysis.

4.1.7.2 Data Set Used

Specify the data you have used—including the period of the data, source of data, and features/fields of the data used. This reconfirms to the management that the right data has been used. You also emphasize the assumptions made.

4.1.7.3 Data Cleaning Steps Carried Out

Specify the issues encountered in the data. You note which nonreliable data had to be removed and which had to be substituted. The rationale for such decisions is presented along with the possible downsides, if any.

4.1.7.4 Method Used to Create the Model

Present the method, technique, or algorithm used to create the model and present your reasoning for this choice. Your aim is to convince the management that the approach is moving the business in the right direction to solve the problem. Also, you present how the model solves the problem, and you explain how the model was evaluated and optimized.

4.1.7.5 Model Deployment Prerequisites

Present the prerequisites for model deployment, including the data and preprocessing requirements. You also present the model deployment hardware and software requirements.

4.1.7.6 Model Deployment and Usage

Here you present how the model will be deployed for use and how it will be used by end users. You can provide reports and their interpretation to specify the do's and don'ts related to the model's use. You also emphasize the importance of not losing sight of the problem you are solving by using the model.

4.1.7.7 Handling Production Problems

Present the ideal process for recording the problems (bugs) observed and the ways they will be reported, analyzed, and addressed. You also emphasize how this step may lead to the optimization of the model over a period of time, as these may indicate the changes happening to the basic assumptions and structure.

4.1.8 Phase 8: Deploy the Model

This is a challenging phase of the project. The model is now deployed for end users and is in a production environment, analyzing the live data. Depending on the deployment scenario and business objectives, the model can perform predictive analytics or simple descriptive analytics and reporting. The model also may behave differently in the production environment, as it may see a totally different set of data from the actual model. In such scenarios, the model is revisited for improvements. In many cases, the analysis is carried out by end customers, and they need to be trained properly.

The success of the deployment depends on the following:

- Proper sizing of the hardware, ensuring required performance
- Proper programming to handle the capabilities of the hardware
- Proper data integration and cleaning
- Effective reports, dashboards, views, decisions, and interventions to be used by end users or end-user systems
- Effective training to the users of the model

The following are the typical issues observed during the deployment:

- Hardware capability (for example, memory or CPUs) is not in tune with the requirements of data crunching or with effective and efficient model usage (for example, the data takes more time to crunch than the time window available or takes so much time that the opportunity for using the generated analysis is reduced). This may also impede the use of real-time data and real-time decision-making.
- Programs used are not effective in using the parallelism and hence reduce the possibility of effectively using the results.

- Data integration from multiple sources in multiple formats including real-time data.
- Data sanitization or cleaning before use.
- Not recognizing the changes in data patterns, nullifying or reducing the model's usability, or resulting in making wrong decisions.
- Not having standby mechanisms (redundancy) to the hardware or network and not having backup or replication for the data.
- Changes to the data sources or data schema, leading to issues with usability of the model.
- Lack of effective user training, leading them to use the model in a defective or wrong way, resulting in wrong decisions.
- Wrong setup of the systems by the administrators, leading to suboptimal or inefficient use of system resources.

4.2 Chapter Summary

In this chapter, we focused on the processes involved in business analytics, including identifying and defining a business problem, preparing data, collecting data, modeling data, evaluating model performance, and reporting to the management on the findings.

You learned about various methods involved in data cleaning, including normalizing, transforming variables, handling missing values, and finding outliers. We also delved into data exploration, which is the most important process in business analytics.

Further, you explored supervised machine learning, unsupervised machine learning, and how to choose different methods based on business requirements. We also touched upon the various metrics to measure the performance of different models including both regression and classification models.

CHAPTER 5

Exploratory Data Analysis

The process of collecting, organizing, and describing data is commonly called *descriptive analytics*. Statistics summarize various measures such as mean, median, mode, and other measures computed from samples. Multiple graphs can be plotted to visualize this data. Data exploration helps in identifying hidden patterns in the data and provides meaningful information to make business decisions. In this chapter, we discuss the fundamentals of data analysis, including various graphs to explore data for further analysis.

5.1 Exploring and Visualizing the Data

Exploratory data analysis (EDA) is the process of exploring data to understand its structure, types, different features, distribution of individual features, and relationships among the variables. EDA relies heavily on representations using graphs and tables. Visual analysis is easily understood and can be quickly absorbed. In this process, data reveals hidden patterns that can help determine not only business decisions but also the selection of techniques and algorithms for further analysis. Therefore, exploring data by using data visualization is a critical part of the overall business analytics process.

Figure 5-1 summarizes the data collection and data preprocessing steps described in the previous chapters, which must occur before the data exploration and data visualization phase. Data is gathered from various sources, including databases, data warehouses, the Web, and logs from various devices, among others. It is consolidated based on particular business problems and requirements. During the integration process, data may have null values, duplicate values, and missing values, and sometimes may require data transformation. All these are explored during EDA using the right visualization and nonvisualization techniques.

U. R. Hodeghatta and U. Nayak, *Practical Business Analytics Using R and Python*,
https://doi.org/10.1007/978-1-4842-8754-5_5

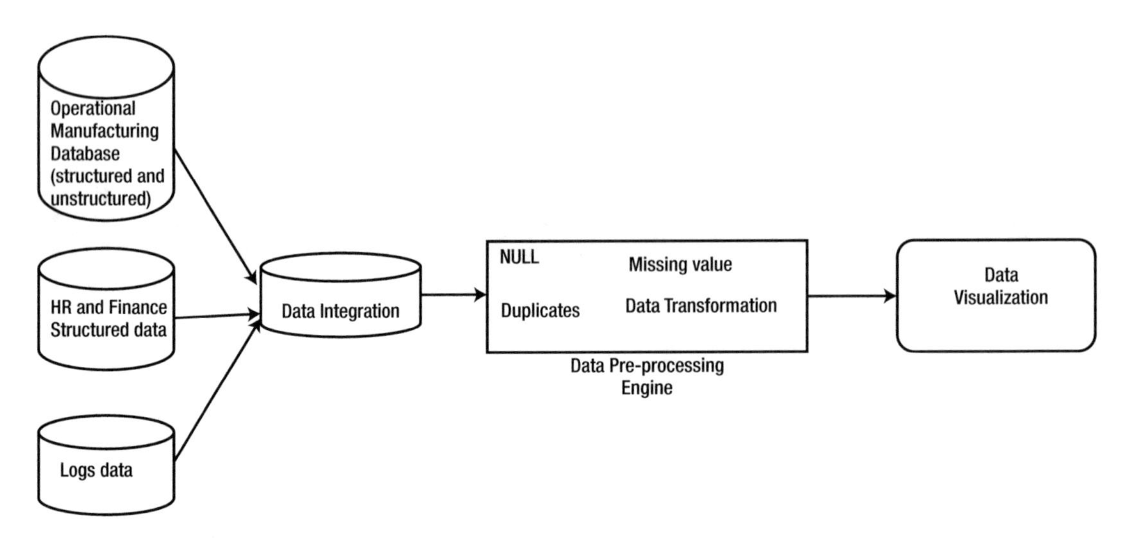

Figure 5-1. *Workflow of data visualization*

The goals of exploratory analysis are as follows:

- To determine the distribution and other statistics of the data set
- To determine whether the data set or a feature set needs normalization
- To determine whether the data set can answer the business problem you are trying to solve
- To come up with a blueprint to solve the business problem

The following sections describe the tables and graphs required for carrying out good data analysis. The basic concepts are explained in each section and are demonstrated using R. Then we demonstrate this using Python. We will try to avoid explaining the concepts twice.

5.1.1 Tables

The easiest and most common tool available for looking at data is a table. Tables contain rows and columns. Raw data is displayed as rows of observations and columns of variables. Tables are useful for smaller sets of samples, so it can be difficult to display the whole data set if you have many records. By presenting the data in tables, you can gain insight into the data, including the type of data, the variable names, and the way the data is structured. However, it is not possible to identify relationships between variables by looking at tables. In R, we use the `View()` function, and in Python, we can use the `display()` and `style()` functions to display the dataframe in the tabular format, as shown in Figure 5-2 and Figure 5-3.

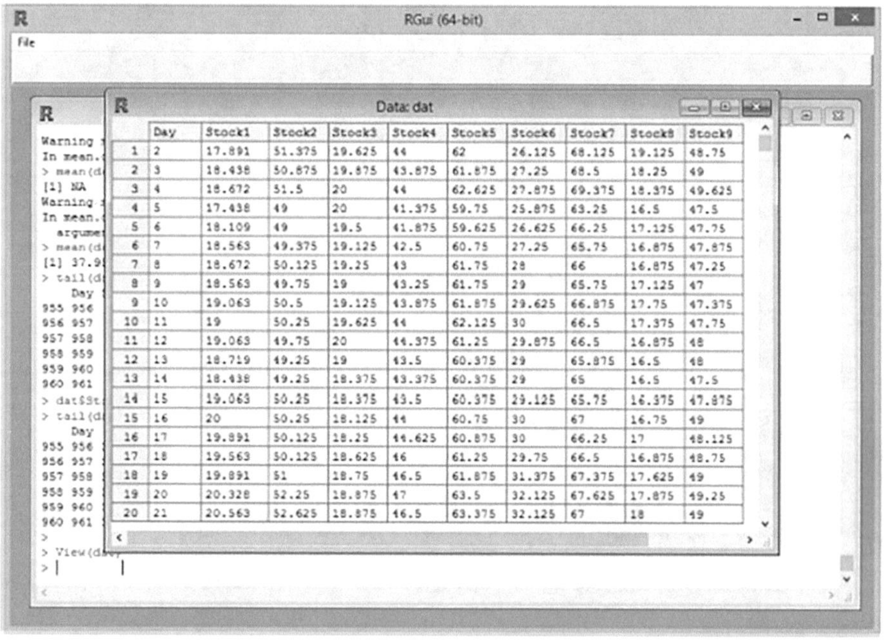

RGui (64-bit)

Data: dat

	Day	Stock1	Stock2	Stock3	Stock4	Stock5	Stock6	Stock7	Stock8	Stock9
1	2	17.891	51.375	19.625	44	62	26.125	68.125	19.125	48.75
2	3	18.438	50.875	19.875	43.875	61.875	27.25	68.5	18.25	49
3	4	18.672	51.5	20	44	62.625	27.875	69.375	18.375	49.625
4	5	17.438	49	20	41.375	59.75	25.875	63.25	16.5	47.5
5	6	18.109	49	19.5	41.875	59.625	26.625	66.25	17.125	47.75
6	7	18.563	49.375	19.125	42.5	60.75	27.25	65.75	16.875	47.875
7	8	18.672	50.125	19.25	43	61.75	28	66	16.875	47.25
8	9	18.563	49.75	19	43.25	61.75	29	65.75	17.125	47
9	10	19.063	50.5	19.125	43.875	61.875	29.625	66.875	17.75	47.375
10	11	19	50.25	19.625	44	62.125	30	66.5	17.375	47.75
11	12	19.063	49.75	20	44.375	61.25	29.875	66.5	16.875	48
12	13	18.719	49.25	19	43.5	60.375	29	65.875	16.5	48
13	14	18.438	49.25	18.375	43.375	60.375	29	65	16.5	47.5
14	15	19.063	50.25	18.375	43.5	60.375	29.125	65.75	16.375	47.875
15	16	20	50.25	18.125	44	60.75	30	67	16.75	49
16	17	19.891	50.125	18.25	44.625	60.875	30	66.25	17	48.125
17	18	19.563	50.125	18.625	46	61.25	29.75	66.5	16.875	48.75
18	19	19.891	51	18.75	46.5	61.875	31.375	67.375	17.625	49
19	20	20.328	52.25	18.875	47	63.5	32.125	67.625	17.875	49.25
20	21	20.563	52.625	18.875	46.5	63.375	32.125	67	18	49

Figure 5-2. *Output of the View() function in R*

The table helps us to quickly check the contents of the data and browse the features we have in the data set. Looking at the data table provides an understanding of the feature name, whether feature names are meaningful, and how the name is related to other features; it also allows us to check the response variable labels (if you are performing predictive analytics) and identify data types.

```
In [126]: stocks.style
```

Out[126]:

	Day	Stock1	Stock2	Stock3	Stcok4	Stock5	Stcok6	Stock7	Stock8	Stock9	Stock10	Ratings
0	1	17.219000	50.500000	18.750000	43.000000	60.875000	26.375000	67.750000	19.000000	48.750000	34.875000	High
1	2	17.891000	51.375000	19.625000	44.000000	62.000000	26.125000	68.125000	19.125000	48.750000	35.625000	Low
2	3	18.438000	50.875000	19.875000	43.875000	61.875000	27.250000	68.500000	18.250000	49.000000	36.375000	Medium
3	4	18.672000	51.500000	20.000000	44.000000	62.625000	27.875000	69.375000	18.375000	49.625000	36.250000	High
4	5	17.438000	49.000000	20.000000	41.375000	59.750000	25.875000	63.250000	16.500000	47.500000	35.500000	Low
5	6	18.109000	49.000000	19.500000	41.875000	59.625000	26.625000	66.250000	17.125000	47.750000	34.375000	Low
6	7	18.563000	49.375000	19.125000	42.500000	60.750000	27.250000	65.750000	16.875000	47.875000	34.000000	Low
7	8	18.672000	50.125000	19.250000	43.000000	61.750000	28.000000	66.000000	16.875000	47.250000	34.625000	Medium
8	9	18.563000	49.750000	19.000000	43.250000	61.750000	29.000000	65.750000	17.125000	47.000000	34.875000	Low
9	10	19.063000	50.500000	19.125000	43.875000	61.875000	29.625000	66.875000	17.750000	47.375000	36.000000	Medium
10	11	19.000000	50.250000	19.625000	44.000000	62.125000	30.000000	66.500000	17.375000	47.750000	35.625000	Medium

```
In [124]: display(stocks)
```

	Day	Stock1	Stock2	Stock3	Stcok4	Stock5	Stcok6	Stock7	Stock8	Stock9	Stock10	Ratings
0	1	17.219	50.500	18.750	43.000	60.875	26.375	67.750	19.000	48.750	34.875	High
1	2	17.891	51.375	19.625	44.000	62.000	26.125	68.125	19.125	48.750	35.625	Low
2	3	18.438	50.875	19.875	43.875	61.875	27.250	68.500	18.250	49.000	36.375	Medium
3	4	18.672	51.500	20.000	44.000	62.625	27.875	69.375	18.375	49.625	36.250	High
4	5	17.438	49.000	20.000	41.375	59.750	25.875	63.250	16.500	47.500	35.500	Low
...	...	...	...	...	...	...	...	...	...	...	...	...
945	946	50.375	46.250	19.375	52.250	61.875	23.500	78.625	26.625	41.875	44.375	Medium
946	947	50.750	46.375	19.625	50.875	64.625	23.250	77.625	26.500	40.750	45.000	High
947	948	50.625	46.625	19.625	50.875	64.625	23.250	75.000	26.250	41.250	44.125	Low
948	949	50.125	47.000	19.875	50.750	62.750	22.875	74.500	25.250	40.625	43.875	Medium
949	950	49.000	47.000	19.500	49.500	60.875	22.750	75.625	25.500	40.500	43.375	High

950 rows × 12 columns

***Figure 5-3.** Output function in Python*

5.1.2 Describing Data: Summary Tables

We already discussed statistics and their significance in Chapter 2. Statistics are the foundation for business analytics; they provide a common way of understanding data. These statistics can be represented as summary tables. Summary tables show the number of observations in each column and the corresponding descriptive statistics for each column. The following statistics are commonly used to understand data:

Minimum: The minimum value

Maximum: The maximum value

Mean: The average value

Median: The value at the midpoint

Sum: The sum of all observations in the group

Standard deviation: A standardized measure of the deviation of a variable from the mean

First quartile: Number between the minimum and median value of the data set

Third quartile: Number between the median and the maximum value of the data set

The following output is the descriptive statistics of the stock price data set using the `summary()` function in R. The output `summary()` provides the statistical mean, variance, first quartile, third quartile, and the other measures described earlier. See Figure 5-4.

```
> stocks3<-read.csv(header=TRUE,"stocks3.csv")
> summary(stocks3)
      Day              Stock1           Stock2
 Min.   :  1.0   Min.   :17.22   Min.   :19.25
 1st Qu.:238.2   1st Qu.:27.78   1st Qu.:35.41
 Median :475.5   Median :38.92   Median :49.06
 Mean   :475.5   Mean   :37.93   Mean   :43.96
 3rd Qu.:712.8   3rd Qu.:46.88   3rd Qu.:53.25
 Max.   :950.0   Max.   :61.50   Max.   :60.25
     Stock3          Stcok4           Stock5
 Min.   :12.75   Min.   :34.38   Min.   :27.75
 1st Qu.:16.12   1st Qu.:41.38   1st Qu.:49.66
 Median :19.38   Median :43.94   Median :61.75
 Mean   :18.70   Mean   :45.35   Mean   :60.86
 3rd Qu.:20.88   3rd Qu.:48.12   3rd Qu.:71.84
 Max.   :25.12   Max.   :60.12   Max.   :94.12
     Stcok6          Stock7           Stock8
 Min.   :14.12   Min.   :58.00   Min.   :16.38
 1st Qu.:18.00   1st Qu.:65.62   1st Qu.:21.25
 Median :25.75   Median :68.62   Median :22.50
 Mean   :24.12   Mean   :70.67   Mean   :23.29
 3rd Qu.:28.88   3rd Qu.:76.38   3rd Qu.:26.38
 Max.   :35.25   Max.   :87.25   Max.   :29.25
     Stock9         Stock10          Ratings
 Min.   :31.50   Min.   :34.00   High  :174
 1st Qu.:41.75   1st Qu.:41.38   Low   :431
 Median :44.75   Median :46.69   Medium:345
 Mean   :44.21   Mean   :46.99
 3rd Qu.:47.62   3rd Qu.:52.12
 Max.   :53.00   Max.   :62.00
```

Figure 5-4. *Summary() statistics of data set in R*

In Python, the descriptive statistics of the data can be explored using the `pandas.describe()` function, as shown next. The first step is reading the data set, and the second step is calling the function `describe()`.

Here is the input:

```
stocks=pd.read_csv("stocks3.csv")

stocks.head()

stocks.describe()
```

Figure 5-5 shows the input and the output.

```
In [25]: stocks.head()
Out[25]:
```

	Day	Stock1	Stock2	Stock3	Stcok4	Stock5	Stcok6	Stock7	Stock8	Stock9	Stock10
0	1	17.219	50.500	18.750	43.000	60.875	26.375	67.750	19.000	48.750	34.875
1	2	17.891	51.375	19.625	44.000	62.000	26.125	68.125	19.125	48.750	35.625
2	3	18.438	50.875	19.875	43.875	61.875	27.250	68.500	18.250	49.000	36.375
3	4	18.672	51.500	20.000	44.000	62.625	27.875	69.375	18.375	49.625	36.250
4	5	17.438	49.000	20.000	41.375	59.750	25.875	63.250	16.500	47.500	35.500

```
In [26]: stocks.describe()
Out[26]:
```

	Day	Stock1	Stock2	Stock3	Stcok4	Stock5	Stcok6	Stock7	Stock8	Stock9	Stock10
count	949.000000	949.000000	949.000000	949.000000	949.000000	949.000000	949.000000	949.000000	949.000000	949.000000	949.000000
mean	475.000000	37.913922	43.952318	18.703635	45.348525	60.863541	24.124078	70.668203	23.291886	44.218124	46.997893
std	274.097002	10.759421	11.382877	2.731681	5.608236	14.306952	5.533698	6.774761	2.970854	4.270644	6.541811
min	1.000000	17.219000	19.250000	12.750000	34.375000	27.750000	14.125000	58.000000	16.375000	31.500000	34.000000
25%	238.000000	27.781000	35.375000	16.125000	41.375000	49.625000	18.000000	65.625000	21.250000	41.750000	41.375000
50%	475.000000	38.922000	49.125000	19.375000	43.875000	61.750000	25.750000	68.625000	22.500000	44.750000	46.750000
75%	712.000000	46.875000	53.250000	20.875000	48.125000	71.875000	28.875000	76.375000	26.375000	47.625000	52.125000
max	949.000000	61.500000	60.250000	25.125000	60.125000	94.125000	35.250000	87.250000	29.250000	53.000000	62.000000

Figure 5-5. *Summary() statistics of data set in Python*

5.1.3 Graphs

Graphs represent data visually and provide more details about the data, enabling you to identify outliers in the data, see the probability distribution for each variable, provide a statistical description of the data, and present the relationship between two or more variables. Graphs include bar charts, histograms, box plots, and scatter plots. In addition, looking at the graphs of multiple variables simultaneously can provide more insights into the data. There are three types of graphical analysis: univariate, bivariate, and multivariate.

Univariate analysis analyzes one variable at a time. It is the simplest form of analyzing data. You analyze a single variable, summarize the data, and find the patterns in the data. You can use several visualization graphs to perform univariate data analysis, including bar charts, pie charts, box plots, and histograms.

Bivariate and *multivariate data analysis* is used to compare relationships between two or more variables in the data. The major purpose of bivariate analysis is to explain the correlations of two variables for comparisons and suspect causal relationships, if any, contingent on the values of the other variables and the relationships between the two variables.

If more than two variables are involved, then *multivariate analysis* is applied. Apart from the two x- and y-axes, the third and fourth dimensions are distinguished using the color or shape or size of the variable. Beyond four or five dimensions is almost impossible to visualize.

Also, visualization is limited to the size of the output device. If you have a huge amount of data and plot a graph, it may become challenging to interpret the results as they are cluttered within the plot area. For example, if your plot area is only 10 inches by 10 inches, your data must fit within this area. It may be difficult to understand the plot. Hence, it is recommended to "zoom" the plot to get a better understanding of it.

5.1.3.1 Histogram

A *histogram* represents the frequency distribution of the data. Histograms are similar to bar charts but group numbers into ranges. Also, a histogram lets you show the frequency distribution of continuous data. This helps analyze the distribution (for example, normal or Gaussian, uniform, binomial, etc.) and any skewness present in the data. Figure 5-6 describes the probability density graph of the first variable of the stock price data, Figure 5-7 describes the histogram, and Figure 5-8 shows both the histogram and distribution in a single graph.

Here is the input:

```
> ##Plotting histogram and density function
> x<-stocks3$Stock1
> plot(density(x), main="Density Function",
+              xlab="Stock price 1")
> hist(x, freq=FALSE,
+      main="Histogram",
+       xlab="Stock price 1",
+      xlim=c(0,70))
> plot(density(x), main="Histogram and Density
Function",
+       xlab="Stock price 1")
> hist(x, freq=FALSE,add=TRUE,
+       main="Histogram",
+       xlab="Stock price 1",
+       xlim=c(0,70))

>
```

Figure 5-6 shows the output.

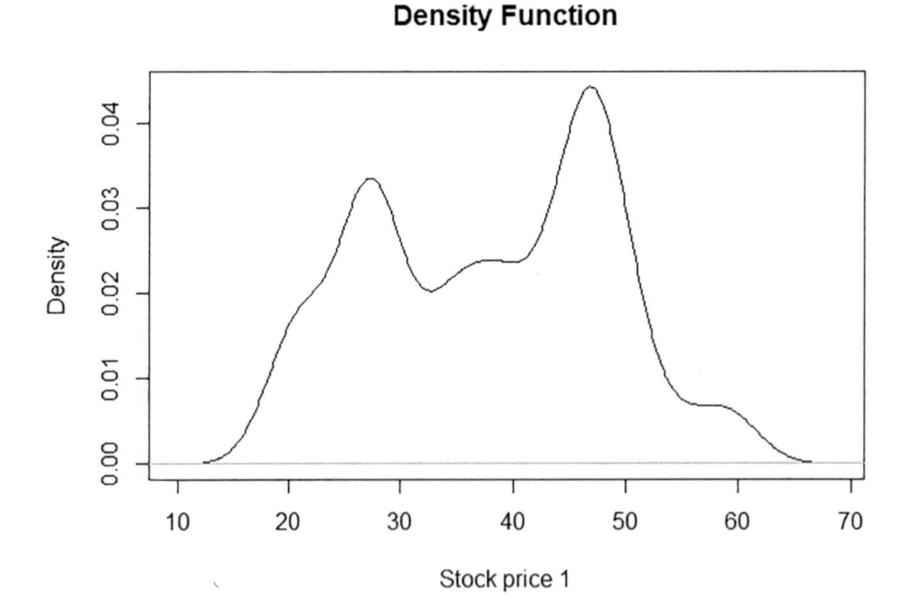

***Figure 5-6.** Stock price variable density function*

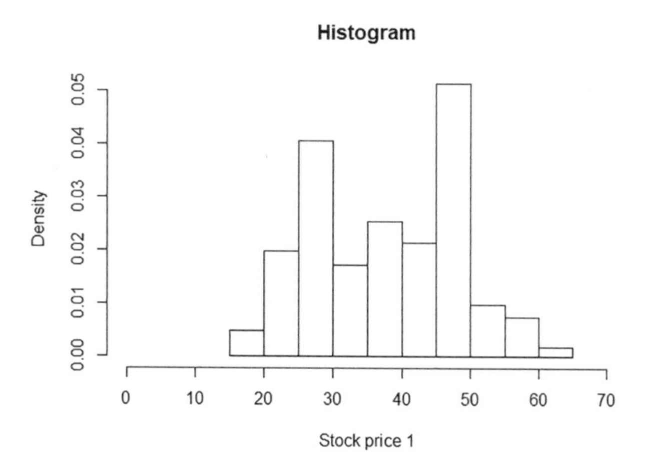

Figure 5-7. *Stock price variable density function*

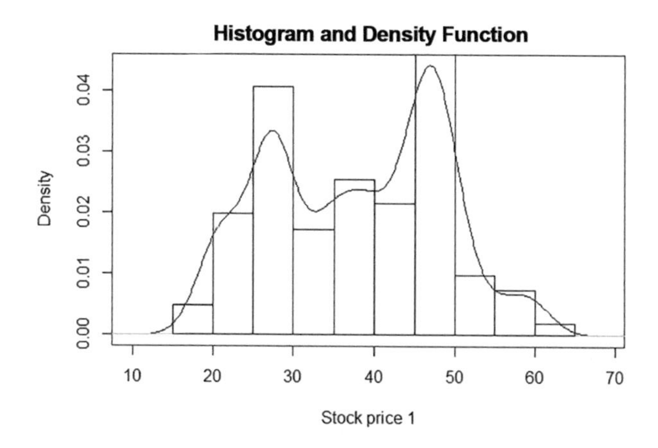

Figure 5-8. *Stock price variable histogram and density function*

In Python, we can plot both the histogram and the probability density function of a single variable to check the distribution. Unlike in R, we use two different functions in Python. The `pandas.hist()` function plots the histogram, and then we use the `pandas.kde()` function to just plot the `density()` function, as shown in Figure 5-9. Both should provide the same information, and both figures should look the same.

Here is the input:

```
#Histograms

stocks.Stock1.plot.hist(by=None, bins=15)

stocks.Stock1.plot.kde(bw_method='silverman', ind=12
00)

import matplotlib.pyplot as plt
import seaborn as sns
sns.histplot(stocks.Stock1, kde=True, bins=15)
plt.show()
```

Figure 5-9 shows the input and the output.

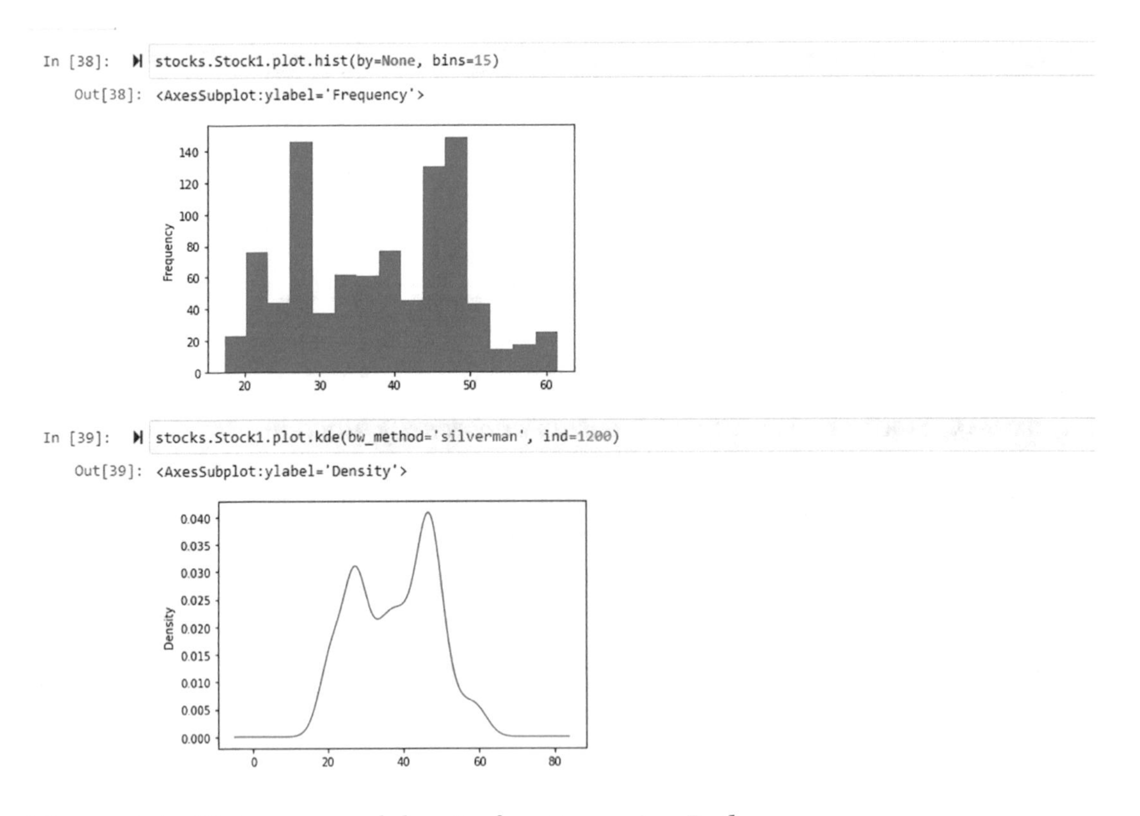

***Figure 5-9.** Histogram and density function using Python*

There may be several ways to plot the density functions and histograms. The one shown here is a basic and simple method supported by the `matplot()` library. Though there are many other visualization libraries in Python, we will not be able to discuss all of them here as that is not the purpose of this book. The objective is to explain the concepts and provide an understanding of the different plots that are available to explain the data through visualization and not to explore different libraries.

For example, using the `seaborn()` library, we can combine both histogram and density functions, as shown in Figure 5-10.

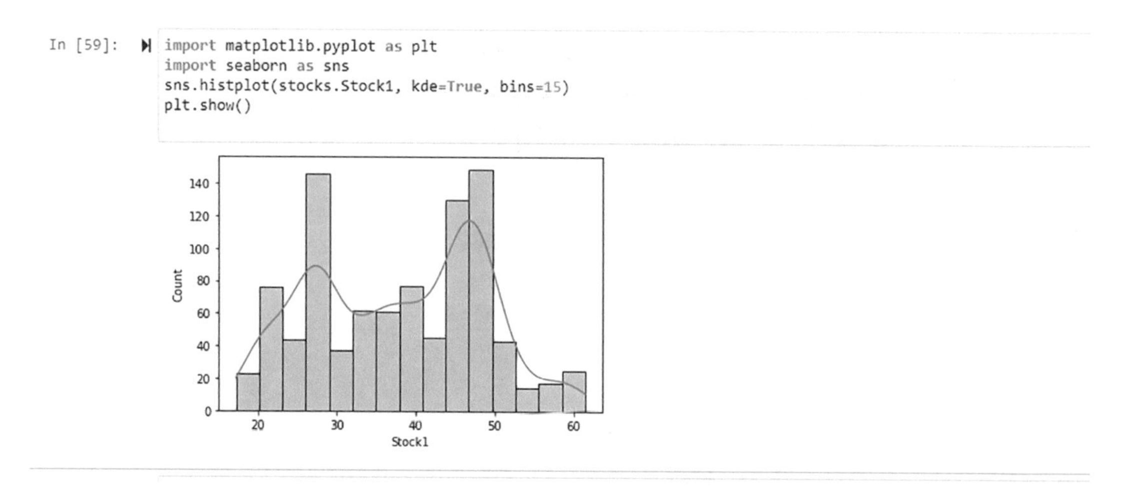
In [59]:
import matplotlib.pyplot as plt
import seaborn as sns
sns.histplot(stocks.Stock1, kde=True, bins=15)
plt.show()
Count
140
120
100
80
60
40
20
0
20
30
40
50
60
Stock1

Figure 5-10. *Histogram and density function using the seaborn() library in Python*

In the previous example, the histogram provides how the data is spread, and in this case, it is a "bimodal" as it has two distinct peaks. The stock prices have reached over 120 two times. The data is not a normal distribution.

Depending on the data set, if the data is normally distributed, the shape can be "bell" shaped. If data is "uniform," then the spread has the same height throughout. If the data has more peaks, it is called a *multimodal* distribution. Data also can be *skewed*, with most of the data having only high values compared to the others; there is a significant difference between high-value and low-value data. Depending on the data set, the data can be left skewed or right skewed, as shown in Figure 5-11 and Figure 5-12. Sometimes, data distribution can be just random with no proper distinct shape.

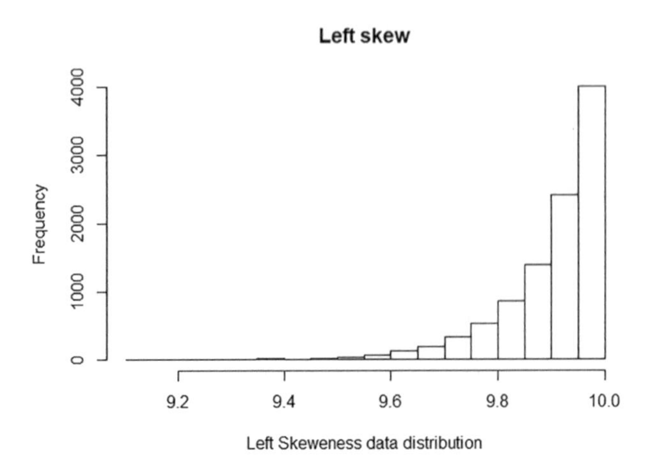

Figure 5-11. *Left skew distribution*

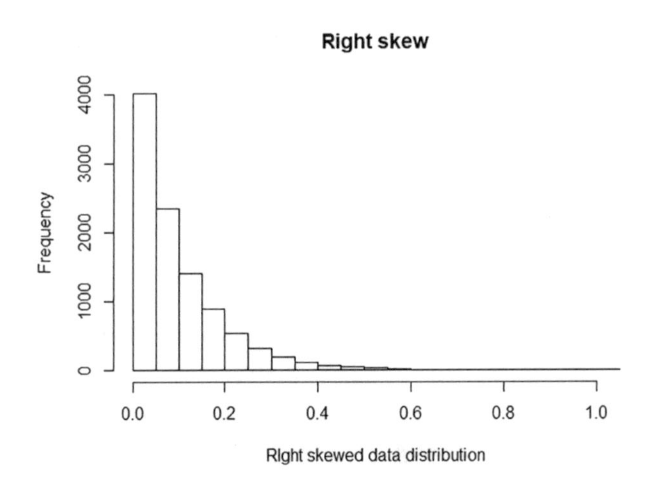

Figure 5-12. *Right skew distribution*

5.1.3.2 Box Plots

A box plot or whisker plot is also a graphical description of data. Box plots, created by John W. Tukey, show the distribution of a data set based on a five-number summary: minimum, maximum, median, first quartile, and third quartile. Figure 5-13 explains how to interpret a box plot and its components. It also shows the central tendency; however, it does not show the distribution like a histogram does. In simple words, a box plot is a graphical representation of the statistical summary, the data spread within the IQR (*interquartile range*), and the outliers above the maximum value and below the maximum value (whiskers). Knowing outliers in the data is useful information for the analytical model exercise. Outliers can have a significant influence and impact on the model. Box plots are also beneficial if we have multiple variables to be compared. For example, if you have a data set that has sales of multiple brands, then box plots can be used to compare the sales of different brands.

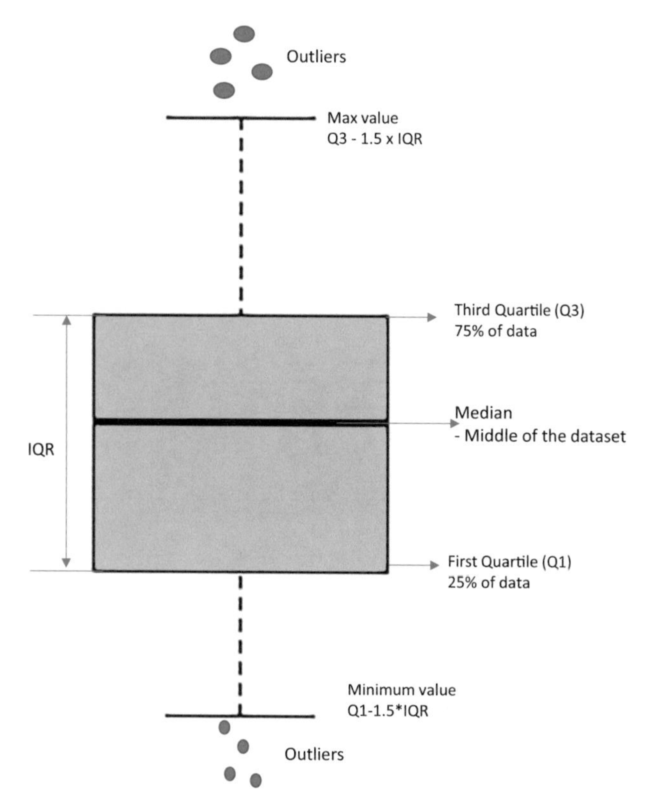

Figure 5-13. *Box plot*

Parts of Box Plots

The central rectangle spans the first and third quartiles (*interquartile range*, or IQR). The line inside the rectangle shows the median. The lines, also called *whiskers*, that are above and below the rectangle show the maximum and minimum of the data set.

Typical data sets do not have a surprisingly high maximum value or low minimum value. Such data is known as an *outlier* and is generally present outside the two whisker lines.

Tukey (Tukey, 1977) has provided the following definitions for outliers:

> *Outliers*—2/3 IQR above the third quartile or 2/3 IQR below the first quartile
>
> *Suspected outliers*—1.5 IQR above the third quartile or 1.5 IQR below the first quartile.

If the data is normally distributed, then $IQR = 1.35\ \sigma$, where σ is the population standard deviation.

The box plot distribution will explain how tightly the data is spread across and whether it is symmetrical or skewed. Figure 5-14 describes the box plot with respect to a normal distribution and how symmetrically data is distributed.

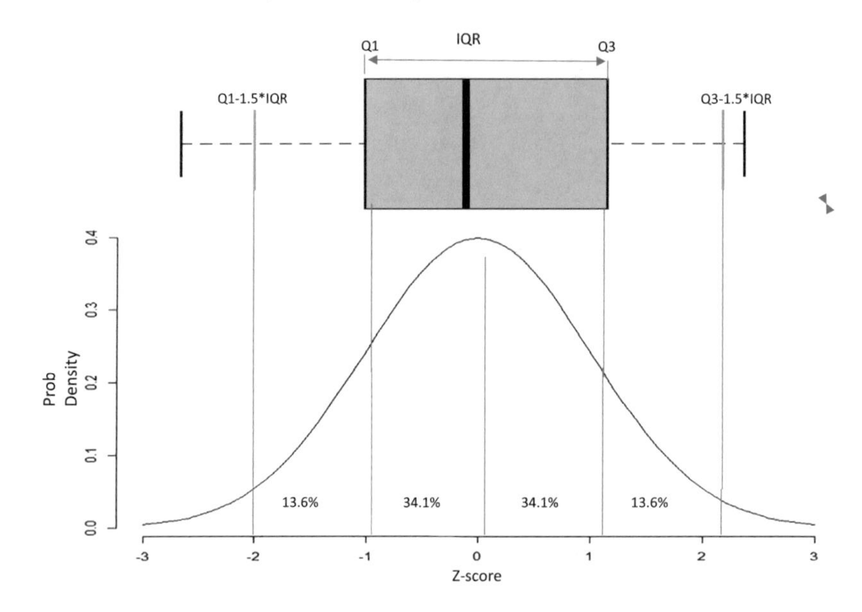

Figure 5-14. *Box plot on a normal distribution*

A box plot is positively skewed if the distance from the median to the maximum is greater than the distance from the median to the minimum. Similarly, a box plot is negatively skewed if the distance from the median to the minimum is greater than the distance from the median to the maximum.

One of the commonly used applications of box plot, apart from finding the spread and outlier, is that you can plot multiple variables in box plots and compare the data of each variable side-by-side. Figure 5-15 shows an example. As you can see from the box plots, all three have a different data spread, their medians are different, and 50 percent of the data for each variable is different. It is clear from the plot that there is a correlation between the three variables.

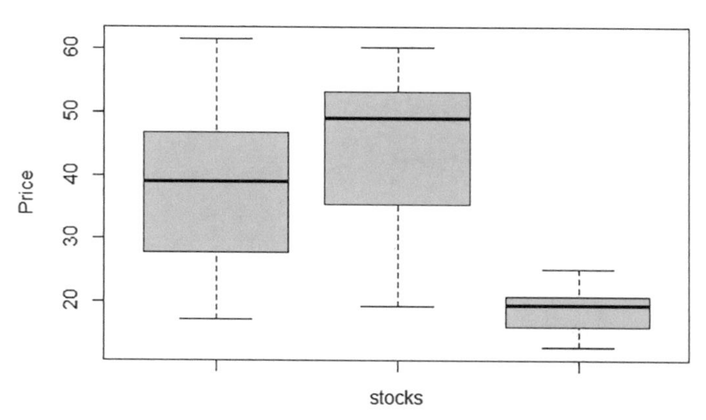

Figure 5-15. *Side-by-side box plot*

Box Plots Using Python

We can plot box plots on a Pandas dataframe, as shown in Figure 5-16. The first box plot is the Stock1 variable, and the second plot is for comparing four different variables. The plots shown here are simple plots using the default arguments available in the Python and Pandas libraries. Fancier plots are supported by different libraries, which are not covered here.

Box plots are also very beneficial if we have multiple variables to be compared. For example, if you have a data set that has sales of multiple brands, then box plots can be used to compare the sales of different brands. In this case, the four stock prices compared have no relationship with each other.

Here is the input:

```
stocks.Stock1.plot.box()

stocks.boxplot(column=['Stock1','Stock2','Stock3',
'Stock5'])
```

Figure 5-16 shows the input and the output.

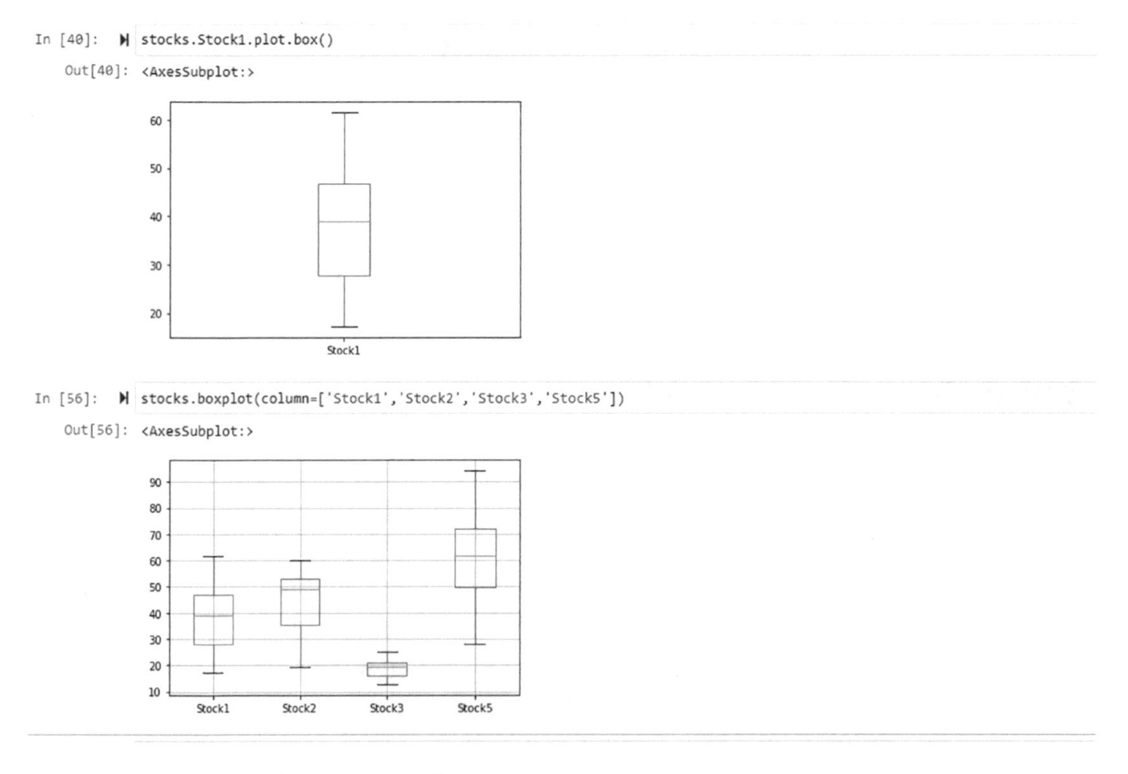

Figure 5-16. *Box plot using Python*

5.1.3.3 Bivariate Analysis

The most common data visualization tool used for bivariate analysis is the *scatter plot*. Scatter plots can be used to identify the relationships between two continuous variables. Each data point on a scatter plot is a single observation. All the observations can be plotted on a single chart.

5.1.3.4 Scatter Plots

Figure 5-17 shows a scatter plot of the number of employees versus revenue (in millions of dollars) of various companies. As you can see, there is a strong relationship between the two variables that is almost linear. However, you cannot draw any causal implications without further statistical analysis. The example shows the scatter plot of the number of employees on the x-axis and revenues on the y-axis. For every point on the x-axis, there is a corresponding point on the y-axis. As you can see, the points are spread in proportion and have a linear relationship. Though not all the points are aligned proportionally, most points are.

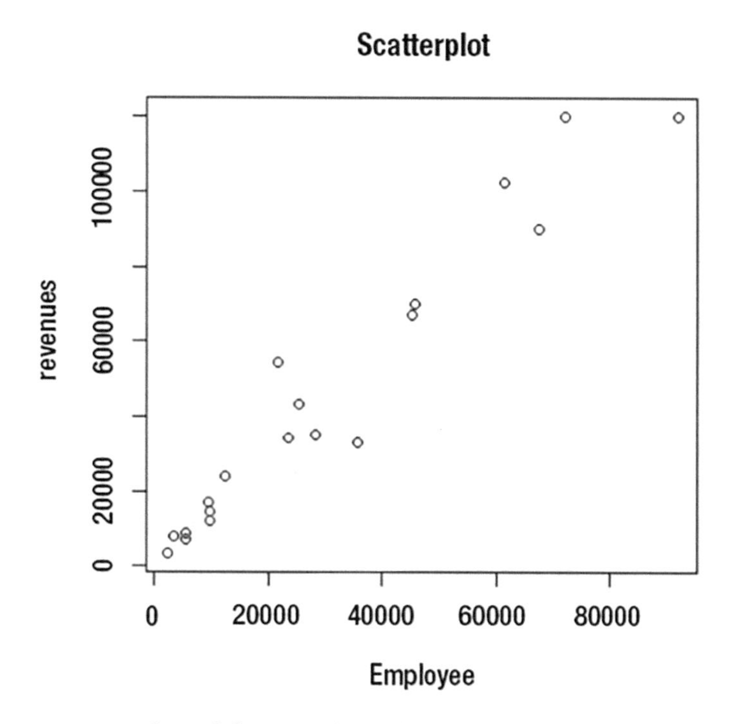

Figure 5-17. *A scatter plot of the number of employees versus revenue (in millions of dollars)*

Unfortunately, scatter plots are not always useful for finding relationships. In the case of our Stocks data set example, it is difficult to interpret any relationship between Stock1 and Stock2, as shown in Figure 5-18. In Python, we use the `plot.scatter()` function for plotting scatter plots.

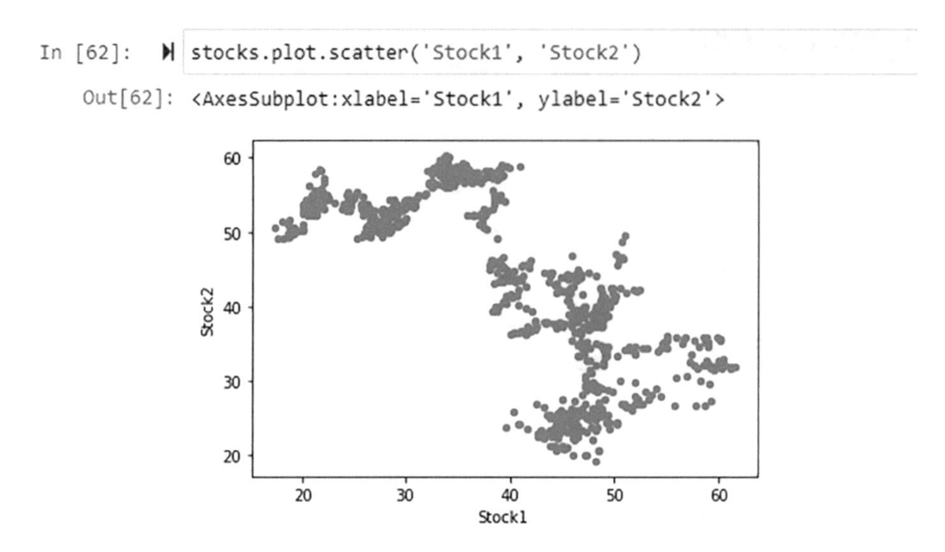

Figure 5-18. *Scatter plot of Stock1 versus Stock2 using Python*

5.1.4 Scatter Plot Matrices

A scatter plot matrix graph plots pairwise scatter plots to look for patterns and relationships of all the variables in the data set. A scatter plot matrix graph takes two variables pairwise and selects all the variables in the data set plots as a single graph so that you can get a complete picture of your data set and understand the patterns and relationships between pairs of variables.

In R, we use the `pairs()` command to plot a scatter matrix graph. And in Python we use the `scatter_matrix()` function of the `plotting()` library, as shown in Figure 5-19 and Figure 5-20.

```
> setwd("E:/Umesh-MAY2022/Personal-May2022/BA2ndEditi
n/Book Chapters/Chapter 5 - EDA")
> stocks<-read.csv(header=TRUE,"stocks.csv")
> pairs(stocks)
```

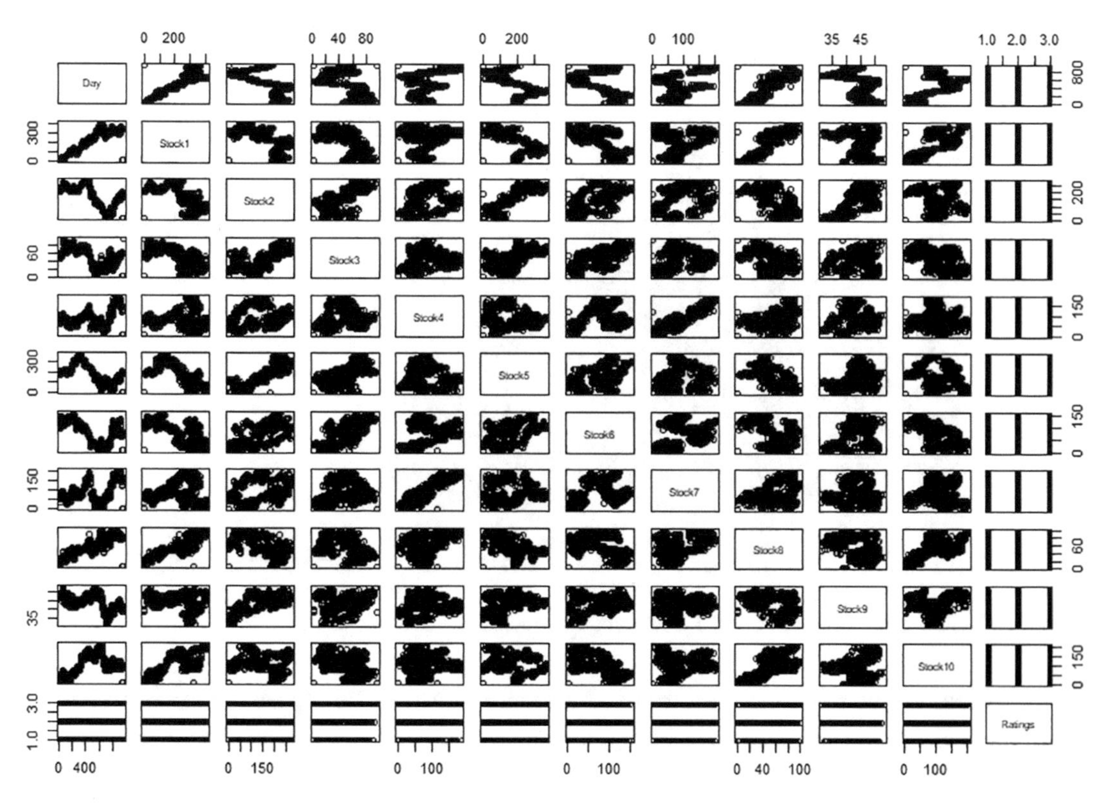

***Figure 5-19.** Scatter matrix plot using R*

Here is the input:

```
import matplotlib.pyplot  as plt
from pandas.plotting  import scatter_matrix

scatter_matrix(stocks,alpha =0.5, figsize=(10,6), ax=N
one, grid=False, color ='red', hist_kwds={'bins':10, 'col
or':'red'})
```

Figure 5-20 shows the input and the output.

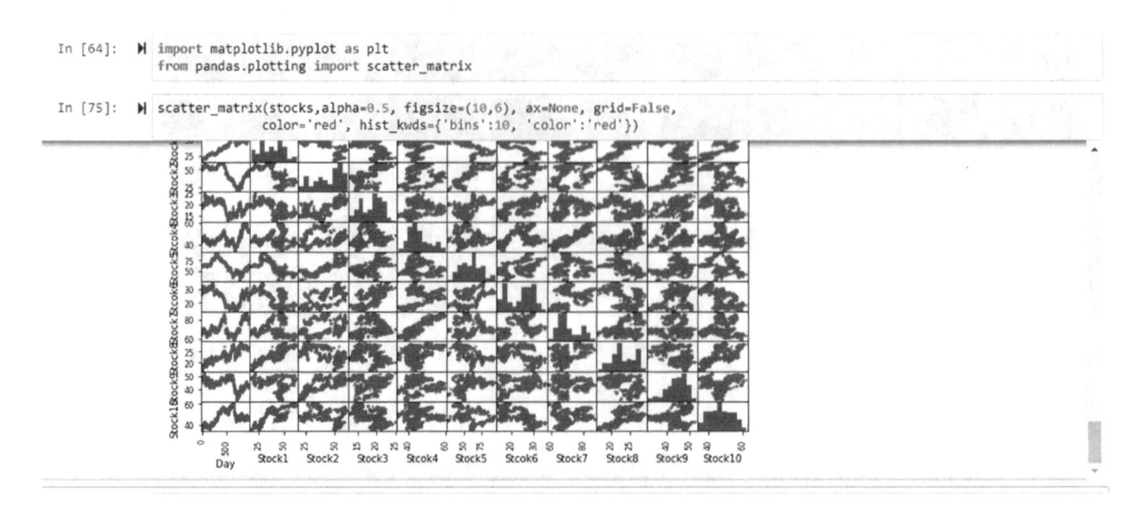
In [64]:
import matplotlib.pyplot as plt
from pandas.plotting import scatter_matrix
In [75]:
scatter_matrix(stocks,alpha=0.5, figsize=(10,6), ax=None, grid=False,
color='red', hist_kwds={'bins':10, 'color':'red'})
Day
Stock1
Stock2
Stock3
Stock4
Stock5
Stock6
Stock7
Stock8
Stock9
Stock10

Figure 5-20. *Scatter_matrix() plot using Python*

In this example, the variables are on the diagonal, from the top left to the bottom right. Each variable is plotted against the other variables. For example, the plot that is to the right of Stock1 and Stock2 represents a plot of Stock1 on the x-axis and Stock2 on the y-axis. Similarly, the plot of Stock8 versus Stock9, plotting Stock8 on the x-axis and Stock9 on the y-axis. If there is any correlation relationship between the two variables, it can be seen from the plot. For example, if two variables have linear relationships, then all the data points would be scattered within a straight line.

5.1.4.1 Correlation Plot

Correlation is a statistical measure that expresses the relationship between two variables. The correlation coefficient, referred to as Pearson correlation coefficient (r), quantifies the strength of the relationship. When you have two variables, X and Y, if the Y variable tends to increase corresponding to the increase in the X variable, then we say we have a *positive correlation* between the variables. When the Y variable tends to decrease as the X variable increases, we say there is a *negative correlation* between the two variables. One way to check the correlation pattern is to use a scatter plot, and another way is to use a correlation graph, as shown in Figure 5-21.

The positive values of ***r*** indicate a positive correlation.

The negative values of ***r*** indicate a negative correlation.

The closer-to-zero values of ***r*** indicate there is a weaker linear relationship.

```
> ##Correlation plot
> library(corrplot)
> cor(stocks3$Stock1, stocks3$Stock2)
[1] -0.7405013
> corel<-cor(stocks3[,2:10])
> corrplot(corel, method="circle")
> head(corel)
             Stock1      Stock2      Stock3       Stcok4
Stock5
Stock1  1.0000000 -0.7405013 -0.6941962  0.22404314 -0.67
743812
Stock2 -0.7405013  1.0000000  0.7277338  0.11507052  0.82
781584
Stock3 -0.6941962  0.7277338  1.0000000  0.10310911  0.64
034833
```

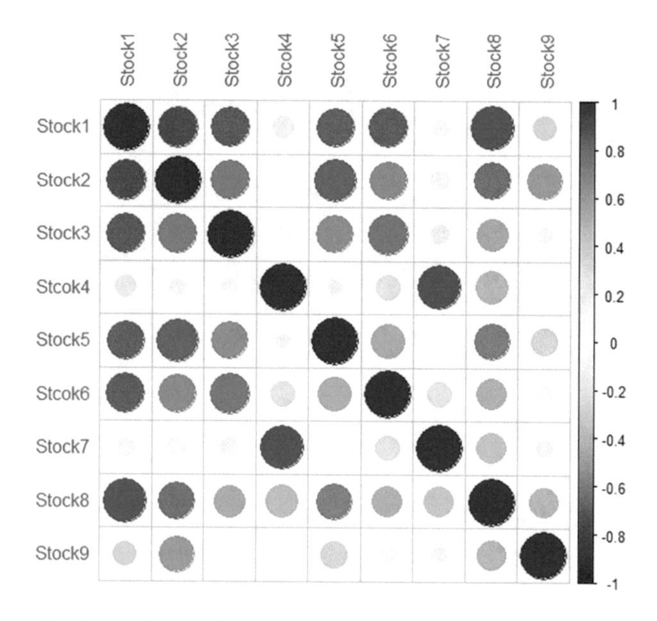

Figure 5-21. *Correlation plot using R*

In this example, a blue dot represents a positive correlation, and red represents a negative correlation. The larger the dot, the stronger the correlation. The diagonal dots (from top left to bottom right) are positively correlated because each dot represents the correlation of each attribute with itself.

In Python the same can be drawn with the help of the `matplotlib()` library, as shown in Figure 5-22. As you can see from the figure, the Pearson coefficient, r, is 0.82 between Stock5 and Stock2 and 0.882 between stock7 and stock4.

Here is the input:

```
Import matplotlib.pyplot as plt

corel=stocks.corr()

corel.style.background_gradient(cmap='coolwarm',set_precision(3)
```

Figure 5-22 shows the input and the output.

```
In [87]:  import matplotlib.pyplot as plt
          corel=stocks.corr()
          corel.style.background_gradient(cmap='coolwarm').set_precision(3)

Out[87]:
```

	Day	Stock1	Stock2	Stock3	Stcok4	Stock5	Stcok6	Stock7	Stock8	Stock9	Stock10
Day	1.000	0.901	-0.716	-0.601	0.447	-0.629	-0.537	0.420	0.900	-0.466	0.495
Stock1	0.901	1.000	-0.741	-0.695	0.223	-0.678	-0.689	0.159	0.878	-0.268	0.712
Stock2	-0.716	-0.741	1.000	0.728	0.115	0.828	0.666	0.169	-0.639	0.592	-0.256
Stock3	-0.601	-0.695	0.728	1.000	0.103	0.640	0.755	0.177	-0.449	0.130	-0.510
Stcok4	0.447	0.223	0.115	0.103	1.000	-0.092	0.294	0.882	0.477	-0.008	0.079
Stock5	-0.629	-0.678	0.828	0.640	-0.092	1.000	0.519	0.047	-0.588	0.351	-0.216
Stcok6	-0.537	-0.689	0.666	0.755	0.294	0.519	1.000	0.295	-0.427	0.155	-0.671
Stock7	0.420	0.159	0.169	0.177	0.882	0.047	0.295	1.000	0.439	-0.124	0.074
Stock8	0.900	0.878	-0.639	-0.449	0.477	-0.588	-0.427	0.439	1.000	-0.397	0.527
Stock9	-0.466	-0.268	0.592	0.130	-0.008	0.351	0.155	-0.124	-0.397	1.000	0.250
Stock10	0.495	0.712	-0.256	-0.510	0.079	-0.216	-0.671	0.074	0.527	0.250	1.000

Figure 5-22. *Correlation plot using Python*

5.1.4.2 Density Plots

A probability density function of each variable can be plotted as a function of the class. Density plots are used to show the distribution of data. They can also be used to compare the separation by class. For example, in our stock price example, each stock is rated as high, low, or medium. You can compare the probability density of each stock price with respect to each class. Similar to scatter plot matrices, a density `featureplot()` function can illustrate the separation by class and show how closely they overlap each other, as shown in Figure 5-23. In this example (of the stock price data set) shown in Figure 5-23, some stock prices overlap very closely and are hard to separate.

```
> ##Density function by ratings
> library(caret)
Loading required package: lattice
Loading required package: ggplot2
Warning message:
package 'caret' was built under R version 3.6.3
> x_dens<-stocks3[,2:11]
> y_dens<-stocks3[,12]
> featurePlot(x=x_dens,y=y_dens, plot="density")


>
```

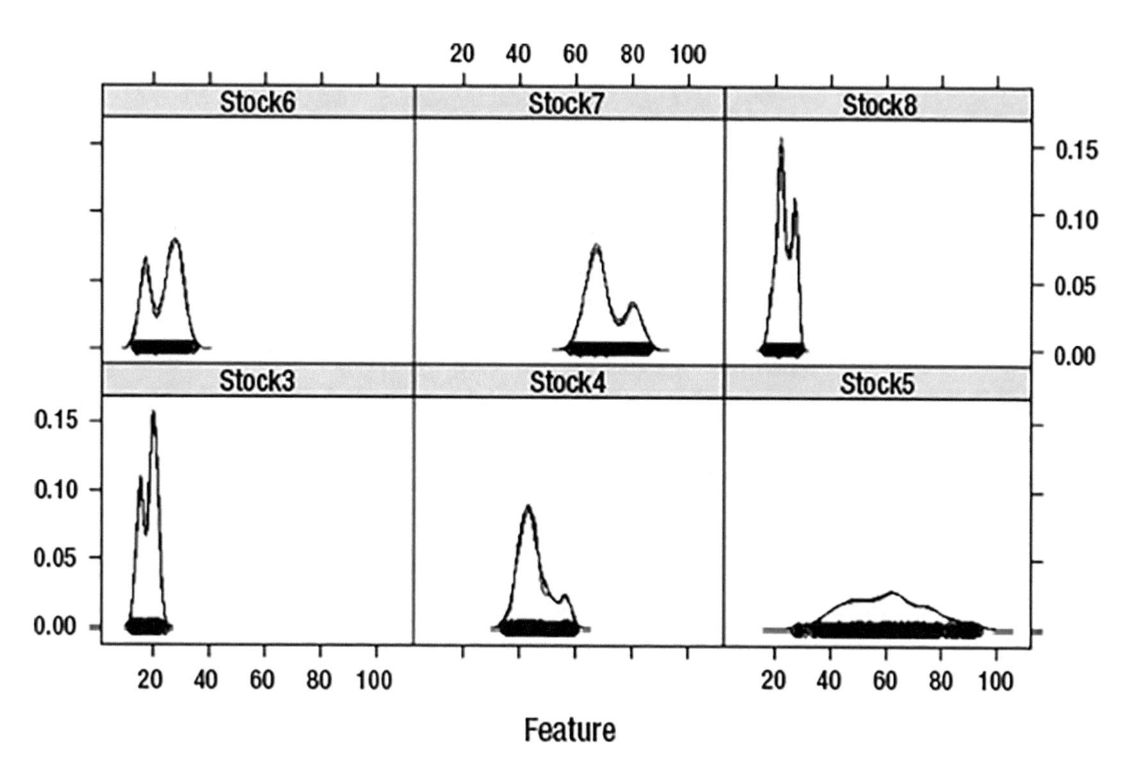

Figure 5-23. *Density plot of stock prices and ratings using R*

We can also plot density plots in Python, as shown in Figure 5-24.

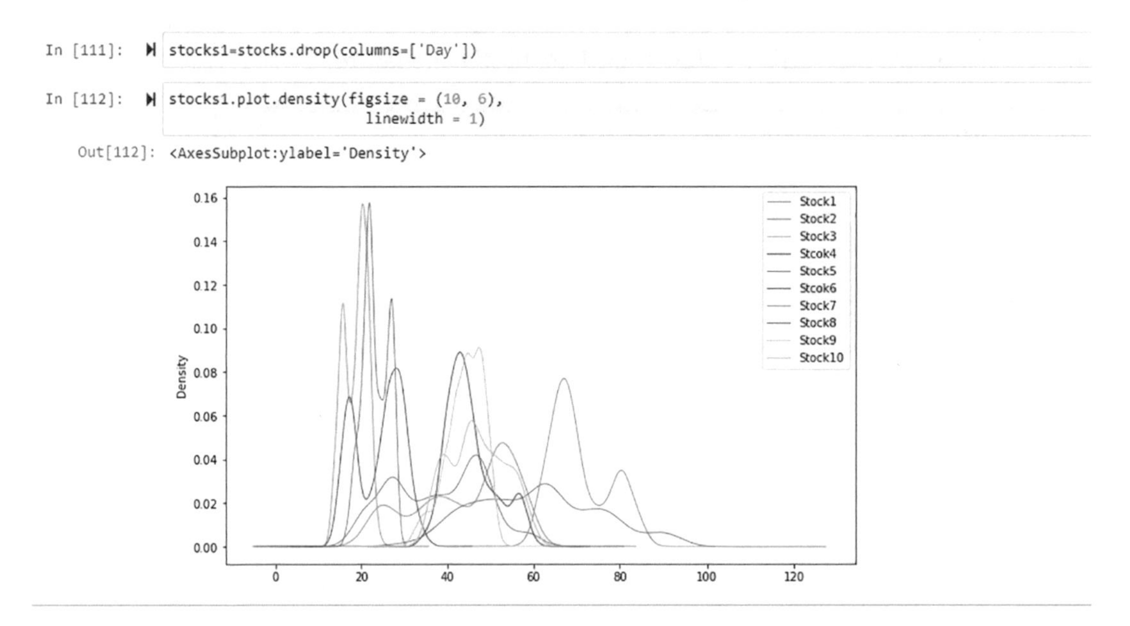

Figure 5-24. *Density plot using Python*

5.2 Plotting Categorical Data

If the data is categorical data, then we use bar chart as a standard visualization method. Each bar in the graph represents the size of the variable. A bar chart describes the comparisons of different discrete categories within the data. In this example, stock ratings is a categorical data, and we can visualize the number of categories in this variable using a bar chart, as shown in Figure 5-25. There are three categories: low, medium, and high. The height, on the y-axis, indicates the total number of stocks with a medium rating. The bar plots can be plotted horizontally or vertically. Both Python and R support `barplot()` functions and are shown here:

```
##Plotting bar plots
> barplot(table(stocks3$Ratings),
+         xlab="Ratings",
+         ylab="Count",
+         border="green",
+         col="red",
+         density=20)

>
```

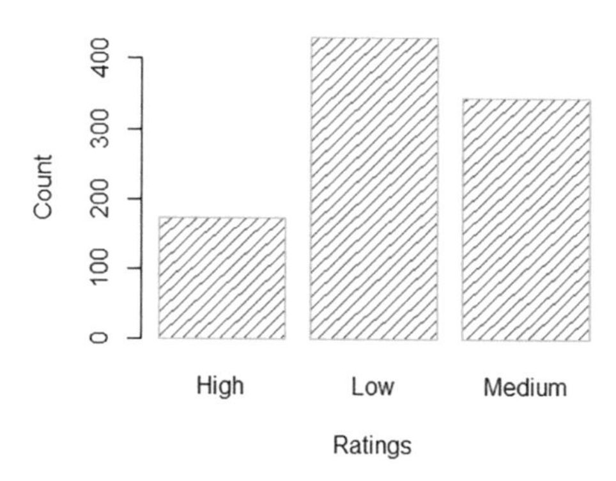

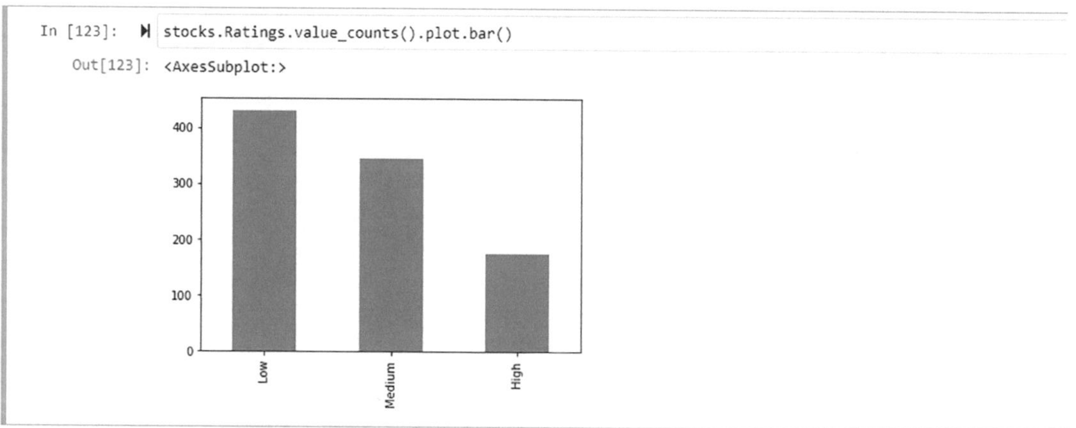

Figure 5-25. *Bar plots*

There are many libraries and packages available to support plots. There are many extensions of the basic plots to get more details, but essentially they all have similar characteristics of describing data. For example, `ggplot()` libraries support graphical description of data in three or four dimensions with the help of different objects (triangle, circle, square, etc.), colors, and also the size of the objects as an indication of different dimensions. It is not in the scope of this book to cover all the libraries available and all the different representations. Our intention is to provide enough insights on fundamental graphical tools available to perform analytics that facilitate developing better models. Depending on the data you are analyzing and depending on the characteristics of the variable, you can select a graphical method. No single technique is considered as the standard.

The process of EDA can be summarized with the help of Figure 5-26. Depending on the type of variables, you can choose different techniques, as shown in Figure 5-26. Data can be of two types, either numerical or categorical. If you are performing analysis on one variable, then it is a univariate analysis; otherwise, it is a bivariate analysis. For the numerical variable, you will describe the various statistical parameters and also plot different graphs such as histograms, box plots, etc. If the variable is categorical, then you have a bar plot, pie chart, etc., and can use simple measures such as counting different categories and checking their probability in percent.

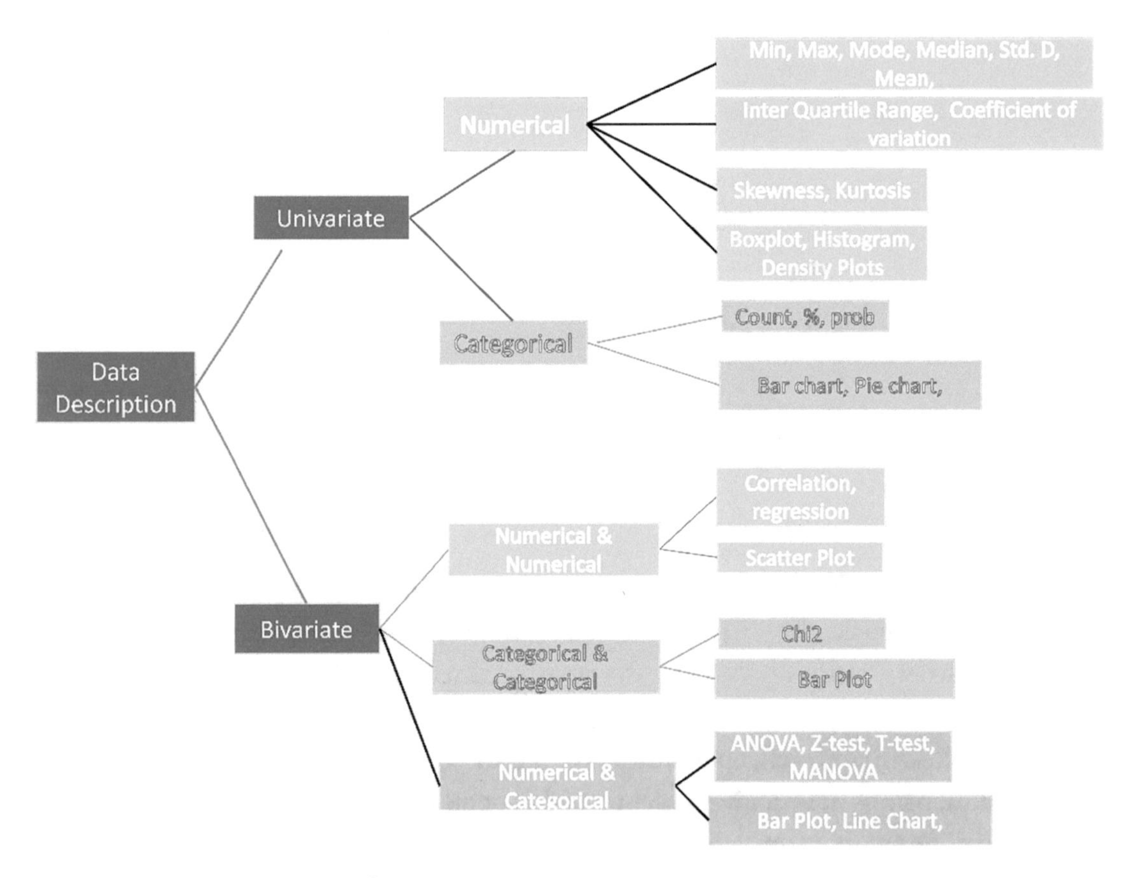

Figure 5-26. *Summary of EDA*

Similarly, in the case of bivariate analysis, you have either numerical-numerical, categorical-categorical, or numerical-categorical combinations. Some of the suggested methods for numerical-numerical are scatter plots and correlations, and for categorical-categorical, they are the Chi-square tests and bar plots. For the categorical-numerical mixed types, we have ANOVA, t-test, bar plots, and line charts, among others.

5.3 Chapter Summary

In this chapter, we focused on the fundamental techniques of exploring data and its characteristics. We discussed both graphical methods and tabulated data.

We covered both univariate and bivariate analysis including histograms, box plots, density plots, scatter plots, correlation plots, and density plots.

We also discussed how to visualize categorical data and its characteristics using bar charts.

CHAPTER 6

Evaluating Analytics Model Performance

There are several ways to measure the performance of different models. This chapter discusses the various measures to test a model's performance, including regression, classification, and clustering methods.

6.1 Introduction

After we create a model, the next step is to measure its performance. We have different measures for classification models and different measures for regression models. Evaluating a model's performance is a key aspect of understanding how accurately your model can predict when applying the model to new data. Though several measures have been used, we will cover only the most commonly used and popular measures. When we have to predict a numerical value, we use regression. When we predict a class or category, we use the classification model.

6.2 Regression Model Evaluation

A *regression model* has many criteria for measuring performance and testing the assumption of normality. The following are some of the evaluation methods for regression. A regression line predicts the y values for a given x value. The difference between the actual and predicted value is called an *error*. The prediction error for an observation k is defined as the difference between the actual value y_k and $\hat{y}_k$ (predicted value of $\hat{y}_k$); $e_k = (y_k - \hat{y}_k)$. Some other popular measures are covered next.

U. R. Hodeghatta and U. Nayak, *Practical Business Analytics Using R and Python*,
https://doi.org/10.1007/978-1-4842-8754-5_6

6.2.1 Root-Mean-Square Error

The *root-mean-square error* (RSME) is yielded by the following formula:

$$\text{RMSE}=\sqrt{\frac{\sum_{k=0}^{n}\left(\hat{y}_k - y_k\right)^2}{n}}$$

Here, y_k is the actual value for the k samples; n is the total number of samples.

$$\hat{y}_k \text{ Is the predicted values}$$

If the model has a good fit, then the error should be less. This is measured by RMSE, and the lower the value of RMSE, the better the model "fit" is to the given data. The RMSE value can vary from one data set to another, and there is no standard RMSE value to say this value is "good" or "bad."

6.2.2 Mean Absolute Percentage Error

Mean absolute percentage error (MAPE) is measured in percentage. MAPE gives the percentage of how predictions deviate from actual values. It is calculated by using the following formula:

$$\frac{1}{\text{n}}\sum_{k=0}^{n}\frac{\text{e}^{\text{k}}}{\text{y}_{\text{k}}}$$

6.2.3 Mean Absolute Error (MAE) or Mean Absolute Deviation (MAD)

This gives the average absolute error. It is calculated by using the following formula:

$$\frac{1}{\text{n}}\sum_{k=0}^{n}\text{ABS}\left|\text{e}^{\text{k}}\right|$$

6.2.4 Sum of Squared Errors (SSE)

The total sum of squared error is another common measure, which is the sum of the error square yielded by this equation:

$$\frac{1}{\mathrm{n}} \sum_{k=0}^{n} \mathrm{e_k}^2$$

6.2.5 R^2 (R-Squared)

R^2 is called the *coefficient of determination* and is a measure that explains how close your predicted values are to the actual values. In other words, it explains the variations of the fitted values to your regression line. R^2 describes the proportion of variance. The R^2 value varies from 0 to 1. The higher the value of R^2, the better the fit is, and the variance is low for a given model. If the regression model is perfect, SSE is zero, and R^2 is 1. R^2 is calculated as the ratio of the comparison of distance of the actual values to the mean and the estimated value to the mean. It is given by the following formula:

$$R^2 = \frac{\mathrm{SSR}}{\mathrm{SST}} = \frac{\mathrm{SST} - \mathrm{SSE}}{\mathrm{SST}}$$

Here, SST is the total sum of squares, SSE is the total sum of error, and SSR is the total sum of residuals. You can refer to any statistics book for the derivation of the equations.

6.2.6 Adjusted R^2

The problem with R^2 is that its value can increase by just having more data points. The more data points, the better the regression fit, and hence you are always tempted to add more data; the same is true if you add more variables. Adjusted R^2 is an adjustment to the R^2 to overcome this situation. Adjusted R^2 considers the number of variables in a data set and penalizes the points that do not fit the model.

$$\mathrm{R}^2_{\mathrm{Adj}} = 1 - \left[\frac{(1 - R^2)(n-1)}{n-k-1} \right]$$

Here, n is the number of samples, and k is the number of predictor variables.

All the previous measures can generate outliers. To keep the outlier or not is something you have to decide as part of exploratory data analysis. These measures are used to compare models and assess the degree of prediction accuracy. It need not be the best model to fit the actual training data perfectly.

6.3 Classification Model Evaluation

In this section we will discuss the methods of evaluating performance of a classification model. We will only discuss the techniques that are commonly used and supported by the opensource tools and popular among data science community.

6.3.1 Classification Error Matrix

The simplest way of measuring the performance of a classifier is to analyze the number of prediction mistakes. A classification error occurs when the observation belongs to one class and the model predicts the class as a different class. A perfect classifier predicts the class with no errors. In the real world, such classifiers cannot construct a model due to noise in the data and not having all the information needed to classify the observations precisely. To uncover misclassification errors, we construct a truth table, also known as a *confusion matrix*. A classification truth table matrix, referred to as a *confusion matrix*, estimates true and predicted values. Figure 6-1 demonstrates a simple 2×2 confusion matrix for two classes (positive and negative). Accuracy measures the quality of classification, which records correctly and incorrectly recognized values for each class.

		Predicted Class	
		Positive (C_0)	Negative (C_1)
Actual Class	Positive (C_0)	80	30
	Negative (C_1)	40	90
	Total Predicted	120	120

Figure 6-1. *Classification error matrix (also called a confusion matrix)*

In Figure 6-1, the total predicted positive class is 120, and the total predicted negative class is 120. However, the actual positive and negative classes are different. The actual positive class is only 110, and the negative class is 130. Therefore, the predictive classification model has incorrectly predicted a class of 10 values and thus has resulted in a classification error.

Further, if the actual class is yes and the predicted class is also yes, then it is considered a *true positive* (TP); if the actual class is yes and the predicted class is no, then it is a *false negative* (FN). Similarly, if the actual class is no and predicted class is also no, then it is referred to as a *true negative* (TN). Finally, if the actual class is no and the predicted is yes, it is called a *false positive* (FP).

Using the contingency table shown in Figure 6-2, we can calculate how well the model has performed. If you want to calculate the accuracy of this model, then just add up the true positives and true negatives and divide by the total number of values. If "a" is true positive, "b" is false negative, "c" is false positive, and "d" is true negative, as shown in Figure 6-2, then the accuracy is calculated as follows:

$$\text{Accuracy} = \frac{a+d}{a+b+c+d} = \frac{TP+TN}{TP+TN+FP+FN}$$

	PREDICTED CLASS		
		Class=Yes	Class=No
ACTUAL CLASS	Class=Yes	a (TP)	b (FN)
	Class=No	c (FP)	d (TN)

a: TP (true positive)
b: FN (false negative)
c: FP (false positive)
d: TN (true negative)

Figure 6-2. *True negative*

For the previous example, the accuracy of the model is = (80+90)/240 * 100 = 70.8%.

6.3.2 Sensitivity Analysis in Classification

As we observed from the previous confusion matrix, there are four possible outcomes. If the actual value is a positive class and is classified as positive by the model, it is counted as a TP; if it is classified as negative, then it is referred to as an FN. Suppose the actual value is negative, and it is predicted as a negative class. In that case, it is counted as TN, and if it is predicted as a positive class instead, it is counted as FP. This can be explained using the confusion matrix shown in Figure 6-2. This matrix forms the basis for many other common measures used in machine learning classifications.

The true positive rate (also called the *hit rate* or *recall*) is estimated as follows:

$$\text{tp rate (recall)} = \frac{\text{TP}}{\text{TP} + \text{FP}}$$

Similarly, the false positive rate (fp rate), also referred to as *false alarm rate*, is estimated as follows:

$$\text{fp rate} = \frac{\text{Negatives incorrectly classified}}{\text{Total negative}}$$

Sensitivity is the metric that measures a model's ability to predict true positives of each available category.

Specificity is the metric that measures a model's ability to predict true negatives of each available category.

$$\text{Specificity} = \frac{\text{True negatives}}{\text{False positives} + \text{True negatives}} = \frac{\text{TN}}{\text{FP} + \text{TN}}$$
$$= 1 - \text{fp rate}$$

Both sensitivity and specificity metrics apply to any classification model.

We also use the terms precision and recall and F-score which are defined as follows:

Precision is to determine the number of selected items which are relevant

$$\text{Precision} = \frac{\text{TP}}{\text{TP} + \text{FP}}$$

Recall is the determination of relevant items which are selected

$$\text{Recall} = \frac{\text{TP}}{\text{TP} + \text{FN}}$$

F-score (also known as F-measure, F1 measure):

An F-score is the harmonic mean of a system's precision and recall values.

It can be calculated by the following formula:

$$\text{F-score} = 2 \times \left[(\text{Precision} \times \text{Recall}) \;/\; (\text{Precision} + \text{Recall}) \right]$$

6.4 ROC Chart

A receiver operating characteristics (ROC) graph represents the performance of a classier. It is a technique to visualize the performance and accordingly choose the best classifier. The ROC graph was first adopted in machine learning by Spackman (1989), who demonstrated how ROC can be used to evaluate and compare different classification algorithms. Now, most of the machine learning community is adopting this technique to measure the performance of a classifier, not just relying on the accuracy of the model (Provost and Fawcett, 1997; Provost et al., 1998). ROC shows a relation between the sensitivity and the specificity of the classification algorithm. A receiver operating characteristic (ROC) graph is a two-dimensional graph of TP rate versus FP rate. It is a plot of the true-positive rate on the y-axis and the false-positive rate on the x-axis.

A true positive rate should be higher for a good classifier model, as shown in Figure 6-3. Area under curve (AUC) provides how the classifier is performing. As we can see from the graph, for a classifier to perform well, it should have a higher true positives rate than a false positives rate. The false positives rate should stabilize over the test instances. In Figure 6-3, we have plotted ROC for the three different models, and the AUC for the first classifier is the highest. Typically, AUC should fall between 0.5 and 1.0. In ideal conditions, when the separation of the two classes is perfect and has no overlapping of the distributions, the area under the ROC curve reaches to 1. An AUC less than 0.5 might indicate that the model is not performing well and needs attention.

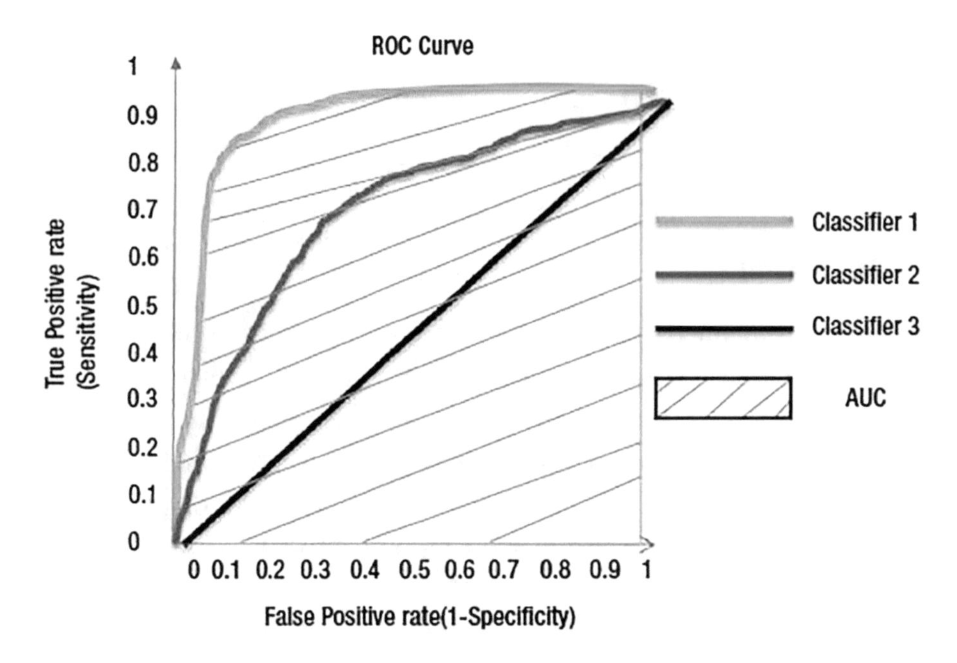

Figure 6-3. *ROC chart*

6.5 Overfitting and Underfitting

Overfitting and underfitting are two important factors that could impact the performance of machine learning models. This section will briefly touch upon the concept of overfitting and underfitting and how to overcome them. *Overfitting* occurs when the model performs well with training data and poorly with test data. *Underfitting* occurs when the model is so simple that it performs poorly with both training and test data.

Overfitting can occur when the model captures the noise in the training data and fits too well for the training data but performs poorly on test data. When this happens, the model will tend to have significant variance. The model does not perform well for training or test data, or sometimes it performs well on test data but poorly on training data. This situation is called *underfitting*. One way to measure the performance of the model in classification models, as explained earlier, is to use the contingency table. The results for a sample data of training set is given in Table 6-1, and the test results are shown in Table 6-2. As we can see from these tables, the accuracy for the training data is higher than the test data. In this case, the model works well on training data but not on test data. This gives a false confidence that your model has performed well and you will tend to take far more risk using the model, which can leave you in a vulnerable situation.

Table 6-1. *Contingency Table for Training Sample*

Predicted	Actual: 1	Actual: 0
1	80	20
0	15	85

Training Data

Table 6-2. *Contingency Table for Test Sample*

Predicted	Actual: 1	Actual: 0
1	30	70
0	80	20

Test Data

6.5.1 Bias and Variance

Bias normally occurs when the model is underfitted and has failed to learn enough from the training data. *Variance* is a prediction error due to different sets of training samples. Ideally, the model error should not vary when tested from one training sample to another. In addition, the model should be stable enough to handle hidden variations between input and output variables.

If either bias or variance is high, the model can be very far off from reality. In general, there is a trade-off between bias and variance. The goal of the machine-learning algorithm is to achieve a low bias and a low variance such that it gives a good prediction performance. In reality, it is hard to calculate the real bias and variance error because of so many other hidden parameters in the model. The bias and variance provide a measure to understand the behavior of the machine-learning algorithm so that the model can be adjusted to provide good prediction performance.

When an overfitting or underfitting situation arises, you must revisit the data split (training and test data) and reconstruct the model. Also, use a k-fold validation (explained under section 6.6) mechanism and take the average or the best model. To reduce the bias error, one should repeat the model-building process by resampling the data. The best way to avoid overfitting is to test the model on data that is entirely outside the scope of your training data. This gives you confidence that you have a representative sample that is part of the production data. In addition to this, it is always good to revalidate the model periodically to determine whether your model is degrading or needs improvement.

Bias and variance are tricky situations. If we minimize bias, then we may end up overfitting the data and end with high variance. If we minimize the variance, then we end up underfitting the model with high bias. Thus, we need to make a compromise between the two. Figure 6-4 depicts the bias-variance trade-off. As the model complexity increases, the error must be reduced to balance the bias and variance.

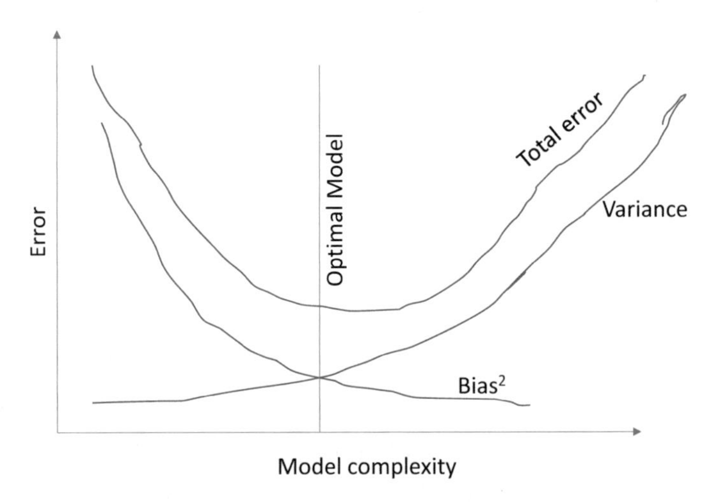

Figure 6-4. *Bias-variance trade-off*

High bias means the model has a high prediction error. This results in unexpected model behavior for the test data (underfitting). Bias can be reduced by the following:

- Add more features to your model.
- Try polynomial functions instead of linear functions.
- In the case of regression with regularization, decrease lambda and penalty values so you can fit the data better.

High variance means the model is picking up noise and the model is overfitting. The following techniques may optimize the model and reduce the variance:

- Increase the training data. A larger data set is more likely to generalize and reduce overfitting.
- Reduce the number of features.
- Increase the lambda, in case of regularization. The higher the lambda, the more the regularization applies.

6.6 Cross-Validation

The goal of the supervised machine learning model is to predict the new data as accurately as possible. To achieve this, we divide the sample data into two sets. The first data set is called the *training data set* and is used to train the model. The second data set, called *test data* (sometimes referred to as *validation set*), is used to test the model performance. The model predicts the class on the test data. The test data already has the actual class that is compared to what is being predicted by the model. The difference between the actual and predicted values gives an error and thus measures the model's performance. The model should perform well on both training and test data set. Also, the model clean up should have the same behavior even if it is tested on a different set of data. Otherwise, as explained in earlier sections, this will result in overfitting or underfitting. To overcome this problem, we use a method called *cross-validation*.

Cross-validation is a sampling technique that uses different portions of the data to train and test the model on different iterations. The goal of the cross-validation is to test the model's ability to predict on unseen data that was not used in constructing the model before and give insight into how the model can behave on independent data. The cross-validation process involves dividing the sample data into k-sets rather than just two sets. For example, for a 10-fold data set, the first model is created using 9-folds of smaller training sets, and the model is tested with 1-fold test data set. In the next iteration, the test data set is shuffled with the other nine sets of training data, and one of the training data sets becomes test data. This process repeats until all the k-sets are covered iteratively. This process gives the entire sample data a chance to be part of the model-building exercise, thus avoiding the overfitting and underfitting problems. The final model would be the average of all the k-models or take the best model out of k-iterations.

Figure 6-5 describes the k-fold validation technique. Typically, the k-value can range anywhere from 5 to 10.

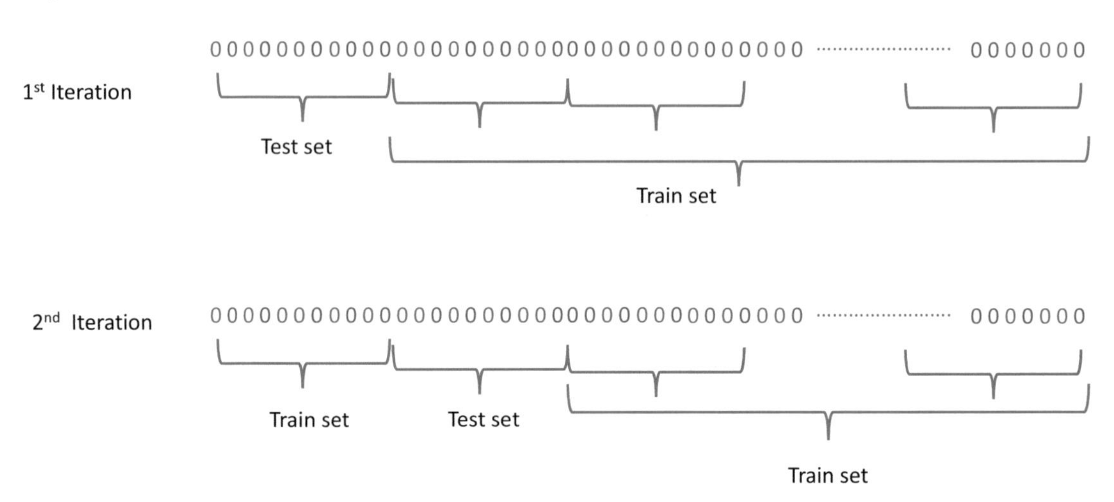

Figure 6-5. *K-fold cross validation*

The first step is to divide the data randomly into k-folds. In the first iteration, select one of the k-fold as test data and the remaining (k-1) sets for training the model. In the next iteration, select a different k-fold for the test from the previous (k-1) set, include the previous test set as part of the new training data, and build the model. Finally, repeat this process on the entire k-fold data. This process allows the entire data to be part of model building exercise, thus reducing any overfit or underfit problems.

The process of k-fold validation is as follows:

1. Split the data set into k folds. The suggested value is k = 10.
2. For each k fold in the data set, build your model on k – 1 folds and test the model to check the effectiveness for the left-out fold.
3. Record the prediction errors.
4. Repeat the steps k times so that each of the k folds are part of the test set.
5. The average of the error recorded in each iteration of k is called the cross-validation error, and this will be the performance metric of the model.

6.7 Measuring the Performance of Clustering

In case of clustering, the performance of the clustering model is based on how well the two clusters are separated and the clusters are clearly distinguished from one another. There are several measures discussed in many scholarly articles. The most common measures are separation index, inter-cluster density (ID), partition index (PI), and so on. The principles behind all of them are the same, and each tool supports a different measure.

6.8 Chapter Summary

In this chapter, we discussed how to measure the performance of a regression model and a classification model.

In regression, we discussed what R^2, adjusted R^2, and RMSE are. How we measure each one is important. We also discussed mean absolute percentage error and mean absolute error.

In classification, we discussed precision, recall, F-score, calculating the accuracy of the model, and sensitivity analysis. We also discussed ROC curves and the area under curve.

We discussed the overfitting and underfitting of the model and the trade-off between the two. We also talked about techniques to solve such challenges when we build the models.

Finally, we mentioned cross validation and the various measures available for the clustering analysis.

PART II

Supervised Learning and Predictive Analytics

CHAPTER 7

Simple Linear Regression

7.1 Introduction

Imagine you are a business investor and want to invest in startup ventures that are likely to be highly profitable. What are all the factors you evaluate in the companies you want to invest in? Maybe it's the innovativeness of the products of the startups, or maybe it's the past success records of the promoters. In this case, we say the profitability of the venture is dependent on or associated with the innovativeness of the products and past success records of the promoters. The innovativeness of the product itself may be associated with various aspects or factors, such as usefulness of the product, competition in the market, and so on. These factors may be a little difficult to gauge, but the past success record of the company can easily be found from publicly available market data. If the promoter had started 10 ventures and 8 were successful, we can say that the success rate of the promoter is 80 percent.

Now imagine you want to increase the sales of the products of your organization. You may want to attract more sales personnel to have a presence in or the capability to service more territories or markets, or you may require more marketing efforts in these areas. All these factors are associated with the quantum of sales or impact the quantum of sales. Imagine the attrition of the employees at any company or industry. There are various factors that might affect attrition rate such as the work environment of the organization, compensation and benefits structure of the organization, how well known the company is in the industry or market, and so forth. Work environment may be how conducive the internal environment is for people to implement their own ideas, how much guidance or help the experienced workers in the organization provide, and the current technological or product landscape of the organization (e.g., whether they are working on the latest technology). It may even be overall satisfaction level of the

U. R. Hodeghatta and U. Nayak, *Practical Business Analytics Using R and Python*,
https://doi.org/10.1007/978-1-4842-8754-5_7

employees. The compensation and benefits structure may include salary structure—that is, how good the salaries are compared to those in other similar organizations or other organizations in the industry, whether there are bonus or additional incentives for higher performance or additional perks, etc. To drive home the point, there may be multiple factors that influence a particular outcome or that are associated with a particular outcome. Again, each one of these may in turn be associated with or influenced by other factors. For example, salary structure may influence the work environment or satisfaction levels of the employees.

Imagine you are a developer of the properties as well as a builder. You are planning to build a huge shopping mall. The prices of various inputs required such as cement, steel, sand, pipes, and so on, vary a lot on a day-to-day basis. If you have to decide on the sale price of the shopping mall or the price of rent that you need to charge for the individual shops, you need to understand the likely cost of building. For this information, you may have to consider the periodic change in the costs of these inputs (cement, steel, sand, etc.) and what factors influence the price of each of these in the market.

You may want to estimate the profitability of the company, arrive at the best possible cost of manufacturing of a product, estimate the quantum of increase in sales, estimate the attrition of the company so that you can plan well for recruitment, decide on the likely cost of the shopping mall you are building, or decide on the rent you need to charge for a square foot or a square meter. In all these cases you need to understand the association or relationship of these factors with the ones that influence, decide, or impact them. The relationship between two factors is normally explained in statistics through *correlation* or, to be precise, the *coefficient of correlation* (i.e., R) or the *coefficient of determination* (i.e., R^2).

The regression equation depicts the relationship between a response variable (also known as the *dependent variable*) and the corresponding independent variables. This means that the value of the dependent variable can be predicted based on the values of the independent variables. When there is a single independent variable, then the regression is called *simple regression*. When there are multiple independent variables, then the regression is called *multiple regression*. Again, the regressions can be of two types based on the relationship between the response variable and the independent variables (i.e., linear regression or nonlinear regression). In the case of linear regression, the relationship between the response variable and the independent variables is explained through a straight line, and in the case of a nonlinear relationship, the relationship between the response variable and independent variables is nonlinear (polynomial like quadratic, cubic, etc.).

Normally we may find a linear relationship between the price of the house and the area of the house. We may also see a linear relationship between salary and experience. However, if we take the relationship between rain and the production of grains, the production of the grains may increase with moderate to good rain but then decrease if the rain exceeds the level of good rain and becomes extreme rain. In this case, the relationship between the quantum of rain and the production of food grains is normally nonlinear; initially food grain production increases and then reduces.

Regression is a supervised method because we know both the exact values of the response (i.e., dependent) variable and the corresponding values of the independent variables. This is the basis for establishing the model. This basis or model is then used for predicting the values of the response variable where we know the values of the independent variable and want to understand the likely value of the response variable.

7.2 Correlation

As described in earlier chapters, correlation explains the relationship between two variables. This may be a cause-and-effect relationship or otherwise, but it need not always be a cause-and-effect relationship. However, variation in one variable can be explained with the help of the other parameter when we know the relationship between two variables over a range of values (i.e., when we know the correlation between two variables). Typically, the relationship between two variables is depicted through a scatter plot as explained in earlier chapters.

Attrition is related to the employee satisfaction index. This means that attrition is correlated with "employee satisfaction index." Normally, the lower the employee satisfaction, the higher the attrition. Also, the higher the employee satisfaction, the lower the attrition. This means that attrition is inversely correlated with employee satisfaction. In other words, attrition has a negative correlation with employee satisfaction or is negatively associated with employee satisfaction.

Normally the profitability of an organization is likely to go up with the sales quantum. This means the higher the sales, the higher the profits. The lower the sales, the lower the profits. Here, the relationship is that of positive correlation as profitability increases with the increase in sales quantum and decreases with the decrease in sales quantum. Here, we can say that the profitability is positively associated with the sales quantum.

Normally, the fewer defects in a product or the faster the response related to issues, the higher the customer satisfaction will be of any company. Here, customer satisfaction is inversely related to defects in the product or negatively correlated with the defects in the product. However, the same customer satisfaction is directly related to or positively correlated with the speed of response.

Correlation explains the extent of change in one of the variables given the unit change in the value of another variable. Correlation assumes a very significant role in statistics and hence in the field of business analytics as any business cannot make any decision without understanding the relationship between various forces acting in favor of or against it.

Strong association or correlation between two variables enables us to better predict the value of the response variable from the value of the independent variable. However, a weak association or low correlation between two variables does not help us to predict the value of the response variable from the value of the independent variable.

7.2.1 Correlation Coefficient

Correlation coefficient is an important statistical parameter of interest that gives us a numerical indication of the relationship between two variables. This will be useful only in the case of linear association between the variables. This will not be useful in the case of nonlinear associations between the variables.

It is easy to compute the correlation coefficient. To compute it, we require the following:

- Average of all the values of the independent variable
- Average of all the values of the dependent variable
- Standard deviation of all the values of the independent variable
- Standard deviation of all the values of the dependent variable

Once we have these values, we need to convert each value of each variable into standard units. This is done as follows:

- (Each value minus the average of the variable) / (Standard deviation of the variable); i.e., ([variable value - mean(variable)] / sd(variable)). For example, if a particular value among the values of the independent variable is 18 and the mean of this independent variable is 15 and the standard deviation of this independent variable is 3, then the value of this independent variable converted into standard units will be (18 – 15)/3 = 1. This is also known as the *z-score* of the variable.

Once we have converted each value of each variable into standard units, the correlation coefficient (normally depicted as r or R) is calculated as follows:

- Average of [(independent variable in standard units) × (dependent variable in standard units)]; i.e., (mean[Σ(z-score of x) * (z-score of y)])

The correlation coefficient can be also found out using the following formula:

- R = [covariance(independent variable, dependent variable)] / [(Standard deviation of the independent variable) × (Standard deviation of the dependent variable)]

In this formula, covariance is [sum(the value of each independent variable minus average of the independent variable values) * (the value of each dependent variable minus the average of the dependent variable values)] divided by [n minus 1].

In the R programming language, the calculation of the correlation coefficient is very simple, as shown in Figure 7-1A. Figure 7-1B shows the corresponding scatter plot.

```
#The following is an example of "attrition" vs. "employee satisfaction"
#The following is the attrition data in terms of % of attrition
attrition <- c(4,5,6,8,10,12,15,18,21,25)
#The following is the employee satisfaction data 1 month
#... prior to the attrition data
empsat <- c(10,9,8,7,6,5,4,3,2,1)
#The employee satisfaction data is collated every quarter
#and the attrition is calculated as the average of the next 3 months
#after employee satisfaction data collation
correl_attri_empsat <- cor(attrition, empsat)
correl_attri_empsat
```

Output: [1] -0.9830268

Figure 7-1A. *Calculating the correlation coefficient in R*

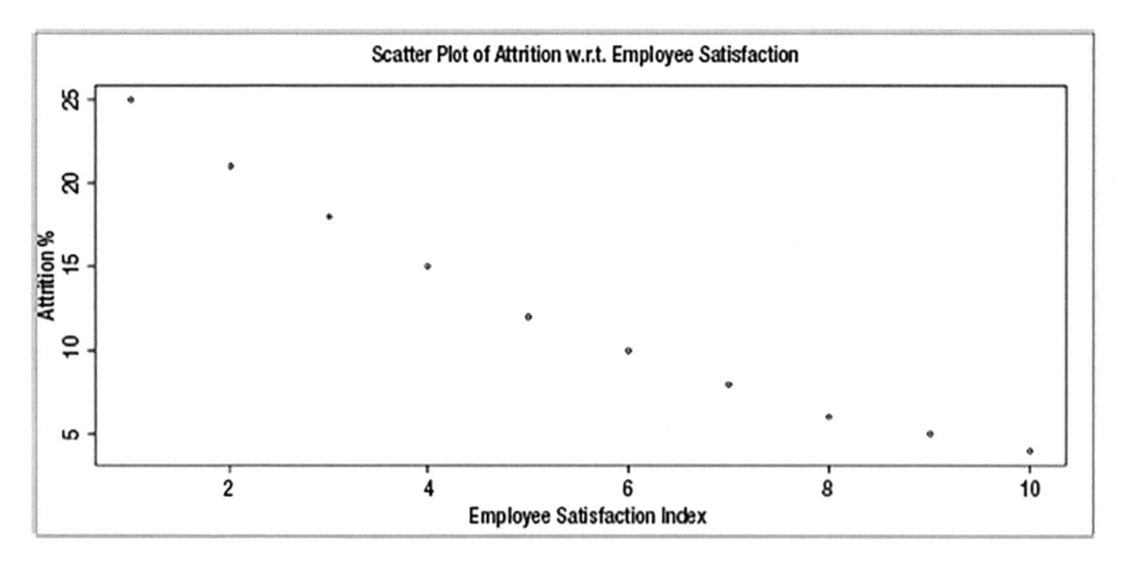

Figure 7-1B. *Scatter plot between employee satisfaction index and attrition*

As shown in the scatter plot in Figure 7-1B, even though the relationship is not linear, it is near linear. This is shown by the correlation coefficient of -0.983. As you can see, the negative sign indicates the inverse association or negative association between attrition percentage and employee satisfaction index. The previous plot shows that the deterioration in the employee satisfaction leads to an increased rate of attrition.

Further, the correlation test shown in Figure 7-2B confirms that there is an excellent statistically significant correlation between attrition and the employee satisfaction index. Figure 7-2A provides the code used to generate this.

```
#We will now test and confirm the if there exists statistically

#significant correlation between "attrition" and "employee

#satisfaction" using Spearman's rank correlation

cor.test(attrition, empsat, method="spearman")
```

Figure 7-2A. *Code to test correlation using Spearman's rank correlation rho*

Figure 7-2B shows the output of the previous code/test.

```
        Spearman's rank correlation rho

data:  attrition and empsat
S = 330, p-value < 2.2e-16
alternative hypothesis: true rho is not equal to 0
sample estimates:
rho
 -1
```

Figure 7-2B. *Test to find out the significance of correlation in R*

Spearman's rho is the measure of the strength of the association between two variables. The Spearman correlation shows whether there is significant correlation (significantly monotonically related). The null hypothesis for the Spearman's rank correlation test is "true rho is equal to zero," and the alternative hypothesis is "true rho is not equal to zero." In the previous test, notice that the p-value is very low and less than the significance level of 0.05. Hence, there is a significant relationship between these two specified variables, and we can reject the null hypothesis and uphold that the true rho is not equal to zero. The estimated rho value thrown out of the test is -1, which specifies a clear and perfect negative correlation between the two variables, i.e., attrition and empsat.

Please note, the previous data is illustrative only and may not be representative of a real scenario. It is used for the purpose of illustrating the correlation. Further, in the case of extreme values (outliers) and associations like nonlinear associations, the correlation coefficient may be very low and may depict no relationship or association. However, there may be a real and good association among the variables.

7.3 Hypothesis Testing

At this point in time, it is apt for us to briefly touch upon hypothesis testing. This is one of the important aspects in statistics. In hypothesis testing we start with an assertion or claim or status quo about a particular population parameter of one or more populations. This assertion or claim or status quo is known as the *null hypothesis*, or H_0. An example of the null hypothesis may be a statement like the following: the population mean of population 1 is equal to the population mean of population 2. There is another statement known as the *alternate hypothesis*, or H_1, which is opposite to the null hypothesis. In our example, the alternate hypothesis specifies that there is a significant difference between the population mean of population 1 and the population mean of population 2. A level of significance or Type I error of normally 0.05 is specified. This is just the possibility that the null hypothesis is rejected when actually it is true. This is represented by the symbol α. The smaller the value of α, the smaller the risk of a Type I error.

Then we decide the sample size required to reduce the errors.

We use test statistics to either reject the null hypothesis or not reject the null hypothesis. When we reject the null hypothesis, it means that the alternate hypothesis is true. However, if we do not reject the null hypothesis it does not mean that the alternate hypothesis is true. It only shows that we do not have sufficient evidence to reject the null hypothesis. Normally, the t-value is the test statistic used.

Then we use the data and arrive at the sample value of the test statistic. We then calculate the p-value on the basis of the test statistic. The p-value is the probability that the test statistic is equal to or more than the sample value of the test statistic when the null hypothesis is true. We then compare the p-value with the level of significance (i.e., α). If the p-value is less than the level of significance, then the null hypothesis is rejected.

This also means that the alternate hypothesis is accepted. If the p-value is greater than or equal to the level of significance, then we cannot reject the null hypothesis.

The p-value is used (among many other uses in the field of statistics) to validate the significance of the parameters to the model in the case of regression analysis. If the p-value of any parameter in the regression model is less than the level of significance (typically 0.05), then we reject the null hypothesis that there is no significant contribution of the parameter to the model, and we accept the alternate hypothesis that there is significant contribution of the parameter to the model. If the p-value of a parameter is greater than or equal to the level of significance, then we cannot reject the null hypothesis that there is no significant contribution of the parameter to the model. We include in the final model only those parameters that have significance to the model.

7.4 Simple Linear Regression

As we mentioned in the introduction to this chapter, simple linear regression depicts the linear relationship between two associated variables—a response (i.e., dependent) variable and an independent variable, which can be depicted through a regression equation. This regression equation can be used for the prediction of the unknown response variable when we know the value of the independent variable. This is so, as the regression describes, how one variable (i.e., the response variable) depends upon another variable (i.e., the independent variable).

Note that the regression equation guarantees predicting the values of the response variable correctly when the independent variable for which the response variable is required to be predicted is within the range of the data of the independent variable used to create the regression equation. In other cases, the regression equation's predicted value may not be reliable. This is because we do not know whether the regression equation holds well beyond the values of the independent variable we had in hand and used to create the regression equation. However, each predicted value may have some residual value. Also, the regression equation may have a small amount of residual value.

7.4.1 Assumptions of Regression

There are four assumptions of regression. These need to be fulfilled if we need to rely upon any regression equation.

- Linear association between the dependent variable and the independent variable.
- Independence of the errors around the regression line between the actual and predicted values of the response variable.
- Normality of the distribution of errors.
- Equal variance of the distribution of the response variable for each level of the independent variable. This is also known as *homoscedasticity*.

7.4.2 Simple Linear Regression Equation

The regression line is a smoothed graph of averages. It is basically the smoothing of the averages of the response variable for each value of the independent variable. The regression line is drawn in such a way that it minimizes the error of the fitted values with respect to the actual values. This method is known as the *least squares method*, and it minimizes the sum of the squared differences between the actual values of the response variable and the predicted values of the response variable. As an alternative to the least squares method, we can use methods such as quantile regression or least absolute deviation. The simple linear regression equation takes the following form:

$$Y_1 = \beta_0 + \beta_1 x_1$$

In the previous equation, β_0 is known as the *intercept*, and β_1 is known as the *slope* of the regression line. Intercept is the value of the response variable when the value of the independent variable (i.e., x) is zero. This depicts the point at which the regression line touches the y-axis when x is zero. The slope can be calculated easily using the following formula: (R × Standard deviation of the response variable) / (Standard deviation of the independent variable).

From this, you can see that when the value of the independent variable increases by one standard deviation, the value of the response variable increases by R × one standard deviation of the response variable, where R is the coefficient of correlation.

7.4.3 Creating a Simple Regression Equation in R

It is easy to create a regression equation in R. We need to use the following function:

- `lm(response variable ~ independent variable, data = dataframe name)`

`lm()` stands for "linear model" and takes the response variable, the independent variable, and the dataframe as input parameters.

Let's look at a simple example to understand how to use the previous command. Assuming the competence or capability of the sales personnel is equal, keeping the sale restricted to one single product, we have a set of data that contains the number of hours of effort expended by each salesperson and the corresponding number of sales made. Figure 7-3A shows the code to import the data from a file; it lists the contents of the dataframe and the summary of the contents of the dataframe.

```
cust_df <- read.table("C:/Users/kunku/OneDrive/Documents/Book
Revision/cust1.txt", header = TRUE, sep=",")

#In the above read statement you can alternatively use read.csv() or
read.csv2()

#we will now print the contents of the imported file

cust_df

summary(cust_df)
```

Figure 7-3A. *Code to read data, list the contents of the dataframe and display the summary*

Figure 7-3B shows the output of the code provided in Figure 7-3A.

```
   Sales_Effort Product_Sales
1           100            10
2            82             8
3            71             7
4           111            11
5           112            11
6            61             6
7            62             6
8           113            11
9           101            10
10           99            10
11           79             8
12           81             8
13           51             5
14           50             5
15           49             5
16           30             3
17           31             3
18           29             3
19           20             2
20           41             4
21           39             4
> summary(cust_df)
  Sales_Effort     Product_Sales
 Min.   : 20.00   Min.   : 2.000
 1st Qu.: 41.00   1st Qu.: 4.000
 Median : 62.00   Median : 6.000
 Mean   : 67.24   Mean   : 6.667
 3rd Qu.: 99.00   3rd Qu.:10.000
 Max.   :113.00   Max.   :11.000
```

Figure 7-3B. *Creating a data frame from a text file (data for the examples)*

In Figure 7-3, we have imported a table of data containing 21 records with the `Sales_Effort` and `Product_Sales` from a file named `cust1.txt` into a data frame named `cust_df`. The `Sales_Effort` is the number of hours of effort put in by the salesperson during the first two weeks of a month, and the `Product_Sales` is the number of product sales closed by the salesperson during the same period. The summary of the data is also shown.

In this data we can treat `Product_Sales` as the response variable and `Sales_Effort` as the independent variable as the product sales depend upon the sales effort put in place by the salespeople.

We will now split the data into `Train_Data` used for the model generation and `Test_Data` used for validating the model generated. Then, we will run the simple linear regression to model the relationship between `Product_Sales` and `Sales_Effort` using the `lm(response variable ~ independent variable, data = dataframe name)` command of R. We will use the `Train_Data` as the input dataframe. Figure 7-4A shows the code.

```
#Now, we will split the data into two separate sets Train_Data,
Test_Data

#One for the model generation purposes, another for testing the model
generated

#As our data set has very limited records we will be utilizing 90% of it for

#the model generation

size = sort(sample(nrow(cust_df),nrow(cust_df)*0.9))

Train_Data = cust_df[size, ]

Test_Data = cust_df[-size, ]

#Printing Train_Data

Train_Data

#Printing Test_Data

Test_Data

#Generating the model using the Train_Data as input

mod_simp_reg = lm(Product_Sales ~ Sales_Effort, data=Train_Data)
```

Figure 7-4A. *Code to split the dataset into two separate dataframes and to generate the linear regression model*

Figure 7-4B shows the output.

```
> Train_Data
   Sales_Effort Product_Sales
1           100            10
2            82             8
3            71             7
4           111            11
5           112            11
6            61             6
7            62             6
8           113            11
11           79             8
13           51             5
14           50             5
15           49             5
16           30             3
17           31             3
18           29             3
19           20             2
20           41             4
21           39             4
> #Printing Test_Data
> Test_Data
   Sales_Effort Product_Sales
9           101            10
10           99            10
12           81             8
> #Generating the model using the Train_Data as input
> mod_simp_reg = lm(Product_Sales ~ Sales_Effort, data=Train_Data)
```

Figure 7-4B. *Splitting the data into train and test sets and generating a simple linear regression model in R*

The output shows clearly the split of data into two sets of data, i.e., `Train_Data` and `Test_Data`, with 18 and 3 records, respectively. We then generated the simple linear regression model named `mod_simp_reg`. The command `lm()` has generated the model but has not thrown any output of the model generated as it has been assigned to `mod_simp_reg`. Now, running the `summary(mod_simp_reg)` command outputs the summary of the simple linear regression model.

Figure 7-5A shows the code.

```
#Generating the summary of the model generated

summary(mod_simp_reg)
```

Figure 7-5A. *Code to generate the summary of the simple linear regression model generated*

Figure 7-5B shows the output.

```
> summary(mod_simp_reg)

Call:
lm(formula = Product_Sales ~ Sales_Effort, data = Train_Data)

Residuals:
     Min       1Q   Median       3Q      Max
-0.14061 -0.07888 -0.02474  0.08353  0.19455

Coefficients:
              Estimate Std. Error t value Pr(>|t|)
(Intercept)  0.0688518  0.0575076   1.197    0.249
Sales_Effort 0.0979316  0.0008271 118.399   <2e-16 ***
---
Signif. codes:  0 '***' 0.001 '**' 0.01 '*' 0.05 '.' 0.1 ' ' 1

Residual standard error: 0.1044 on 16 degrees of freedom
Multiple R-squared:  0.9989,    Adjusted R-squared:  0.9988
F-statistic: 1.402e+04 on 1 and 16 DF,  p-value: < 2.2e-16
```

Figure 7-5B. *Output of the summary (mod_simp_reg)*

The initial part of the summary shows which element of the data frame is regressed against which other element and the name of the data frame that contained the data to arrive at the model.

Residuals depict the difference between the actual value of the response variable and the value of the response variable predicted using the regression equation. The maximum residual is shown as 0.19455. The spread of residuals is provided here by specifying the values of min, max, median, Q1, and Q3 of the residuals. In this case, the spread is from -0.14061 to +0.19455. As the principle behind the regression line and regression equation is to reduce the error or difference, the expectation is that the median value should be very near to 0. As you can see, here the median value is -0.02474, which is almost equal to 0. The prediction error can go up to the maximum value of the residual. As this value (i.e., 0.19455) is very small, we can accept this residual.

The next section specifies the coefficient details. Here β_0 is specified by the intercept estimate (i.e., 0.0688518), and β_1 is specified by the `Sales_Effort` estimate (0.0979316). Hence, the simple linear regression equation is as follows:

$$\text{Product\_Sales}_i = 0.0688518 + 0.0979316\ \text{Sales\_Effort}_i$$

The value next to the coefficient estimate is the *standard error* of the estimate. This specifies the uncertainty of the estimate. Then comes the "t" value of the standard error. This specifies how large the coefficient estimate is with respect to the uncertainty. The next value is the probability that the absolute(t) value is greater than the one specified, which is due to a chance error. Ideally, the "Pr" (or Probability value, or popularly known as p-value) should be very small (like 0.001, 0.005, 0.01, or 0.05) for the relationship between the response variable and the independent variable to be significant. The p-value is also known as the *value of significance*. As the probability of the error of the coefficient of `Sales_Effort` here is less (i.e., almost near 0, <2e-16), we reject the null hypothesis that there is no significance of the parameter to the model and accept the alternate hypothesis that there is significance of the parameter to the model. Hence, we conclude that there is a significant relationship between the response variable `Product_Sales` and the independent variable `Sales_Effort`. The number of asterisks (*s) next to the p-value of each parameter specifies the level of significance. Please refer to "Signif. codes" in the model summary as given in Figure 7-5B.

The next section shows the overall model quality-related statistics.

- The *degrees of freedom* specified here is nothing but the number of rows of data minus the number of coefficients. In our case, it is 18 – 2 = 16. This is the residual degrees of freedom. Ideally, the number of degrees of freedom should be large compared to the number of coefficients for avoiding the overfitting of the data to the model. We have dealt with overfitting and underfitting of the data to the model in one of the chapters. Let's remember for the time being that the overfitting is not good. Normally, according to the rule of thumb, 30 rows of data for each variable is considered good for the training sample. Further, we cannot use the ordinary least squares method if the number of rows of data is less than the number of independent variables.

- The *residual standard error* shows the sum of the squares of the residuals as divided by the degrees of freedom (in our case 16) as specified in the summary. This is 0.1044 and is very low, as required by us.

- *The multiple R-squared* value shown here is just the square of the *correlation coefficient* (i.e., R). *Multiple R-squared* is also known as the *coefficient of determination*. However, the *adjusted R-squared value* is the one that is the adjusted value of R-squared adjusted to avoid overfitting. Here again, we rely more on adjusted R-squared value than on multiple R-squared. The value of multiple R-squared is 0.9989, and the adjusted R-squared is 0.9988, which are very high and show the excellent relationship between the response variable `Product_Sales` and the independent variable `Sales_Effort`.
- Finally, the *F-statistic* is based on the *F-test*. As in the case of coefficients, here also we want the *p-value* to be very small (like 0.001, 0.005, 0.01, 0.05) for the model to be significant. As you can see, in our case it is < 0.001 (i.e., <2.2e-16). Hence, the model is significant.

As shown, there is an excellent and significant association between the response variable `Product_Sales` and the independent variable `Sales_Effort`.

7.4.4 Testing the Assumptions of Regression

Before accepting the model as usable, it is essential to validate that the regression assumptions are true in respect to the fitted model.

7.4.4.1 Test of Linearity

To test the linearity, we plot the residuals against the corresponding fitted values. Figure 7-6B depicts this.

Figure 7-6A shows the simple code.

```
#Test of Linearity

plot(mod_simp_reg)

#This provides many graphs
```

Figure 7-6A. *Code to generate the plot of the model to test linearity*

Figure 7-6B shows the output in terms of the plot.

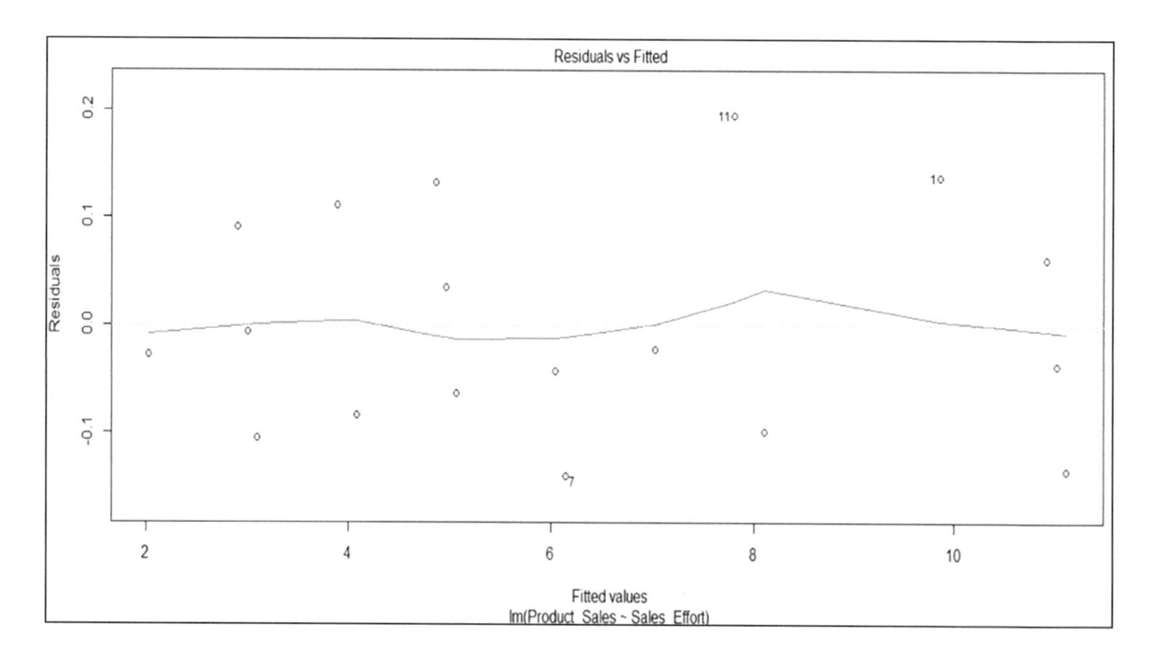

***Figure 7-6B.** Residuals versus fitted plot to test the linearity*

For the model to pass the test of linearity, we should not have any pattern in the distribution of the residuals, and they should be randomly placed around the 0.0 residual line. That is, the residuals will be randomly varying around the mean of the value of the response variable. In our case, as we can see there are no patterns in the distribution of the residuals. Hence, it passes the condition of linearity.

7.4.4.2 Test of Independence of Errors Around the Regression Line

This test is not required to be conducted when we know that each data point is independent of the other. In our case, the data is collected over the same two-week period but pertains to different individuals, and hence this test is not required. This test is primarily conducted to check the autocorrelation that is introduced when the data is collected over a close period of time or one data record is related to other data record, where there is a possibility of auto correlation. Further, as shown in Figure 7-6B, the residuals are distributed randomly around the mean value of the response variable. If we need to test for the autocorrelation, we can use the Durbin-Watson test.

Figure 7-7A gives the code.

```
#Test of Independence of errors around the Regression Line
#using Durbin-Watson Test
library(lmtest)
dwtest(mod_simp_reg)
```

Figure 7-7A. *Code to test the independence of errors around the regression line*

Figure 7-7B shows the output.

```
> library(lmtest)
> dwtest(mod_simp_reg)

        Durbin-Watson test

data:  mod_simp_reg
DW = 2.3242, p-value = 0.6991
alternative hypothesis: true autocorrelation is greater than 0
```

Figure 7-7B. *Durbin-Watson test in R to check autocorrelation*

In the case of the Durbin-Watson test, the null hypothesis (i.e., H_0) is that there is no autocorrelation, and the alternative hypothesis (i.e., H_1) is that there is autocorrelation. If the p-value is < 0.05, then we reject the null hypothesis—that is, we conclude that there is autocorrelation. In the previous case, the p-value is greater than 0.05, which means that there is no evidence to reject the null hypothesis that there is no autocorrelation. Hence, the test of independence of errors around the regression line passes. Alternatively for this test, you can use the `durbinWatsonTest()` function from `library(car)`.

7.4.4.3 Test of Normality

As per this test, the residuals should be normally distributed. To check on this, we will look at the normal Q-Q plot (one among the many graphs created using the `plot(model name)` command), as shown in Figure 7-8.

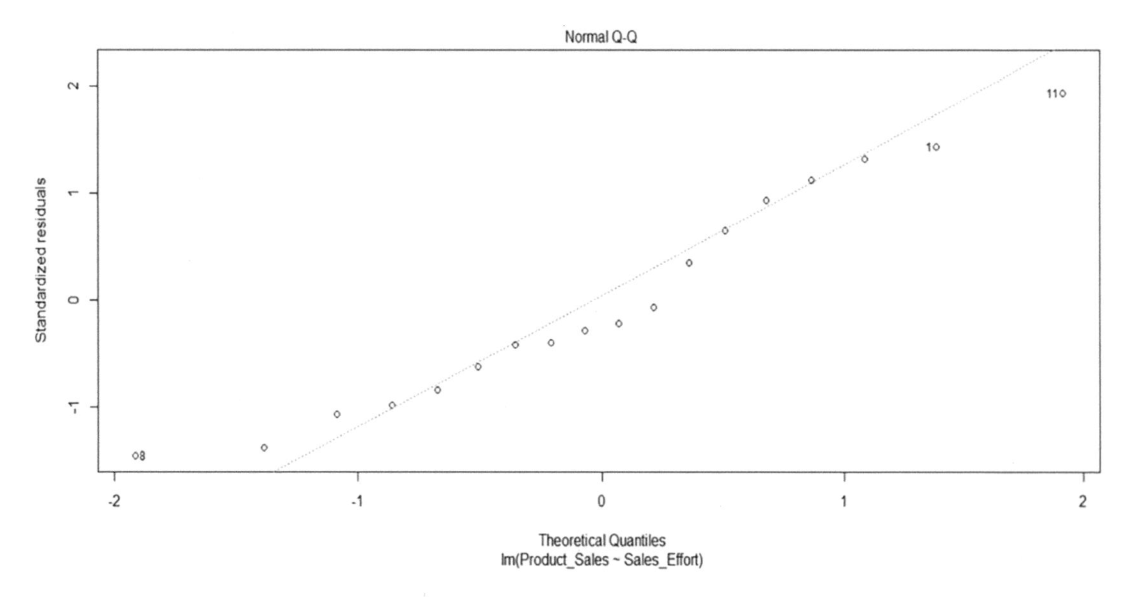

Figure 7-8. *Test for assumption of "normality" using normal Q-Q plot*

Figure 7-8 shows that the residuals are almost on the straight line in the above normal Q-Q plot. This shows that the residuals are normally distributed. Hence, the normality test of the residuals passes.

7.4.4.4 Equal Variance of the Distribution of the Response Variable

We can use the same plot of the residuals we used for testing the linearity. As we can see from the plot (in Figure 7-6B in the case of test of linearity), there is no significant difference in the variation of the residuals for different levels of the independent variable's values. Hence, we consider this test as passed.

7.4.4.5 Other Ways of Validating the Assumptions to Be Fulfilled by a Regression Model

There are many other ways of validating whether the assumptions of the regression are fulfilled by the regression model. These are in addition to the ones mentioned earlier.

Using the gvlma Library

The other easy way to validate whether a regression model has fulfilled the assumptions of the regression model is by using `library(gvlma)`, which performs the validation of the regression model assumptions as well as the evaluation of other related aspects such as skewness, kurtosis, link function, and heteroscedasticity.

Figure 7-9A shows how to use this library (i.e., the code), and Figure 7-9B shows the output from R.

```
#Other ways of validating the model assumptions

library(gvlma)

gv_model <- gvlma(mod_simp_reg)

summary(gv_model)
```

Figure 7-9A. *Using the gvlma() function to validate the model assumptions*

```
> summary(gv_model)

Call:
lm(formula = Product_Sales ~ Sales_Effort, data = Train_Data)

Residuals:
     Min       1Q   Median       3Q      Max
-0.14061 -0.07888 -0.02474  0.08353  0.19455

Coefficients:
              Estimate Std. Error t value Pr(>|t|)
(Intercept)  0.0688518  0.0575076   1.197    0.249
Sales_Effort 0.0979316  0.0008271 118.399   <2e-16 ***
---
Signif. codes:  0 '***' 0.001 '**' 0.01 '*' 0.05 '.' 0.1 ' ' 1

Residual standard error: 0.1044 on 16 degrees of freedom
Multiple R-squared:  0.9989,	Adjusted R-squared:  0.9988
F-statistic: 1.402e+04 on 1 and 16 DF,  p-value: < 2.2e-16


ASSESSMENT OF THE LINEAR MODEL ASSUMPTIONS
USING THE GLOBAL TEST ON 4 DEGREES-OF-FREEDOM:
Level of Significance =  0.05

Call:
 gvlma(x = mod_simp_reg)

                     Value p-value                Decision
Global Stat        1.32363  0.8574 Assumptions acceptable.
Skewness           0.40996  0.5220 Assumptions acceptable.
Kurtosis           0.77517  0.3786 Assumptions acceptable.
Link Function      0.09841  0.7537 Assumptions acceptable.
Heteroscedasticity 0.04009  0.8413 Assumptions acceptable.
```

Figure 7-9B. *Output of the gvlma() with the linear regression assumptions validated*

In Figure 7-9B we have given the output of the `gvlma()` function from R on our model. The Global Stat line clearly shows that the assumptions related to this regression model are acceptable. Here, we need to check whether the p-value is greater than 0.05. If the p-value is greater than 0.05, then we can safely conclude as shown above that the assumptions are validated. If we have p-value less than 0.05, then we need to revisit the regression model.

Using the Scale-Location Plot

The scale-location graph is the one of the graphs generated using the command `plot(regression model name)`. In our case, this is `plot(mod_simp_reg)`. Figure 7-10 shows this graph.

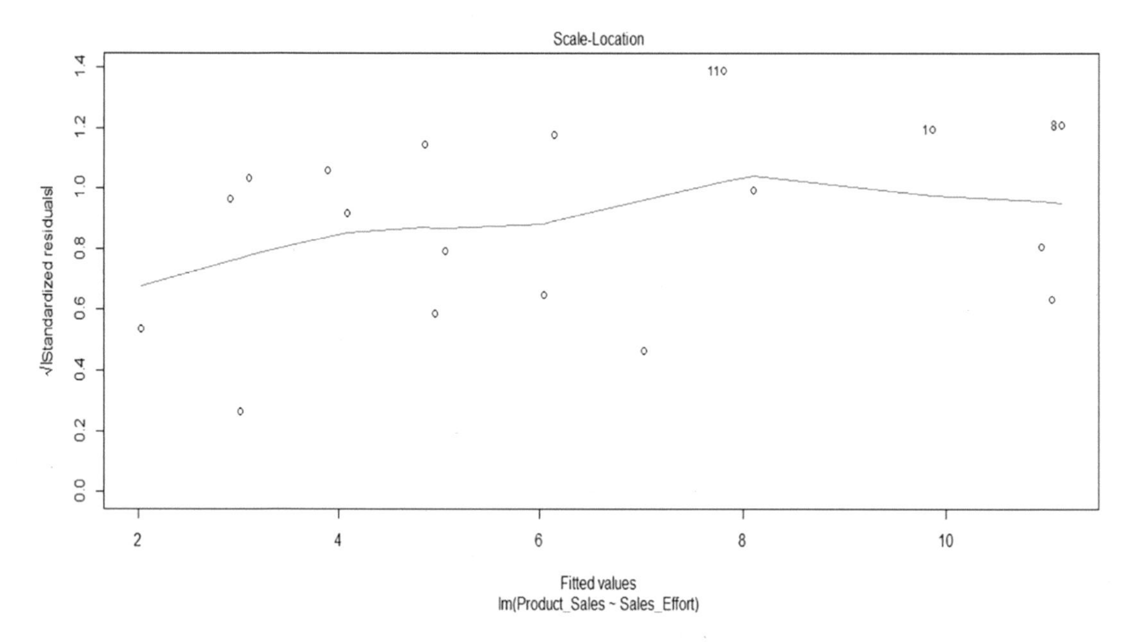

Figure 7-10. *Scale-location plot generated in R to validate homoscedasticity*

In Figure 7-10, as the points are spread in a random fashion around the near horizontal line, this assures us that the assumption of constant variance (or homoscedasticity) is fulfilled.

Using the crPlots(model name) Function from library(car)

We can use the `crPlots(mod_simp_reg)` command to understand the linearity of the relationship represented by the model. Nonlinearity requires us to re-explore the model. The graph in Figure 7-11 shows that the model we created, like earlier (i.e., `mod_simp_reg`), is linear.

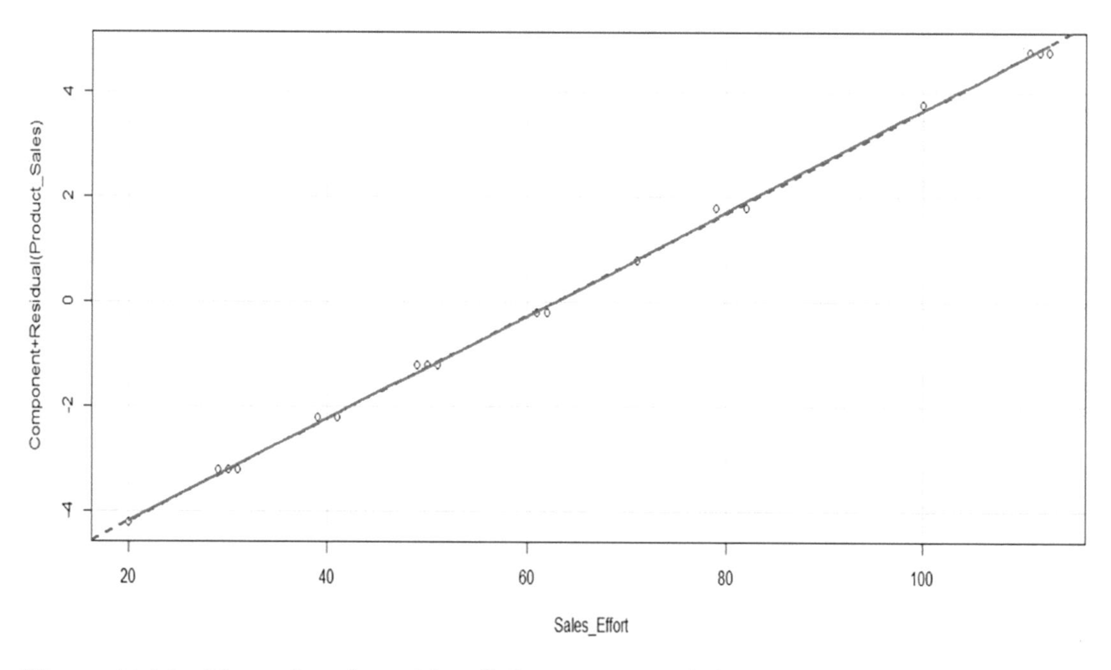

Figure 7-11. *Plot using the crPlots() function to validate the linearity assumption*

7.4.5 Conclusion

As shown, the simple linear regression model fitted using the R function `lm(response variable ~ independent variable, data = dataframe name)` representing the simple linear regression equation, namely, $Product\_Sales_i = 0.0688518 + 0.0979316\ Sales\_Effort_i$, is a good model as it passes the tests to validate the assumptions of the regression too. A note of caution here is that there are various ways the regression equation may be created and validated.

7.4.6 Predicting the Response Variable

The fitted model not only explains the relationship between the two variables, namely, the response variable and the independent variable, but also provides a mechanism to predict the value of the response variable from the value of the new independent variable. Here, we are using `Test_Data` for the prediction that we had earmarked for the validation of the model.

Figure 7-12A shows the code used.

```
#Using the model mod_simp_reg on the Test_Data

#to check if the model works well in prediction

predicted <- predict(mod_simp_reg, newdata=Test_Data,
interval="prediction")

predicted
```

Figure 7-12A. *Code to predict using the generated model*

Figure 7-12B shows the output.

```
> predicted
        fit      lwr       upr
9  9.959946 9.722815 10.197076
10 9.764083 9.527918 10.000247
12 8.001313 7.771603  8.231024
```

Figure 7-12B. *Prediction using the model on the Test_Data*

This is done using the function `predict(model name, newdata)` where the model name is the name of the model arrived at from the input data (i.e., `Train_Data` in our case) and `newdata` contains the independent variable data for which the response variable has to be predicted. In our case, we used the `Test_Data` we had earmarked for the model validation. However, we can create any new dataframe with new `Sales_Effort` variables and use it as an input `newdata` in the previous equation to obtain the prediction.

As shown, we have obtained a prediction interval, which specifies the range of the distribution of the prediction also with the additional parameter `interval = "prediction"` on the `predict()` function. This uses by default the confidence interval as 0.95.

As you can see from the predicted values, if we round off the `Sales_Effort` values predicted to the nearest numbers, they are as per the `Test_Data`, as shown in Figure 7-4B. This suggests that the model is working well on the prediction.

7.4.7 Additional Notes

It may be observed from the model fitted earlier that the intercept is not zero, but it is 0.0688518. Actually, when there is no sales effort, ideally, there should not be any sales. But this may not be so; there may be some walk-in sales possible because of the other means such as advertisements, websites, etc. Similarly, there cannot be partial product sales like 3.1. However, the sales efforts put in would have moved the salesperson toward the next potential sale partially. If we are interested in arriving at the model without intercept (i.e., no product sales when there is no sales effort), then we can do so as shown by forcing the intercept to zero value.

Figure 7-13A shows the code used.

```
#Generating the model without intercept

mod_simp_reg_wo_intercept <- lm(cust_df$Product_Sales ~
cust_df$Sales_Effort + 0)

summary(mod_simp_reg_wo_intercept)
```

Figure 7-13A. *Model without the intercept*

Figure 7-13B shows the output.

```
> summary(mod_simp_reg_wo_intercept)

Call:
lm(formula = cust_df$Product_Sales ~ cust_df$Sales_Effort + 0)

Residuals:
      Min        1Q    Median        3Q       Max
-0.190158 -0.060146 -0.001823  0.097205  0.196233

Coefficients:
                      Estimate Std. Error t value Pr(>|t|)
cust_df$Sales_Effort 0.0990279  0.0003201   309.4   <2e-16 ***
---
Signif. codes:  0 '***' 0.001 '**' 0.01 '*' 0.05 '.' 0.1 ' ' 1

Residual standard error: 0.1079 on 20 degrees of freedom
Multiple R-squared:  0.9998,	Adjusted R-squared:  0.9998
F-statistic: 9.573e+04 on 1 and 20 DF,  p-value: < 2.2e-16
```

Figure 7-13B. *Generating a simple linear regression model without an intercept*

However, if we have to believe this model and use this model, we have to validate the fulfilment of the other assumptions of the regression.

7.5 Using Python to Generate the Model and Validating the Assumptions

In the previous sections of this chapter, we used the R programming language to generate the model, predict the response variable for the new predictor value, and also validate the regression assumptions.

Here, we are going to use Python and the packages available within the Anaconda framework like Jupyter Notebook, scikit-learn, statsmodels, pandas, NumPy, SciPy, etc. (as relevant), to generate the model, make the predictions using the generated model, and also validate the regression assumptions. To keep it simple and to explain the details in the context of each input and output, the comments are embedded along with the code instead of writing the content separately. The extract from the Jupyter Notebook we used to carry out the this work is attached in the following logical pieces.

7.5.1 Load Important Packages and Import the Data

Figure 7-14A shows the code.

```
#import all the necessary packages

import numpy as np

import pandas as pd

import sklearn
```

Figure 7-14A. *Code to load important packages*

These commands load the packages into the memory for usage.

Now, we will read the file into the working environment for further processing. This is done using the code given in Figure 7-15A.

```
#import the text file with data from the local machine

#this is a text file with headers and hence the

#headers are read from the file itself

df = pd.read_csv("C:/Users/kunku/OneDrive/Documents/Book Revision/cust1.txt", header=0)

print(df)
```

Figure 7-15A. *The code to read the file and create a dataframe*

Figure 7-15B shows the output.

```
    Sales_Effort  Product_Sales
0            100             10
1             82              8
2             71              7
3            111             11
4            112             11
5             61              6
6             62              6
7            113             11
8            101             10
9             99             10
10            79              8
11            81              8
12            51              5
13            50              5
14            49              5
15            30              3
16            31              3
17            29              3
18            20              2
19            41              4
20            39              4
```

Figure 7-15B. *Output showing the contents of the dataframe created*

In the previous dataframe, we have 21 records with one response variable and one predictor variable (row index starting from 0).

7.5.2 Generate a Simple Linear Regression Model

There are many different ways you can generate a simple linear regression model. Here, we will walk you through one of those ways.

We first create two separate dataframes, one with only the data of the response variable and another with only the data of the predictor variable.

Figure 7-16A shows the code related to the creation of a separate dataframe with only the response variable values.

```
#separating the response variable data to a separate dataframe
y = df["Product_Sales"]
#displaying top 5 values from this dataframe
y.head(5)
```

Figure 7-16A. *Code to create a separate dataframe with response variables only*

Figure 7-16B shows the output.

```
0    10
1     8
2     7
3    11
4    11
Name: Product_Sales, dtype: int64
```

Figure 7-16B. *Output showing Top 5 records of the response variable dataframe created*

Figure 7-17A shows the code related to the creation of a separate dataframe with only the predictor variable.

```
#other than the Product_Sales other data is carried over to a separate dataframe

X = df.drop(['Product_Sales'], axis=1)

#print out the data types

print(X.dtypes)

#displaying top 5 values from this dataframe

X.head(5)
```

Figure 7-17A. *Code to create a separate dataframe with only predictor variables*

Figure 7-17B shows the output.

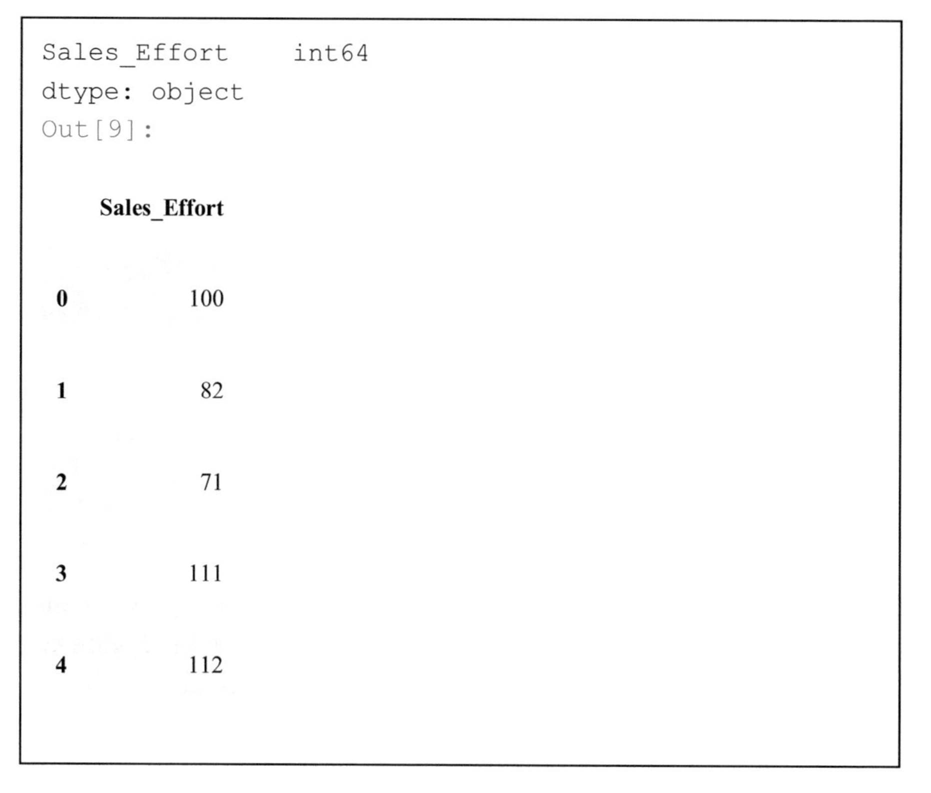

```
Sales_Effort    int64
dtype: object
Out[9]:
```

	Sales_Effort
0	100
1	82
2	71
3	111
4	112

Figure 7-17B. *Output showing the types and five values of X*

In the following section, we have split the data set into two separate data sets, i.e., the training data set and test data set, and have generated the simple linear regression model using the scikit-learn package.

Figure 7-18A shows the code.

```
#The data is split as training set and test set
#There are multiple means - Here we are doing it manually
train_samples = (15)
from sklearn.model_selection import train_test_split
X_train, X_test, y_train, y_test = train_test_split(X, y,
            train_size=train_samples, test_size=6)
#importing necessary packages
from sklearn.linear_model import LinearRegression
#running the LinearRegression
Line_Reg = LinearRegression()
Line_Reg.fit(X_train, y_train)
#Line_Reg is the model generated
```

Figure 7-18A. *Code to split the data into training data and test data and also to generate the model using training data*

The output of the previous is the creation of four separate dataframes, i.e., `X_train`, `X_test`, `y_train`, and `y_test`, and also the generation of the model. We have used the scikit-learn package to split the data as well as generate the linear regression model named `Line_Reg`. However, the output generated by Python in Jupiter is shown in Figure 7-18B, which suggests a linear regression model has been built.

```
Out[11]:  LinearRegression()
```

Figure 7-18B. *Output related to train-test split and linear regression model generation*

In the following section, we have checked for the coefficient value of the predictor and model accuracy based on the test data set.

Figure 7-19A shows the code.

```
#print regression coefficients

print(Line_Reg.coef_)

#print the regression score using the test data

#this provides the regression model accuracy

Line_Reg.score(X_test, y_test)
```

Figure 7-19A. *Code to get the coefficient of the model as well as accuracy score of the model*

Figure 7-19B shows the output.

```
[0.09826593]
0.9990992557160832
```

Figure 7-19B. *Output related to obtaining the regression coefficient and the regression model accuracy score*

Figure 7-19B shows the coefficient of the `Sales_Effort` as 0.9826593 and the `Line_Reg` regression model accuracy score as 99.91 percent.

In the following section, we have checked on the mean square error of the model. This is one of the measures of the model error, as you are aware. Figure 7-20A shows the code related to this.

```
#finding the mean square error on the test predictions

np.mean((Line_Reg.predict(X_test)-y_test)**2)
```

Figure 7-20A. *Code to generate the mean square error of the model*

Figure 7-20B shows the output.

```
0.007806450460612922
```

Figure 7-20B. *Output of the mean square error*

As you can observe from Figure 7-20B, the mean square error is very small. It is nearly zero. Hence, we have obtained a fit model.

In the following section, we have predicted the values of the response variable on the predictor from the `X_test` data set. This depicts high agreement between the predicted (i.e., predicted) and the actual values (i.e., y_test). Of course, there are some residuals (i.e., the difference between the fitted and actual values). Figure 7-21A shows the code.

```
#run the model on the test data to arrive at the predicted values

predicted = Line_Reg.predict(X_test)

#printing the predicted and actual response values

print(predicted)

print(y_test)
```

Figure 7-21A. *Code to predict on the test data and to print predicted and original response variable values*

Figure 7-21B shows the output.

```
[4.08978733  9.98574284 11.06666801  6.05510583  7.82389248  3.00886215]
19     4
8     10
4     11
5      6
10     8
15     3
Name: Product_Sales, dtype: int64
```

Figure 7-21B. *Predicted value values versus actual values*

7.5.3 Alternative Way for Generation of the Model

As discussed, there are multiple ways you can generate a simple linear regression model. In Figure 7-22A, we are exploring another way by using the statsmodels package. Figure 7-22A shows the code.

```
#generate linear regression model using statsmodels package
#this will also enable us to test for the regression
#assumption related to autocorrelation
from statsmodels.formula.api import ols
#fit linear regression model
Line_Reg_1 = ols('Product_Sales ~ Sales_Effort', data=df).fit()
#print model summary
print(Line_Reg_1.summary())
```

Figure 7-22A. *Code related to linear regression model generation using statsmodels*

As output of the previous code, an OLS model named `Line_Reg_1` is generated, and the `print(Line_Reg_1.summary)` prints the output given in Figure 7-22B.

```
                            OLS Regression Results
==============================================================================
Dep. Variable:          Product_Sales   R-squared:                       0.999
Model:                            OLS   Adj. R-squared:                  0.999
Method:                 Least Squares   F-statistic:                 1.530e+04
Date:                Sun, 04 Sep 2022   Prob (F-statistic):           4.42e-29
Time:                        13:48:54   Log-Likelihood:                 17.877
No. Observations:                  21   AIC:                            -31.75
Df Residuals:                      19   BIC:                            -29.66
Df Model:                           1
Covariance Type:            nonrobust
================================================================================
                  coef    std err          t      P>|t|      [0.025      0.975]
--------------------------------------------------------------------------------
Intercept       0.0501      0.059      0.856      0.403      -0.072       0.173
Sales_Effort    0.0984      0.001    123.703      0.000       0.097       0.100
==============================================================================
Omnibus:                        1.042   Durbin-Watson:                   1.507
Prob(Omnibus):                  0.594   Jarque-Bera (JB):                0.941
Skew:                           0.328   Prob(JB):                        0.625
Kurtosis:                       2.196   Cond. No.                         182.
==============================================================================
```

Figure 7-22B. *Summary of the OLS model generated using the statsmodels package*

The R-Squared and Adj. R-Squared value both are 0.999 showing a high correlation (i.e., almost perfect) between `Product_Sales` and `Sales_Effort`. The p-value (P>|t|) of 0.000 in the case of `Sales_Effort` shows that `Sales_Effort` is significant to the model. The probability of the F-statistic of a very low value (i.e., 4.42e-29) shows that the model generated is significant. The Durbin_Watston test value of 1.507 (which is between 1.50 to 2.50) shows that there is no autocorrelation. `Prob(Jarque_Bera)` of 0.625 shows that we cannot reject the null hypothesis that the residuals are normal.

7.5.4 Validation of the Significance of the Generated Model

The model summary of the OLS model discussed in section 7.5.3 shows the P > |t| value for the `Sales_Effort` predictor as 0.000. Hence, the `Sales_Effort` predictor is significant to the model as a predictor. Further, the model's F-statistic has a very low value (i.e., 4.42e-29). Hence, we can safely conclude that the model generated is significant and can be used for prediction.

7.5.5 Validating the Assumptions of Linear Regression

We now need to validate that the model generated fulfils the assumptions of the linear regression. For this purpose, we will use the model `Line_Reg_1` to predict on the full dataframe (X). Figure 7-23A shows the code.

```
#We will use the entire data set to carry out the
#prediction using the regression model generated
predicted_full = Line_Reg_1.predict(X)
#printing the predicted and actual response values
print(predicted_full)
print(y)
```

Figure 7-23A. *Code related to the prediction on the entire dataframe*

Figure 7-23B shows the output.

```
0      9.890614
1      8.119317
2      7.036858
3     10.973073
4     11.071478
5      6.052805
6      6.151210
7     11.169883
8      9.989019
9      9.792208
10     7.824101
11     8.020912
12     5.068751
13     4.970346
14     4.871940
15     3.002239
16     3.100644
17     2.903833
18     2.018185
19     4.084698
20     3.887887
dtype: float64
0     10
1      8
2      7
3     11
4     11
5      6
6      6
7     11
8     10
9     10
10     8
11     8
12     5
13     5
14     5
15     3
16     3
17     3
18     2
19     4
20     4
Name: Product_Sales, dtype: int64
```

Figure 7-23B. *Predicted y values and the actual y values*

Now, we will first plot the scatter plot between the predicted values and the residuals of the model, through code shown in Figure 7-24A.

```
#creating the scatter plot between predicted values
#and the residuals of the model
import matplotlib.pyplot as plt
plt.scatter(predicted_full, (predicted_full-y))
plt.xlabel("fitted")
plt.ylabel("residuals")
plt.show()
```

Figure 7-24A. *Plotting of a scatter plot between actual response variable values versus the predicted values*

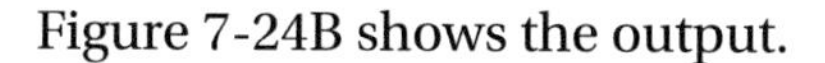

Figure 7-24B shows the output.

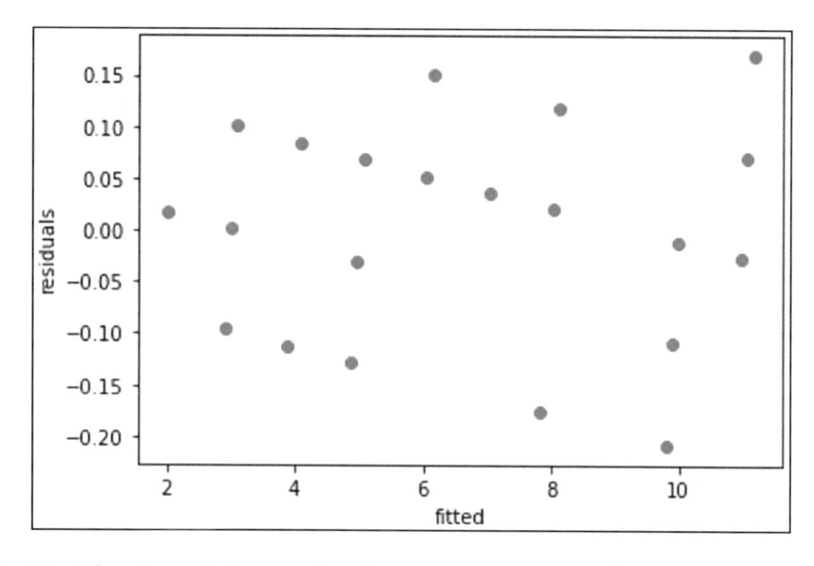

Figure 7-24B. *Plotting of the residuals versus predicted (fitted) values*

- *Validation of the regression assumption, linearity*: As you can see in the previous graph, the residuals are very small. Further, the residuals are randomly spread around the 0.0 residuals. Hence, this depicts the linearity assumption is validated.
- *Validation of the regression assumption, test of equal variance (homoscedasticity)*: As we can see from the plot in Figure 7-24B, there is no significant variation of the residuals for the different levels of the independent variable's values. Hence, we consider that the test of equal variance of the distribution of the response variable as passed.
- *Validation of the regression assumption, autocorrelation, i.e., test of independence of the errors around the regression line*: From the summary of the model `Line_Reg_1`, we have observed that the Durbin-Watson test has returned the result of ~1.51, which is between 1.50 and 2.50. This clearly suggests that there is no autocorrelation. We also know from our data collection effort that each data point is related to different person's sales effort.
- *Validation of the regression assumption, normal distribution of the residuals*: For this purpose, we plot the ordered values of the residuals against the theoretical quantiles as follows. The expectation is that all the residuals should fall on a 45-degree straight line. Figure 7-25A shows the code.

```
#plotting for the validation of normal distribution of the residuals
import scipy
from scipy import stats
stats.probplot(y-predicted_full, plot=plt)
plt.show()
```

Figure 7-25A. *The code to validate the assumption of normal distribution of the residuals*

Figure 7-25B shows the output.

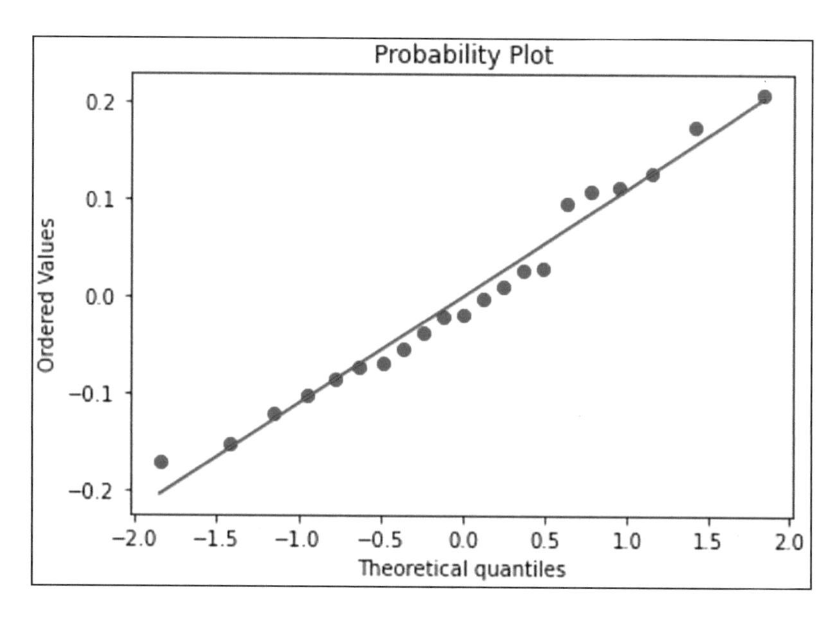

Figure 7-25B. *The validation of the assumption of normal distribution of the residuals*

The graph in Figure 7-25B confirms the assumption that the residuals are normally distributed.

As you can see, all the assumptions of the linear regression are validated. Hence, we can use the model for the predictions.

7.5.6 Predict Using the Model Generated

Now, we use the model generated to predict the response variable in case of a set of new predictor variables. Figure 7-26A shows the code.

```
#Now let us create a dataframe with predictors to check that
#the model predicts the values of response variables accurately
#(even though we have tested the model against the test data
data = {'Sales_Effort': [111, 79]}
df_new_predictors = pd.DataFrame(data=data)
predicted_resp_variable = Line_Reg_1.predict(df_new_predictors)
print(predicted_resp_variable)
```

Figure 7-26A. *Code to create a dataframe with new predictor variables and carry out the prediction using the model*

Figure 7-26B shows the output.

```
0     10.973073
1      7.824101
dtype: float64
```

Figure 7-26B. *Values predicted on the new predictors*

The value of the `Product_Sales` predicted (after rounding off to the nearest number) are 11 and 8.

7.6 Chapter Summary

- In this chapter, you went through some examples as to how the relationship between various aspects influence other factors. Understanding these relationships helps us not only to understand what can happen to the other associated factors but also to predict the value of others. You learned how the regression model or regression equation explains the relationship between a response variable and the independent variable(s). You also learned about the linear or nonlinear relationship as well as simple regression and multiple regression.

- You explored the concept of correlation with examples. You explored the uses of correlation, strong correlation, positive correlation, negative correlation, and so on. You also learned how to calculate the correlation coefficient (R).
- You learned about simple linear regression by highlighting that it is a smoothed line of averages. You also explored the four important assumptions of regression: linearity, equivalence of errors, normality, and homoscedasticity. You learned about simple linear regression, what intercept and slope are, and how to calculate them.
- You explored examples of how, using R, you can arrive at the best-fit simple linear regression model and the simple linear regression equation. You looked at examples and learned how to validate the best-fit model and the model arrived at for the fulfillment of the regression assumptions.
- You explored how the simple linear regression model arrived at can be used to predict the value of the response variable when it is not known but when the related independent variable value is known.
- You looked at how, using R, you can arrive at the simple linear regression model without intercept and the usage of the same.
- We finally demonstrated to you how to carry on all those analyses we carried out using R in Python also.

CHAPTER 8

Multiple Linear Regression

In Chapter 7, you explored simple linear regression, which depicts the relationship between the response variable and one predictor. You saw that the expectation is that the response variable is a continuous variable that is normally distributed. If the response variable is a discrete variable, you use a different regression method. If the response variable can take values such as yes/no or multiple discrete variables (for example, views such as *strongly agree, agree, partially agree*, and *do not agree*), you use logistic regression. You will explore logistic regression in Chapter 11. When you have more than one predictor—say, two predictors or three predictors or *n* predictors (with *n* not equal to 1)—the regression between the response variable and the predictors is known as *multiple regression*, and the linear relationship between them is expressed as *multiple linear regression* or a *multiple linear regression equation*. In this chapter, you will see examples of situations in which many factors affect one response, outcome, or dependent variable.

Imagine that you want to construct a building. Your main cost components are the cost of labor and the cost of materials including cement and steel. Your profitability is positively impacted if the costs of cement, steel, and other materials decrease while keeping the cost of labor constant. Instead, if the costs of materials increase, your profitability is negatively impacted while keeping the cost of labor constant. Your profitability will further decrease if the cost of labor also increases.

It is possible in the market for one price to go up or down or all the prices to move in the same direction. Suppose the real estate industry is very hot, and there are lots of takers for the houses, apartments, or business buildings. Then, if there is more demand for the materials and the supply decreases, the prices of these materials are likely to increase. If the demand decreases for the houses, apartments, or business buildings, the prices of these materials are likely to decrease as well (because of the demand being less than the supply).

U. R. Hodeghatta and U. Nayak, *Practical Business Analytics Using R and Python*,
https://doi.org/10.1007/978-1-4842-8754-5_8

Now let's presume that the selling prices are quite fixed because of the competition, and hence the profitability is decided and driven primarily by the cost or cost control. We can now collect data related to the cost of cement, steel, and other materials, as well as the cost of labor as predictors or independent variables, and profitability (in percent) as the response variable. Such a relationship can be expressed through a multiple linear regression model or multiple linear regression equation.

In this example, suppose we find that the relationship of the cost of other materials (one of the predictors) to the response variable is dependent of the cost of the cement. Then we say that there is a *significant interaction* between the cost of other materials and the cost of the cement. We include the interaction term *cost of other materials:cost of cement* in the formula for generating the multiple linear regression model while also including all the predictors. Thus, our multiple linear regression model is built using the predictors *cost of cement, cost of steel, cost of other materials,* and the interaction term, *cost of other materials:cost of cement* versus the *profitability* as the response variable.

Now imagine that you are a human resources (HR) manager. You know that the compensation to be paid to an employee depends on their qualifications, experience, and skill level, as well as the availability of other people with that particular skill set versus the demand. In this case, compensation is the response variable, and the other parameters are the independent variables, or the predictor variables. Typically, the higher the experience and higher the skill levels, the lower the availability of people compared to the demand, and the higher the compensation should be. The skill levels and the availability of those particular skills in the market may significantly impact the compensation, whereas the qualifications may not impact compensation as much as the skill levels and the availability of those particular skills in the market.

In this case, there may be a possible relationship between experience and skill level; ideally, more experience means a higher skill level. However, a candidate could have a low skill level in a particular skill while having an overall high level of experience—in which case, experience might not have a strong relationship with skill level. This feature of having a high correlation between two or more predictors themselves is known as *multicollinearity* and needs to be considered when arriving at the multiple linear regression model and the multiple linear regression equation.

Understanding the *interactions between the predictors* as well as *multicollinearity* is very important in ensuring that we get a correct and useful multiple regression model. When we have the model generated, it is necessary to validate it on all *four assumptions of regression*:

- Linearity between the response variable and the predictors (also known as *independent variables*)
- Independence of residuals
- Normality of the distribution of the residuals
- Homoscedasticity, an assumption of equal variance of the errors

The starting point for building any multiple linear regression model is to get our data in a dataframe format, as this is the requirement of the `lm()` function in R. The expectation when using the `lm()` function is that the response variable data is distributed normally. However, independent variables are not required to be normally distributed. Predictors can contain factors too.

Multiple regression modeling may be used to model the relationship between a response variable and two or more predictor variables to *n* number of predictor variables (say, 100 or more variables). The more features that have a relationship with the response variable, the more complicated the modeling will be. For example, a person's health, if quantified through a health index, might be affected by the quality of that person's environment (pollution, stress, relationships, and water quality), the quality of that person's lifestyle (smoking, drinking, eating, and sleeping habits), and genetics (history of the health of the parents). These factors may have to be taken into consideration to understand the health index of the person.

8.1 Using Multiple Linear Regression

Now, let's discuss how to arrive at multiple linear regression.

8.1.1 The Data

To demonstrate multiple linear regression, we have created data with three variables: Annual Salary, Experience in Years, and Skill Level. These are Indian salaries, but for the sake of convenience, we have converted them into U.S. dollars and rounded them to thousands. Further, we have not restricted users to assess and assign skill levels in decimal points. Hence, in the context of this data, even Skill Level is represented as continuous data. This makes sense, as in an organization with hundreds of employees, it is not fair to categorize all of them, say, into five buckets; it's better to differentiate them with skill levels such as 4.5, 3.5, 2.5, 0.5, and 1.5. In this data set, all the variables are continuous variables.

Here, we import the data from the CSV file `sal1.txt` to the dataframe `sal_data_1`. Figure 8-1A shows the code.

```
#Reading a file using read.csv() command

#Alternatively you can use read.table() or read.csv2()

sal_data_1 <- read.csv("C:/Users/kunku/OneDrive/Documents/Book
Revision/Sal1.txt", header=TRUE, sep=',')

#Understand the details of the dataframe

str(sal_data_1)
```

Figure 8-1A. *Code to read the file and to display the details of the dataframe created*

Figure 8-1B shows the output.

```
> str(sal_data_1)
'data.frame':    48 obs. of  3 variables:
 $ Annu_Salary: int  4000 6000 8000 10000 12000 14000 16000 14000 18000 20000 ...
 $ Expe_Yrs   : num  0 1 2 3 4 4.5 5 4 6 7 ...
 $ Skill_lev  : num  0.5 1 1.5 2 2.5 3.5 4 4 4 4.5 ...
```

Figure 8-1B. *The output of the str(sal_data_1)*

If you use the `head()` and `tail()` functions on the data, you will get an idea of what the data looks like. Figure 8-2A shows the code.

```
#Understand the data

head(sal_data_1)

tail(sal_data_1)
```

Figure 8-2A. *Code to read the first few and last few records of the dataframe sal_data_1*

Figure 8-2B shows the output.

```
> head(sal_data_1)
  Annu_Salary Expe_Yrs Skill_lev
1        4000      0.0       0.5
2        6000      1.0       1.0
3        8000      2.0       1.5
4       10000      3.0       2.0
5       12000      4.0       2.5
6       14000      4.5       3.5
> tail(sal_data_1)
   Annu_Salary Expe_Yrs Skill_lev
43       15000     5.50       3.0
44       15000     5.00       3.5
45       19000     6.50       4.0
46       19000     6.30       4.2
47        5000     0.50       0.5
48        6000     0.75       1.0
```

Figure 8-2B. *The display of first few and last few rows of the dataframe*

Please note that we have not shown all the data, as the data set has 48 records. In addition, this data is illustrative only and may not be representative of a real scenario. The data was collected at a certain point in time.

8.1.2 Correlation

We explained correlation in Chapter 7. *Correlation* specifies the way that one variable relates to another variable. This is easily done in R by using the `cor()` function.

Figure 8-3A shows the coefficient of correlation R between each pair of these three variables (Annual Salary, Experience in Years, Skill Level).

```
#Understanding the correlation between each field to the other fields

cor(sal_data_1)
```

Figure 8-3A. *Code to provide the correlation between each field to the other fields of the dataframe*

Figure 8-3B shows the output.

```
> cor(sal_data_1)
            Annu_Salary  Expe_Yrs Skill_lev
Annu_Salary   1.0000000 0.9888955 0.9414029
Expe_Yrs      0.9888955 1.0000000 0.8923255
Skill_lev     0.9414029 0.8923255 1.0000000
```

Figure 8-3B. *Correlation between each field to the other fields of the dataframe*

As you can see from Figure 8-3B, there is a very high correlation of about 0.9888 between Annual Salary and Experience in Years. Similarly, there is a very high correlation of about 0.9414 between Annual Salary and Skill Level. Also, there is a very high correlation of about 0.8923 between Experience in Years and Skill Level. Each variable is highly correlated to the other variable. All the correlation is positive correlation.

The relationship between two pairs of variables is generated visually by using the R command, as shown in Figure 8-4A.

```
#Visually depicting the relationship between a pair of fields

library(caret)

featurePlot(x=sal_data_1[,c("Expe_Yrs","Skill_lev")],
y=sal_data_1$Annu_Salary, plot="pairs")
```

Figure 8-4A. *Code to visually depict the relationship between the fields*

The output, i.e., the visual relationship between the fields, is provided in Figure 8-4B through a scatter plot matrix. A scatter plot matrix explains the relationship between each variable in the data set to the other variables. In our case, we have three variables, i.e., `Annu_Salary`, i.e., represented as y, `Expe_Yrs`, and `Skill_lev`. Hence, we will have a 3 × 3 scatter plot matrix. The label in each row specifies the y-axis for the row and the x-axis for the column. In our scatter plot matrix shown in Figure 8-4B, we have three labels (`Annu_Salary`, `Skill_lev`, `Expe_Yrs`) each representing one of the variables in the data set. The first plot in Figure 8-4B has y as the y-axis and `Expe_Yrs` as the x-axis, and the second plot in the first row has y as the y-axis and `Skill_lev` as the x-axis. Similarly, the second row in the first plot has `Skill_lev` as the y-axis and `Expe_Yrs` as the x-axis, and the third plot in the second row has `Skill_lev` as the y-axis and y as the x-axis. Similarly, the second plot in the third row has `Expe_Yrs` as the y-axis and `Skill_lev` as the x-axis, and the third plot in the third row has `Expe_Yrs` as the y-axis and y as the x-axis.

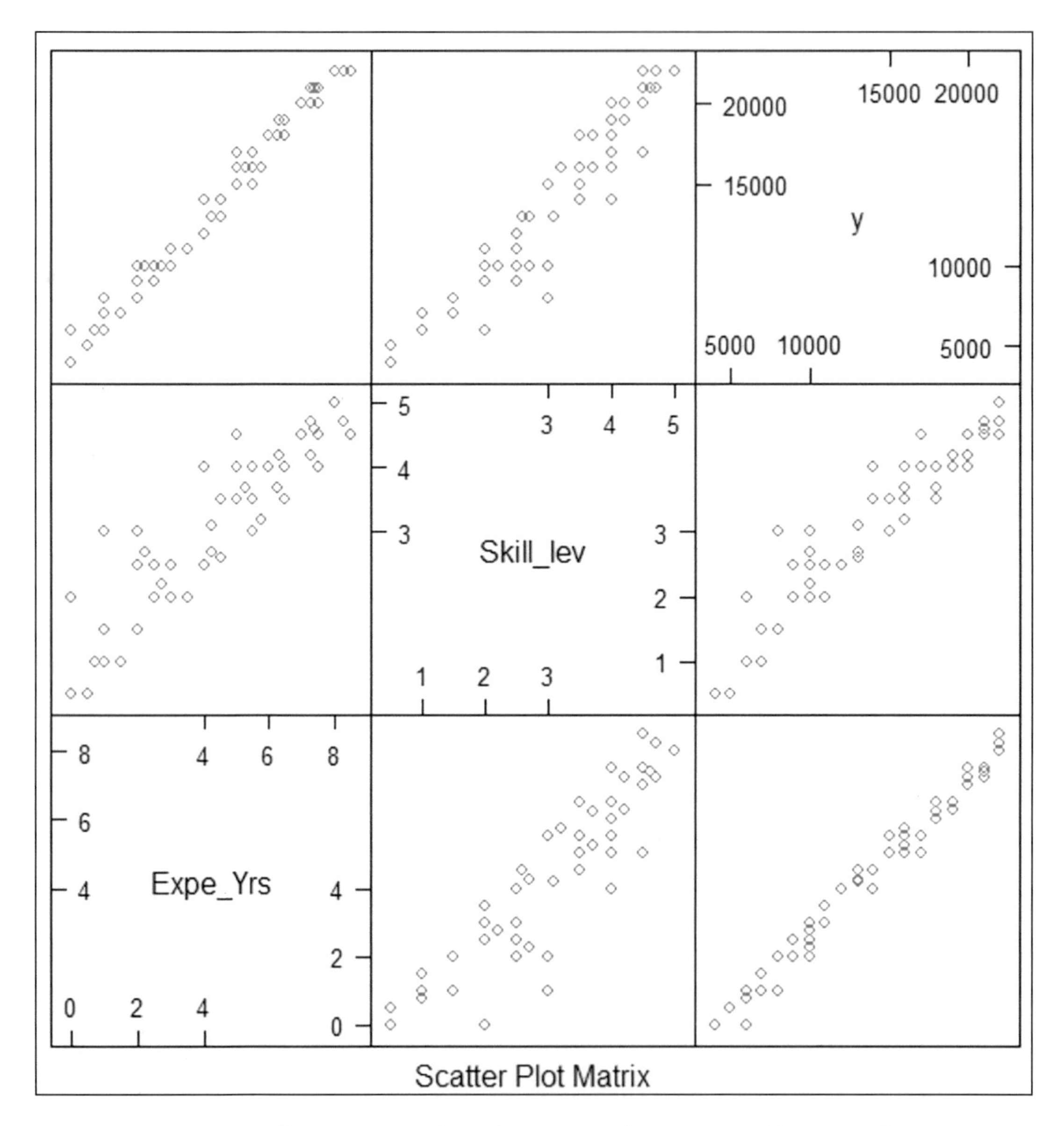

***Figure 8-4B.** Visually depicting the relationship between fields using the featurePlot() function*

Here, we use the `caret` package and the `featurePlot()` function. The response variable is plotted as y, and the predictor variables are plotted with their respective variable names.

8.1.3 Arriving at the Model

Now, let's create the multiple linear regression model by using the `lm()` function of R. To start, let's generate the model by using all the parameters, as they are highly correlated with Annual Salary as the response (or dependent) variable, and we want to use this model to predict, or set, the salary later in tune with this model. The other variables, Experience in Years and Skill Level, will be predictor (or independent) variables. Figure 8-5A shows the command in R to generate the multiple linear regression model.

```
#Generating the Multiple Linear Regression Model using lm()

sal_model_1 <- lm(Annu_Salary ~ Expe_Yrs + Skill_lev, data =
sal_data_1)
```

Figure 8-5A. *Code to generate the model using lm()*

The model gets created, but there will not be any output in R. To see the details of the model created using the `lm()` function, we need to use `summary(model name)`. Figure 8-5B shows the command along with the output.

```
> summary(sal_model_1)

Call:
lm(formula = Annu_Salary ~ Expe_Yrs + Skill_lev, data = sal_data_1)

Residuals:
   Min      1Q Median      3Q     Max
-605.6 -318.5  -23.8  354.3  666.0

Coefficients:
            Estimate Std. Error t value Pr(>|t|)
(Intercept)  3011.66     162.58   18.52  < 2e-16 ***
Expe_Yrs     1589.68      50.86   31.26  < 2e-16 ***
Skill_lev    1263.65     102.03   12.38 4.25e-16 ***
---
Signif. codes:  0 '***' 0.001 '**' 0.01 '*' 0.05 '.' 0.1 ' ' 1

Residual standard error: 377.5 on 45 degrees of freedom
Multiple R-squared:  0.995,     Adjusted R-squared:  0.9948
F-statistic:  4469 on 2 and 45 DF,  p-value: < 2.2e-16
```

Figure 8-5B. *Summary of the model sal_model_1 displayed*

You can see in this summary of the multiple regression model that both independent variables are significant to the model, as the p-value for both is less than 0.05. Further, the overall model p-value is also less than 0.05. Further, as you can see, the adjusted R-squared value of 99.48 percent indicates that the model explains 99.48 percent of the variance in the response variable. Further, the residuals are spread around the median value of -23.8, very close to 0.

You can explore the individual aspects of this model by using specific R commands. You can use `fitted(model name)` to understand the values fitted using the model. The fitted values are shown in Figure 8-6. Similarly, you can use `residuals(model name)` to understand the residuals for each value of Annual Salary fitted versus the actual Annual Salary per the data used. You can use `coefficients(model name)` to get the details of the coefficients (which is part of the summary data of the model shown previously).

```
> fitted(sal_model_1)
        1         2         3         4         5         6         7         8         9        10        11        12        13        14        15
 3643.481  5864.982  8086.484 10307.985 12529.487 14587.973 16014.636 14424.959 17604.314 19825.815 22047.316  9981.956  9350.132  9513.147  8392.279
       16        17        18        19        20        21        22        23        24        25        26        27        28        29        30
10939.809 13179.636  6496.806  6659.821 11102.824 13450.690 16646.460 16809.475 20620.654 20475.964 15545.827 15382.812  5538.953 20588.051 13605.611
       31        32        33        34        35        36        37        38        39        40        41        42        43        44        45
22065.641 22210.331 16032.961 16177.651 16195.976 17622.638 17767.328 19988.830 19844.140 10144.971 10163.295 10000.281 15545.827 15382.812 18399.152
       46        47        48
18333.946  4438.320  5467.563
```

Figure 8-6. *Values fitted using the sal_model_1*

8.1.4 Validation of the Assumptions of Regression

You can use `plot(model name)` to explore the model further. This command generates diagnostic plots to evaluate the fit of the model. Three of the important plots generated are shown in Figures 8-7 to 8-9.

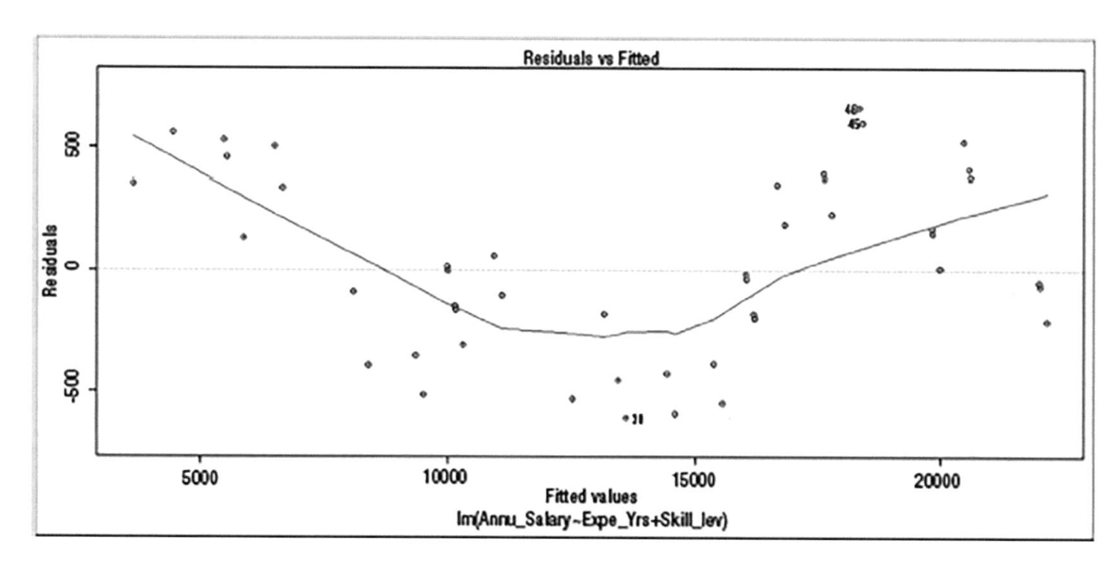

Figure 8-7. *Residuals versus fitted*

Here the residuals versus fitted plot seem to show that the residuals are spread randomly around the dashed line at 0. If the response variable is linearly related to the predictor variables, there should be no relationship between the fitted values and the residuals. The residuals should be randomly distributed. Even though there seems to be a pattern, in this case we know clearly from the data that there is a linear relationship between the response variable and the predictors. This is also shown through a high correlation between the response variable and each predictor variable. Hence, we cannot conclude that the linearity assumption is violated. The linear relationship between the response variable and predictors can be tested through the `crPlots(model name)` function, as shown in Figure 8-10. As both the graphs in Figure 8-10 show near linearity, we can accept that the model `sal_model_1` fulfils the test of linearity.

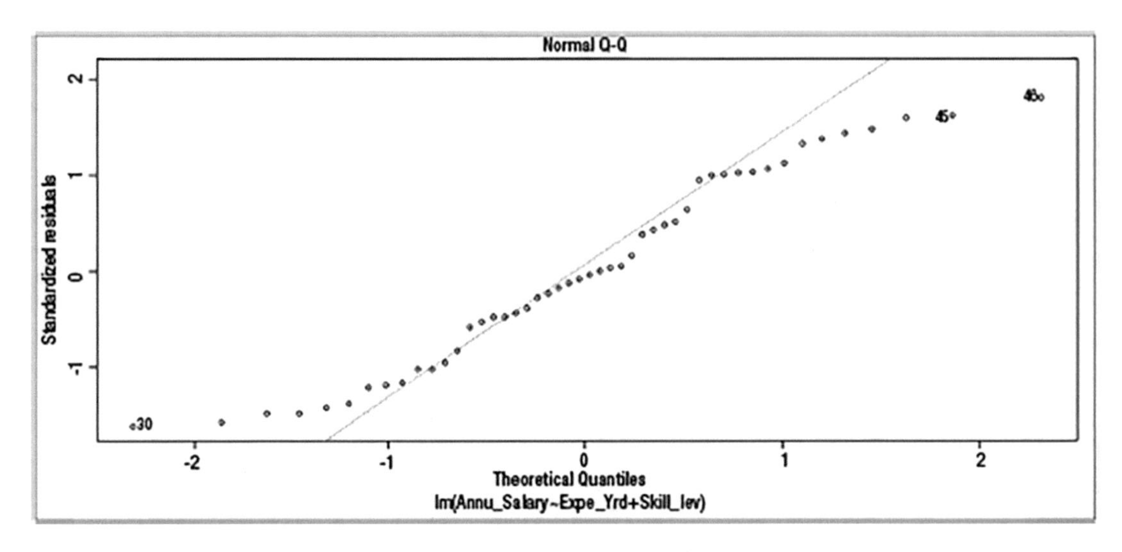

Figure 8-8. *Normal Q-Q plot*

The normal Q-Q plot seems to show that the residuals are not normally distributed. However, the visual test may not always be appropriate. We need to ensure normality only if it matters to our analysis and is really important, because, in reality, data may not always be normal. Hence, we need to apply our judgment in such cases. Typically, as per the central limit theorem and rule of thumb, we do not require validating the normality for huge amounts of data, because it has to be normal. Furthermore, if the data is very small, most of the statistical tests may not yield proper results. However, we can validate the normality of the model (that is, of the residuals) through the Shapiro-Wilk normality test by using the `shapiro.test(residuals(model name))` command. The resultant output is as follows:

```
> shapiro.test(residuals(sal_model_1))

        Shapiro-Wilk normality test

data:  residuals(sal_model_1)
W = 0.9551, p-value = 0.06401
```

Here, the null hypothesis is that the normality assumption holds good. We reject the null hypothesis if the p-value is < 0.05. As the p-value is > 0.05, we cannot reject the null hypothesis that normality assumption holds good.

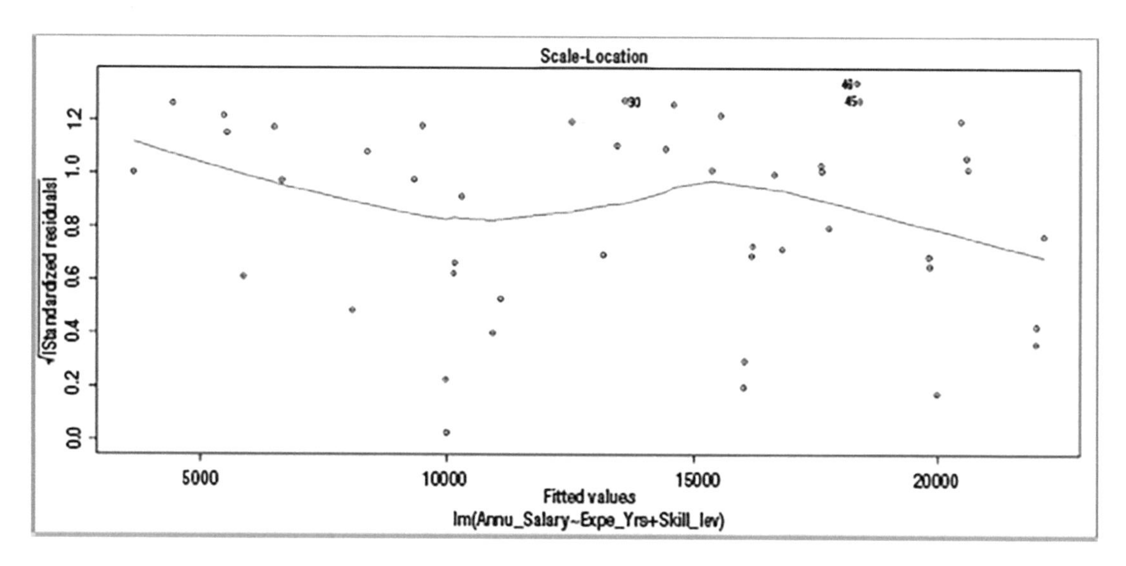

Figure 8-9. *Scale-location plot*

Further, the scale-location plot (Figure 8-9) shows the points distributed randomly around a near-horizontal line. Thus, the assumption of constant variance of the errors is fulfilled. We validate this understanding with `ncvTest(model name)` from the `car` library. Here, the null hypothesis is that there is constant error variance and the alternative hypothesis is that the error variance changes with the level of fitted values of the response variable or linear combination of predictors. As the p-value is > 0.05, we cannot reject the null hypothesis that there is constant error variance. See the following Figure.

```
> ncvTest(sal_model_1)
Non-constant Variance Score Test
Variance formula: ~ fitted.values
Chisquare = 0.1810408, Df = 1, p = 0.67048
```

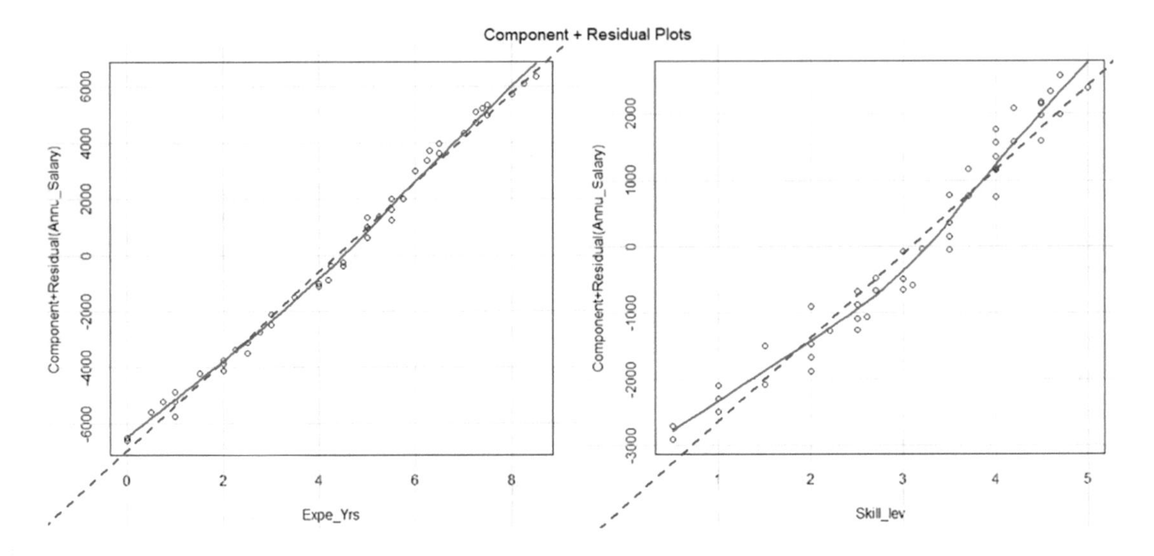

Figure 8-10. *crPlots(model name) plot showing the near linearity*

From the earlier discussion and the previous graphs that show near linearity, we can accept that the model `sal_model_1` fulfils the test of linearity.

Another way to visually confirm the assumption of normality (in addition to the discussion earlier in this regard) is by using `qqPlot(model name, simulate = TRUE, envelope = 0.95)`. Figure 8-11 shows the resultant plot. As you can see, most of the points are within the confidence interval of 0.95.

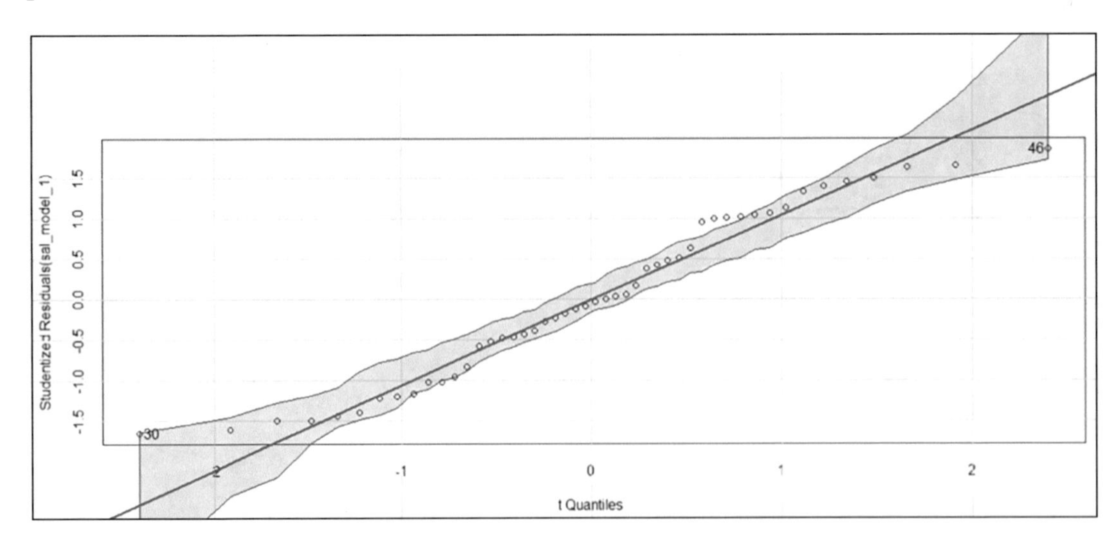

Figure 8-11. *Plot of normality generated through qqPlot(model name, simulate = TRUE, envelope = 0.95)*

Now, we need to validate the independence of the residuals or the independence of the errors, or lack of autocorrelation. However, in this case, because we know that each data entry belongs to a different employee and that no dependencies exist between the data of one employee and another, we can safely assume that there is independence of the residuals and independence of errors. From an academic-interest point of view, however, you can run a Durbin-Watson test to determine the existence of autocorrelation, or lack of independence of errors. The command in R to do this and the corresponding output is as follows:

```
> durbinWatsonTest(sal_model_1)
 lag Autocorrelation D-W Statistic p-value
   1       0.2987001      1.338564   0.012
 Alternative hypothesis: rho != 0
```

As the p-value is < 0.05, the Durbin-Watson test rejects the null hypothesis that there is no autocorrelation. Hence, the Durbin-Watson test holds up the alternate hypothesis that there exists autocorrelation, or the lack of independence of errors. However, as mentioned previously, we know that the data used does not lack independence. Hence, we can ignore the Durbin-Watson test.

As you can see, all the tests of the regression model assumptions are successful.

8.1.5 Multicollinearity

Multicollinearity is another problem that can happen with multiple linear regression methods. Say you have both date of birth and age as predictors. You know that both are the same in a way, or, in other words, that one is highly correlated with the other. If two predictor variables are highly correlated with each other in this way, there is no point in considering both of these predictors in a multiple linear regression equation. We usually eliminate one of these predictors from the multiple linear regression model or equation, because multicollinearity can adversely impact the model.

Multicollinearity can be determined in R easily by using the `vif(model name)` function. VIF stands for *variance inflation factor*. VIF for any two predictors is calculated using the simple formula, $VIF = 1/(1-R^2)$, where R is the correlation coefficient between these two predictors. The correlation coefficient R between the predictor `Expe_Yrs` and `Skill_lev` in our case is 0.8923255 (refer to Figure 8.3B); i.e., R^2 is 0.7962447979. So, VIF = 1/(1-0.7962447979) = ~ 4.91. Typically, for this test of multicollinearity to pass, the VIF value should be greater than 5. Figure 8-12 is the test showing the calculation of VIF.

```
> #Testing the Multicollinearity
> #We use the Variance Inflation Factor (VIF)
> vif_value <- vif(sal_model_1)
> #VIF value below 5 suggests lack of multicollinearity
> #Outputting the calculated VIF value
> vif_value
 Expe_Yrs Skill_lev
 4.907852  4.907852
```

Figure 8-12. *Variance inflation factor calculation in R*

As the VIF is less than 5, we can assume that there is no significant multicollinearity. However, we cannot rule out moderate multicollinearity.

Multicollinearity typically impacts the significance of one of the coefficients and makes it nonsignificant. However, in this case, we do not see such an impact. Instead, we see that the coefficients of both predictors are significant. Further, the existence of multicollinearity does not make a model not usable for prediction. Hence, we can use the preceding model for prediction even though we presume multicollinearity (according to some schools of statistical thought that suggest multicollinearity when the VIF value is greater than 4).

Figure 8-13A shows how to use the model to predict the Annual Salary for the new data of Experience in Years and Skill Level.

```
#Using the Model to predict on the new values of the predictors

#We first create a data frame with new predictor values

predictor_newdata <- data.frame(Expe_Yrs = 3, Skill_lev = 3)

#We will check the predicted value of the Annu_Salary using the Model

predict(sal_model_1, newdata=predictor_newdata)
```

Figure 8-13A. *The code for the prediction using the model on the new values of the predictors*

Figure 8-13B shows the output.

```
> predict(sal_model_1, newdata=predictor_newdata)
       1
11571.63
```

***Figure 8-13B.** Value predicted using the sal_model_1 on the predictor_newdata*

Now, let's take another set of values for the predictor variables and check what the model returns as Annual Salary. Figure 8-14 shows the code used and the prediction made by `sal_model_1`.

```
> #Predicting on another set of predictor values
> predictor_newdata1 <- data.frame(Expe_Yrs = 5, Skill_lev = 5)
> predict(sal_model_1, newdata=predictor_newdata1)
       1
17278.28
```

***Figure 8-14.** Value predicted using the sal_model_1 on the predictor_newdata1*

If we see significant multicollinearity between the predictor variables, one approach to building the model is to drop one of the predictor variables and proceed with the other. Similarly, you can try leaving out the other predictor and building the model with the first predictor. After both models are built, check which of these models is more significant statistically. One approach to compare the models is to compute the Akaike information criterion (AIC) value. AIC specifies how well a model fits the data from which it was generated. The AIC value is calculated using the number of predictor variables and also the estimate of the maximum likelihood of the model. Any model with a lesser AIC value is a more significant model. A lower AIC value means the complexity of the model is lower and the model better explains the variations.

In this case, if we eliminate one of the variables (say, Skill Level), we tend to get a regression of only one predictor variable (Experience in Years) against the response variable (Annual Salary), which is nothing but a case of simple linear regression. Figure 8-15A shows the code.

```
#Let us drop one of the predictor variable Skill_lev

#as it normally changes with the Expe_Yrs

#Let us now get the model with only Expe_Yrs as the predictor

sal_model_2 <- lm(Annu_Salary ~ Expe_Yrs, data = sal_data_1)

#We will now generate the summary of sal_model_2 model generated

summary(sal_model_2)
```

Figure 8-15A. *Code to generate the model after dropping Skill_lev and displaying the summary of the model*

Figure 8-15B shows the output.

```
> summary(sal_model_2)

Call:
lm(formula = Annu_Salary ~ Expe_Yrs, data = sal_data_1)

Residuals:
     Min       1Q   Median       3Q      Max
-1279.40  -591.36   -93.04   584.68  1796.49

Coefficients:
            Estimate Std. Error t value Pr(>|t|)
(Intercept)  4444.63     237.22   18.74   <2e-16 ***
Expe_Yrs     2151.78      47.68   45.13   <2e-16 ***
---
Signif. codes:  0 '***' 0.001 '**' 0.01 '*' 0.05 '.' 0.1 ' ' 1

Residual standard error: 783.9 on 46 degrees of freedom
Multiple R-squared:  0.9779,	Adjusted R-squared:  0.9774
F-statistic:  2037 on 1 and 46 DF,  p-value: < 2.2e-16
```

Figure 8-15B. *Summary of the model without Skill_lev as a predictor*

This is also a significant model with the response variable Annual Salary and the predictor variable Experience in Years, as the p-value of the model as well as the p-value of the predictor are less than 0.05. Further, the model explains about 97.74 percent of the variance in the response variable.

Alternatively, if we remove the Experience in Years predictor variable, we use the code in Figure 8-16A to get the model shown in Figure 8-16B.

```
#Let us now drop the other variable Expe_Yrs and
#build the model only with the predictor variable Skill_lev
sal_model_3 <- lm(Annu_Salary ~ Skill_lev, data = sal_data_1)
#We will now generate the summary of sal_model_3 model generated
summary(sal_model_3)
```

***Figure 8-16A.** Code to generate the model after dropping Expe_Yrs and displaying the summary of the model*

```
> summary(sal_model_3)

Call:
lm(formula = Annu_Salary ~ Skill_lev, data = sal_data_1)

Residuals:
    Min      1Q  Median      3Q     Max 
-5528.9  -597.1   505.6  1389.2  2416.5 

Coefficients:
            Estimate Std. Error t value Pr(>|t|)    
(Intercept)   1201.2      716.0   1.678      0.1    
Skill_lev     4109.2      217.1  18.930   <2e-16 ***
---
Signif. codes:  0 '***' 0.001 '**' 0.01 '*' 0.05 '.' 0.1 ' ' 1

Residual standard error: 1779 on 46 degrees of freedom
Multiple R-squared:  0.8862,	Adjusted R-squared:  0.8838 
F-statistic: 358.4 on 1 and 46 DF,  p-value: < 2.2e-16
```

***Figure 8-16B.** Summary of the model without Expe_Yrs as a predictor*

This is also a significant model with the response variable Annual Salary and the predictor variable Skill Level, as the p-value of the model as well as the p-value of the predictor are less than 0.05. Further, the model explains about 88 percent of the variance in the response variable as shown by the R-squared value.

However, when we have various models available for the same response variable with different predictor variables, one of the best ways to select the most useful model is to choose the model with the lowest AIC value. Here we have made a comparison of the AIC values of three models: `sal_model_1` with both Experience in Years and Skill Level as predictors, `sal_model_2` with only Experience in Years as the predictor variable, and `sal_model_3` with only Skill Level as the predictor variable. Figure 8-17 shows both the code and the output related to this.

```
> #Checking all the 3 models generated so far as to which is the
> #best model using the comparison of AIC value
> AIC(sal_model_1)
[1] 710.7369
> AIC(sal_model_2)
[1] 779.9501
> AIC(sal_model_3)
[1] 858.63
```

Figure 8-17. *Code for generation of the AIC values of the models along with the corresponding output*

If you compare the AIC values, you find that the model with both Experience in Years and Skill Level as predictors is the best model.

8.1.6 Stepwise Multiple Linear Regression

Other ways to build models and select the best model include carrying out a forward stepwise regression or a backward stepwise regression. In the forward stepwise regression, we start with one predictor and at each run keep on adding another predictor variable until we exhaust all the predictor variables. In the case of backward stepwise regression, we start with all the predictors and then start removing the parameters one by one. Both will yield the same result. Let's do a backward stepwise regression now. We will use AIC as shown previously to select the best model. The model with the lowest AIC is the model we'll select.

We need to have the package MASS and then use this library to run the `stepAIC(model name)` function. Figure 8-18A provides the code related to stepwise multiple linear regression model generation.

```
#Now, we will build the Multiple Linear Regression Stepwise

library(MASS)

sal_model <- lm(Annu_Salary ~ Expe_Yrs + Skill_lev, data = sal_data_1)

#First we will check from backward i.e., last model first

stepAIC(sal_model, direction = "backward")
```

Figure 8-18A. *Code showing the stepwise model generation*

Figure 8-18B shows the output.

```
> stepAIC(sal_model, direction = "backward")
Start:  AIC=572.52
Annu_Salary ~ Expe_Yrs + Skill_lev

            Df Sum of Sq       RSS    AIC
<none>                     6411961 572.52
- Skill_lev  1  21857249  28269210 641.73
- Expe_Yrs   1 139199116 145611077 720.41

Call:
lm(formula = Annu_Salary ~ Expe_Yrs + Skill_lev, data = sal_data_1)

Coefficients:
(Intercept)     Expe_Yrs    Skill_lev
       3012         1590         1264
```

Figure 8-18B. *Output showing the stepwise model generation*

This confirms our understanding as per the discussion in the prior section of this chapter. The downside for the effective use of this stepwise approach is that as it drops predictor variables one by one, it does not check for the different combinations of the predictor variables.

8.1.7 All Subsets Approach to Multiple Linear Regression

Another alternative is to use the all subsets regression approach, which evaluates all the possible subsets out of the predictors.

We need the leaps package and need to use the `regsubsets()` function. We have used the R-squared option in the plot function as follows to understand which of the predictors we need to include in our model. We should select the model that has the highest adjusted R-squared value. Figure 8-19A shows the code to do this.

```
#All Subsets approach to Multiple Linear Regression

library(leaps)

#We use regsubsets() function from the leaps library

#This is like lm() except that lm() is replaced here by regsubsets()

get_leaps <- regsubsets(Annu_Salary ~ Expe_Yrs + Skill_lev, data =
sal_data_1, nbest = 2)

#Best model is the one with the highest Adjusted R-Squared value or R-
Squared Value

#First, we will check the R-Squared Value

plot(get_leaps, scale = "r2")
```

Figure 8-19A. *Code for generating all subsets of multiple linear regression*

Figure 8-19B shows the output.

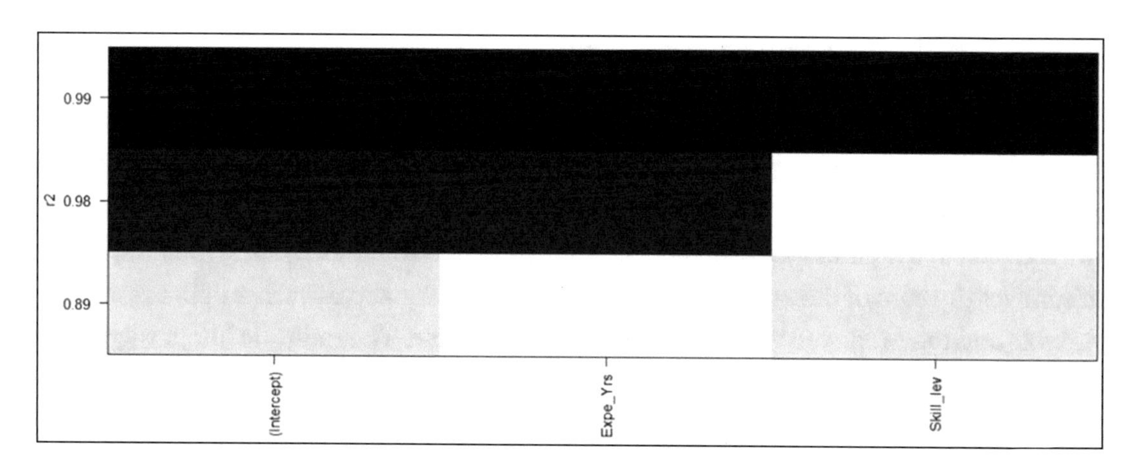

Figure 8-19B. *Plot of the multiple linear regression model generated in R using the all subsets approach scaled with R-squared*

If we use the adjusted R-squared value instead of the R-squared value, we get the plot shown in Figure 8-20. To get this, we use "adjr2" instead of "r2" in the code given in Figure 8-19A.

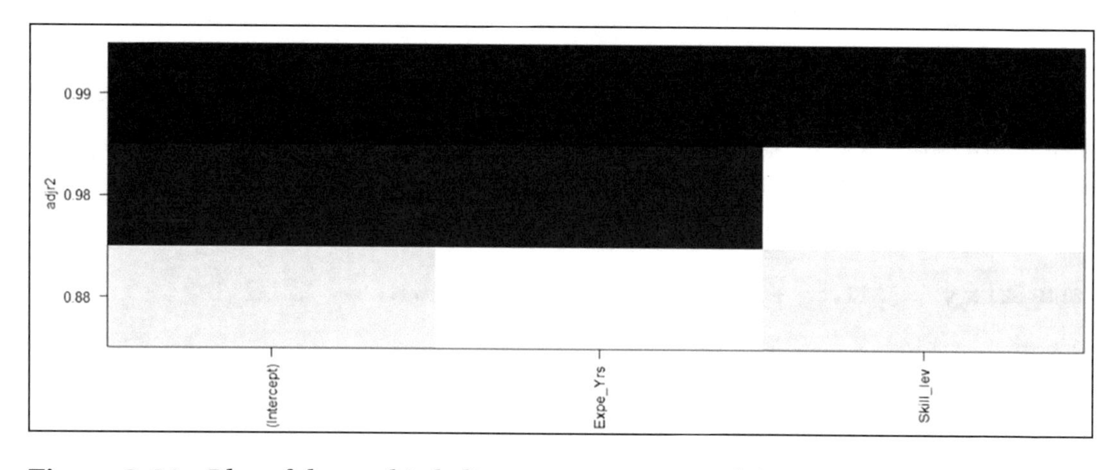

Figure 8-20. *Plot of the multiple linear regression model generated in R using the all subsets approach scaled with adjusted R-squared value*

As you can see, we select both the predictors Experience in Years and Skill Level, as this model has the highest R-squared value of 0.99 and also the one with the highest adjusted R-squared value. If the adjusted R-squared value is higher than the one with R-squared, we use the model recommended by it as it adjusts for the degrees of freedom.

In this example, the best model selected is the one with both the predictors.

8.1.8 Multiple Linear Regression Equation

The regression line is a smoothed graph of averages. It is basically a smoothing of the averages of the response variable for each value set of the independent variables. The regression line is drawn in such a way that it minimizes the error of the fitted values with respect to the actual values. This method, known as the *least squares method,* minimizes the sum of the squared differences between the actual values of the response variable and the predicted values of the response variable. The multiple linear regression equation takes the following form:

$$Y_i = \beta_0 + \beta_1 x_1 + \beta_2 x_2 + \beta_3 x_3 + \dots$$

In this equation, β_0 is known as the *intercept,* and β_i is known as the *slope* of the regression line. The intercept is the value of the response variable when the values of each independent variable are 0. This depicts the point at which the regression line touches the y-axis when x_1 and x_2, etc., are 0. The slope can be calculated easily by using this formula:

(R × Standard deviation of the response variable) / Standard deviation of the independent variable

From this formula, you can see that when the value of the independent variable increases by one standard deviation, the value of the response variable increases by R × one standard deviation of the response variable, where R is the coefficient of correlation.

In our example, the multiple linear regression equation is as follows:

```
Annu_Salary = 3011.66 + 1589.68 × Expe_Yrs + 1263.65 × Skill_lev
```

where β_0 = intercept = 3011.66

β_1 = coefficient of Experience in Years = 1589.68

β_2 = coefficient of Skill_levl = 1263.65

8.1.9 Conclusion

From our data, we got a good multiple linear regression model that we validated for the success of the assumptions. We also used this model for predictions.

8.2 Using an Alternative Method in R

As we mentioned in the introduction of this chapter, the expectation for the use of the `lm()` function is that the response variable is normally distributed. As we know from our sample data, Annual Salary is normally distributed, or follows Gaussian distribution. Hence, we can alternatively use the generalized linear model using the R function `glm()` with the frequency distribution as belonging to the `gaussian` family with the parameter `link = "identity"`. Figure 8-21A shows the code to generate this and to display the summary.

```
#Using alternative method glm() to generate the Multiple Linear
Regression Model

glm_sal_model <- glm(Annu_Salary ~ Expe_Yrs + Skill_lev, data =
sal_data_1, family = gaussian(link="identity"))

#Display the summary of glm_sal_model

summary(glm_sal_model)
```

Figure 8-21A. *Code to generate the multiple linear regression using the alternative method glm()*

Figure 8-21B shows the output.

```
> summary(glm_sal_model)

Call:
glm(formula = Annu_Salary ~ Expe_Yrs + Skill_lev, family = gaussian(link = "identity"),
    data = sal_data_1)

Deviance Residuals:
   Min       1Q  Median       3Q      Max
-605.6   -318.5   -23.8    354.3    666.0

Coefficients:
            Estimate Std. Error t value Pr(>|t|)
(Intercept)  3011.66     162.58   18.52  < 2e-16 ***
Expe_Yrs     1589.68      50.86   31.26  < 2e-16 ***
Skill_lev    1263.65     102.03   12.38 4.25e-16 ***
---
Signif. codes:  0 '***' 0.001 '**' 0.01 '*' 0.05 '.' 0.1 ' ' 1

(Dispersion parameter for gaussian family taken to be 142488)

    Null deviance: 1.280e+09  on 47  degrees of freedom
Residual deviance: 6.412e+06  on 45  degrees of freedom
AIC: 710.74

Number of Fisher Scoring iterations: 2
```

Figure 8-21B. *Output of the model generated using glm()*

As expected, you will find that the result is the same as that obtained through the **lm()** function.

8.3 Predicting the Response Variable

The fitted model not only explains the relationship between the response variable and the independent variables but also provides a mechanism to predict the value of the response variable from the values of the new independent variables. Figure 8-22 shows the code required to do this and the predicted value output.

```
> #Predicting the response variable on the new predictor variables using
> #glm_sal_model
> predictor <- data.frame(Expe_Yrs = 5, Skill_lev = 4)
> predicted <- predict(glm_sal_model, newdata=predictor)
> predicted
       1
16014.64
```

Figure 8-22. *Code to generate prediction using glm_sal_model and the predicted value output*

As you can see from the data used for building the model, the predicted value is almost in tune with the related actual values in our data set.

Please note: we have used `glm_sal_model` instead of `sal_model_1` generated using the `lm()` function, as there is no difference between the two.

This is done using the function `predict(model name, newdata)`, where `model name` is the name of the model arrived at from the input data, and `newdata` contains the data of independent variables for which the response variable has to be predicted.

This model may work in predicting the Annual Salary for Experience in Years and Skill Level beyond those in the data set. However, we have to be very cautious in using the model on such an extended data set, as the model generated does not know or has no means to know whether the model is suitable for the extended data set. We also do not know whether the extrapolated data follows the linear relationship. It is possible that after a certain number of years, while experience increases, the Skill Level and the corresponding Annual Salary may not go up linearly, but the rate of increase in salary may taper off, leading to a slowly tapering down in the slope of the relationship.

8.4 Training and Testing the Model

We can also use separate subsets of the data set for training the model (basically, to arrive at the model) and testing the model. Ideally, such data sets have to be generated randomly by splitting the data set. Typically, 75 percent to 80 percent of the data set is taken to train the model, and another 25 percent to 20 percent of the data set is used to test the model.

Let's use the same data set that we used previously to do this. For this, we have to use `install.packages("caret")` and `library(caret)`. Figure 8-23A shows the splitting of the data set into two subsets—training data set and test data set—using R code.

```
#Training and Testing the model by splitting the Original Data

#We use library(caret) for this purpose

library(caret)

#Setting the seed to ensure the repeatability of the results with
different runs

set.seed(45)

#Now we are going to partition the data into Training_Data and
Test_Data sets

Data_Partition <- createDataPartition(sal_data_1$Annu_Salary, p=0.80,
list=FALSE)

Train_Data <- sal_data_1[Data_Partition, ]

summary(Train_Data)

#Creating Test_Data set out of the remaining records from the original
data

Test_Data <- sal_data_1[-Data_Partition, ]

summary(Test_Data)
```

Figure 8-23A. *Code to split the data into two separate datframes, namely, one for training and another for testing*

Figure 8-23B shows the output of the previous code showing `summary(Train_Data)` and `summary(Test_Data)`.

```
> summary(Train_Data)
  Annu_Salary        Expe_Yrs        Skill_lev
 Min.   : 4000   Min.   :0.000   Min.   :0.50
 1st Qu.:10000   1st Qu.:2.500   1st Qu.:2.50
 Median :14500   Median :4.750   Median :3.35
 Mean   :13975   Mean   :4.406   Mean   :3.15
 3rd Qu.:18000   3rd Qu.:6.263   3rd Qu.:4.00
 Max.   :22000   Max.   :8.250   Max.   :5.00
> #Creating Test_Data set out of the remaining records from the original data
> Test_Data <- sal_data_1[-Data_Partition, ]
> summary(Test_Data)
  Annu_Salary        Expe_Yrs        Skill_lev
 Min.   : 5000   Min.   :0.500   Min.   :0.500
 1st Qu.: 6750   1st Qu.:1.000   1st Qu.:1.375
 Median :14000   Median :4.875   Median :2.850
 Mean   :13250   Mean   :4.206   Mean   :2.725
 3rd Qu.:18000   3rd Qu.:5.975   3rd Qu.:4.125
 Max.   :22000   Max.   :8.500   Max.   :4.600
```

Figure 8-23B. *Output showing the summary(Train_Data) and summary(Test_Data) of the two separate dataframes created*

This split of the entire data set into two subsets, `Train_Data` and `Test_Data`, has been done randomly. Now, we have 40 records in `Train_Data` and 8 records in `Test_Data`.

We will now train our `Train_Data` using the machine-learning concepts to generate a model. Figure 8-24A provides the code.

```
#Now we will generate the model using only the Train_Data

Trained_Model <- train(Annu_Salary ~ Expe_Yrs + Skill_lev, data =
Train_Data, method = "lm")

summary(Trained_Model)
```

Figure 8-24A. *Model generation using the Train_Data dataframe*

Figure 8-24B shows the output.

```
> summary(Trained_Model)

Call:
lm(formula = .outcome ~ ., data = dat)

Residuals:
    Min      1Q  Median      3Q     Max 
-579.94 -288.70  -29.87  338.51  644.57 

Coefficients:
            Estimate Std. Error t value Pr(>|t|)    
(Intercept)  2881.98     190.75   15.11  < 2e-16 ***
Expe_Yrs     1606.49      53.22   30.19  < 2e-16 ***
Skill_lev    1274.42     109.45   11.64 6.19e-14 ***
---
Signif. codes:  0 '***' 0.001 '**' 0.01 '*' 0.05 '.' 0.1 ' ' 1

Residual standard error: 372.4 on 37 degrees of freedom
Multiple R-squared:  0.9947,	Adjusted R-squared:  0.9944 
F-statistic:  3453 on 2 and 37 DF,  p-value: < 2.2e-16
```

Figure 8-24B. *Summary of the Trained_Model output*

As you can see, the model generated, `Trained_Model`, is a good fit with the p-values of the coefficients of the predictors being < 0.05 as well as the overall model p-value being < 0.05.

We will now use the `Trained_Model` arrived at previously to predict the values of `Annual_Salary` in respect to Experience in Years and Skill Level from the `Test_Data`. Figure 8-25A shows the code.

```
#Now we will use the model to predict the Annu_Salary on the
Test_Data

predicted_Annu_Salary <- predict(Trained_Model, newdata =
Test_Data, interval = "prediction")

summary(predicted_Annu_Salary)
```

Figure 8-25A. *Predicting on the Test_Data earmarked for testing*

Figure 8-25B shows the output of the previous code showing the summary of the predicted values. The output shows the minimum, maximum, first quartile, median, mean, and third quartile of the predictions made.

```
> summary(predicted_Annu_Salary)
  Min. 1st Qu.  Median    Mean 3rd Qu.    Max.
  4322    6241   14346   13112   17770   22272
```

Figure 8-25B. *The output showing the summary of the predicted_Annu_Salary, which is the prediction on the Test_Data*

Now, let's see whether the values of Annual Salary we predicted using the model generated out of `Training_Data` and actual values of the Annual Salary in the `Test_Data` are in tune with each other. Figure 8-26 shows the code and the output showing actual values and predicted values.

```
> #Now, let us check the values predicted against the actual Annu_Salary from Test_Data
> Test_Data$Annu_Salary
[1]  6000 13000  7000 17000 21000 22000 15000  5000
> predicted_Annu_Salary
        2        17        18        23        29        32        43        47
 5762.892 13150.495  6400.105 16815.356 20632.335 22272.027 15540.931  4322.437
```

Figure 8-26. *Shows both the Actual Annu_Salary and the Predicted Annu_Salary for comparison*

You can see here that both the Actual Annual Salary from `Test_Data` and the Annual Salary from the model generated out of `Training_Data`, i.e., `predicted_Annu_Salary`, match closely with each other. Hence, the model generated out of `Training_Data`, i.e., `Trained_Model`, can be used effectively.

8.5 Cross Validation

Cross validation like k-fold cross validation is used to validate the model generated. This will be useful when we have limited data and the data is required to be split into a small test set (like 20 percent of data) and a relatively bigger training set (like 80 percent of data), and this makes the validation of the model relatively difficult or not feasible because the data in the training set and test set may be split in such a way that both may not be very representative of the entire data set.

This problem is eliminated by cross validation methods like k-fold cross validation. Here, we actually divide the data set into k-folds and use k-1 sets as training data and the remaining 1 set as the test data and repeatedly do this for k times. Each k-fold will have almost an equal number of data points (depends upon k-value) randomly drawn from the entire data set. In our case, as we have 48 records in total for k=10, each fold has 4 or 5 records. None of the elements of one fold is repeated in the other fold. For each fold, the model is validated, and the value predicted by the model (`Predicted`) and by the cross validation is tabulated as the output along with the difference between actual value and the cross validated prediction (`cvpred`) as CV residual. Using the difference between the `Predicted` values and the `cvpred`, we can arrive at the root mean square error of the fit of the linear regression model. This cross validation method is also useful in the case of a lot of data as every data point is used for the training as well as for testing by rotation and in none of the folds the same data point is taken again for consideration.

Let's look at the example of the multiple linear regression we used earlier. Let's validate this model using the k-fold validation. In our example, we will be using K = 10. This is for the sake of convenience so that every time we have 90 percent of the data in the training set and another 10 percent of the data in the test set. This way, the data once in the training set will move some other time to the test set. Thus, in fact, by rotating, we will ensure that all the points are used for model generation as well as the testing. For this cross validation, we will use `library(DAAG)` and use either the `cv.lm(data set, formula for the model, m=number of folds)` or `CVlm(data set, formula for the model, m=number of folds)` function. Here `m` is the number of folds (i.e., k) we have decided on. Figure 8-27A shows the code to generate the `cv_model`.

```
#Cross Validation

library(DAAG)

cv_Model <- CVlm(sal_data_1, Annu_Salary ~ Expe_Yrs + Skill_lev, m =
10)

plot(cv_Model)
```

Figure 8-27A. *Code to carry out cross validation and to plot the model generated*

Figure 8-27B shows the output generated by the previous code.

```
> cv_Model <- CVlm(sal_data_1, Annu_Salary ~ Expe_Yrs + Skill_lev, m = 10)


fold 1
Observations in test set: 4
                      16         32         34         46
Predicted    10939.80927 22210.3309 16177.6508 18333.9463
cvpred       10932.31932 22208.6958 16170.9502 18319.6737
Annu_Salary  11000.00000 22000.0000 16000.0000 19000.0000
CV residual     67.68068  -208.6958  -170.9502   680.3263

Sum of squares = 540202.5    Mean square = 135050.6    n = 4

fold 2
Observations in test set: 5
                      6         21         24         27          33
Predicted    14587.9734 13450.6901 20620.6536 15382.812 16032.96106
cvpred       14613.1484 13473.9446 20637.4377 15406.233 16055.93163
Annu_Salary  14000.0000 13000.0000 21000.0000 15000.000 16000.00000
CV residual   -613.1484  -473.9446   362.5623  -406.233   -55.93163

Sum of squares = 900179.4    Mean square = 180035.9    n = 5
```

Figure 8-27B. *Partial view of the k-fold cross validation*

The previous code carves out 10 (as m=10) separate data sets from the original data. As you can see, in each data set carved out there will be either 4 or 5 rows (as we have only 48 records in total in our original data set). Each fold shows the number of observations in the set. As you can see in Figure 8-27B, fold 1 has four observations, and fold 2 has five observations. The other eight folds are not shown in Figure 8-27B in order to avoid the clutter. The index of the rows selected for each fold is also shown. The output also shows, for each fold, the `Predicted` value, which shows the value predicted using all the observations; the `cvpred` value, which shows the value predicted using cross validation; the actual response variable value, `CV Residual`, which is the difference between the `cvpred` value and the actual value of the response variable. At the end, the sum of squares error and the mean square error of the cross validation residuals are also shown.

Figure 8-27C shows the plot generated by `plot(cv_Model)`.

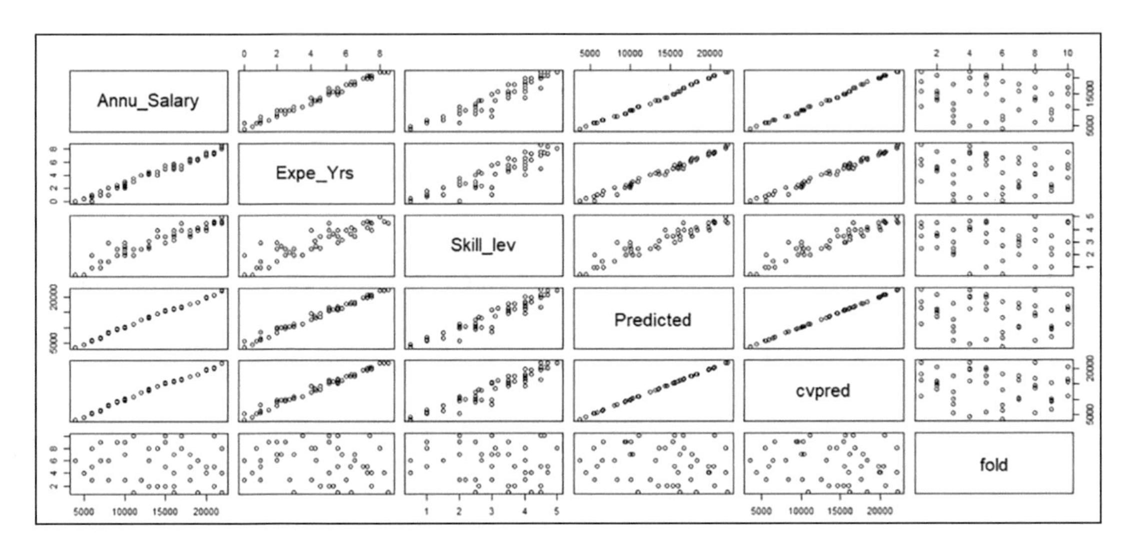

Figure 8-27C. *Plot of the cross validation results*

The values `Predicted` (predicted using all the values) and `cvpred` (predicted using each fold) are almost similar as is evident from the previous plot. Also, we can see from Figure 8-27B, the predictions made by `cv_Model` (i.e., cvpred) are almost similar to the Predicted values as well as the actual values of Annu_Salary. Now, let's check the root mean square error between the model generated using the linear regression model and the cross validation model. Figure 8-28 shows the code and output.

```
> #Calculating the Root Mean Square Error of the cv_Model
> rmse <- sqrt(mean(cv_Model$cvpred - cv_Model$Predicted)^2)
> rmse
[1] 0.2554861
```

Figure 8-28. *Root mean square error calculated between linear regression model and cross validation model*

If we check the root mean square error between the values predicted by multiple linear regression models and cross validation models, as shown in Figure 8-28, it is very negligible, i.e., 0.25. Hence, k-fold cross validation validates the multiple linear regression model arrived at earlier.

Instead of k-fold cross validation, we can use other methods like repeated k-fold cross validation, bootstrap sampling, and leave-one-out cross validation.

Note We need to use the `set.seed()` command when carrying out the model generation or splitting the data so that the result for each run is consistent.

8.6 Using Python to Generate the Model and Validating the Assumptions

In the previous sections of this chapter, we used the R programming language to generate the model, predict the response variable for the new predictor values, and also validate the regression assumptions.

Here, we are going to use Python and the packages available within the Anaconda framework like Jupyter Notebook, scikit-learn, statsmodels, pandas, NumPy, SciPy, etc. (as relevant), to generate the model, make the predictions using the generated model, and also validate the regression assumptions. To keep it simple and to explain the details in the context of each input and output, the comments are embedded along with the code instead of writing the content separately. The extract from the Jupyter Notebook used are provided in the following sections in logical pieces.

8.6.1 Load the Necessary Packages and Import the Data

First, the necessary packages are loaded into the memory using the `import` statements. There is no specific output from the code segment in Figure 8-29.

```
#import all the necessary packages

import numpy as np

import pandas as pd

import sklearn
```

Figure 8-29. *Code to load necessary packages*

Figure 8-30A shows the code to read the data.

```
#import the text file with data from the local machine

#the data is ',' separated

#this is a text file with headers and hence the header row is specified

df = pd.read_csv('E:/Book_Revision/sal1.txt', sep=',', header=0)

print(df)
```

Figure 8-30A. *Code to import the data into a dataframe from the text file*

Figure 8-30B shows the output of the code to import the data, i.e., the listing of the data imported.

```
    Annu_Salary  Expe_Yrs  Skill_lev
0          4000      0.00        0.5
1          6000      1.00        1.0
2          8000      2.00        1.5
3         10000      3.00        2.0
4         12000      4.00        2.5
5         14000      4.50        3.5
6         16000      5.00        4.0
7         14000      4.00        4.0
8         18000      6.00        4.0
9         20000      7.00        4.5
10        22000      8.00        5.0
11        10000      2.00        3.0
12         9000      2.00        2.5
13         9000      2.50        2.0
14         8000      1.00        3.0
15        11000      3.00        2.5
16        13000      4.25        2.7
17         7000      1.00        1.5
18         7000      1.50        1.0
```

Figure 8-30B. *Partial view of the of the data of the dataframe created*

In the previous dataframe we have 48 records with one response variable and two predictor variables (only the partial view is shown here).

8.6.2 Generate Multiple Linear Regression Model

Here, we will create the multiple linear regression model using the statsmodels package/tool. Figure 8-31A provides the code used to generate the multiple linear regression model and print the summary.

```
#Thanks to: Seabold, Skipper, and Josef Perktold. "statsmodels: Econometric and statistical

#modeling with python." Proceedings of the 9th Python in Science Conference. 2010.

#Generate multiple linear regression model using statsmodels package

from statsmodels.formula.api import ols

#fit Multiple linear regression model

Mul_Line_Reg_1 = ols('Annu_Salary ~ Expe_Yrs + Skill_lev', data=df).fit()

#print model summary

print(Mul_Line_Reg_1.summary())
```

Figure 8-31A. *Code to generate the multiple linear regression model using statsmodels and to print the summary of the model created*

Figure 8-31B shows the output of the previous code.

```
                            OLS Regression Results
==============================================================================
Dep. Variable:            Annu_Salary   R-squared:                       0.995
Model:                            OLS   Adj. R-squared:                  0.995
Method:                 Least Squares   F-statistic:                     4469.
Date:                Wed, 10 Aug 2022   Prob (F-statistic):           1.76e-52
Time:                        11:26:15   Log-Likelihood:                -351.37
No. Observations:                  48   AIC:                             708.7
Df Residuals:                      45   BIC:                             714.4
Df Model:                           2
Covariance Type:            nonrobust
==============================================================================
                 coef    std err          t      P>|t|      [0.025      0.975]
------------------------------------------------------------------------------
Intercept   3011.6568    162.584     18.524      0.000    2684.195    3339.119
Expe_Yrs    1589.6773     50.860     31.256      0.000    1487.239    1692.115
Skill_lev   1263.6482    102.028     12.385      0.000    1058.154    1469.142
==============================================================================
Omnibus:                        8.967   Durbin-Watson:                   1.339
Prob(Omnibus):                  0.011   Jarque-Bera (JB):                2.649
Skew:                           0.053   Prob(JB):                        0.266
Kurtosis:                       1.854   Cond. No.                         20.1
==============================================================================
```

Figure 8-31B. *Output of the OLS regression results*

Here is the interpretation of the model output: The intercept value is 3011.6568, the coefficient of Expe_Yrs is 1589.6773, and that of Skill_lev is 1263.6482. As you can see, the P>|t| value is 0.000 in the case of both the predictors, i.e., `Expe_Yrs` and `Skill_lev`, which is less than the significance level of 0.05. This informs us that both the predictor variables are significant to the model. Further, the Prob (F-statistic) value (i.e., 1.76e-52) is very low and is less than the significance level of 0.05. Moreover, both the R-squared and adjusted R-squared values are 0.995, which signifies significant explanation of the model by the predictors. Hence, the model generated is significant. The AIC value of 708.7 is also significantly low. The Durbin-Watson value shown in the model is 1.339, which shows minimal auto correlation (a value between ~1.5 to 2.5 shows no autocorrelation). The Jarque-Bera test checks for the normality of the data. The null hypothesis here is that the data is normally distributed. The Prob(JB) value is 0.266, which is not significant. Hence, we cannot reject the null hypothesis. This means that it confirms the normality of the data.

8.6.3 Alternative Way to Generate the Model

There are many alternative ways you can generate a multiple linear regression model. You have already seen one of the ways in the previous section. Here, we will walk you through another way.

We first create two separate dataframes, one with only the data of the response variable and another with only the data of the predictor variables.

Figure 8-32A shows the code to create a separate dataframe with only the response variable.

```
#The following section provides another way of generating the model

#separating the response variable data to a separate dataframe

y = df["Annu_Salary"]

#displaying top 5 values from this dataframe

y.head(5)
```

Figure 8-32A. *Code to create a separate dataframe y with only the response variable data*

The check for whether the dataframe has been created properly is done through `y.head(5)`, given earlier. Figure 8-32B shows the output.

```
0      4000
1      6000
2      8000
3     10000
4     12000
Name: Annu_Salary, dtype: int64
```

Figure 8-32B. *Output: top 5 records of the dataframe y*

Now, we will create another dataframe with only the predictor variable data. Figure 8-33A shows the code.

```
#other than the Annu_Salary other predictor data is carried over to a separate dataframe

X = df.drop(['Annu_Salary'], axis=1)

print(X.dtypes)

#displaying top 5 values from this dataframe

X.head(5)
```

***Figure 8-33A.** The code to create a separate dataframe X with only the predictor variable data*

In the previous code, `X.head(5)` is added to check if the dataframe has been populated properly. The output of the code given in Figure 8-33A is provided in Figure 8-33B.

```
Expe_Yrs      float64
Skill_lev     float64
dtype: object
Out[6]:
```

	Expe_Yrs	Skill_lev
0	0.0	0.5
1	1.0	1.0
2	2.0	1.5
3	3.0	2.0
4	4.0	2.5

***Figure 8-33B.** Output of the code given in Figure 8-33A*

In the following section, we have split the data set into two separate data sets, i.e., the training data set and test data set, and have generated the multiple linear regression model using the scikit-learn package. Even the `train_test_split` has been accomplished using the scikit-learn package. Figure 8-34A shows the code.

```
#The data is split as training set and test set

#There are multiple means - Here we are doing it manually

train_samples = (35)

from sklearn.model_selection import train_test_split

X_train, X_test, y_train, y_test = train_test_split(X, y,
train_size=train_samples, test_size=12)

#importing necessary packages

from sklearn.linear_model import LinearRegression

#running the LinearRegression

Mul_Line_Reg = LinearRegression()

Mul_Line_Reg.fit(X_train, y_train)

#Mul_Line_Reg is the model generated
```

Figure 8-34A. *Code to carry out the split of the dataframes into separate training and test datasets (both X, y) and to generate the model using the training data*

Figure 8-34B shows the output.

```
LinearRegression(copy_X=True, fit_intercept=True, n
_jobs=1, normalize=False)
```

Figure 8-34B. *Output of the code given in Figure 8-34A*

The previous output only shows that a linear regression model has been generated.

In the following section, we have checked for the coefficient values of the predictors, model accuracy based on the test data set, and also the mean square error on the predictions from the test data set. Figure 8-35 provides the code and output.

```
#Code:
#print regression coefficients
print(Mul_Line_Reg.coef_)
Output: [ 1628.07306079  1176.49916823]

#Code:
#print the regression score using the test data
#this provides the regression model accuracy
Mul_Line_Reg.score(X_test, y_test)
Output: 0.99391939705948518
#Code
#the above result shows an accuracy of 99.39%
#finding the mean square error on the test predictions
np.mean((Mul_Line_Reg.predict(X_test)-y_test)**2)

Output: 158940.20463956954
```

Figure 8-35. *Code and output showing the coefficients of the model, the accuracy of the model, and the mean square error on the test predictions*

In the following section, we have predicted the values of the response variable on the predictors from the test data set. This depicts high agreement between the predicted and actual values. Of course, there are some residuals (i.e., difference between the fitted and actual values). Figure 8-36A shows the code.

```
#the above shows very less overall error in the prediction

#run the model on the test data to arrive at the predicted values

predicted = Mul_Line_Reg.predict(X_test)

#printing the predicted and actual response values

print(predicted)

print(y_test)
```

Figure 8-36A. *Code to predict on the X_test data using the model trained on training data*

Figure 8-36B shows the output.

```
[ 6527.4644841   22267.50994474  20639.43688395  12588.18283471
  15618.54201002  10371.8601898   22041.72299846  11185.8967202
  18423.11423904  16260.86010886  15392.75506373   8155.53754489]
17     7000
31    22000
23    21000
4     12000
25    15000
3     10000
10    22000
19    11000
44    19000
34    16000
43    15000
2      8000
Name: Annu_Salary, dtype: int64
```

Figure 8-36B. *Output of the prediction on the X_test and actual values from y_test*

8.6.4 Validating the Assumptions of Linear Regression

We now need to validate that the model generated fulfils the assumptions of the linear regression. We will take you through how we do this along with the detailed explanations embedded within the portions of the Jupyter Notebook. Please note that we have used only a portion of the dataset in most places, i.e., the test dataset, to validate the assumptions. Ideally, you should use the model to carry out the prediction on the entire dataset and use the entire dataset to validate the assumptions.

To enable us to validate the assumptions at first, we will create a scatter plot between the fitted values and the residuals. Figure 8-37A shows the code.

```
#creating the scatter plot between response variable
#actual values and predicted values
import matplotlib.pyplot as plt
plt.scatter(predicted, (predicted-y_test))
plt.xlabel("fitted")
plt.ylabel("residuals")
plt.show()
```

***Figure 8-37A.** Code to create a scatter plot of fitted values to the residuals*

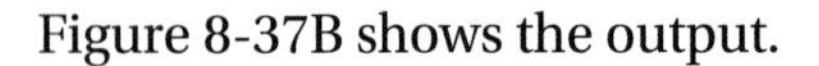

Figure 8-37B shows the output.

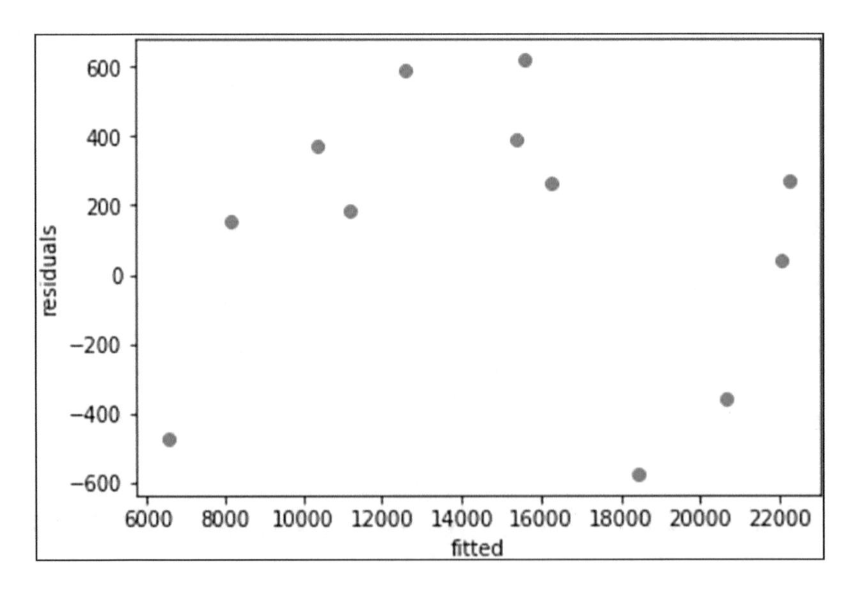

***Figure 8-37B.** Output: scatter plot of fitted values to the residuals*

- *Regression Assumption: Linearity: Validation*: As you can observe in the previous graph, the residuals are very small compared to the actual values. Further, residuals are randomly spread around the

0.00 residuals. Hence, this depicts that the linearity assumption is validated.

- *Regression Assumption: Test of equal variance of the distribution of the response variable: Validation*: As we can observe from the previous plot, there is no significant difference in variation of the residuals for the different levels of the independent variable's values. Hence, we consider that the test of equal variance of the distribution of the response variable as validated.
- *Regression Assumption: No autocorrelation: Validation*: The Durbin-Watson test in the model shows the value of 1.339. This indicates minor positive autocorrelation. Any value between 1.50 to 2.50 for this test shows no autocorrelation. However, from the data collection process we know that each data pertains to different people. Hence, there is no possibility of autocorrelation, and we consider the test of no autocorrelation validated.

We will now plot a scatter plot of ordered residuals against the theoretical quantiles using the SciPy package. This will help us to understand the normality of the distribution of the residuals. Figure 8-38A shows the code.

```
#Validating the regression assumption: Normality

#plotting for the validaton of normal distribution of the residuals

import scipy

from scipy import stats

stats.probplot(y_test-predicted, plot=plt)

plt.show()
```

Figure 8-38A. *Code to plot ordered residuals against the theoretical quantiles to validity the assumption of normality*

Figure 8-38B shows the output.

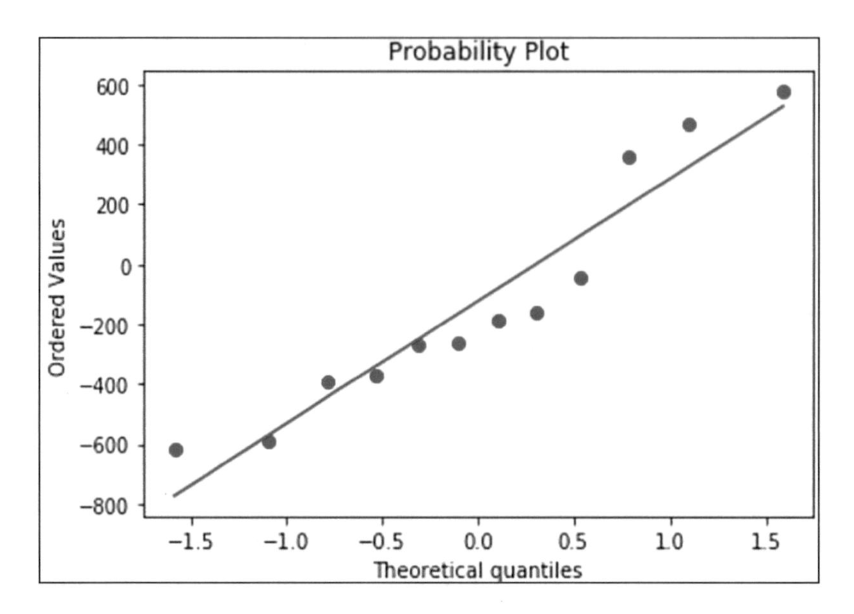

Figure 8-38B. *The plot of the ordered values of the residuals against the theoretical quantiles for validating the assumption of the normality*

- *Regression Assumption: Normality: Validation*: As most of the residuals are almost on the 45-degree line, we can assume that the test of the normality is passed. Further, as we discussed as part of the model output interpretation, the Prob(Jarque-Bera) is more than the significance level of 0.05, and hence we cannot reject the null hypothesis that the data is normally distributed.

We will also test the assumption of the linearity in another way. For this purpose, we will use the Seaborn package to create a `pairplot()`, plotting the relationship between each pair of the variables. Figure 8-39A shows the code.

```
#Alternative way of validating the linear relationship assumption

#plot the relationship between the data elements in the dataframe

#We use the seaborn package here

import seaborn as sns

sns.pairplot(df)

plt.draw()

plt.show()
```

Figure 8-39A. *Code to create a pairplot() on the original dataframe showing the relationship between each pair of the variables*

Figure 8-39B shows the output.

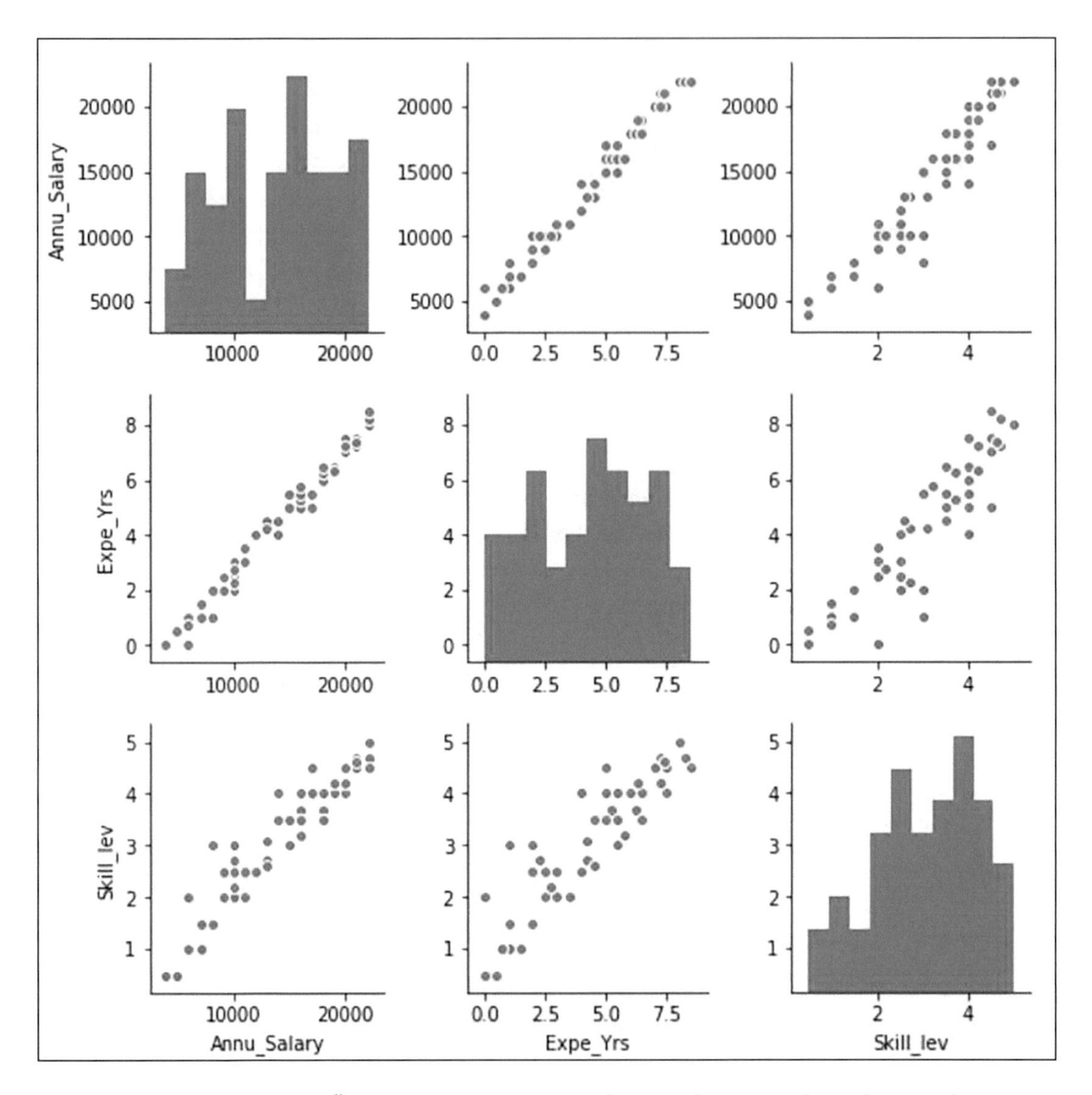

Figure 8-39B. *pairplot() on the original dataframe showing the relationship between each pair of the variables*

From the previous graph, we can confirm that there is almost a linear relationship between each variable to the other variable. Hence, we can conclude that the assumption of linearity holds.

As you can see from the previous plot, all the assumptions of the linear regression are validated. Hence, we can use the model for the predictions.

8.6.5 Predict Using the Model Generated

Here, we use the model generated to predict the response variable in the case of a set of predictor variables. Figure 8-40A shows the code used in this regard to get the new values of the predictors into a new dataframe and to predict using the multiple linear regression model already created, i.e., `Mul_Line_Reg`.

```
#Now let us create a df with predictors to check that
#the model predicts the values accurately (even though
#we have tested the model against the test data
data = {'col1': [3.0, 5.0], 'col2': [2.0, 4.0]}
df_new_predictors = pd.DataFrame(data=data)
predicted_resp_variable = Mul_Line_Reg.predict(df_new_predictors)
print(predicted_resp_variable)
```

Figure 8-40A. *Code to create a new dataframe with the new predictor values and to predict using these values as inputs*

Figure 8-40B shows the output.

```
[ 10371.8601898   15981.00464785]
```

Figure 8-40B. *Predicted Response Variable Values*

You can observe from the previous listed output that the predicted values are very close to the actual values from the initial data set, i.e., 10000 and 16000, respectively.

8.7 Chapter Summary

In this chapter, you saw examples of multiple linear relationships. When you have a response variable that is continuous and is normally distributed with multiple predictor variables, you use multiple linear regression. If a response variable can take discrete values, you use logistic regression.

You also briefly looked into significant interaction, which occurs when the outcome is impacted by one of the predictors based on the value of the other predictor. You learned about multicollinearity, whereby two or more predictors may be highly correlated or may represent the same aspect in different ways. You saw the impacts of multicollinearity and how to handle them in the context of the multiple linear regression model or equation.

You then explored the data we took from one of the entities and the correlation between various variables. You saw a high correlation among all three variables involved (the response variable as well as the predictor variables). By using this data in R, you learned how to arrive at the multiple linear regression model and to validate if it is a good model with significant predictor variables.

You learned various techniques to validate that the assumptions of the regression are met. Different approaches can lead to different interpretations, so you have to proceed cautiously in that regard.

In exploring multicollinearity further, you saw that functions such as `vif()` enable you to understand the existence of multicollinearity. You briefly looked at handling multicollinearity in the context of multiple linear regression. Multicollinearity does not reduce the value of the model for prediction purposes. Through the example of Akaike information criterion (AIC), you learned to compare various models and that the best one typically has the lowest AIC value.

You then explored two alternative ways to arrive at the best-fit multiple linear regression models: stepwise multiple linear regression and the all subsets approach to multiple linear regression. You depicted the model by using the multiple linear regression equation.

Further, you explored how the `glm()` function with the frequency distribution `gaussian` with `link = "identity"` can provide the same model as that generated through the `lm()` function, as we require normality in the case of a continuous response variable.

Further, you saw how to predict the value of the response variable by using the values of the predictor variables.

We also made you explore how to split the data set into two subsets, training data and test data. You learned how to use the training data to generate a model for validating the response variable of test data, by using the `predict()` function on the model generated using the training data.

We finally demonstrated to you how to carry on all those analyses we carried out using R in Python also.

CHAPTER 9

Classification

In this chapter, we will focus on classification techniques. *Classification is the task of predicting the value of a categorical variable.* A *classification model* predicts the category of class, whereas *regression models* predict a continuous value. In this chapter, we will focus on classification techniques used in predicting a class based on a learning algorithm. We will explain some classification methods such as naïve Bayes, decision trees, etc. We will demonstrate the classification techniques using R libraries, and then we will use Python library packages to perform the same tasks.

9.1 What Are Classification and Prediction?

Classification is a data mining technique used to predict the dependent variables, also referred to as the target variable, given a set of independent variables in a data set. The classification algorithm learns the data from the training data and predicts the class of the new data, as shown in Figure 9-1.

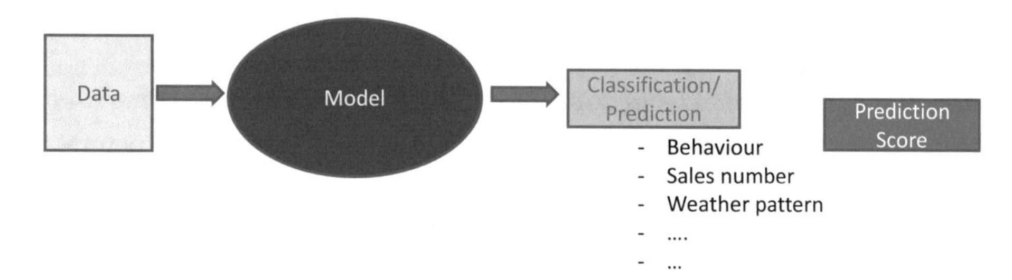

Figure 9-1. *Classification and prediction model*

U. R. Hodeghatta and U. Nayak, *Practical Business Analytics Using R and Python*,
https://doi.org/10.1007/978-1-4842-8754-5_9

Assume that you are applying for a mortgage loan. You fill out a long application form with all your personal details, including income, age, qualification, location of the house, valuation of the house, and more. You are anxiously waiting for the bank's decision on whether your loan application has been approved. The bank has to make a decision about whether the loan should be approved or not. How does the bank decide? The bank reviews various parameters provided in your application form, and then—based on similar applications received previously and the experience the bank has had with those customers—the bank decides whether the loan should be approved or denied. The bank may be able to review and classify ten or twenty applications in a day but may find it extremely difficult if it receives 1000s of applications at a time. Instead, you can use classification algorithms to perform this task. Similarly, assume that a company has to make a decision about launching a new product in the market. Product performance depends on various parameters. This decision may be based on similar experiences a company has had launching similar products in the past, based on numerous market parameters.

There are numerous examples of classification tasks in business; some examples include the following:

- Classifying a specific credit card transaction as legitimate or fraudulent
- Classifying a home loan to be approved or not
- Classifying a tumor as benign or malignant
- Classifying an email message as spam or not spam

Prediction is the same as classification except that the results you are predicting are representation of the future. Examples of prediction tasks in business include the following:

- Predicting the value of a stock price for the next three months
- Predicting weather for tomorrow's golf game
- Predicting which basketball team will win this year's game based on the past data
- Predicting the percentage decrease in traffic deaths next year if the speed limit is reduced

Classification modeling is a *supervised* machine learning model. In supervised machine learning, a machine creates a model with the help of a training data set. The training data set consists of data that has class labels prepared through a manual process by an expert of that particular domain, as shown in Figure 9-2. The model is based on an iterative process. The learning stops when the algorithm achieves an acceptable level of performance.

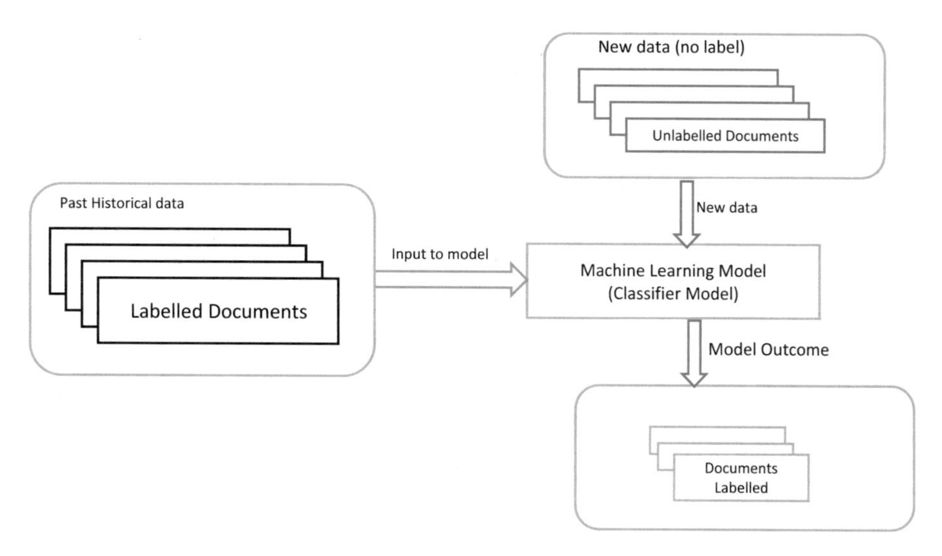

***Figure 9-2.** Supervised machine learning*

Classification is a two-step process. In the first step, a model is constructed by analyzing the data and the set of attributes (X) that define the class variable Y. The class Y is determined by a set of input variables, $\{x_1, x_2, x_3, \ldots\}$, and a set of classes, $\{c_1, c_2, c_3, \ldots\}$. The task of the classification model is to determine the relationship between the class variable Y and the input variables. Once this relationship is determined, the second step is to use the model to predict the class of the new data for which the class is unknown.

In the example, as shown in Figure 9-3, the classifier model is to predict the class Yes or No for an unknown data set. The training data already has a class label, which is manually annotated into different classes. The classification algorithm learns data patterns and characteristics and creates a model. The model once created is able to classify and predict the class of any new data whose labels are unknown. Common classification algorithms include logistic regression, decision tree, naïve Bayes, support vector machines, and k-nearest neighbor. The selection of the algorithm is based on the input characteristics of the data. Each algorithm's performance may vary. Though there are certain guidelines for selecting the algorithms given by different scholars and data scientists, there is no standard method. We recommend you try different algorithms and select the best model and algorithm that provides the best accuracy.

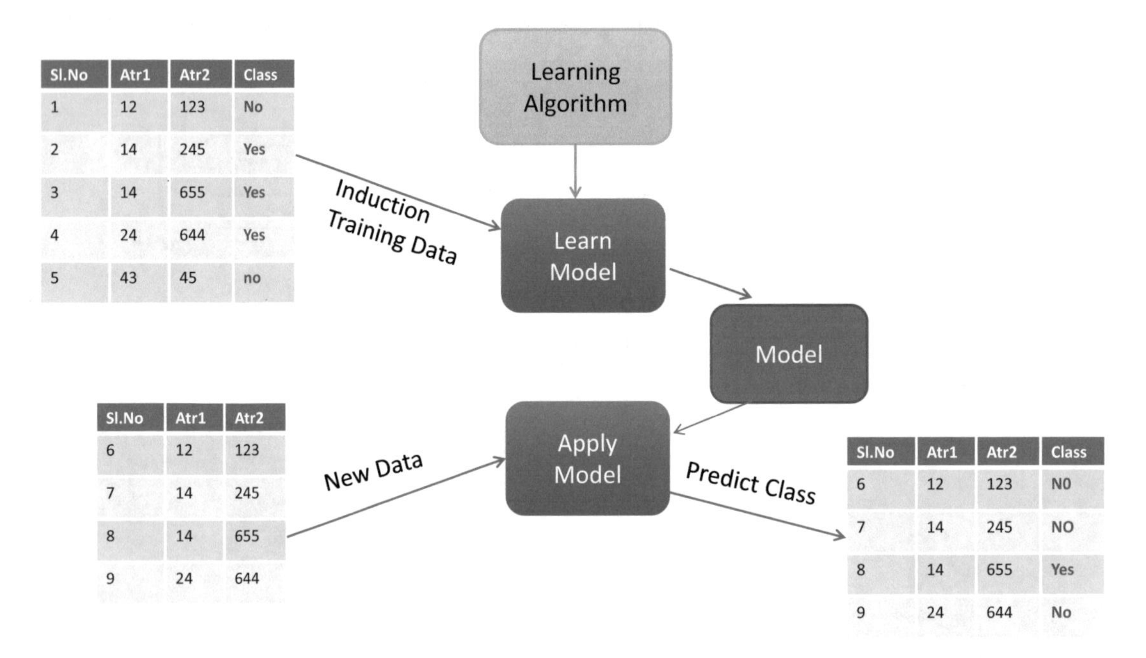

Figure 9-3. *Typical classification model*

We begin our discussion with a simple classifier and then discuss the other classifier models such as decision trees, naïve Bayes, and random forest.

9.1.1 K-Nearest Neighbor

We begin our discussion with the k-nearest neighbor (KNN) algorithm, one of the simpler algorithms for classification based on the distance measure. We will explore the concepts behind the nearest neighbor with an example and demonstrate how to build the KNN model using both the R and Python libraries.

The nearest neighbor algorithm is commonly known as a *lazy learning algorithm* because it does not have any complicated rules defined to learn the data; instead, it memorizes the training data set. You will find many applications of KNN in the real world for categorizing a voter as Democrat or Republican or grouping news events into different category.

The *KNN* classifier is based on learning numeric attributes in an *n*-dimensional space. All of the training samples are stored in an *n*-dimensional space with a distinguished pattern. When a new sample arrives with an unknown class, the classifier algorithm searches for the pattern that is closest to the training sample pattern and then labels the class based on the k-pattern space (called the *k-nearest neighbor*). The "closeness" is defined in terms of distance measured. The unknown sample is assigned the nearest class among the k-nearest neighbor pattern. The idea is to look for the records that are similar to, or "near," the record to be classified in the training records that have values close to $X = (x_1, x_2, x_3, \ldots)$. These records are grouped into classes based on the "closeness," and the unknown sample will look for the class (defined by k) and identify itself to that class that is nearest in the k-space.

Though there are different distance measures, Euclidean distance measure is more popular than other measures.

The Euclidean distance between two points, $X = (x_1, x_2, x_3, \ldots x_n)$ and $Y = (y_1, y_2, y_3, \ldots y_n)$, is the square root of the summation of the square of the difference between the two points. Mathematically, it is defined as follows:

$$\text{Euclidian Distance} = \sqrt{\sum_{i=1}^{k} (x_i - y_i)^2}$$

Similarly, the Manhattan distance for the two points, $X = (x_1, x_2, x_3, \ldots x_n)$ and $Y = (y_1, y_2, y_3, \ldots y_n)$ is defined as follows:

$$\text{Manhattan Distance} = \sum_{i=1}^{k} |x_i - y_i|$$

And the Minkowski distance for the two points, $X = (x_1, x_2, x_3, \ldots x_n)$ and $Y = (y_1, y_2, y_3, \ldots y_n)$, is defined as follows:

$$\text{Minkowski} = \left(\sum_{i=1}^{k} (|x_i - y_i|)^q \right)^{1/q}$$

If q=1, then it is equivalent to the Manhattan distance, and for the case q=2, it is equivalent to the Euclidian distance. Although q can be any real value, it is typically set to a value between 1 and 3.

9.1.2 KNN Algorithm

K-nearest neighbor does not assume any relationship among the predictors (X) and class (Y). Instead, it draws the conclusion of the class based on the similarity measures between predictors and records in the data set. Though there are many potential measures, KNN uses Euclidean distance between the records to find the similarities to label the class. For the algorithm to work, you need the following:

1. Set of sample data (training data).
2. Select the distance measure between the records. Different tools may support different distance measures, and you can choose the one appropriate for your data after reading the definition of the distance measure.

When new data arrives, to classify an unknown record, do the following:

a. Compute the distance from all the other records in the training data.

b. Identify the k-nearest neighbors.

c. Classify the unknown data "class" by taking a majority vote of the class labels of the nearest neighbors.

Let's demonstrate this with the following example. In this loan approval example, we have a target class variable, `Approval`, which depends on two independent variables, `Age` and `PurchaseAmount`, as shown in Figure 9-4. The data already has a class label of Yes or No.

Age	PurchaseAmoun	Approval
44	204	No
35	183	Yes
41	221	No
40	158	Yes
40	280	No
32	362	No
36	436	No
45	457	Yes
38	479	Yes
33	492	No
38	523	No
41	623	No
32	686	Yes
35	695	Yes

New Sample Data of "Joe"	
Age	PurchaseAmount
34	200

Figure 9-4. *Loan approval training data and new data for KNN*

Given the new data, with a variable `Age` of 34 and a `PurchaseAmount` of 200, we calculate the distance from all the training records. We will use the Euclidian distance measure for our distance calculation. As shown in Figure 9-5, a distance of (34,200) is 10.77033 units from (44, 204), 17.029386 from (35, 183), 22.135944 from (41,221), and so on. Once the distance measure is calculated, now choose the k-value and classify the new unknown data class by taking the majority vote. For example, if k=5, then you have 3 No and 2 Yes. The majority is No, and hence the new data is classified as No. If a new record has to be classified, it finds the nearest match to the record and tags to that class. For example, if it looks like a mango and tastes like a mango, then it's probably a mango.

Age	PurchaseAmoun	Approval	DistMeas
44	204	No	10.77033
35	183	Yes	17.029386
41	221	No	22.135944
40	158	Yes	42.426407
40	280	No	80.224684
32	362	No	162.01235
36	436	No	236.00847
45	457	Yes	257.2353
38	479	Yes	279.02867
33	492	No	292.00171
38	523	No	323.02477
41	623	No	423.05792
32	686	Yes	486.00412
35	695	Yes	495.00101
New Sample Data of "Joe"			
Age	PurchaseAmount		
34	200		

Figure 9-5. *Classifying unknown data using KNN*

After computing the distances between records, we need a rule to put these records into different classes (k). A higher value of k reduces the risk of overfitting due to noise in the training set. Ideally, we balance the value of k such that the misclassification error is minimized. Generally, we choose the value of k to be between 2 and 10; for each iteration of k, we calculate the misclassification error and select the value of k that gives the minimum misclassification error.

Please note that the predictor variables should be standardized to a common scale before computing the Euclidean distances and classifying. We can use any of the standardization techniques we discussed in Chapter 4, such as min-max or z-score transformation.

9.1.3 KNN Using R

In this example, we will investigate the loan approval problem. The data consists of three parameters: the dependent variable, the target class to determine whether the loan is approved based on the `Age` value of the applicant, and the `PurchaseAmount`, as shown in Figure 9-6. This data consists of 53 training records and 10 test data records. The class 0 means the loan is not approved, and 1 means the loan is approved. We will use the `class` package in R to run the KNN algorithm function.

Age	PurchaseAmount	Approval
44	204	0
35	183	1
41	221	0
40	158	1
40	280	0
32	362	0
36	436	0
45	457	1
38	479	1
33	492	0
38	523	0
41	623	0

***Figure 9-6.** Training data for the KNN classifier*

The following section explains the step-by-step process of creating a KNN model and testing the performance.

Step 1: Read the training data and test data.

```
> # Read the Training data
> knndata<-read.csv("knn-data.csv",header=TRUE, sep=",")

> ##Read the Test data
> knntest<-read.csv("knntestdata.csv",header=TRUE,sep=",")
```

Step 2: Preprocess the data if required. Check the data types of each variable and transform them to the appropriate data types.

```
> ##The Objective of this KNN model is to classify and predict
  Loan Approval
> ##Approval is the target(response) variable and Age, PurchaseAmount
  are independent variable
> ## Data has 3 columns, Age, Purchase Amount and APproval
> ##Checking the data type to make sure if any transformation is required
> ##APproval variable is a categorical and R automatically reads it as
  a "factor"
> ## In R Factor is categorical variable
> str(knndata)
```

```
 'data.frame': 52 obs. of  3 variables:
$ Age           : int  44 35 41 40 40 32 36 45 38 33 ...
$ PurchaseAmount: int  204 183 221 158 280 362 436 457 479 492 ...
$ Approval      : int  0 1 0 1 0 0 0 1 1 0 ...
```

Step 3: Prepare the data as per the library requirements. In this case, read the KNN function R documentation and accordingly prepare your training and test data.

```
> ## Read KNN library documentation to see the input parameters
  required using help(knn).
> ## For this library function, input training data should not have the
  response variable
> # Hence separate the response variable
> train_data=knndata[,-3]
> # Seprate the response variable from the test data
> test_data = knntest[,-3]

> ## Number of categories/classes
> cls<-factor(knndata[,3])
> cls   ## There are Two classes (Yes and NO)
 [1] 0 1 0 1 0 0 0 1 1 0 0 0 1 1 1 0 1 0 0 1 1 1 0 0 1 0 0 0
[29] 1 1 1 1 1 0 0 0 0 1 0 0 0 0 0 0 1 1 0 1 0 1 0 0
Levels: 0 1

>

> ##Separating test data categories/class
> actual<-factor(knntest[,3])
> actual
 [1] 0 0 1 1 1 1 1 1 0 0 1 0 1 1
Levels: 0 1
```

Step 4: Create the KNN model using the knn() function from the class package. If you have not installed the class package, you must install the package first.

Here is the description of the function from the official R documentation (https://www.rdocumentation.org/packages/class/versions/7.3-20/topics/knn):

knn {class} R Documentation

k-Nearest Neighbour Classification

Description

k-nearest neighbour classification for test set from training set. For each row of the test set, the k nearest (in Euclidean distance) training set vectors are found, and the classification is decided by majority vote, with ties broken at random. If there are ties for the kth nearest vector, all candidates are included in the vote.

Usage

knn(train, test, cl, k = 1, l = 0, prob = FALSE, use.all = TRUE)

`train`	matrix or data frame of training set cases.
`test`	matrix or data frame of test set cases. A vector will be interpreted as a row vector for a single case.
`cl`	factor of true classifications of training set
`k`	number of neighbours considered.
`l`	minimum vote for definite decision, otherwise `doubt`. (More precisely, less than `k-l` dissenting votes are allowed, even if k is increased by ties.)
`prob`	If this is true, the proportion of the votes for the winning class are returned as attribute `prob`.
`use.all`	controls handling of ties. If true, all distances equal to the kth largest are included. If false, a random selection of distances equal to the kth is chosen to use exactly k neighbours.

Value

Factor of classifications of test set. `doubt` will be returned as NA.

References

Ripley, B. D. (1996) *Pattern Recognition and Neural Networks.* Cambridge.

Venables, W. N. and Ripley, B. D. (2002) *Modern Applied Statistics with S.* Fourth edition. Springer.

```
> ## Creating KNN Model using knn() function
> ##The Objective of this KNN model is to classify and predict Loan
  Approval (Approval)
> ##Approval is the target(response) variable and Age, PurchaseAmount
  are independent variable
```

```
> library(class)
> knnmod<-knn(train=train_data,test=test_data,cl=cls,k=3,
+               prob=FALSE)
```

The model has training data and test data as input, and it uses the training data to build the model and the test data to predict. At this stage, we do not know the exact value of k and assume that k is 3. In step 6, we will demonstrate how to find the value of k.

Step 5: Our KNN model has predicted the test data target class. We must check the model performance by comparing the predicted class values with the actual values. We will write a small function called `check_error()`, as shown next, which compares the actual values and the predicted values by calculating the mean error.

```
> ##Predict the new test data
> predicted<-knnmod
> ## Calculate the classification error
> check_error = function(actual, predicted) {
+   mean(actual != predicted)
+ }
> check_error(actual,predicted)
[1] 0.5
```

Please note that the confusion matrix is a truth table of actual versus predicted in a matrix form. The predicted values are displayed vertically, and the actual values are displayed horizontally. For example, in this case, there are four cases where the model-predicted output value is 0, and the actual value is also 0. Similarly, there are eight cases where the model-predicted value as 1 but the actual value is 0. This matrix helps in determining the accuracy of the model, precision, and recall of the model. The actual formula and calculations are explained in Chapter 6. We urge you to refer to this chapter and calculate manually for better understanding of the matrix and the output produced by the code.

```
> # Caclulate accuracy and other parameters using caret functions
> library(e1071)
> confusionMatrix (actual, predicted)
```

```
Confusion Matrix and Statistics

          Reference
Prediction 0 1
         0 4 1
         1 8 1

               Accuracy : 0.3571
                 95% CI : (0.1276, 0.6486)
    No Information Rate : 0.8571
    P-Value [Acc > NIR] : 1.0000

                  Kappa : -0.0678

 Mcnemar's Test P-Value : 0.0455

            Sensitivity : 0.3333
            Specificity : 0.5000
         Pos Pred Value : 0.8000
         Neg Pred Value : 0.1111
             Prevalence : 0.8571
         Detection Rate : 0.2857
   Detection Prevalence : 0.3571
      Balanced Accuracy : 0.4167

       'Positive' Class : 0
```

The accuracy of the model is around 35.71 percent with a confidence of 95 percent. The results also provide sensitivity and specificity measures.

Step 6: We will determine the exact value of k by calculating errors for each value of k from 1 to 10. Normally, the k-values lie between 1 to 10. We will not consider value 1 because of the inherent nature of KNN for the majority voting count. We will write a small function that will loop 10 times for different values of k and calculate the error. Once we have errors for each value of k, we will plot the graph of k-values versus error. The final model will be with the k-value, which has a minimum error. In this example, we have not demonstrated the normalizing parameters. As part of fine-tuning model performance, you should consider normalizing the variables.

```
> #Try different values of k
> ks = 1:10 ##Try k values from 1 to 10
> ##Storing errors in an array
> store_error = rep(x=0, times = length(ks))
> for (i in seq_along(ks)) {
+   predicted = knn(train = train_data,
+                   test = test_data,
+                    cl = cls,
+                    k = ks[i])
+ store_error[i] = check_error(actual, predicted)
+ }

# ##Plot Error Vs K-values graph
> plot(store_error,
+        type='b',
+        col = "blue",
+        cex = 1, pch = 20,
+        xlab = "K - values",
+        ylab = "classification error",
+        main = "Error rate vs K-Values")
```

Figure 9-7 shows the k-values versus the error plot. This is referred to as an *elbow plot*. This method is normally used to determine the optimum value of k. In this example, an error is minimal for any values of k between 2 and 6. Since this is a small data set, we recommend a k-value of 3.

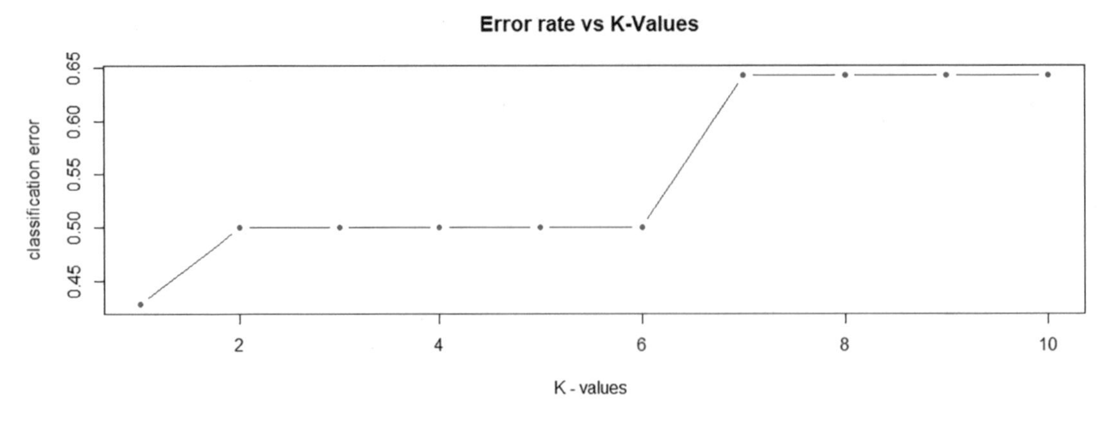

Figure 9-7. *Error versus k-values*

9.1.4 KNN Using Python

We will use the sklearn library to build our KNN classifier. The sklearn library function for KNN is `KNeighborsClassifier` from `sklearn.neighbors`. The process of building the model in Python is exactly the same as before. We are using the same data set. Please note that you can download the entire `jupyternotebook.ipynb` from the GitHub repository and execute it on your device.

Step 1: Load all the Python libraries required to create the KNN model.

Step 2: Read the training data and test data.

Here is the input:

```
raw_data = pd.read_csv("knn-data.csv")
raw_test = pd.read_csv("knntestdata.csv")
```

Step 3: Prepare the data and preprocess the data.

```
import os # Operating System functionality libraries
import re # Regular expression library
import pprint #provides a capability to "pretty-print" arbitrary Python
data structures

import numpy as np  # Library numpy array
import pandas as pd  # pandas library

import matplotlib.pyplot as plt  # Plotting library
%matplotlib inline
## Splitting dataset and normalizing data
from sklearn.model_selection import train_test_split
from sklearn.preprocessing import scale
##KNN
from sklearn.neighbors import KNeighborsClassifier
##Measuring Accuracy of the Model
from sklearn.metrics import roc_auc_score
from sklearn.metrics import roc_curve
from sklearn.metrics import classification_report
from sklearn.metrics import confusion_matrix
```

In this step, we will split the data into `X_train`, `y_train`, `X_test`, and `y_test` by separating the `Approval` variable as shown. Then we will check the data types and convert them to the appropriate data types. For example, the Approval variable should be converted to the categorical variable using the astype() function.

Check the data types of the variables in the data set and take the necessary actions to convert the data types. In this case, as you can see, Approval is the target variable, and it should be categorical.

```
#drop the target variable from the X and keep Y separately for the KNN
API() function
X_train = raw_data.drop(columns='Approval')
y_train = raw_data['Approval']
X_test = raw_test.drop(columns='Approval')
y_test = raw_test['Approval']
```

Here is the input:

```
# check data type
raw_data.dtypes
```

Here is the output:

```
Out[47]: Age                int64
         PurchaseAmount     int64
         Approval           int64
         dtype: object
```

Convert the Approval variable to categorical using the astype() function.

Here is the input:

```
#convert to 'categorical' variable
y_train=y_train.astype('category')
y_test = y_test.astype('category')
```

Step 4: Create the KNN model using the KNeighborsClassifier() function. We can use any values of k, but to start with, we will choose k=2, and then we will use k=2 and find the optimal value of k using the elbow method using misclassification errors as a measure for different values of k.

Here is the input:

```
knn_model = KNeighborsClassifier(n_neighbors = 2)
knn_model.fit(X_train, y_train)

knn_model.classes_
```

Here is the actual output along with input commands.

```
In [63]:  knn_model = KNeighborsClassifier(n_neighbors = 2)
          knn_model.fit(X_train, y_train)

Out[63]:  KNeighborsClassifier(n_neighbors=2)

In [72]:  knn_model.classes_

Out[72]:  array([0, 1], dtype=int64)
```

Step 5: Predict the class of the new data using the KNN model.

Here is the input:

```
predicted = knn_model.predict(X_test)
predicted
```

Here is the output:

```
In [73]:  predicted

Out[73]:  array([0, 0, 0, 1, 0, 0, 0, 0, 0, 0, 0, 0, 0, 0], dtype=int64)
```

Step 6: Using the `classification_report()` and `confusion_matrix()` functions, we will calculate the accuracy of the model, including precision, recall, and f1-score. This model resulted in an overall accuracy of 43 percent. It has also given the precision and recall measures for both 0 and 1 (`Approval` or `Denied`).

Here is the input:

```
print(confusion_matrix(y_test, predicted))
print(classification_report(y_test, predicted))
```

Here is the output:

```
[[5 0]
 [8 1]]
```

```
In [75]: print(classification_report(y_test, predicted))
              precision    recall  f1-score   support

           0       0.38      1.00      0.56         5
           1       1.00      0.11      0.20         9

    accuracy                           0.43        14
   macro avg       0.69      0.56      0.38        14
weighted avg       0.78      0.43      0.33        14
```

Step 7: The final step is to calculate the error for different k-values and select the k based on the minimum error. For this, we will write a small function to loop through different k-values, create the model, and calculate the error. We will also plot the error versus k-values graph to choose the k-value.

Here is the input:

```
err = []
for i in np.arange(1,10):
    knn_new = KNeighborsClassifier(n_neighbors=i)
    knn_new.fit(X_train, y_train)
    new_predicted = knn_new.predict(X_test)
    err.append(np.mean(new_predicted != y_test))
plt.plot(err)
```

Here is the output:

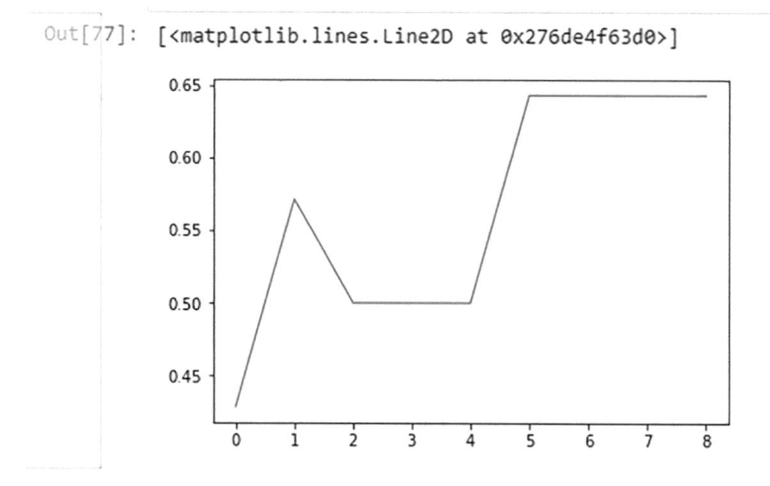

As you can see, the misclassification error is least for the k between 2 and 4. For the optimal model performance, we can choose k=3 (odd number) since the algorithm is based on the majority vote method.

The results obtained for R and Python are slightly different. We used the open-source libraries, and we have not looked at all the parameters of the input API function as well as the same distance measure. We are confident that if you go through the documentation and apply the right parameters in both R and Python, you will be able to achieve the same results. We strongly recommend reading the documentation, using the right parameters, and rerunning the code.

9.2 Naïve Bayes Models for Classification

The *naïve Bayes* classifier is a simple probabilistic classifier based on the Bayes theorem. It has shown good performance in many common applications such as email spam detection, sentiment analysis, weather predictions, etc. It is widely used as its performance surprisingly outperforms more sophisticated classification methods. Naïve Bayes classification is based on the approximating joint distribution, with an assumption of independence between the predictors, and then decomposing the probability into a product of conditional probability.

A conditional probability is the probability that a random variable takes on a particular value given that the outcome of another variable is known. Let X and Y be the pair of random variables. Their joint probability, P(X=x, Y=y), is the likelihood of two events occurring at the same time, i.e., the probability of event Y occurring at the same time as the event X. A conditional probability is the likelihood of an event occurring given the other event has already occurred. The conditional probability of event A, given event B occurred, is denoted by P(A|B).

$$P(X,Y) = P(Y|X) * P(X) = P(X|Y) * P(Y)$$

By rearranging these terms we get:

$$P(Y|X) = \frac{P(X|Y) * P(Y)}{P(X)}$$

Applying the same thing in the context of classification, Bayes' theorem provides a formula for calculating the probability of a given record belonging to a class. Suppose you have *m* classes, C_1, C_2, C_3, ... C_m, and the probability of classes is $P(C_1)$, $P(C_2)$, ... $P(C_m)$. If you know the probability of occurrence of x_1, x_2, x_3, ... attributes within each class, then by using Bayes' theorem, you can calculate the probability of the record xi, belonging to class C_i:

$$P_{(C|X)} = \frac{P(x|c) * P(c)}{P(x)} \qquad (1)$$

where:

P (c|x) is the posterior probability of a class given a predictor.

P(x|c) is the likelihood of occurrence, for instance, of x given class C.

P(c) is the probability of class C.

P(x) is the prior probability of the predictors.

The naïve Bayes classifier assumes what is called *conditional independence*; that is, the effect of the value of a predictor (x) on a given class (c) is independent of the values of the other predictors. The previous equation simplifies to the following:

$$P(c|x) = P(x_1|c) * P(x_2|c)\ P(x_3|c) \ldots P(X_n|c) * P(C) \qquad (2)$$

$P(C_i)$ is the prior probability of belonging to class C_i in the absence of any other attributes. $P(C_i|X_i)$ is the posterior probability of X_i belonging to class C_i. To classify a record using Bayes' theorem, we compute the probability of each training record belonging to each class C_i and then classify based on the highest probability score calculated using the naïve Bayes formula.

9.2.1 Naïve Bayes Classifier Model Example

In this section, let's look at an example data set and apply naïve Bayes to build our predictive modeling. Table 6-1 presents a training set of data tuples for a bank credit-card approval process. The data samples in Table 9-1 consist of four attributes: Purchase Frequency, Credit Rating, Age, and Loan Approval. Loan Approval is the target class we are predicting using the naïve Bayes model, which depends on three independent variables: Age, Purchase Frequency, and Credit Rating. There are only 14 records in this data set; however, in real-life scenarios, the data set is large, and the naive Bayes model performs well with large data sets and training records.

Table 9-1. *Sample Training Set*

ID	Purchase Frequency	Credit Rating	Age	Loan Approval (Yes = Approved No = Denied)
1	Medium	OK	< 35	No
2	Medium	Excellent	< 35	No
3	High	Fair	35–40	Yes
4	High	Fair	> 40	Yes
5	Low	Excellent	> 40	Yes
6	Low	OK	> 40	No
7	Low	Excellent	35–40	Yes
8	Medium	Fair	< 35	No
9	Low	Fair	< 35	No
10	Medium	Excellent	> 40	No
11	High	Fair	< 35	Yes
12	Medium	Excellent	35–40	No
13	Medium	Fair	35–40	Yes
14	High	OK	< 35	No

We will calculate the prior probability and class probability for each class. To calculate these probabilities, we can construct a frequency table for each attribute against the target class. Then, from the frequency table, calculate the likelihood; finally, apply the naïve Bayes equation to calculate the posterior probability for each class. Then the class with the highest posterior probability decides the prediction class.

Figure 9-8 is the frequency table for the Purchase Frequency attribute. Using this table we can calculate the probabilities:

Purchase Frequency / Loan Approval

	Yes	NO
High	3	1
Medium	1	5
Low	2	2

P(C) = p(Yes)
= 6/14 = 0.42

Loan Approval

Likelihood	Yes	No	
High	3/6	1/8	4/14
Medium	1/6	5/8	6/14
Low	2/6	2/8	4/14
	6/14	8/14	

P(X) = P(high)
= 4/14 = 0.28

P(x|c) = P(High|Yes)
= 3/6 = 0.5

Figure 9-8. *Frequency table to calculate probabilities, Purchase Frequency attribute*

$P(high) = 4/14 = 0.28$ (out of 14 training records, we have 4 high purchases). The probability would be 4/14, i.e, 0.28. Similarly, we can calculate all the other Purchase Frequency prior probabilities.

$P(Yes) = 6/14$

$P(No) = 8/14.$

$P(High|Yes) = 3/6 = 0.5$; there are 3 High purchases out of 6 Yes.

Now, given all theses probabilities, we can calculate the posterior probability using the naïve Bayes theorem (assuming the conditional independence).

$P(C|X) = (PYes|high) = P(High|Yes) * P(Yes) / P(High)$

$= (0.5 * 0.42) / 0.28 = 0.75$

Similarly, we can come up with frequency tables to calculate the likelihood for other attributes, as shown in Figure 9-9.

Credit Rating	Yes	No	
Excellent	2, 2/6	3, 3/8	5/14
OK	0, 0/6	3, 3/8	3/14
Fair	4, 4/6	2, 2/8	6/14

Age	Yes	No	
<35	1, 1/6	5, 5/8	6/14
35-40	3, 3/6	1, 1/8	4/14
>40	2, 2/6	2, 2/8	4/14

Figure 9-9. *Frequency table to calculate probabilities, Credit Rating and Age attributes*

Assume you are given a new set of data, shown here:

X = where (Age >40, Purchase Frequency = Medium, Credit Rating = Excellent)

We will demonstrate how to apply the naïve Bayes model to predict the loan status (Approved or Denied). We will use the previous frequency table to perform all our probability calculations.

To classify a record, first compute the probability of a record belonging to each class by computing $P(C_i|X_1,X_2, \ldots X_p)$ from the training record. Then decide on the class based on the highest probability.

> We already have the probability of the Approval/Denied class, P(C):
>
> P(Application Approval = Yes) = 6/14 = 0.428
>
> P(Application Approval = No) = 8/14 = 0.571

Let's compute $P(X|C_i)$, for i =1, 2:

> P(Age > 40 | Approval = Yes) = 2/6 = 0.333
>
> P(Age > 40 | Approval = No) = 2/8 = 0.25
>
> P(Purchase Frequency = Medium | Approval = Yes) =1/6 = 0.167
>
> P(Purchase Frequency = Medium | Approval = No) = 5/8 = 0.625
>
> P(Credit Rating = Excellent | Approval = Yes) = 2/6 = 0.333
>
> P(Credit Rating = Excellent | Approval = No) = 3/8 = 0.375

Using these probabilities, apply naïve Bayes from equation (3):

$$P(c|x) = P(x_1|c) * P(x_2|c)\ P(x_3|c) \ldots P(X_n|c) * P(C) \quad (3)$$

Using these probabilities, you can obtain the following:

For the new data, X = where (Age >40, Purchase Frequency = Medium, Credit Rating = Excellent).

= P(Age>40|yes) * P(Purchase Frequency=Medium|Yes) * P(CreditRatng=Excellent|Yes) * P(Yes)

= 0.333 × 0.167 × 0.333 * 0.428 = 0.0079

= P(Age>40|No) * P(PurchaseFrequency=Medium|NO) * P(CreditRating=Excellent|NO) * P(NO)

=P(X | Approval = No) = 0.25 × 0.625 × 0.375 * 0.571 = 0.0334

The naïve Bayesian classifier predicts Approval = No for the given set of sample X.

9.2.2 Naïve Bayes Classifier Using R (Use Same Data Set as KNN)

Let's try building the model by using R. We'll use the same example. The data sample sets have the attributes Age, Purchase Frequency, and Credit Rating. The class label attribute has two distinct classes: Approved or Denied. The objective is to predict the class label for the new sample, where Age > 40, Purchase Frequency = Medium, and Credit_Rating = Excellent.

The first step is to read the data from the file.

Here is the input:

```
➢ credData
➢ str(credData)
```

Here is the output:

```
> credData
   ID PurchaseFrequency CreditRating   Age Approval
1   1            Medium           OK   <35       No
2   2            Medium    Excellent   <35       No
3   3              High         Fair 35-40      Yes
4   4              High         Fair   >40      Yes
5   5               Low    Excellent   >40      Yes
6   6               Low           OK   >40       No
7   7               Low    Excellent 35-40      Yes
8   8            Medium         Fair   <35       No
9   9               Low         Fair   <35       No
10 10            Medium    Excellent   >40       No
11 11              High         Fair   <35      Yes
12 12            Medium    Excellent 35-40       No
13 13            Medium         Fair 35-40      Yes
14 14              High           OK   <35       No
> |
```

```
> str(credData)
'data.frame':   14 obs. of  5 variables:
 $ ID               : int  1 2 3 4 5 6 7 8 9 10 ...
 $ PurchaseFrequency: Factor w/ 3 levels "High","Low","Medium": 3
3 1 1 2 2 2 3 2 3 ...
 $ CreditRating     : Factor w/ 3 levels "Excellent","Fair",..: 3
1 2 2 1 3 1 2 2 1 ...
 $ Age              : Factor w/ 3 levels "<35",">40","35-40": 1 1
3 2 2 2 3 1 1 2 ...
 $ Approval         : Factor w/ 2 levels "No","Yes": 1 1 2 2 2 1 2
 1 1 1 ...
>
```

Once we read the data, divide the data into test and train.

Here is the input:

```
# Step 5: Divide data into training and test datasets.
# We will use caret() library package to perfrom this task

library(caret)
set.seed(1234)
data_partition<-createDataPartition(y=data_df$Approval,
                        p=0.8,
                        list=FALSE)
train<-data_df[data_partition,]
test<-data_df[-data_partition,]
```

The next step is to build the classifier (naïve Bayes) model by using the `mlbench` and `e1071` packages.

Here is the input:

```
#Step 6: Create a Naive Bayes model using training data
# We will use 'e107' library
library(e1071)
nb_model <- naiveBayes(Approval~Income+CreditRating+Age, data=train)
```

For the new sample data X = (Age > 40, Purchase Frequency = Medium, Credit Rating = Excellent), the naïve Bayes model has predicted Approval = No.

Here is the input:

```
#Step 7: Predict test data
nb_model
nb_pred<-predict(nb_model,test)
nb_pred
```

Here is the R output with input commands:

```
> nbmodel

Naive Bayes Classifier for Discrete Predictors

Call:
naiveBayes.default(x = X, y = Y, laplace = laplace)

A-priori probabilities:
Y
       No       Yes
0.5714286 0.4285714

Conditional probabilities:
     CreditRating
Y      Excellent      Fair        OK
  No   0.3750000 0.2500000 0.3750000
  Yes  0.3333333 0.6666667 0.0000000

     PurchaseFrequency
Y           High       Low    Medium
  No   0.1250000 0.2500000 0.6250000
  Yes  0.5000000 0.3333333 0.1666667

     Age
Y            <35       >40     35-40
  No   0.6250000 0.2500000 0.1250000
  Yes  0.1666667 0.3333333 0.5000000
```

```
> ctest
  ID PurchaseFrequency CreditRating Age
1  1            Medium     Exellent >40
>
>
> pred<-predict(nbmodel,ctest)
> pred
[1] No
Levels: No Yes
```

Here is the input:

```
> nb_pred<-predict(nb_model,test)
> nb_pred
```

Here is the R output with input commands:

```
> nb_pred
 [1] No  Yes Yes No  No  Yes No  No  Yes No  Yes No  Yes No
Levels: No Yes
```

The next step is to measure the performance of the NB classifier and how well it has predicted. We use the `caret()` libraries to print the confusion matrix. As discussed in the earlier sections, the confusion matrix is a truth table and provides the table of what is being predicted by the model versus the actual values. The function `confusionMatrixReport()` calculates the accuracy of the model and also provides a sensitivity analysis report. As you can see from the output results, our NB model achieved an accuracy of 67 percent.

Here is the input:

```
> #Step 8: Meausure the performance of your model
> library(caret)
> table(nb_pred, test$Approval)
```

Here is the output:

```
nb_pred No Yes
    No   5   3
    Yes  2   4
```

Here is the input:

```
> confusionMatrix(as.factor(nb_pred), test$Approval)
```

Here is the R output with input commands:

```
> confusionMatrix(as.factor(nb_pred), test$Approval)
Confusion Matrix and Statistics

          Reference
Prediction No Yes
       No   5   3
       Yes  2   4

               Accuracy : 0.6429
                 95% CI : (0.3514, 0.8724)
    No Information Rate : 0.5
    P-Value [Acc > NIR] : 0.212

                  Kappa : 0.2857

 Mcnemar's Test P-Value : 1.000

            Sensitivity : 0.7143
            Specificity : 0.5714
         Pos Pred Value : 0.6250
         Neg Pred Value : 0.6667
             Prevalence : 0.5000
         Detection Rate : 0.3571
   Detection Prevalence : 0.5714
      Balanced Accuracy : 0.6429

       'Positive' Class : No
```

9.2.3 Advantages and Limitations of the Naïve Bayes Classifier

The naïve Bayes classifier is the simplest classifier of all. It is computationally efficient and gives the best performance when the underlying assumption of independence is true. The more data you have, the better naïve Bayes performs.

The main problem with the naïve Bayes classifier is that the classification model depends on posterior probability, and when a predictor category is not present in the training data, the model assumes zero probability. This can be a big problem if this attribute value is important. With a zero value, the conditional probability values calculated could result in wrong predictions. *When the preceding calculations are done using computers, sometimes it may lead to floating-point errors.* The class with the highest log probability is still the most significant class. If a particular category does not appear in a particular class, its conditional probability equals 0. Then the product becomes 0. If we use the second equation, log(0) is infinity. To avoid this problem, we use add-one or Laplace smoothing by adding 1 to each count. Laplace smoothing tackles the zero probability problem. By adding 1, the likelihood is pushed toward a value of nonzero and thus optimizes the overall model. This is more relevant in text classification problems. Also, for numeric attributes, normal distribution is assumed. However, if we know the attribute is not a normal distribution and is likely to follow some other distribution, you can use different procedures to calculate estimates—for example, kernel density estimation does not assume any particular distribution (Kernel Density Estimation (KDE) is a method to estimate the probability density function of a continuous random variable). Another possible method is to discretize the data. Although conditional independence does not hold in real-world situations, naïve Bayes tends to perform well.

9.3 Decision Trees

A *decision tree* builds a classification model by using a tree structure. A decision tree builds the tree incrementally by breaking down the data sets into a smaller subset and incrementally builds the tree structure. The final decision tree structure consists of a root node, branches, and leaf nodes called *decision nodes*. The root node is the topmost node attribute, the branches are the next set of attributes selected based on the decisions of the root node, and the leaf nodes are the final class of the tree model.

Figure 9-10 demonstrates a simple decision tree model. Once the root node is decided, the decision happens at each node for an attribute on how to split. Selection of the root node attribute and the criterion for the split node attribute is based on various methods such as gini index, entropy, misclassification errors, etc. The next section explains the techniques in detail. In this example, the root node is Purchase Frequency, which has two branches, High and Low. If Purchase Frequency is High, the next decision node would be Age, and if Purchase Frequency is Low, the next decision node is Credit Rating. The next nodes are chosen based on the information and purity of each node. The leaf node represents the classification decision; in this case it is Yes or No. The final decision tree consists of a set of rules, more of an 'If Else' rule used to classify the new data.

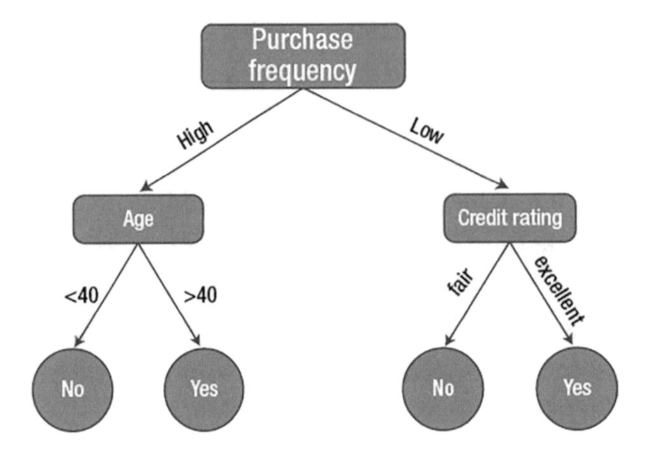

Figure 9-10. *Example of a decision tree*

The decision tree algorithm is based on the divide-and-conquer rule. Let x_1, x_2, and x_3 be independent variables and Y denote the dependent variable. The X variables can be continuous, binary, or ordinal. The first step is selecting one of the variables, x_i, as the root node to split. Depending on the type of the variable and the values, the split decision is made. After the root attribute is split, the next attribute is selected at each branch, and the algorithm proceeds recursively by splitting at each child node. The splitting continues until the decision class is reached. We call the final leaf *homogeneous*, meaning the final terminal node contains only one class.

The following are the basic steps involved in creating a decision tree structure:

1. The tree starts by selecting the root attribute from the training set, based on the gini index and entropy.

2. The root node is branched, and the decision to split is made based on the attribute characteristics and the split measures.

3. At each branch, the attributes of the next node are selected based on the information gain and entropy measure.

4. This process continues until all the attributes are considered for the decision.

The growth of the tree stops when the following occur:

- No more attributes remain in the samples data for further partitioning.
- There are no samples for the attribute branch.
- All the samples belong to the same class.

The decision tree can handle both categorical and numerical variables.

9.3.1 Decision Tree Algorithm

One of the simplest ways to build the tree is to consider all the attributes, come up with all the possible decision trees and rules that correctly classify the training set, and select the simplest of them. This tree will be very large and hard to interpret. J.R Qinlan (1996) came up with an algorithm to build an optimized tree algorithm called ID3, where the training data is large and has many attributes, without much computation. The ID3 algorithm employs entropy and information gain to decide the split and construct a decision tree. Many other methods to split the tree have been invented, but we will not be able to cover all of them here; however, referring to the appropriate implementation documentation will help you to understand the other methods.

9.3.1.1 Entropy

Entropy is a measure of how messy the data is. That is, it is how difficult it is to separate data in the sample into different classes. For example, let's say you have an urn that has 100 marbles and all of them are red marbles; then whichever you pick, it turns out to be red marble, so there is no randomness in the data. This means the data is in good order and the level of disorder is zero. The entropy is zero.

Similarly, if the urn has red marbles, blue marbles, and green marbles, in random order, then the data is in disorder. The level of randomness is high as you do not know which marble you get. The data is completely random, and the entropy is high. The distribution of the marbles in the urn determines the entropy. If the three marbles are equally distributed, then the entropy is high; if they are distributed as 50 percent, 30 percent, and 10 percent, then entropy is somewhere between low to medium. The maximum value of the entropy can be 1, and the lowest value of entropy can be 0.

Mathematically, entropy is given by the following formula:

$$E(A) = -\sum_{i=1}^{C} -p_i log_2 p_i$$

Where $E(A)$ is the entropy of an attribute
P is the probability
i is the class

For example, let's say we have an attribute `LoanApproval`, with 6 Yes and 8 No, then
Entropy(LoanApproval) = Entropy(8,6).
= E(0.429, 0.571)
= - (0.428 log2 (0.428) – (0.571log2(0.571)
= 0.9854
The entropy of two attributes would be as follows:

$$E(A, X) = -\sum_{c \in x} P(c)E(c)$$

For example, E(LoanApproval, PurchaseFrequency) is calculated with the help of frequency tables, as shown in Figure 9-11.

Loan Approval

Purchase Frequency	Yes	NO
High	3	1
Medium	1	5
Low	2	2

Figure 9-11. *Frequency table of PurchaseFrequency and LoanApproval*

E(LoanApproval, PurchaseFrequency) = P(High), E(3,1) + P(Medium)*E(1,5) + P(Low) *E(2,2)

= 4/14*E(0.75,0.25) + 6/14*E(0.166,0.833) + 4/14 *E(0.5, 0.5)
= 4/14*0.811 + 6/14*0.649 + 4/14* 1
= 0.79593

9.3.1.2 Information Gain

A decision tree consists of many branches and many levels. *Information gain* is the decrease of entropy from one level to the next level. For example, in Figure 9-12, the tree has three levels.

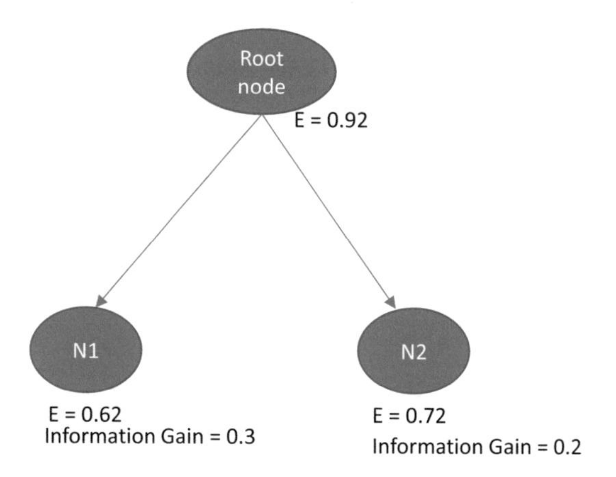

Figure 9-12. *Information gain*

The root node has an entropy of 0.92, node1 has an entropy of 0.62, and node2 has an entropy of 0.72. The entropy has decreased by 0.3 and 0.2, respectively. A higher entropy means the data is uniformly distributed, and a lower entropy means data is more varied. The decision tree algorithm uses this method to construct the tree structure. It is all about finding the attribute that returns the highest information gain (i.e., easier to make the split decision at each branch and node). The purity of the subset partition depends on the value of entropy. The smaller the entropy value, the greater the purity.

In order to select the decision-tree node and attribute to split the tree, we measure the information provided by each attribute. Such a measure is referred to as a measure of the *goodness of split*. The attribute with the highest information gain is chosen as the test attribute for the node to split. This attribute minimizes the information needed to classify the samples in the recursive partition nodes. This approach of splitting minimizes the number of tests needed to classify an object and guarantees that a simple tree is formed. Many algorithms use entropy to calculate the homogeneity of a sample.

Let N be a set consisting of n data samples. Let the k is the class attribute, with m distinct class labels C_i (for $i = 1, 2, 3, \dots m$).

The Gini impurity index is defined as follows:

$$G(N)=1-\sum_{k=1}^{m}(p_k)^2 \quad (1)$$

Where p_k is the proportion of observations in set N that belong to class k. G(N) = 0 when all the observations belong to the same class, and G(N)= (m – 1) / m when all classes are equally represented. Figure 6-2 shows that the impurity measure is at its highest when $p_k = 0.5$.

The second impurity measure is the entropy measure. For the class of *n* samples having distinct *m* classes, C_i (for i = 1, ... *m*) in a sample space of N and class C_i, the expected information needed to classify the sample is represented as follows:

$$I(n_1,n_2,n_3,...n_m)=-\sum_{K=1}^{n}p_k \log_2(p_k) \quad (2)$$

Where p_k is the probability of an arbitrary sample that belongs to class C_i.

Let X be attributes with *n* distinct records A = {a_1, a_2, a_3, ... a_n}. X attributes can be partitioned into subsets as {S_1, S_2, ... S_v}, where S_k contains the a_j values of A. The tree develops from a root node selected from one of these attributes based on the information gain provided by each attribute. Subsets would correspond to the branches grown from this node, which is a subset of S.

The *entropy*, or *expected information*, of attributes A is given as follows:

$$E(A)=\sum_{j=1}^{n}\frac{S_{1j}+S_{2j}+S_{3j}+.....S_{mj}}{S}=I(S_{1j},S_{2j},...S_{mj}) \quad (3)$$

The purity of the subset partition depends the value of entropy. The smaller the entropy value, the greater the purity. The information gain of each branch is calculated on X attributes as follows:

$$\text{Gain}(A)=I(s_1,s_2,s_3,...s_m)-E(A) \quad (4)$$

Gain (A) is the difference between the entropy of each attribute of X. It is the expected reduction on entropy caused by individual attributes. The attribute with the highest information gain is chosen as the root node for the given set S, and the branches are created for each attribute as per the sampled partition.

$$Information\ Gain\ (A,X) = Entropy\ (A) - Entropy(A,X)$$

9.3.2 Building a Decision Tree

Let's illustrate the decision tree algorithm with an example. Table 9-2 presents a training set of data for a retail store credit-card loan approval process. The data samples in this training set have the attributes Age, Purchase Frequency, and Credit Rating. The target attribute has two distinct classes: Approved or Denied. Let C_1 correspond to the class Approved, and let C_2 correspond to the class Denied. Using the decision tree, we want to predict the loan approval decision for the sample X given Age > 40, Purchase Frequency = Medium, and Credit Rating = Excellent.

X = (Age > 40, Purchase Frequency = Medium, Credit Rating = Excellent)

Table 9-2. *Sample Training Set*

ID	Purchase Frequency	Credit Rating	Age	Loan Approval (Yes/No)
1	Medium	OK	< 35	No
2	Medium	Excellent	< 35	No
3	High	Fair	35–40	Yes
4	High	Fair	> 40	Yes
5	Low	Excellent	> 40	Yes
6	Low	OK	> 40	No
7	Low	Excellent	35–40	Yes
8	Medium	Fair	< 35	No
9	Low	Fair	< 35	No
10	Medium	Excellent	> 40	No
11	High	Fair	< 35	Yes

(*continued*)

Table 9-2. *(continued)*

ID	Purchase Frequency	Credit Rating	Age	Loan Approval (Yes/No)
12	Medium	Excellent	35–40	No
13	Medium	Fair	35–40	Yes
14	High	OK	< 35	No

Table 9-3. *Frequency Table of Loan Approval*

Loan Approval	
Yes	No
6	8

We will use the recursive partitioning to build the tree. The first step is to calculate overall impurity measures of all the attributes. Select the root node based on the purity of the node. At each successive stage, repeat the process by comparing this measure for each attribute. Choose the node that has minimum impurity.

In this example, Loan Approval is the target class that we have to predict and build the tree model, which has two class labels: Approved or Denied. There are two distinct classes ($m = 2$). C_1 represents the class Yes, and class C_2 corresponds to No. There are eight samples of class Yes and six samples of class No. To compute the information gain of each attribute, we use equation 1 to determine the expected information needed to classify a given sample.

Entropy(LoanApproval) = Entropy(8,6)

= E(0.429, 0.571)

= - (0.428 * log2 (0.428) - (0.571*log2(0.571))

= 0.9854

Next, compute the entropy of each attribute—Age, Purchase Frequency, and Credit Rating. For each attribute, look at the distribution of Yes and No and compute the information for each distribution. Let's start with the Purchase Frequency attribute. Having a frequency table for each helps the computation.

The first step is to calculate the entropy of each PurchaseFrequency category, as shown in Figure 9-13.

Loan Approval

	Yes	NO
High	3	1
Medium	1	5
Low	2	2

***Figure 9-13.** Frequency table of Purchase Frequency attribute*

For Purchase Frequency = High and Approval = Yes, $C_{11} = 3$, $C_{21} = 2$, PurchaseFrequency = High, and Approval = No:

$E(C_{11}, C_{21}) = E(3,1);$

$= (-\frac{3}{4})\log_2(3/4) - (1/4)\log_2(1/4) = -(0.75 \times -0.41) + (0.25 \times -2)$

$=0.811$

For Purchase Frequency = Medium:

$E(C_{12},C_{22}) = E(1,5) = -1/6 \log(1/6) - 5/6 \log(5/6)$

$= -(0.1666 \times -2.58) - (0.8333 \times -0.2631) = 0.4298 + 0.2192$

$E(C_{12},C_{22}) = 0.6490$

For Purchase Frequency = Low:

$E(C_{13}, C_{23}) = E(2,2) = -(0.5 \times -1)-(0.5 \times -1)$

$= 1$

$E(\text{Purchase Frequency}) = 4/14 \times E(C_{11}, C_{12}) + 6/14 \times E(C_{12},C_{22}) + 4/14 \times E(C_{13},C_{23})$

$E(\text{Purchase Frequency}) = 0.79593$

Gain in information for the attribute Purchase Frequency is calculated as follows:

$\text{Gain}(\text{Purchase Frequency}) = E(C_1, C_2) - E(\text{Purchase Frequency})$

$\text{Gain}(\text{Purchase Frequency}) = 0.9852 - 0.7945 = 0.1907$

Similarly, compute the Gain for other attributes, Gain (Age) and Gain (Credit Rating). Whichever has the highest information gain among the attributes, it is selected as the root node for that partitioning. Similarly, branches are grown for each attribute's values for that partitioning. The decision tree continues to grow until all the attributes in the partition are covered.

For Age < 35, $E(C_{11}, C_{12}) = E(1,5) = 0.6498$

For Age > 40, $E(C_{12}, C_{22}) = E(2,2) = 1$

For Age 35 – 40, $E(C_{13}, C_{23}) = E(3,1) = 0.8113$

$E(Age) = 6/14 \times 0.6498 + 4/14 \times 1 + 4/14 \times 0.8113 = 0.2785 + 0.2857 + 0.2318 = 0.7960$

$Gain (Age) = E(C_1,C_2) - E(Age) = 0.9852 - 0.7960 = 0.1892$

For Credit Rating Fair, $E(C_{11},C_{12}) = E(4,2) = - 4/6 \log(4/6) - 2/6 \log(2/6)$

$= -(0.6666 \times -0.5851) - (0.3333 \times -1.5851 = 0.3900 + 0.5 283 = 0.9183$

For Credit Rating OK, $E(C_{21},C_{22}) = E(0,3) = 0$

For Credit Rating Excellent, $E(C_{13},C_{23}) = (2,3) = -2/5 \log(2/5) - 3/5 \log(3/5)$

$= -(0.4 \times -1.3219) - (0.6 \times -0.7370) = 0.5288 + 0.4422 = 0.9710$

$E(Credit Rating) = 6/14 \times 0.9183 + 3/14 \times 0 + 5/14 \times 0.9710 = 0.3935 + 0.3467$

$E(Credit Rating) = 0.7402$

$Gain (Credit Rating) = E(C_1,C_2) - E(Credit rating) = 0.9852 - 0.7402 = 0.245$

CreditRating has the highest information gain, and it is used as a root node, and branches are grown for each attribute value. The next tree branch node is based on the remaining two attributes, Age and Purchase Frequency. Both Age and Purchase Frequency have almost the same information gain. Either of these can be used as a split node for the branch. The final decision tree looks like Figure 9-14.

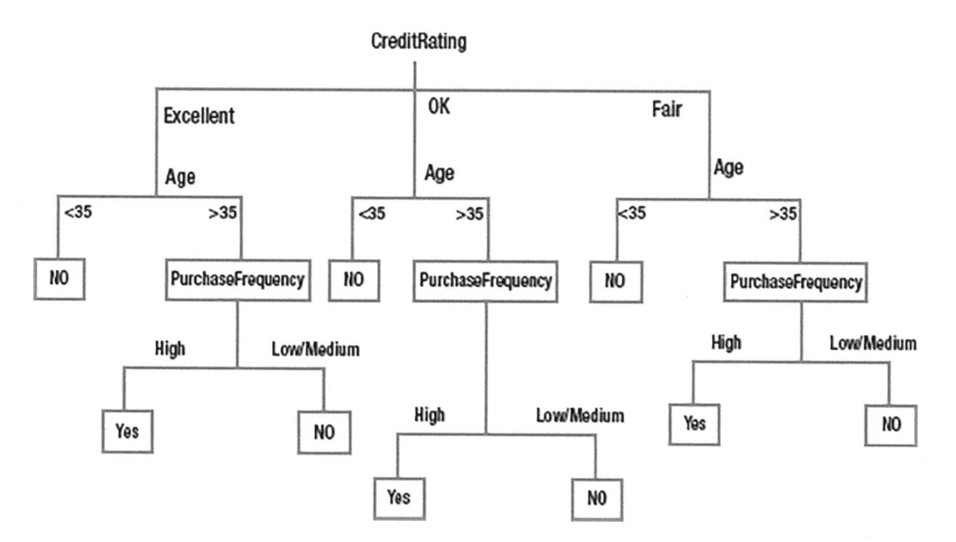

Figure 9-14. *Full-grown decision tree*

9.3.3 Classification Rules from Tree

The decision trees provide an easily understandable rule. Traversing through each tree leaf provides a classification rule. For the previous example, the decision tree algorithm gives us the following rules:

> *Rule 1*: If Credit Rating is Excellent, Age > 35, Purchase Frequency is High, then Loan Approval = Yes.
>
> *Rule 2*: If Age < 35, Credit Rating = OK, Purchase Frequency is Low, then Loan Approval = No.
>
> *Rule 3*: If Credit Rating is OK or Fair, Purchase Frequency is High, then Loan Approval = Yes.
>
> *Rule 4*: If Credit Rating is Excellent, OK, or Fair, Purchase Frequency Low or Medium and Age > 35, then Loan Approval = No.

Applying the previous rules on the sample gives us X = (Age > 40, Purchase Frequency = Medium, Credit Rating = Excellent).

The prediction of the class is: Loan Approval = No.

9.3.3.1 Limiting Tree Growth and Pruning the Tree

If the data is large and has many variables, then the tree can grow big in size, and it becomes difficult to interpret the rules. Also, a full-grown tree is prone to overfitting. Overfitting happens when the tree learns so well that it produces fewer errors in the training data, whereas errors in the test set are significant. Using the entire data set for a full-grown tree leads to complete overfitting of the data, as each and every record and variable is considered to fit the tree. The number of errors for the training set is expected to decrease as the number of tree levels grows; for the new data, the errors decrease until a point where all the predictors are considered, and if we continue building the tree, then the tree starts modeling noise, and thus the overall errors start increasing, as shown in Figure 9-15.

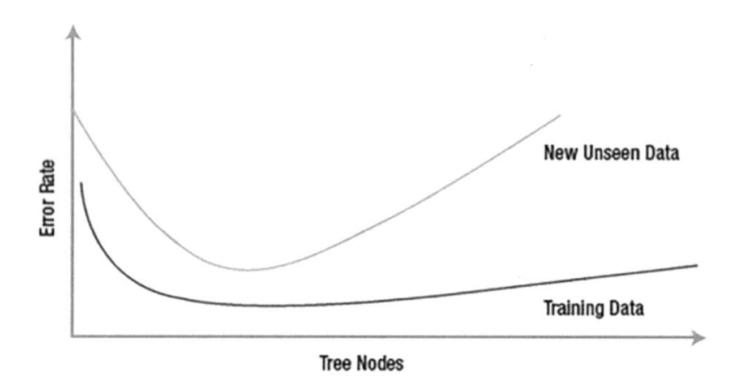

Figure 9-15. *Overfitting error*

There are two ways to limit the overfitting error. One way is to set rules to stop the tree growth at the beginning. The other way is to allow the full tree to grow and then prune the tree to a level where it does not overfit.

One method is to stop the growth of the tree and set some rules at the beginning before the model starts overfitting the data. It is not easy to determine a good point for stopping the tree growth. One popular method used is Chi-Squared Automatic Interaction Detection (CHAID), which has been widely used in many open-source tools. CHAID uses a well-known statistical test called a *chi-squared test* to assess whether splitting a node improves the purity of a node and is statistically significant. At each node split, the variables with the strongest association with the response variable are selected based on the chi-squared test of independence. The tree split is stopped when this test does not show a significant association. This method is more suitable for categorical variables, but it can be adopted by transforming continuous variables into categorical bins.

The other method is to allow the tree to grow fully and then prune the tree. The purpose of pruning is to identify the branches that hardly reduce the error rate and remove them. The process of pruning consists of successively selecting a decision node and redesignating it as a leaf node, thus reducing the size of the tree. The pruning process should reduce misclassification errors and noise but capture the tree patterns.

Most tools provide an option to select the size of the split and the method to prune the tree. If you do not remember the chi-squared test, that's okay. However, it is important to know which method to choose and when and why to choose it. The pruning tree method is implemented in multiple software packages such as SAS, SPSS, C4.5, and other packages. We recommend you read the documentation of the appropriate libraries and packages before selecting the methods.

9.4 Advantages and Disadvantages of Decision Trees

A decision tree is a simple classifier that shows the important predictor variable as a tree's root node. From the end-user perspective, a decision tree requires no transformation of variables or selection of variables to split the tree branch, or even pruning as they are taken care of automatically without any user interference. Furthermore, the construction of trees depends on the observation values and not on the actual magnitude values; therefore, trees are intrinsically robust to outliers and missing values. Also, decision trees are easy to understand because of the simple rules they generate.

However, from a computational perspective, building a decision tree can be relatively expensive because of the need to identify splits for multiple variables. Overfitting is one of the practical challenges of a decision tree algorithm. Further pruning also adds computational time.

9.5 Ensemble Methods and Random Forests

If the data set is large and has too many attributes, then building the large tree, pruning the tree, and avoiding overfitting problems can be challenging. Ensemble methods address this issue. It constructs a set of classifier models from different training sets and aggregates all the models. As a result, the final prediction is based on the aggregate model instead of a single decision tree model, as shown in Figure 9-16.

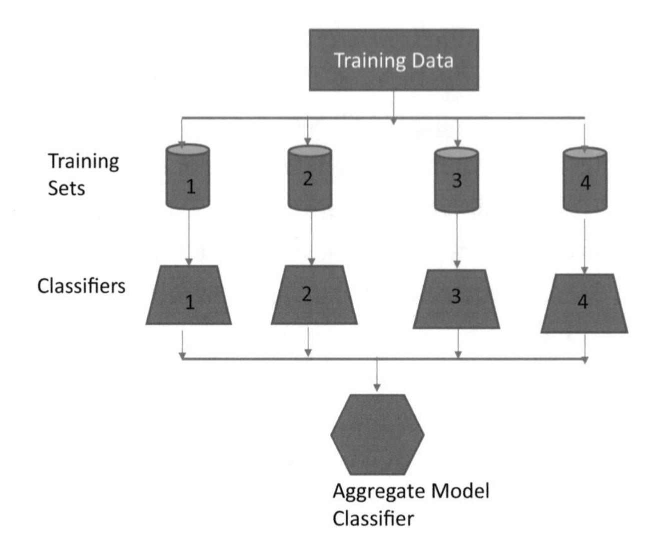

Figure 9-16. *Constructing an ensemble model*

We generally do not have the luxury of having multiple training sets. Instead, we bootstrap by taking repeated samples from the same training set. In this approach, we generate Z different bootstrapped training data sets and then train the model on the bootstrapped training set and average all the predictions. This method is called *bagging*. Though bagging improves the overall predictability and performance, it is often difficult to select the most important variables to the procedure.

A random forest is a small tweak to the bagging tree. A random forest builds trees on a bootstrapped training sample. Each time the tree builds on a random sample of m predictors from a full set of p predictors where m $\approx\sqrt{p}$. The number of predictors considered for each tree split is approximately equal to the square root of the total number of predictors. All predictors have equal chance to participate, and the average model is more reliable with better performance and a low test error overbagging.

Random forests have low bias. By adding more trees, we can reduce variance and thus overfitting. Random forest models are relatively robust to the set of input variables and often do not care about preprocessing of data. Research has shown that they are more efficient in building than other models.

Table 9-4 lists the various types of classifiers and their advantages and disadvantages.

***Table 9-4.** Clasification Algorithms—Advantages and Disadvatages*

SI No	Classification Method	Advantages	Disadvantages
1	Naïve Bayes	Computationally efficient when assumption of independence is true.	Model depends on the posterior probability. When a predictor category is not present in the training data, the model assumes the probability and could result in wrong predictions.
2	Decision tree	Simple rules to understand and easy to comprehend. It does not require any domain knowledge. The steps involved in learning and classification are simple and fast. A decision tree requires no transformation of variables or selecting variables to split the tree branch.	Building a decision tree can be relatively expensive, and further pruning adds computational time. Requires a large data set to construct a good performance classifier.
3	Nearest neighbor	Simple and lack of parametric assumptions. Performs well for large training sets.	Time to find the nearest neighbors. Reduced performance for large number of predictors.
4	Random Forests	Performs well for small and large data sets. Balances bias and variance and provides better performance. More efficient to build than other advanced models.	If a variable is a categorical variable with multiple levels, random forests are biased toward the variable having multiple levels.

9.6 Decision Tree Model Using R

We demonstrate a predictive model using a decision tree algorithm in this example. The data set we are using is the Attrition data set. The company's HR department would like to predict employee attrition. The HR department has captured various parameters of the employees. The data consists of 10 independent variables and target variables. The variables are Attrition, YrsExp, WorkChallenging, WorkEnvir, Compensation, TechExper, maritalstatus, education, children, ownhouse, and loan. We will first demonstrate the decision tree model in R and then repeat it in Python. We will follow the typical process to develop a model as described earlier in the book. The typical process is first to read the data, explore the data, preprocess the data, split data into test and train, then build the model, and test the model performance by measuring the accuracy of the model. We will follow these steps, first in R followed by Python.

Step 1: The first step is to load the data set.

```
>
> #Step1: Set working directory
> setwd("E:/Umesh-MAY2022/Personal-May2022/BA2ndEdition/2ndEdition/Book
  Chapters/Chapter 9 - Pred-Classification/Code/DecisionTree")
> #step 2: Read data
> attrition_df<-read.csv("attrdataDecisionTree.csv")
```

Step 2: Once we read the data, we explore the data set to check the type of the variable, missing values, and null values that may exist in the data. Since the data set is complete and clean, we will continue with the next step. The R tool has read and assigned the correct data types. It has identified categorical and integer variables properly; hence, we do not need to convert any variables' data types.

```
> #Step 2: Data preprocessing and data preparation
> head(attrition_df)
```

```
  Attrition YrsExp WorkChallenging WorkEnvir Compensation
1       Yes    2.5              No       Low          Low
2        No    2.0             Yes Excellent    Excellent
3        No    2.5             Yes Excellent          Low
4       Yes    2.0              No Excellent          Low
5        No    2.0             Yes       Low          Low
6       Yes    2.0              No       Low          Low
  TechExper maritalstatus     education children ownhouse loan
1 Excellent       married undergraduate       no       no   no
2 Excellent       married      graduate       no      yes  yes
3 Excellent        single      graduate       no      yes   no
4 Excellent       married      graduate       no      yes  yes
5       Low       married undergraduate       no      yes   no
6 Excellent        single      graduate       no       no   no

>
```

Step 3: The next step is exploring the data to understand the distribution. You can plot various graphs and tables to check the distribution, skewness, etc. This should indicate how well your assumptions are and how well the model can perform.

The target class is quite balanced. So, you should aim to achieve model performance higher than this distribution. Similarly, all the other variables are also balanced, and the machine can learn quite well; we expect the model to perform beyond 54 percent.

```
> ##Step 3: Explore data. In this case we will look into how data
> # is distributed into different categories.
> table(attrition_df$Attrition)

 No Yes
 96 112
> prop.table(table(attrition_df$Attrition))

       No       Yes
0.4615385 0.5384615
```

```
> prop.table(table(attrition_df$WorkChallenging))

       No       Yes
0.5384615 0.4615385
> prop.table(table(attrition_df$WorkEnvir))

Excellent       Low
0.5384615 0.4615385
> prop.table(table(attrition_df$Compensation))

Excellent       Low
0.4038462 0.5961538

>
```

Step 4: The next step is to divide the sample into two: train and test. We use the `caret()` library in R to perform this operation. Typically we use 80 percent sample data to train the model and 20 percent sample data to test the model. As you can see from the following code, there are 208 sample pieces of data, 80 percent of which is around 167 used to train the model with the remaining 41 samples applied to test the model.

```
> library(caret)
> set.seed(1234)
> data_partition<-createDataPartition(y=attrition_df$Attrition,
+                                     p=0.8, list=FALSE)
> train<-attrition_df[data_partition,]
> test<-attrition_df[-data_partition,]
> nrow(attrition_df)
[1] 208
```

```
> nrow(train)
[1] 167
> nrow(test)
[1] 41
> head(train)
  Attrition YrsExp WorkChallenging WorkEnvir Compensation TechExper maritalstatus
2        No    2.0             Yes Excellent    Excellent Excellent       married
3        No    2.5             Yes Excellent          Low Excellent        single
4       Yes    2.0              No Excellent          Low Excellent       married
5        No    2.0             Yes       Low          Low       Low       married
6       Yes    2.0              No       Low          Low Excellent        single
7        No    2.0              No Excellent    Excellent       Low       married
      education children ownhouse loan
2      graduate       no      yes  yes
3      graduate       no      yes   no
4      graduate       no      yes  yes
5 undergraduate       no      yes   no
6      graduate       no       no   no
7 undergraduate       no      yes   no

>
```

Step 5: Once we have the data split for training and testing, we create the model using the decision tree algorithm. We use the `rpart()` library package to create the decision tree. We use information gain, as discussed earlier, to split the nodes. Please refer to the documentation for more details on the various input parameters used by `rpart()` function. (See `https://cran.r-project.org/web/packages/rpart/rpart.pdf`.)

```
> library(rpart)
> equation = Attrition~YrsExp+WorkChallenging+WorkEnvir+Compens
  ation+Tech
  Exper+maritalstatus+education+children+ownhouse+loan
> attr_tree<-rpart(formula = equation,
+                  data = train,
+                  method = 'class',
+                  minsplit=2,
+                  parms = list(split = 'information')
+                  )

>
```

Step 6: Summarize the model and plot the decision tree structure. The model summary provides information about the split, variables it considered for the tree, and other useful information.

```
> summary(attr_tree)
Call:
rpart(formula = equation, data = train, method = "class", parms =
list(split = "information"), minsplit = 2)
  n= 167

          CP nsplit  rel error    xerror       xstd
1 0.50649351      0 1.00000000 1.0000000 0.08365996
2 0.08441558      1 0.49350649 0.7142857 0.07887537
3 0.06493506      3 0.32467532 0.5844156 0.07446243
4 0.03896104      6 0.11688312 0.1688312 0.04496588
5 0.01948052      7 0.07792208 0.1168831 0.03789665
6 0.01000000     11 0.00000000 0.0000000 0.00000000
```

```
Variable importance
         YrsExp        WorkEnvir WorkChallenging
             20               15              13
  maritalstatus     Compensation       TechExper
             13               12               9
      education        ownhouse         children
              7                7               2
           loan
              2

Node number 1: 167 observations,    complexity param=0.5064935
  predicted class=Yes  expected loss=0.4610778  P(node) =1
    class counts:    77    90
   probabilities: 0.461 0.539
  left son=2 (77 obs) right son=3 (90 obs)
```

R provides many libraries to plot the structure of the final tree. Figure 9-17 is the final tree using the `rpart.plot` function.

```
> library(rpart.plot)
> rpart.plot(attr_tree)

>
```

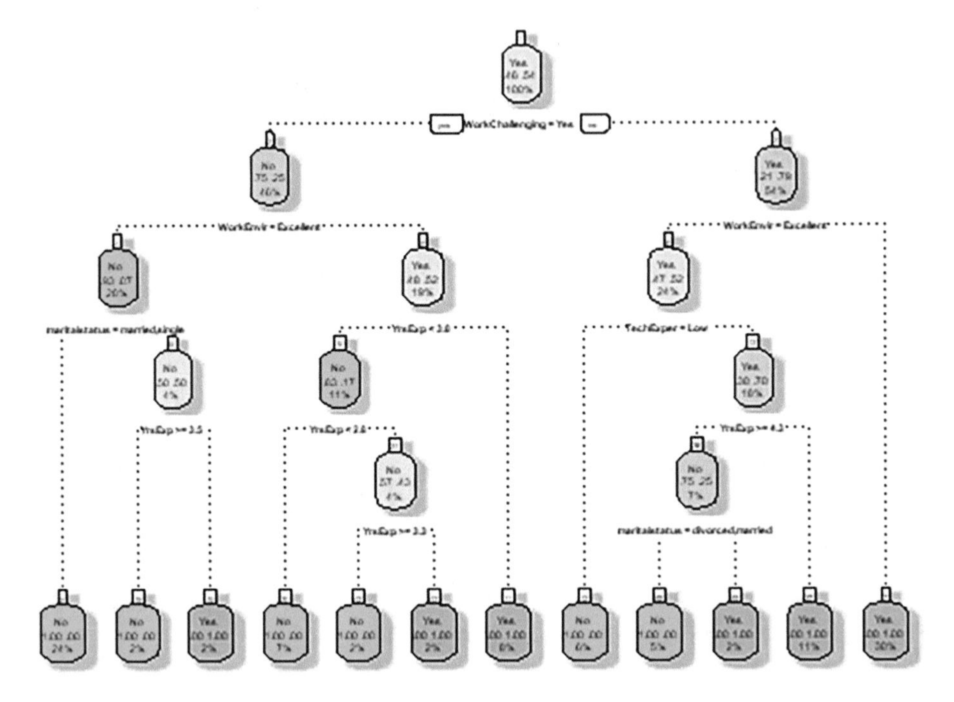

***Figure 9-17.** Full-grown decision tree*

Step 7: Let's apply our fully grown decision tree model to predict the new data (test data). The model's performance is measured by how well the model has predicted the test data. The following code is to predict the test data, and the next step is to measure the performance of the model:

```
> #Step 7: Predict test data
> tree_pred<-predict(attr_tree,newdata=test,
+                      type = 'class', prob = TRUE)
```

Step 8: Let's measure the performance of the model using a contingency table, also called a *confusion matrix*.

```
> #Step : Measure the performance of your model
> library(caret)
> table(tree_pred, test$Attrition)
```

```
tree_pred No Yes
      No  19   0
      Yes  0  22

> confusionMatrix(as.factor(tree_pred), test$Attrition)
Confusion Matrix and Statistics

          Reference
Prediction No Yes
       No  19   0
       Yes  0  22

               Accuracy : 1
                 95% CI : (0.914, 1)
    No Information Rate : 0.5366
    P-Value [Acc > NIR] : 8.226e-12

                  Kappa : 1

 Mcnemar's Test P-Value : NA

            Sensitivity : 1.0000
            Specificity : 1.0000
         Pos Pred Value : 1.0000
         Neg Pred Value : 1.0000
             Prevalence : 0.4634
         Detection Rate : 0.4634
   Detection Prevalence : 0.4634
      Balanced Accuracy : 1.0000

       'Positive' Class : No
```

The confusion matrix function in R provides the accuracy of the model, false-positive rates, specificity, sensitivity, and other measures. In this case, the accuracy of the model is 100 percent. However, the model has considered all the variables. We do not know whether the tree is under fitted or overfitted as we did not measure the performance of training data. This exercise allows you to follow the same steps and functions to calculate the training data prediction performance and determine whether the model is overfitting or underfitting. Also, if you look at the tree, it did not consider all the variables.

Step 9: We assume that the tree is full-grown, and we will demonstrate how to prune. We will also plot the pruned tree (Figure 9-18). We will be using the built-in `prune()` AIP function. It takes a fully grown tree model and applies the pruning method. In this case, this package has an option to use `cp` (complexity parameter) method. You should refer to the `rpart()` documentation for the other methods and options.

(Here is the `rpart` documentation: `https://www.rdocumentation.org/packages/rpart/versions/4.1.16/topics/rpart.`) See Figure 9-18.

```
> ##Now, we will prune the tree
> pruned_tree<-prune(attr_tree,cp=0.05)
> rpart.plot(pruned_tree)

>
```

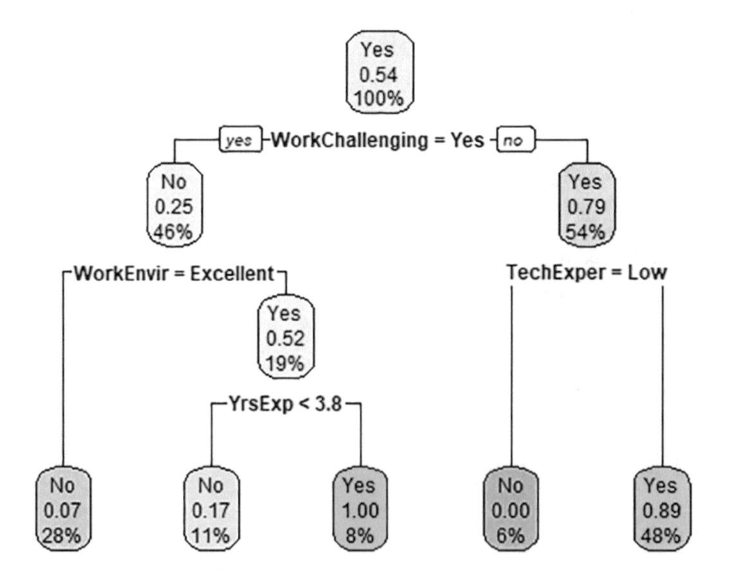

Figure 9-18. *Plot of a pruned tree model*

Step 10: We will predict the test data on the pruned tree model and check the accuracy of the model.

```
> ##Predict using pruned tree
> pred_prun<-predict(newdata=test, pruned_tree, type='class')
> ##Measure performance
> confusionMatrix(as.factor(pred_prun), test$Attrition)
Confusion Matrix and Statistics

          Reference
Prediction No Yes
       No  16   2
       Yes  3  20

               Accuracy : 0.878
                 95% CI : (0.738, 0.9592)
    No Information Rate : 0.5366
    P-Value [Acc > NIR] : 3.487e-06

                  Kappa : 0.7539

 Mcnemar's Test P-Value : 1

            Sensitivity : 0.8421
            Specificity : 0.9091
         Pos Pred Value : 0.8889
         Neg Pred Value : 0.8696
             Prevalence : 0.4634
         Detection Rate : 0.3902
   Detection Prevalence : 0.4390
      Balanced Accuracy : 0.8756

       'Positive' Class : No
```

The pruned tree has an accuracy of 87 percent, and it has considered only four variables for constructing the tree. Further, by working on other parameters of the `rpart()` library, you can improve the performance of the model.

Finally, we will plot the receiver operating curve (ROC) to measure the performance of the model using sensitivity analysis. In R, use the ROCR library to plot the ROC graph and measure the AUC. See Figure 9-19.

```
> ##ROC Curve
> library("ROCR")
> pred_prob <- predict(pruned_tree,newdata=test, type = "prob")[, 2]
> pred2_prob = prediction(pred_prob, test$Attrition)
> plot(performance(pred2_prob, "tpr", "fpr"),col='red',
+       main="ROC: Predicting Attrition")

> > auc = performance(pred2_prob, 'auc')
> slot(auc, 'y.values')

[[1]]
[1] 0.9055024
```

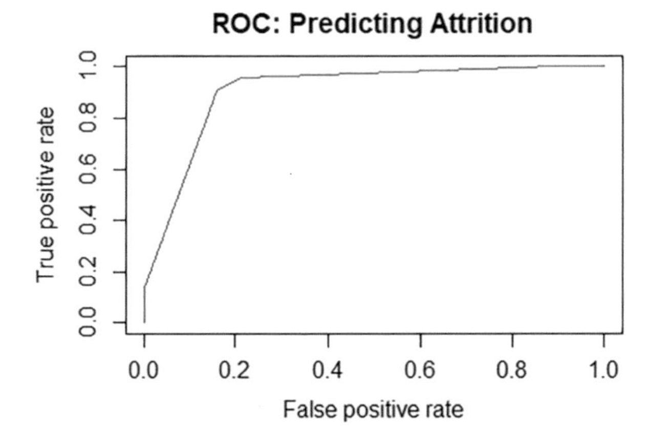

Figure 9-19. *ROC of the decision tree model*

Note To learn more about the `rpart()` library and options, please refer to the documentation: `https://cran.r-project.org/web/packages/rpart/rpart.pdf`.

9.7 Decision Tree Model Using Python

In this example, we demonstrate a predictive model using a decision tree algorithm with Python. We use the same data set and the process we used in our earlier example to create the decision tree model. The data set we are using is the Attrition data set. The company's HR department would like to predict employee attrition. The HR department has captured various parameters of employees. The data consists of 10 independent variables and a target variable. The variables include Attrition, YrsExp, WorkChallenging, WorkEnvir, Compensation, TechExper, maritalstatus, education, children, ownhouse, and loan.

Step 1: Import all the necessary libraries required to create the model. These are the typical libraries we use in Python. The main packages for the data manipulations is the Pandas dataframe, and we use `sklearn()` for creating the model.

```
from __future__ import absolute_import, division, print_function
import codecs
import glob
import logging
import multiprocessing
import os
import pprint
import re

##Load pandas and numpy library
import pandas as pd
import numpy as np

## Loading plot libraries
%matplotlib inline
import matplotlib.pyplot as plt
from pandas.plotting import scatter_matrix
from sklearn.metrics.pairwise import cosine_similarity
from scipy.spatial.distance import cosine
import sklearn as sk
import matplotlib.pyplot as plt
import seaborn as sns
```

```
# Preprocessing libraries
from sklearn.model_selection import train_test_split
from sklearn.preprocessing import scale
from sklearn.model_selection import train_test_split

############# Load Decision Tree library
from sklearn.tree import DecisionTreeClassifier
from sklearn import model_selection

##Measuring Accuracy of the Model
#Import scikit-learn metrics module for accuracy calculation

from sklearn.metrics import roc_auc_score
from sklearn.metrics import roc_curve
from sklearn.metrics import classification_report
from sklearn.metrics import confusion_matrix
from sklearn import metrics
from sklearn.metrics import roc_curve
from sklearn.metrics import auc
#https://scikit-learn.org/stable/modules/generated/
sklearn.tree.DecisionTreeClassifier.html
```

Step 2: Read the data set into a Pandas dataframe. All the data manipulations are performed on the Pandas dataframe.

```
#Set the working directory where the dataset is stored before reading
the dataset
data_dir = 'E:/Code/DecisionTree'
os.chdir(data_dir)
attrData_df = pd.read_csv("attrdataDecisionTree.csv")
attrData_df.head()
```

```
  Attrition  YrsExp WorkChallenging  WorkEnvir Compensation  TechExper  \
0       Yes     2.5              No        Low          Low  Excellent
1        No     2.0             Yes  Excellent    Excellent  Excellent
2        No     2.5             Yes  Excellent          Low  Excellent
3       Yes     2.0              No  Excellent          Low  Excellent
4        No     2.0             Yes        Low          Low        Low

  maritalstatus      education children ownhouse loan
0       married  undergraduate       no       no   no
1       married       graduate       no      yes  yes
2        single       graduate       no      yes   no
3       married       graduate       no      yes  yes
4       married  undergraduate       no      yes   no
##COpy data to another variable(dataframe) as abackup.
X = attrData_df.copy()
X.head()
  Attrition  YrsExp WorkChallenging  WorkEnvir Compensation  TechExper  \
0       Yes     2.5              No        Low          Low  Excellent
1        No     2.0             Yes  Excellent    Excellent  Excellent
2        No     2.5             Yes  Excellent          Low  Excellent
3       Yes     2.0              No  Excellent          Low  Excellent
4        No     2.0             Yes        Low          Low        Low

  maritalstatus      education children ownhouse loan
0       married  undergraduate       no       no   no
1       married       graduate       no      yes  yes
2        single       graduate       no      yes   no
3       married       graduate       no      yes  yes
4       married  undergraduate       no      yes   no
```

Step 3: Explore the data. Check the data type, missing values, etc.

In this case, the data set is clean and there are no missing values. However, in reality, this is not true. As a first step, you should always clean the data set. Unlike R,

Python sklearn() libraries do not accept text values to build the models. Hence, we have to convert any text values to numerical values. In our datset, the variables are categorical, and they are text. We use the LabelEncoder() function to convert them to numerical values. We could also use the Pandas get_dummies() function or sklearn OrdinalEncoder() function. Since the variables are just nominal and has only two categories (YES / NO, Married / Single, Graduate / Undergraduate) the get_dummies(), function also gives the same results. To learn when and how to use dummy variables, you could refer to the sklearn documentation. Typically we use LabelEncoder() to the target class and get_dummies() for the independent variables. Since we have only two classes in our data set variables, we will use the LabelEncoder() function instead of get_dummies as both provide the same end results.

Step 4: This step includes data exploration and preparation.

```
##Check the data types, missing values, etc
##Use LabeEncoder to code Target variable 'Attrition'
# Converting string labels into numbers.
from sklearn.preprocessing import LabelEncoder
encoder = LabelEncoder()
X[['Attrition','WorkChallenging',
         'WorkEnvir','Compensation',
         'TechExper', 'maritalstatus','education',
          'children','ownhouse','loan']] = X[['Attrition',
          'WorkChallenging',
         'WorkEnvir','Compensation',
         'TechExper', 'maritalstatus','education',
          'children','ownhouse','loan']].apply(encoder.fit_transform)
          X.head()
   Attrition  YrsExp  WorkChallenging  WorkEnvir  Compensation  TechExper  \
0          1     2.5                0          1             1          0
1          0     2.0                1          0             0          0
2          0     2.5                1          0             1          0
3          1     2.0                0          0             1          0
4          0     2.0                1          1             1          1
```

```
   maritalstatus  education  children  ownhouse  loan
0              1          1         0         0     0
1              1          0         0         1     1
2              2          0         0         1     0
3              1          0         0         1     1
4              1          1         0         1     0
```

Step 4a: Convert the object data types to categorical data types. Since our data set variables are categorical and Python reads them as an object, we need to convert them to categorical.

```
cat_vars = ['Attrition','WorkChallenging',
   'WorkEnvir','Compensation',
   'TechExper', 'maritalstatus','education',
   'children','ownhouse','loan']
for var in cat_vars:
    X[var] = X[var].astype('category',copy=False)
X.dtypes
Attrition          category
YrsExp              float64
WorkChallenging    category
WorkEnvir          category
Compensation       category
TechExper          category
maritalstatus      category
education          category
children           category
ownhouse           category
loan               category
dtype: object
```

Step 4b: To understand the data distribution and how all the variables influence the target class, explore the data using data visualization or tables. In the following example, we just use the tables as all our data has only two classes.

```
X.Attrition.value_counts()
1    112
0     96
Name: Attrition, dtype: int64
X.WorkChallenging.value_counts()
0    112
1     96
Name: WorkChallenging, dtype: int64
X.TechExper.value_counts()
0    176
1     32
Name: TechExper, dtype: int64
X.Compensation.value_counts()
1    124
0     84
Name: Compensation, dtype: int64
```

Step 4c: In this step, prepare the data as per the `sklearn()` algorithm API function. The `DecisionTree()` function of sklearn accepts inputs as X, only independent variables, Y, only target class. Hence, we separate `Attrition` from the X as shown.

```
##Prepare data for the sklearn decisiontree function(). We need to
separate X and Y parameters
X_NoAttrVar = X.drop(columns='Attrition')
Y = X['Attrition']
X_NoAttrVar.head(2)

   YrsExp WorkChallenging WorkEnvir Compensation TechExper maritalstatus  \
0     2.5               0         1            1         0             1
1     2.0               1         0            0         0             1

  education children ownhouse loan
0         1        0        0    0
1         0        0        1    1
Y.head(2)

0    1
1    0
Name: Attrition, dtype: category
Categories (2, int64): [0, 1]
```

Step 5: Split the data into train and test. We use 80 percent for training and 20 percent data for testing. This is the standard rule used across the industry. If you have lots of data, you can even split 75 percent and 25 percent.

```
# Split dataset into training set and test set. We will use sklearn
train_test_split() function
X_train, X_test, y_train, y_test = train_test_split(X_NoAttrVar, Y,
train_size = 0.8, random_state=1)
# 80% training and 20% test
X_train.shape
(166, 10)
```

```
y_train.shape
(166,)
X_test.shape
(42, 10)
y_test.shape
(42,)
```

Step 6: Create a decision tree model using the sklearn `decisiontreeclassifier()` function. Use the training data to create the model and test data to test the model. Once again, the API supports the number of parameters, and each parameter has a significance and helps in fine-tuning the model performance. In our case, we have used the default values and allowed the tree to grow 10 levels. (Refer to `https://scikit-learn.org/stable/modules/classes.html#module-sklearn.preprocessing` for more information.)

9.7.1 Creating the Decision Tree Model

Here is the decision tree model:

```
dec_model= DecisionTreeClassifier(random_state=10,
                                  criterion="entropy", max_depth=10)
dec_model_fit = dec_model.fit(X_train, y_train) ## Train the model
DecisionTreeClassifier(criterion='entropy', max_depth=10, random_
state=10)
dec_model_fit.get_n_leaves()
11
dec_model_fit.classes_
array([0, 1], dtype=int64)
```

Step 6a: Plot the tree model. Once you have the model, plot the decision tree and see how many levels your model has. This helps you to understand the model and arrive at the rules.

```
# Plot the tree structure of fully grown tree
from sklearn import tree
tree.plot_tree(dec_model_fit)
```

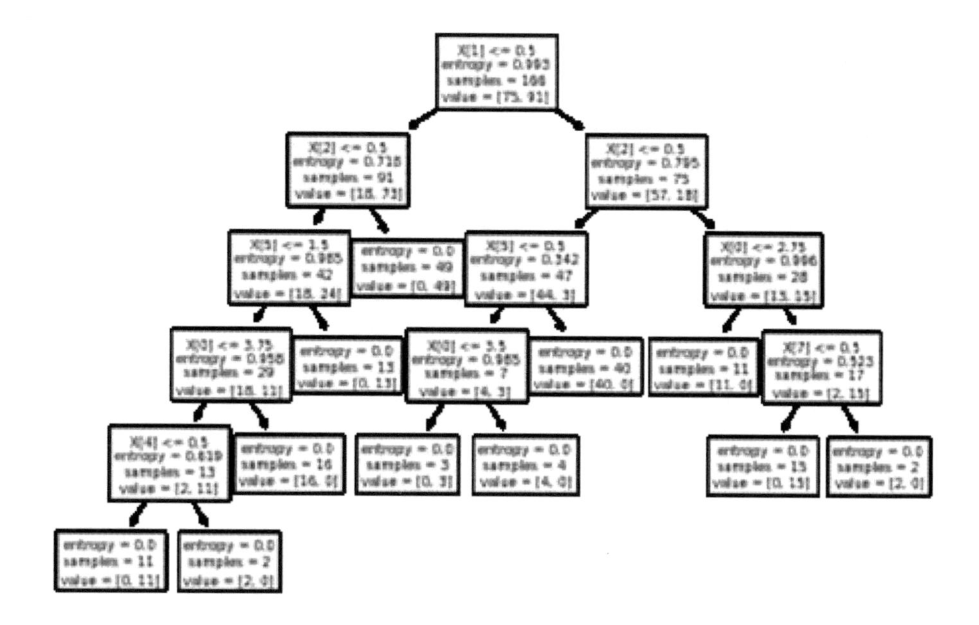

Step 7: The final step is to predict the test data, without the labels, from the decision tree model we just created and measure the model accuracy.

9.7.2 Making Predictions

```
predicted = dec_model_fit.predict(X_test)
```

Step 8: Measure the accuracy of the model using a contingency table (also referred to as a *confusion matrix*). The classification_report() function also provides a sensitivity analysis report including precision, recall, and f-score.

9.7.3 Measuring the Accuracy of the Model

Here is how to measure the accuracy:

```
print(classification_report(y_test, predicted))

              precision    recall  f1-score   support

           0       1.00      1.00      1.00        21
           1       1.00      1.00      1.00        21

    accuracy                           1.00        42
   macro avg       1.00      1.00      1.00        42
weighted avg       1.00      1.00      1.00        42

print(confusion_matrix(y_test, predicted))
[[21  0]
 [ 0 21]]
print("Accuracy:",metrics.accuracy_score(y_test, predicted))
Accuracy: 1.0
```

9.7.4 Creating a Pruned Tree

In this section, we will demonstrate how to prune a fully grown tree.

```
dec_model_2= DecisionTreeClassifier(random_state=0,
                                   criterion="entropy", max_depth=4)
dec_model_fit_2 = dec_model_2.fit(X_train, y_train) ## Train the model
##Predict Test Data
predicted_2 = dec_model_fit_2.predict(X_test)
```

Step 9: Measure the performance of the "pruned tree" and check the accuracy of the model. The accuracy of the pruned tree model is only 95.23 percent. Finally, plot the pruned tree using the plot_tree() function.

```
##Performance of the model
print(classification_report(y_test, predicted_2))
print("Accuracy:",metrics.accuracy_score(y_test, predicted_2))

##Plot the tree
tree.plot_tree(decision_tree = dec_model_fit_2,
               feature_names = X_train.columns,
               class_names = (['yes', 'No']),
               filled = True,
               impurity = False
              )

              precision    recall  f1-score   support

           0       1.00      0.90      0.95        21
           1       0.91      1.00      0.95        21

    accuracy                           0.95        42
   macro avg       0.96      0.95      0.95        42
weighted avg       0.96      0.95      0.95        42

Accuracy: 0.9523809523809523
```

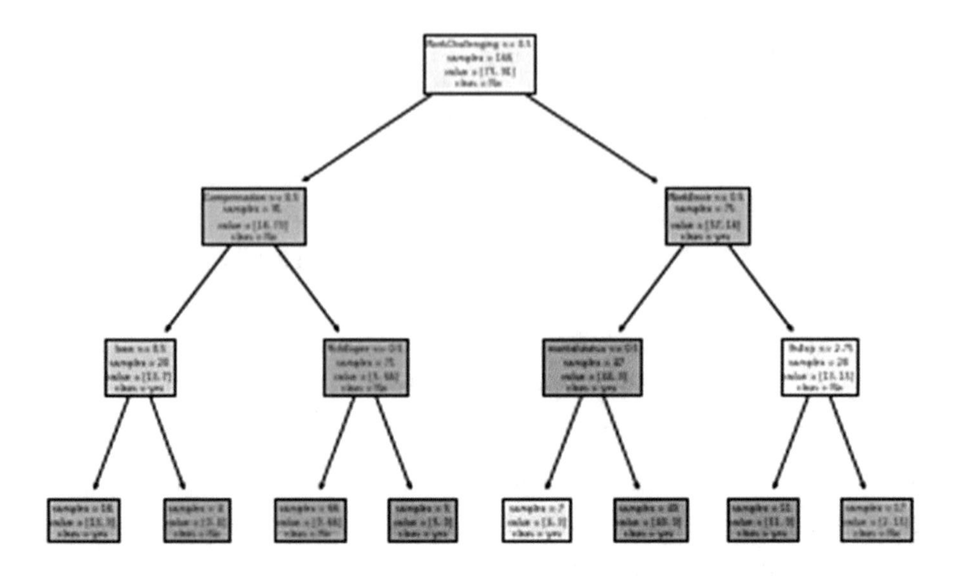

Step 10: Plot the ROC curve and find out the AUC measure as well. See Figure 9-20.

```
roc_auc_score(y_test,dec_model_fit_3.predict_proba(X_test)[:,1])
0.9761904761904762
# Compute fpr, tpr, thresholds
fpr, tpr, thresholds = roc_curve(y_test, predicted_3)
plt.plot(fpr, tpr)
plt.xlabel('False Positive Rate or (1 - Specifity)')
plt.ylabel('True Positive Rate or (Sensitivity)')
plt.title('Receiver Operating Characteristic')
```

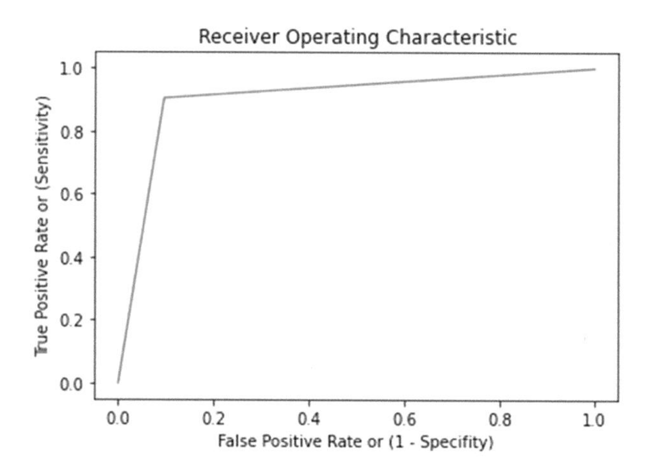

***Figure 9-20.** ROC of the pruned decision tree model*

9.8 Chapter Summary

The chapter explained the fundamental concepts of the classification method in supervised machine learning and the differences between the classification and prediction models.

We discussed various classification techniques such as k-nearest neighbor, naïve Bayes, and decision tree. We also touched upon ensemble models.

This chapter described the decision tree model, how to build the decision tree, how to select the decision tree root, and how to split the tree. You saw examples of building the decision tree, pruning the tree, and measuring the performance of the classification model.

You also learned about the bias-variance concept with respect to overfitting and underfitting.

Finally, you explored how to create a classification model, how to measure the performance of the model, and how to improve the model performance, using R and Python.

CHAPTER 10

Neural Networks

Though neural networks have been around for many years, because of technological advancement and computational power, they have gained popularity recently and now perform better than other machine learning algorithms today. In this chapter, we will discuss using neural networks and associated deep neural network algorithms to solve classification problems.

10.1 What Is an Artificial Neural Network?

Our brain is highly complicated and is a structure that does reasoning, makes connections, and comes to decisions using a network of natural neurons. Let's just take an example of visually recognizing a handwritten number. Our human brains can easily recognize the numbers in Figure 10-1 as 3 even though there is a variation between them.

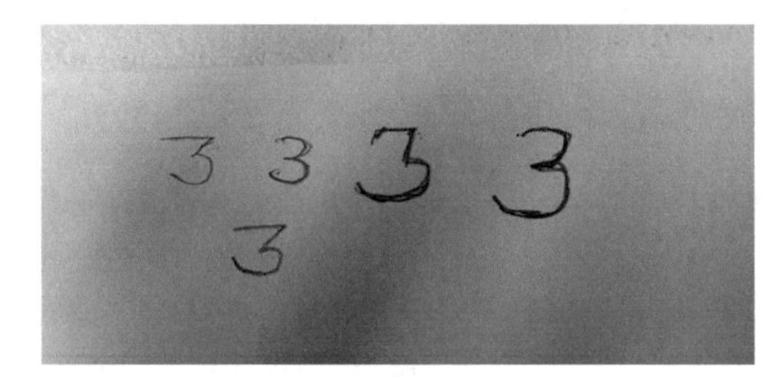

Figure 10-1. *How a human senses images*

This is possible because the human brain can process every detail and variation with the help of millions of neurons based on the past memory of a similar object or pattern and experience. Millions of neurons and billions of connections connect the brain and eyes. This process happens seamlessly, as the brain is like a supercomputer and able to process complex information within microseconds. Hence, you can recognize

U. R. Hodeghatta and U. Nayak, *Practical Business Analytics Using R and Python*,
https://doi.org/10.1007/978-1-4842-8754-5_10

patterns presented to you visually almost instantaneously. Neurons are the structural and functional units of the nervous system. The nervous system is divided into a central and peripheral nervous system. The central nervous system (CNS) is composed of the brain and its neuronal connections. Neurons transmit signals between each other via junctions known as *synapses*. Synapses are of two types, electrical and chemical. The CNS neurons are connected by electrical synapses, as shown in Figure 10-2.

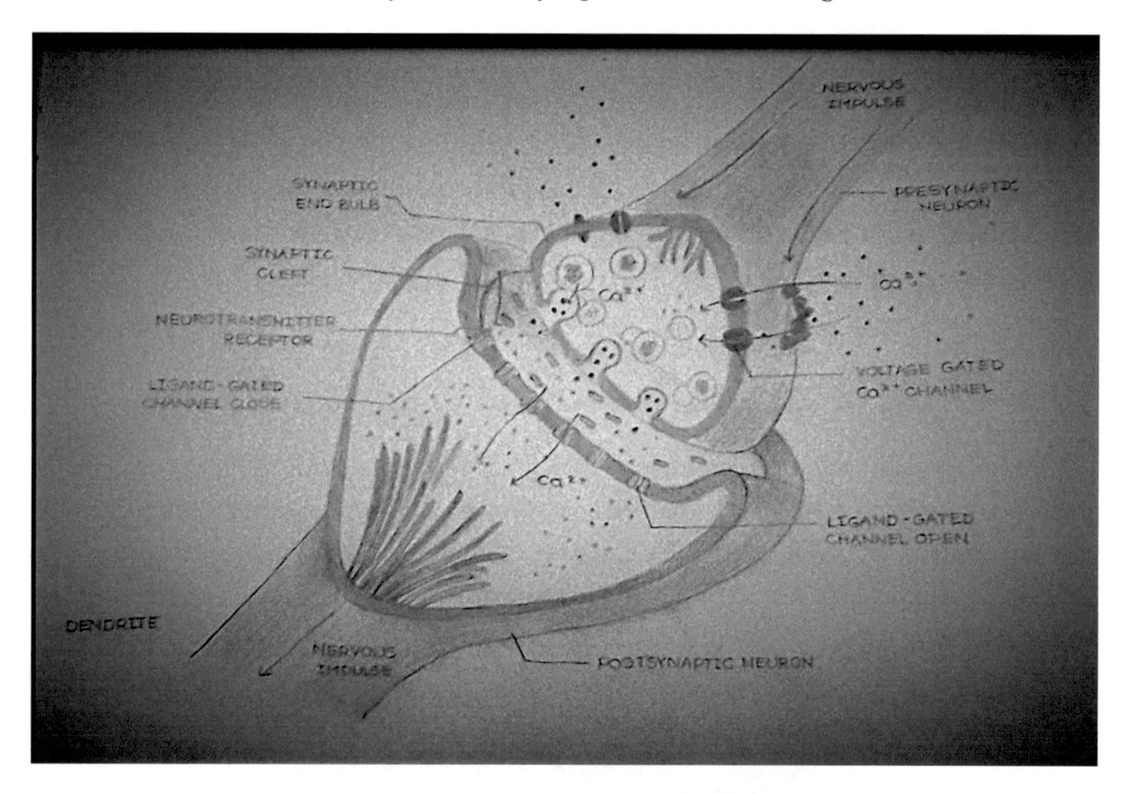

Figure 10-2. *Human nervous system (Drawn By: Mahaswin Sai M)*

The neurons act via action potentials with the help of voltage-gated and ligand-gated ionic channels. Action potentials *activate* the neuron and transmit signals in synapses using neurotransmitters. The neurotransmitters can excite or inhibit other neurons. The main seat of memory in the CNS is the hippocampus. The hippocampus is important for the conversion of short-term memory to long-term memory. This conversion is known as *consolidation* or *long-term potentiation*. Repeated action by revisions strengthens the synopsis between neurons. This long-term memory helps in the future retrieval of memory. Thus, if the hippocampus is lesioned, it can lead to amnesia, popularly known as memory loss.

Computers cannot do this as easily as humans. *Artificial neural networks* (ANNs) emulate how the brain works as far as the learning or cognition is concerned. Layers of artificial neurons are deployed to solve the given problem. The inputs from the input layer pass through and are transformed until the final output is created through the output layer. *Deep learning* is learning through the deep woven architecture of the neural nets. Here, a number of different layers of artificial neurons are typically deployed to ensure effective and high learning from the data. Currently, artificial intelligence is complemented by the huge data processing capabilities of the cloud and a cluster of machines using the distributed and collective power of multiple machines. This has made computer learning possible using the vast amount of data being collected. Based on this learning, computers can now learn from data and compete with humans in games like checkers, chess, etc. These networks not only learn but have the capability to improve on their performance with more data and learning cycles.

10.2 Concept and Structure of Neural Networks

The concept of a neural network was developed in the 1940s with the introduction of simplified neurons by McCulloch and Pitts (McCulloch and Pitts, 1943). Their paper presented neurons as a biological model that could perform complex computational tasks. After a few years of research, in 1962 Rosenblatt published *Principles of neurodynamics: Perceptrons and the theory of brain mechanisms* (Spartan Books, 1962). In 1969, Minsky and Papert published the book *Perceptrons* (Misky and Papert, 1969). They described the single neuron model with mathematical proof of some of the characteristics of layers connecting different neurons with a feed-forward approach. *Perceptrons* was the first systematic study of parallelism in computation and has been a classical work on automata networks for nearly two decades. This marked the historical beginning of artificial intelligence. This is where computer scientists are able to simulate brain-like entities from the networks of neurons, and the evaluation of higher-processing computers with parallelism gave perceptrons a new edge in the field.

10.2.1 Perceptrons

Let's start the discussion by understanding how a perceptron works. Once we understand how a perceptron works, we can learn to combine multiple perceptrons to form a feed-forward neural network. At its core, a *perceptron* is a simple mathematical model that takes a set of inputs and does certain mathematical operations to produce computation results. A perceptron with the activation function is similar to a biological neuron; the neurotransmitters can excite or inhibit other neurons based on the electrical signals by sensors. In this case, the perceptron with its activation function has to decide whether to fire an output or not based on the threshold values.

A perceptron can take several input values and produce a single output. In the example shown in Figure 10-3, the perceptron takes two inputs, x_1 and x_2, and produces a single output. In general, it could have more than two inputs. The output is the summation of weights multiplied by the input x.

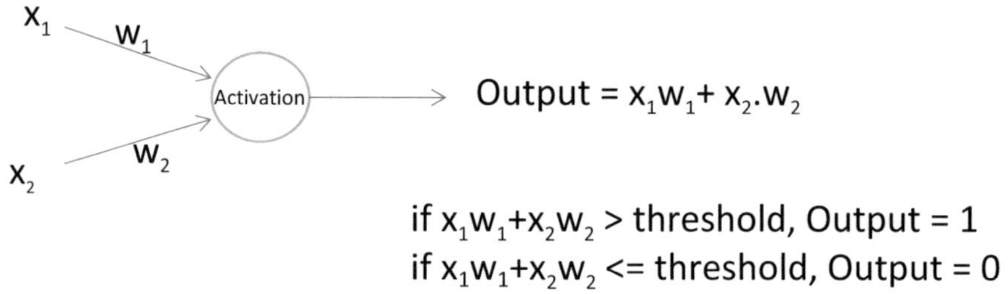

x_1 w_1 x_2 w_2 Activation

Output = $x_1w_1 + x_2.w_2$

if $x_1w_1+x_2w_2$ > threshold, Output = 1
if $x_1w_1+x_2w_2$ <= threshold, Output = 0

Figure 10-3. *Perceptron*

Rosenblatt in his book proposed a function to compute the output. The output Y depends on the inputs x_1 and x_2, and in his book he proposed giving the weightage w_1, w_2 corresponding to x_1 and x_2. The neuron triggers output 1 or 0 determined by the weighted summation of w and x, $W.X = \sum_{k=0}^{n} w_k \cdot x_k$. The neuron triggers output 0 or 1 depending on the threshold value. To put this in a perspective, here it is expressed as an algebraic equation:

$$\text{Output} = \begin{cases} 0 = if \sum_{k=0}^{n} x_k w_k & \leq \text{threshold} \\ 1 = if \sum_{k=0}^{n} x_k w_k & \geq \text{threshold} \end{cases} \tag{1}$$

The output is 0 if the sum of all the weights multiplied by input x is less than some threshold, and the output is 1 if it is greater than some threshold.

The mathematical function is called the *activation* function. In this case, the activation function is the summation function of all the inputs and corresponding weights, and it fires 0 or 1 based on the threshold set, as shown in the previous equation. Since the output of the neuron is binary, it can be treated as a simple classification model. For example, using this model, you could build a model to predict playing golf, based on rain, x_1, and weather outlook, x_2. Depending on the importance of the features, corresponding weights can be assigned. In this case, as shown in Figure 10-4, if you assign 0 to a rainy day, 1 to a sunny day, and 1 if the outlook is sunny, 0 if the outlook is cloudy, and assign $w_1 = 5$, $w_2 = 3$, then if $x_1 = 0$, $x_2 = 0$, you will not play golf. If $x_1 = 1$, $x_2 = 0$, you would play golf, and vice versa.

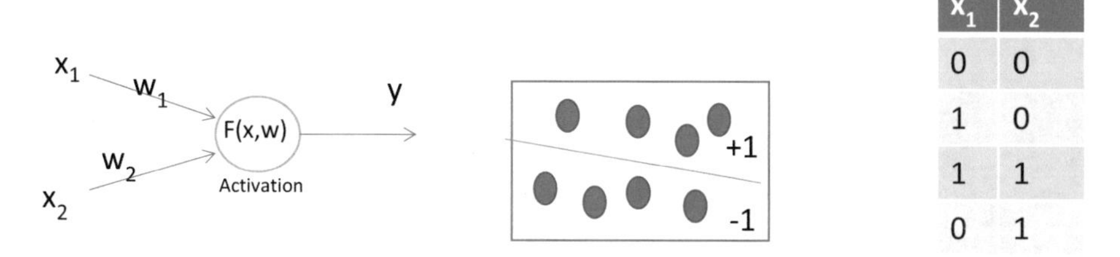

x_1	x_2
0	0
1	0
1	1
0	1

Figure 10-4. *A perceptron as a classifier*

By rearranging the terms and replacing the threshold with bias, b, the equation simplifies as follows:

$$\begin{aligned} \text{Output} = 0 \text{ If } \quad & w.x + b <= 0 \\ \text{Output} = 1 \text{ If } \quad & w.x + b => 0 \end{aligned} \tag{2}$$

Term *b* is called *bias*. Bias determines how easy it is to fire the perceptron output. For a large bias, it is easy for perceptron output to be 1, whereas, for a small bias, it is difficult to fire output. Henceforth, we will not use the term *threshold*; instead, we use bias. The perceptron fires out 0 or 1 based on both weights and bias. For simplicity we can keep the bias fixed, and we can optimize the weights. How do we tune the weights and biases in response to external stimuli without intervention? We can devise a new learning algorithm, which automatically tunes the weights and biases of an artificial neuron. This is the basic idea of the neural network algorithm.

A simple perceptron, as shown earlier, fires binary output, such as 0 or 1, True or False, and Yes or No, and thus this concept can be applied to a linear classification. If x_1 and x_2 are two inputs for the perceptron, then the perceptron output can be used to classify by implementing logical functions like AND, OR, or NOT. Output regions can be separated by a line, called *linearly separable regions*, given the inputs x_1 and x_2.

However, the XOR operator is not linearly separable by a single neuron, as shown in Figure 10-5. But yet we could overcome this problem by using more than one neuron.

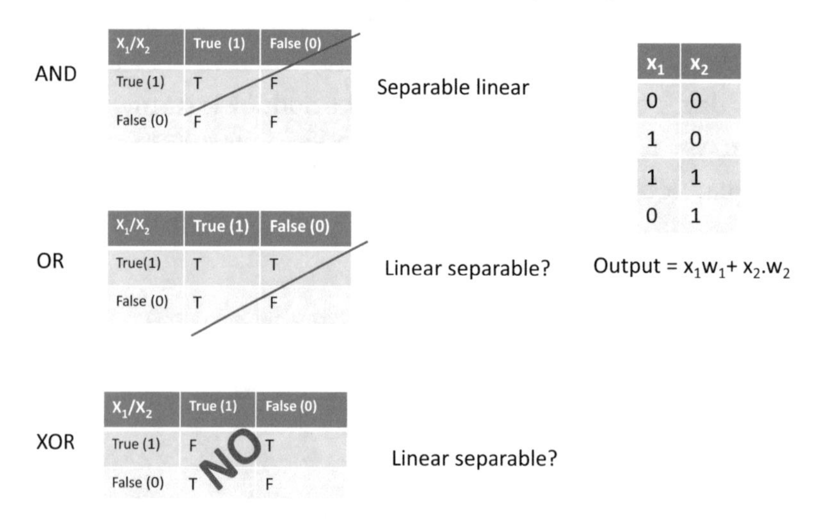

Figure 10-5. *Perceptron as a linear classifier*

Footnote: Those unfamiliar with basic logic operations such as AND, OR, NOT, and XOR, please refer to advanced books on electronics engineering.

The previous example demonstrates how to use a logical operator since this is one of the most common problems used in networks and circuits. However, more complex problems are solved by using multiple neurons and complex functions to stimulate the output by tuning weights and biases without direct intervention by a programmer. In the following section, let's explore the neural network architecture with multiple neurons and how the activation function triggers output. Also, we will explore the algorithm that tunes weights and biases automatically.

10.2.2 The Architecture of Neural Networks

The purpose of the neural network is to capture complicated relationships and patterns by learning input information through a series of neurons. Researchers have studied numerous types of neural network architectures, but the most successful neural networks have been multilayer feed-forward networks. The basic neural network architecture consists of multiple neurons organized in a specific way. It consists of multiple layers, as shown in Figure 10-6. The leftmost layer in this network is called the *input layer*, and the neurons within this layer are called *input neurons*. The middle layer is called a *hidden layer*, as the neurons in this layer have inputs from the previous layer and outputs to the next layer and do not receive direct input or output. There could be multiple hidden layers in the network as the network learning deepens. Figure 10-6 is a three-layer neural network, whereas Figure 10-7 is a four-layer network with two hidden layers.

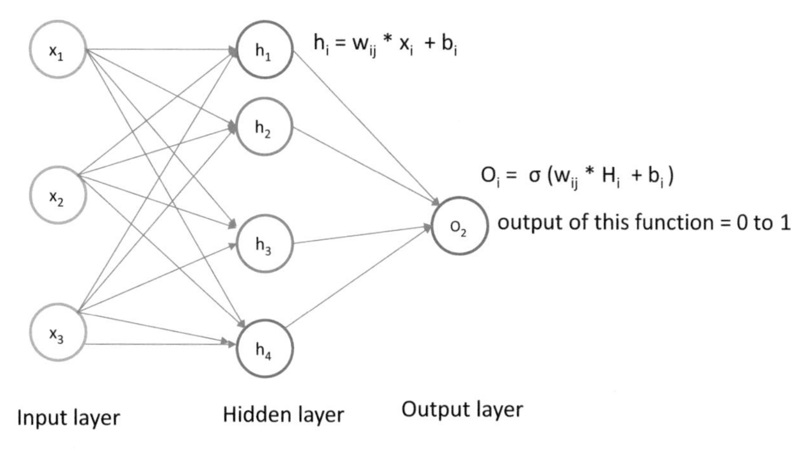

Figure 10-6. *Fully connected neural network architecture*

The last layer is called the *output layer*, which contains the output neuron. The output neuron can be only one, or it can be more than one depending on the input conditions. In general, each neuron is called a *node*, and each link connection is associated with weights and bias. Weights are similar to coefficients in linear regression and subjected to iterative adjustments. Biases are similar to coefficients but are not subjected to iterative adjustments.

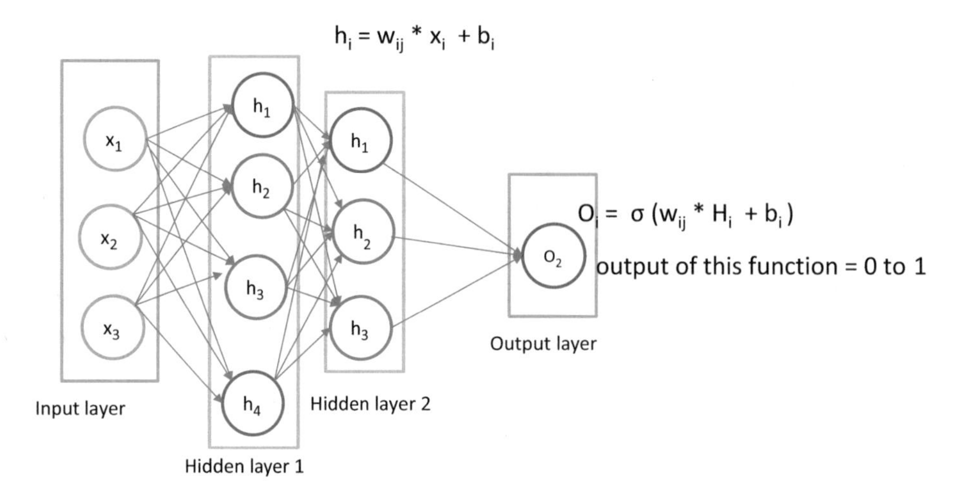

***Figure 10-7.** Fully connected four-layer neural network architecture*

$x_1, x_2, x_3, \ldots x_n$ are the input to the hidden-layer neurons; the output of the hidden neurons is the summation of the weights and bias.

$$h_i = w_{ij}{}^{*} x_i + b_i$$

Similarly, the output is the summation of weights and bias of hidden neurons.

$$O_i = \sigma\left(w_{ij}{}^{*} H_i + b_i\right).$$

These are represented as vectors. The output depends on the activation function. The most common activation function used in the output is a sigmoid. We will discuss different types of activation functions in the next section. Overall, the learning happens from one layer to the next in a sequential pattern. Each layer attempts to learn the minute details of the feature to predict the outcome as accurately as possible.

10.3 Learning Algorithms

The purpose of a neural network algorithm is to capture the complicated relationship between the response variable and the predictors variable much more accurately. For instance, in linear regression we assume that the relationship between the response variable and the predictor variable is normally distributed and is linear. In many cases this is not true, and the relationship may be unknown. To overcome this, we adopt several transformations of the data to make it normal. In the case of neural networks, no such transformation or correction is required. The neural network tries to learn such a relationship by passing through different layers and adjusting the weights. How does learning happen, and how does a neural network capture and predict output? We will illustrate this with an example of a simple data set. This data set is only for the purpose of explaining the concept, but in practice, the features are much more complex, and the data size is also larger. In the example demonstration, we will be using a larger data set to demonstrate how to create the neural network model.

10.3.1 Predicting Attrition Using a Neural Network

In this example, we will consider a small data set, shown in Table 10-1, which has only six training records. The predictor variable `Attrition` depends on the three variables `YrsExp`, `AnnuSalary`, and `SkillLev`. `Attrition` has two classes: 1 or 0; 1 means yes, which means there is attrition in the company, and 0 means no attrition. The objective of the neural network learning model is to predict the attrition given the inputs, i.e., three independent variables.

Table 10-1. *The Example Data Set for Predicting Attrition*

Attrition	YrsExp	AnnuSalary	SkillLev
1	2.5	4000	0.5
0	2	6000	1
0	2.5	8000	1.5
1	2	10000	2
0	2	12000	2.5
1	2	14000	3.5
0	2	16000	4

Let's design a simple neural network model to understand how a neural network predicts attrition. Figure 10-8 is our neural network model consisting of three input neurons, four hidden-layer neurons, and one output neuron. For simplicity, we will name each node. Nodes 1, 2, and 3 are input-layer nodes; nodes 4, 5, 6, and 7 belong to the hidden layer, and node 8 is the output layer. Each hidden neuron is connected by a weights arrow from the input neurons and denoted by the weight vector, w_{ij}, from node i to j, and the additional bias vector is denoted by b_i.

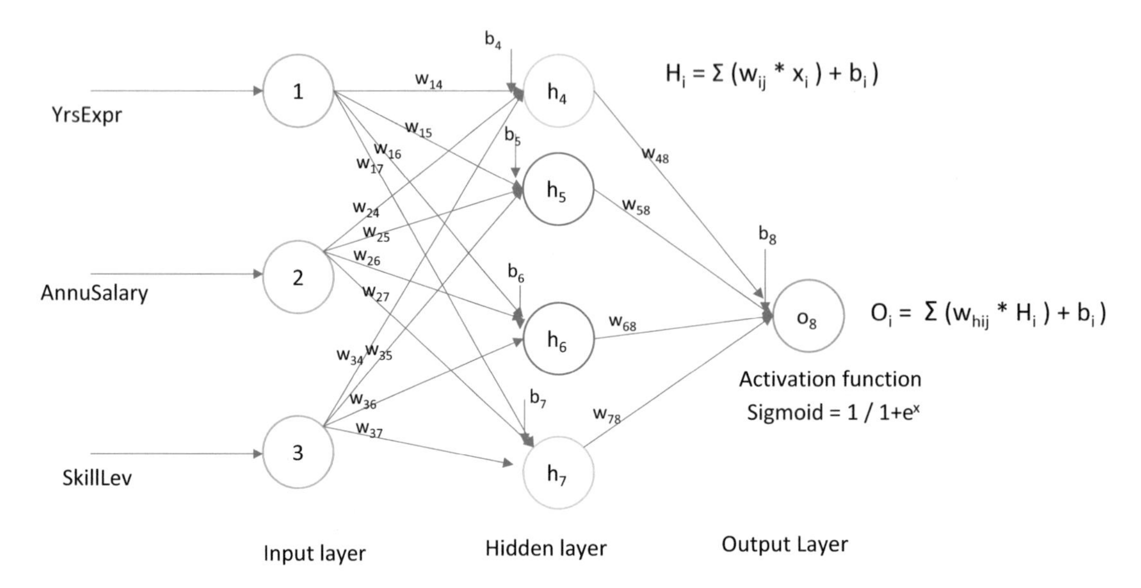

Figure 10-8. *Predicting attrition using a neural network model*

The objective is to predict the output, attrition, which is 0 or 1, based on the input training data. The output neuron triggers 0 or 1 in response to the external stimuli without any intervention. The network is provided with the known class labels, and the network should "guess" the output. After "guessing," the error is computed as follows:

Error = actual – guess

Adjust the weights according to the error until the error is minimized.

Initially, the weights (w_0, w_1, w_2,...w_n) and biases are initialized to some random values.

The error function for this is as follows:

$$E = \sum_i \left[Y_i - f\left(w_i, X_i\right) \right]^2 \tag{3}$$

where Y_i is the actual value of the class, and $f(w,x)$ is the prediction function of weights and input data. This is a linear programming minimization problem where the objective function is the error function of the previous equation 3. The objective is to find the weights that minimize the error.

The output of each hidden node is the sum product of the input values x_1, x_2, x_3 ... x_n and the corresponding weights w_1, w_2, ... w_n and the bias b represented mathematically as follows:

$$h_i = \Sigma\left(w_{ij}{}^{*}x_i\right) + b_i \tag{4}$$

where weights w_1, w_2, ... w_n are initially set to some random values. These weights are automatically adjusted as the network "learns." The output of O_8 is the sum product of input values h_1, h_2, h_3 ... h_n and corresponding weights w_{h11}, w_{h22}, w_{h31}, ... w_{hn1} and the bias b, as shown in figure, represented mathematically as follows:

$$O_i = \Sigma\left(w_{hij}{}^{*}H_i\right) + b_i) \tag{5}$$

If the network is learning, then we should see changes in the output for any small changes in the weights (or bias). If a network is learning that a small change in weight (or bias) causes only a small change in output, then we should continuously use this fact to get the network to behave in such a manner that it should predict as accurately as possible by constantly adjusting weights (or bias). Since the output is in binary, 0 or 1, measuring such a small change to trigger 0 or 1 can be difficult and hence makes it difficult to gradually modify the weights and biases so that the network gets closer to the desired output. This problem can be resolved by introducing what is called a *sigmoid* function as an activation function. So, the output of the sigmoid for the given inputs x_1, x_2, ... x_n, instead of 0 and 1, can have ranges of values between 0 and 1, and you can set the threshold for 0 or 1.

The sigmoid function is given as follows:

$$g(z) = \frac{1}{1 + e^{-z}} \longrightarrow \quad (6)$$

The output of a neuron of a neural network is a function of weights and bias, f(wx,b), given by the following:

$$z = \Sigma\ (w_{ij} * x_i\ ) + b_i\ ) \longrightarrow \quad (7)$$

Substituting z, in equation 7, we have this:

$$g(z) = \frac{1}{1 + \exp\left(- \Sigma\ (w_{ij} * x_i\ ) - b_i\ \right)} \longrightarrow \quad (8)$$

Suppose $z = w.x + b$ is a large number, e^{-z} then ~ 0, so $g(z)$ ~ 1. In other words, when z is large and positive, the output from the sigmoid neuron is 1; on the other hand, if z is very negative, then the output of a neuron is 0. Figure 10-9 shows the shape of the sigmoid function.

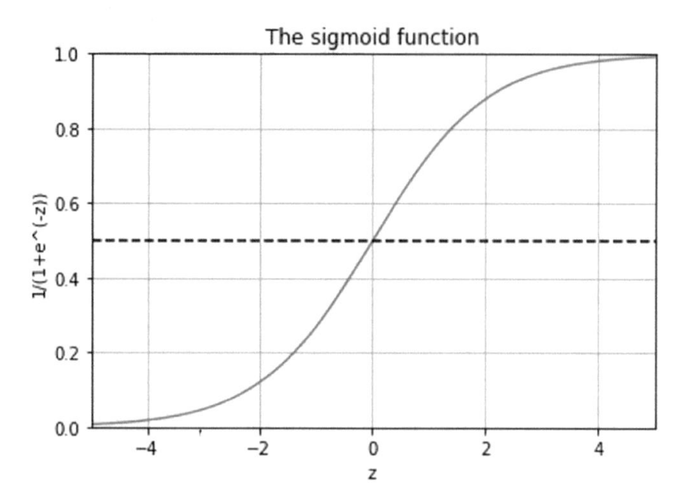

Figure 10-9. *Sigmoid function*

10.3.2 Classification and Prediction Using a Neural Network

The values of bias and weights are initialized to very small values, generally a random value of 0.00 ±0.05. Such small values have no effect on the network, and thus the output it predicts may not have any significance for the first round of training. Let's assume initial weights and bias at nodes 4, 5, 6, and 7, as shown in Figure 10-10, and compute the output of each of the nodes using the sigmoid function. For example, at node 4, we assume that the following weights have been initialized: $b_4 = 0.1$, $w_{14}= 0.05$, $w_{24} = 0.02$, $w_{34}=0.01$ (shown in Figure 10-10). Using the sigmoid function, we can compute the output of node 4 in the hidden layer (using the first observation) as follows:

$$H_4 = \frac{1}{1+\exp\left(-\left(0.05^{*}2.5-0.07^{*}4^{*1}+0.01^{*}0.5-0.01\right)\right)} \tag{9}$$

$$= 0.4600$$

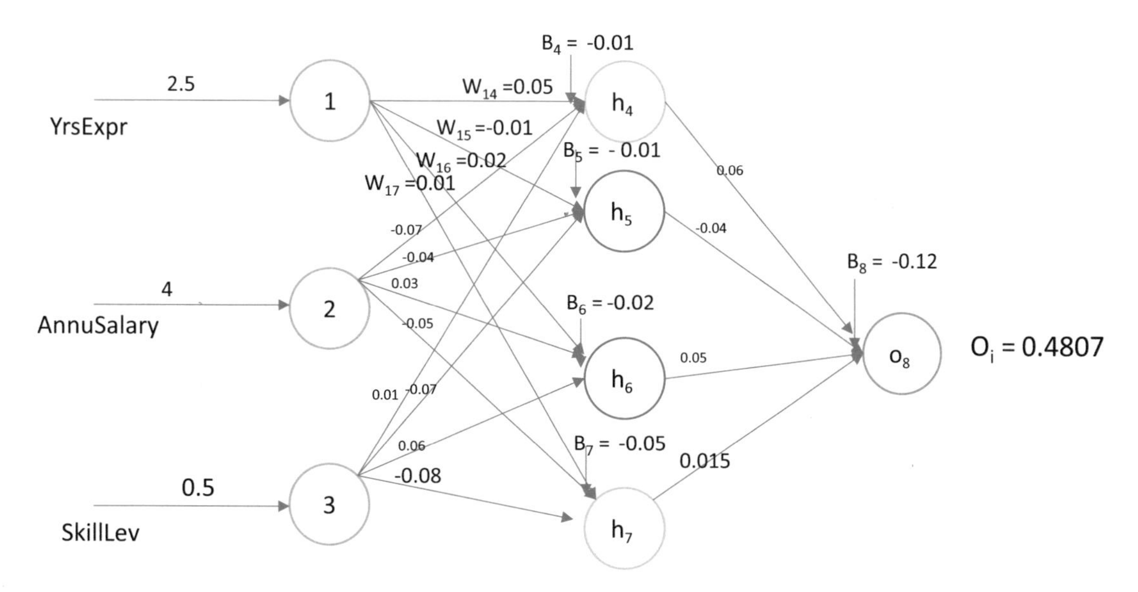

Figure 10-10. *Output of first training round**

* Normalize AnnuSalary data by dividing by 1,000 for better performance.

Similarly, you can apply the same formula and calculate the output of h_5, h_6, and h_7; the corresponding outputs are calculated and listed in Table 10-1.

Table 10-2. *Output of Hidden-Layer Nodes*

h_4	h_5	h_6	h_7
0.460085115	**0.461327**	**0.544878892**	**0.434135**

The output of the output node is calculated using the following formula:

$$O_i = \Sigma\left(w_{hij}{}^{*}H_i\right) + b_i) \tag{10}$$

Just like before, the input to output node 8 is from the hidden-layer neurons. And it is given by the following:

$$O_8 = \frac{1}{1+\exp\left(-\left(0.06^{*}0.4600 - 0.04^{*}0.4613 + 0.05^{*}0.5448 + 0.015^{*}0.4341 - 0.12\right)\right)} \tag{11}$$

$$O_8 = 0.480737$$

Note that if there is more than one hidden layer, the same calculations are applied, except the input values for the subsequent hidden layer would be the output of the previous hidden layer.

Finally, to classify (output prediction) for this training data, we can use some cutoff value, say, 0.5. If the output is less than 0.5, then it is 0, and if the output greater than 0.5, then it is 1. For this model and for the given data, the output is 0 because 0.48037 < 0.5.

We have discussed how the neural network output is generated using inputs from one layer to the next layer. Such networks are called *feed-forward* neural networks; information is always fed forward to the next layer. In the next section, we will discuss how to train the neural network to produce the best prediction results.

10.3.3 Training the Model

Training the model in a neural network means estimating the weights and biases for the best prediction outcome. The process for computing neural network output is repeated for each of the observations in the training data set as described earlier. During each observation, the model compares the predicted results with the actual value of the response variable. If the predicted and actual values vary, this results in an *error*.

10.3.4 Backpropagation

The estimation of weights is based on errors. The errors are computed at the last layer and propagate backward to the hidden layers for estimating (adjusting) weights. This is an iterative process. The initial weights are randomly assigned to each connected neuron, as shown in Figure 10-10, and for the first training record, the output is computed as discussed earlier. Let's say the predicted class is Y^, and the actual class is Y. The error is as follows:

$$\text{Error} = \left(Y_i \wedge - Yi\right) \tag{12}$$

$$Y \wedge \text{for the } k^{th} \text{node is} = wi_k{}^* x_i + b_{ki} \tag{13}$$

For the next iteration, the new weight should be the previous weight plus the small adjustment Δ.

$$\text{newweight}_{i+1} = \text{weight}_{(i-1)} + \Delta\text{weight} \tag{14}$$

The Δ weight is the error multiplied by the input.

$$\Delta\text{weight} = \text{error}^{*}\text{input}_{xi} \tag{15}$$

$$\text{newweight} = \text{weight}_{i} + \text{error}^{*}\text{input}_{xi}^{\;*}(\text{learning rate}) \tag{16}$$

The learning rate (LR) is the fine-tuning factor. If the learning rate is higher, then the new weight will be larger, and vice versa. It is a constant and can take a value from 0 to 1.

$$\text{newweight} = \text{oldweight} + \text{error}^{*}\text{LR} \tag{17}$$

$$\text{newbias} = \text{oldbias} + \text{error} * \text{LR} \tag{18}$$

$$Y\wedge_{i+1} = \text{newweight}^{*}X_{i+1} + b_{ki} \tag{19}$$

$$Y\wedge_{i+1} = \left(\text{weight} + \text{error}^{*}\text{input}_{xi*LR}\right) * X_{i+1} + b_{ki+1} \tag{20}$$

The prediction for the next training record depends on the LR and the input. And this iteration process continues until all the records are processed. In our example, for the first observation the output is 0.48037, and the error is 1 - 0.48037 = 0.51963. This error is used to compute the weights for the next set of training records as in the previous equations 1–6. These weights are updated after the second observation is passed through the network, and the process continues until all the observations in the training set are used. This one complete cycle is called the *epoch* or *iteration* or *cycle*. Typically, there are many iterations (epochs) before achieving optimal weights and biases for the network to perform its best. In case of batches, the entire batch of training data is run through the network before the updates of weights take place.

Figure 10-11 demonstrates the backpropagation algorithm that has been explained. The updating of the weights and the network (and weights) optimization process stops when one of the following conditions are met:

- The error rate has reached the threshold where any updates to weights are making no difference.
- The updating of the new weights is insignificant compared to the previous ones.
- The limits to the number of iterations (epochs) are reached.

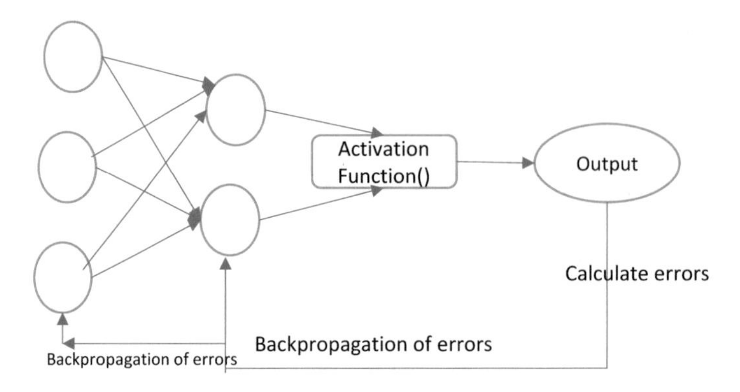

Figure 10-11. *Backpropagation algorithm*

For the previous example of predicting attrition, we calculate the optimized weight using the R computer program. Figure 10-12 shows the optimized weights (and bias). The final weights and bias are labeled for each of the nodes. This process took 1,780 iterations to find the optimized weight.

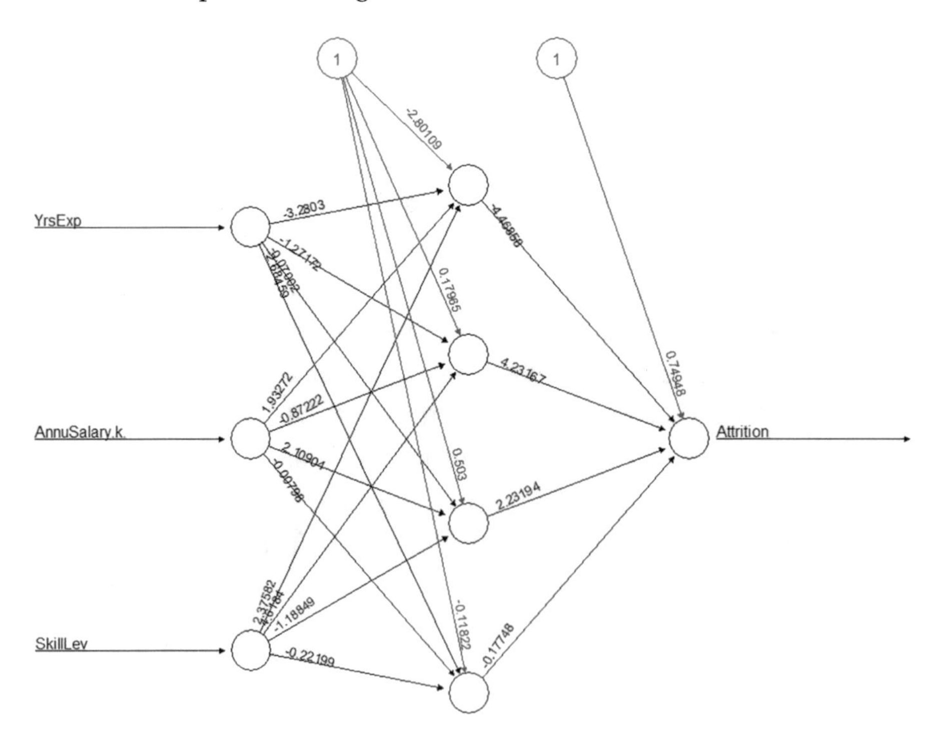

Figure 10-12. *Neural network of the attrition model*

For these optimized weights and bias, the model-predicted output and the actual values are as shown in Figure 10-13.

```
> model_nn$weights
[[1]]
[[1]][[1]]
          [,1]       [,2]        [,3]         [,4]
[1,] -2.801087  0.1796452  0.50299565 -0.118217978
[2,] -3.280298 -1.2717189 -0.07001939 -2.684591064
[3,]  1.932718 -0.8722163  2.10904149 -0.007976765
[4,]  2.375824  4.6183961 -1.18849287 -0.221991349

[[1]][[2]]
           [,1]
[1,]  0.7494801
[2,] -4.4685788
[3,]  4.2316660
[4,]  2.2319444
[5,] -0.1774764
```

Figure 10-13. *Optimized weights for the attrition example*

Since the predicted output is in the probabilities values, we could use a threshold value for deciding the final output values of 0 or 1. We will set the threshold value of 0.5 and higher for 1 and anything less than 0.5 as 0. As you can observe from the truth table in Figure 10-14, the predicted values for five records are correct, and it matches with the actual value. However, the fourth record prediction is wrong. Our neural network has predicted 0, whereas the actual value is 1. This is called a *prediction error*. The model can be further tuned to improve the accuracy. We will learn more about tuning the model in the next section.

```
  Actual Output Neural Net Output
1          TRUE         0.9275408
2         FALSE         0.2246251
3         FALSE         0.2151029
4          TRUE         0.2865495
5         FALSE         0.3612136
6          TRUE         0.8836209
```

Figure 10-14. *Neural network predicted versus actual*

10.4 Activation Functions

In our earlier discussion, we mentioned the activation function. Activation functions are necessary to trigger the response for a given input. At every stage of the neural network layers, the output of the neuron depends on the input from the previous layer's neurons. The output of each layer's neuron is the sum of products of the inputs and their corresponding weights. This is passed through an activation function. The corresponding output of the activation is the input to the next layer's neuron and so on. The sum of products of weights and inputs is simply a linear function, just a polynomial of degree one. If the activation function is not used, then the model acts as a simple linear regression model. However, the data is not always linear. The simple linear regression model will not learn all the details for complex data such as videos, images, audio, speech, text, etc., thus resulting in the poor performance of the model. Because of this reason, we use the activation function. Activation function would take the high dimensional and nonlinear input data and transform them into a more meaningful data pattern output signal, which is fed into the next layers. At every stage, activation functions are applied so that the final output of the model provides accurate prediction performance based on its learning at each layer. Activation functions provide the nonlinear property to the neural networks.

There are different types of activation functions available for a neural network. We will discuss the commonly used activation functions. These activation functions are sometimes referred to as *threshold functions* or *transfer functions* since they force the output signal to a finite value.

10.4.1 Linear Function

Linear is the simplest activation function represented by the following formula:

$$Y = f(x) \ \alpha \ x$$

The output is directly proportional to the input x ranging from ($-\infty$ to ∞). Figure 10-15 shows the linear activation function, and the input and output follow a linear relationship.

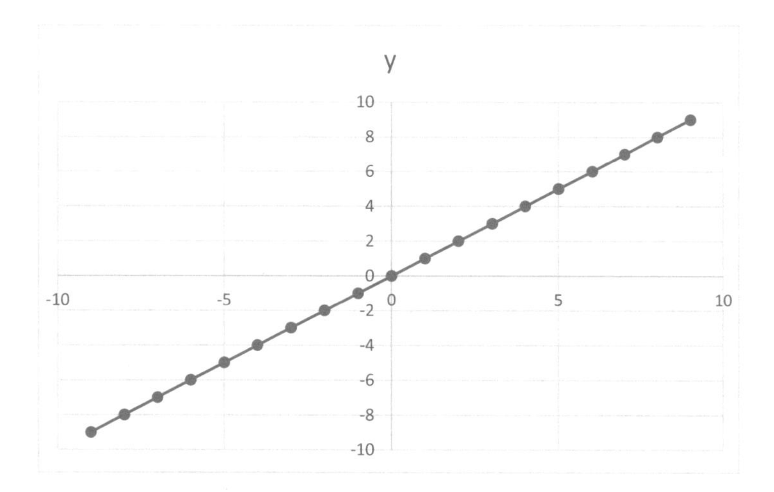

Figure 10-15. *Linear activation function*

10.4.2 Sigmoid Activation Function

The sigmoid function is the most commonly used activation function, particularly in the output layer of classification problems. The sigmoid function transforms the values of probabilities values ranging from 0 to 1. The sigmoid function is a special case of logistic function and is represented by the following formula:

$$f(x) = 1/1 + e - x$$

The output of the sigmoid function is a smooth, "S-shaped" curve, as shown in Figure 10-16. The sigmoid function is not a symmetric function about zero. As you can see from the graph, the values increase slowly initially and then increase exponentially before reaching a saturation point. When the output value is close to 0, the neuron is inactive, and when the output value is 1, the neuron is activated and enables the flow of information.

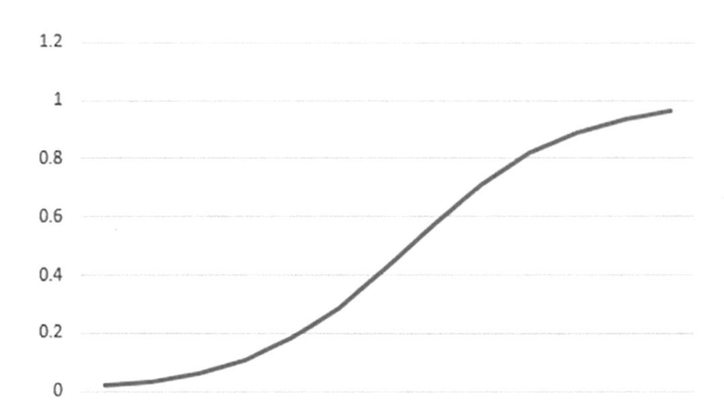

Figure 10-16. *Sigmoid activation function*

10.4.3 Tanh Function

The hyperbolic tangent function (tanh) function is the extension of the sigmoid (logistic) function. The function is similar to the sigmoid function except corrected to have the symmetry around the origin and has gradients that are not restricted to vary in a certain direction, as shown in Figure 10-17.

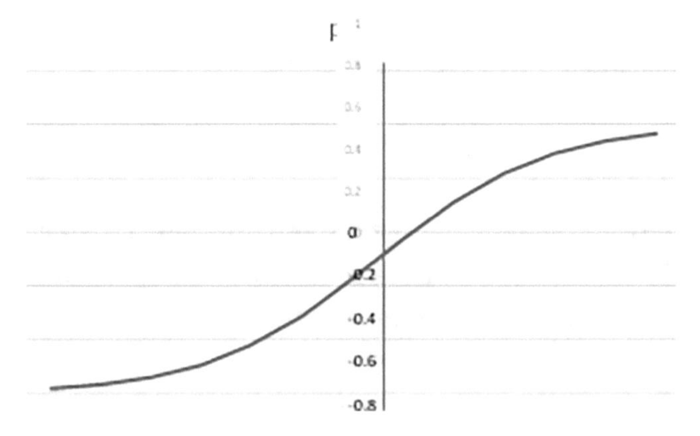

Figure 10-17. *Tanh activation function*

The function is a `tanh()` function defined by the following formula:

$$f(x)=\tanh(x) \text{ for } x\ (-1,1)$$

The tanh function expresses similar behavior to the sigmoid function. The main difference is that the tanh function pushes the input values to 1 and -1 instead of 1 and 0.

10.4.4 ReLU Activation Function

The rectified linear unit (ReLU) function is the most recent activation function whose output is between 0 and infinity. In other words, the threshold of the activation is at zero (Figure 10-18). The ReLU function is defined by the following formula:

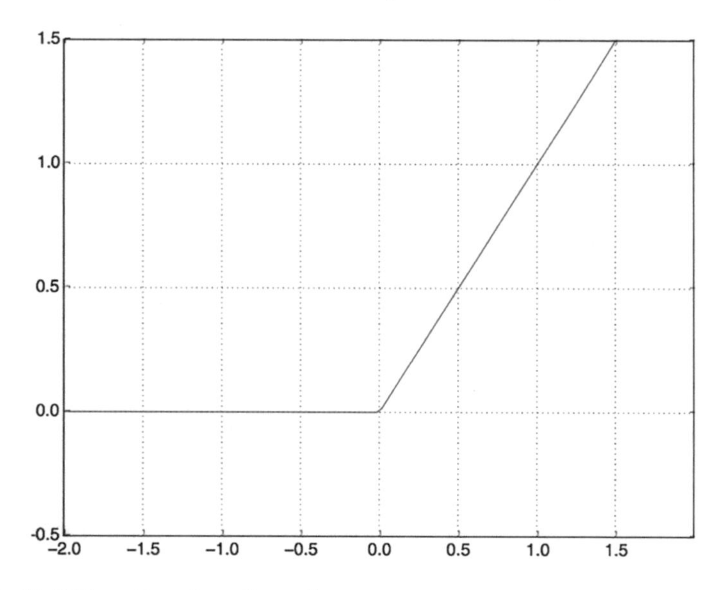

Figure 10-18. *ReLU activation function*

$$f(x)=0 \text{ when } x<0$$
$$f(x)=x \text{ when } x>0$$

As you can see from Figure 10-18, it is similar to a linear function that will output the input directly if it is positive; otherwise, it will output zero.

10.4.5 Softmax Activation Function

The softmax activation function is used for multiclass classification problems. It is similar to the sigmoid function and is expressed with the following formula:

$$f(x)=\frac{e^{xj}}{\sum_{k=0}^{k} e^{xk}} \qquad \text{for } j=1,2\ldots k$$

Where the xj values are the elements of the input vector, it can take any real value. The denominator term at the bottom of the formula is the normalization term to ensure all the function output values will sum to 1 to constitute a valid probability distribution.

When we have a multiclass problem, the neural network model output layer will have the same number of neurons as the number of target classes. The softmax activation function returns the probability for every data point of all the individual classes.

10.4.6 Selecting an Activation Function

Selecting the right activation function depends on the data and the problem you are trying to solve. Though there is no standard definition of which activation should be used and when, research has shown some limitations of using activation functions. Here are some examples:

- Sigmoid and softmax functions are generally used at the output layer for classification problems.
- ReLU is used only inside hidden layers but not at the output layer.
- Tanh has shown better results than sigmoid and can be used in both output and hidden layers.

There are many other activation functions developed recently, which include leakyReLU, swish, exponential ReLU, etc. They are not as common as the others. Also note that not all activation functions will be supported by the library you intend to use.

10.5 Practical Example of Predicting Using a Neural Network

Up until now, we have learned about neural networks, including using activation functions, training the model, improving performance, etc. In this example, we demonstrate the neural network concepts by taking a real data set and creating a neural network model to predict the job attrition from the data provided by a private firm. Table 10-3 shows a snippet of the data set.

***Table 10-3.** Job Attrition Data Set*

1	Attrition	YrsExp	WorkChall	WorkEnvir	Compensa	TechExper
2	Yes	2.5	No	Low	Low	Excellent
3	No	2	Yes	Excellent	Excellent	Excellent
4	No	2.5	Yes	Excellent	Low	Excellent
5	Yes	2	No	Excellent	Low	Excellent
6	No	2	Yes	Low	Low	Low
7	Yes	2	No	Low	Low	Excellent
8	No	2	No	Excellent	Excellent	Low
9	Yes	2.5	No	Low	Excellent	Excellent
10	Yes	2	No	Excellent	Low	Excellent
11	Yes	3	Yes	Low	Excellent	Excellent
12	Yes	3.5	No	Excellent	Low	Excellent

Job attrition may be related to many reasons and factors, but in this specific case, we have data consisting of six variables. `Attrition` represents whether the employee has exited the organization or is still in the organization (Yes for Exit and No for Currently continuing in the organization). `Yrs_Exp` represents the experience of the employee at this point of time (in years), `Work_Challenging` represents whether the work assigned to the employee is challenging or not, `Work_Envir` represents whether the work environment is Excellent or Low, `Compensation` represents whether the compensation is Excellent or Low, and `Tech_Exper` represents whether the employee is an expert (Excellent or Low). The data is for the last two years and pertains to only those employees with experience between 2 to 5 years.

As you can observe from the data set, there are attributes that carry the same information and may be correlated. In a neural network, we do not need to worry about features, removing features, etc., unlike in other algorithms. We will continue to create the neural network model without removing variables. By nature, neural networks should adjust the weights and biases based on what it learns from the data.

10.5.1 Implementing a Neural Network Model Using R

In this section, we will demonstrate how to create a neural network model using R libraries.

10.5.1.1 Exploring Data

The data is extracted from the comma-separated value text file `AttrData_NN.txt`. This data set has only 52 records.

Step 1: Read the data set and check the data summary using the `summary()` function of R.

Here is the input:

```
> data_df<-read.csv("attrData_NN.csv",header=TRUE, sep=",")
> summary(data_df)
```

Figure 10-19 shows the input and the output.

```
> #Read data
> data_df<-read.csv("AttrData_NN.csv",header=TRUE, sep=",")
> summary(data_df)
 Attrition     YrsExp       WorkChallenging     WorkEnvir      Compensation     TechExper
 No :24      Min.   :2.000   No :28          Excellent:28   Excellent:21    Excellent:44
 Yes:28      1st Qu.:2.500   Yes:24          Low      :24   Low      :31    Low      : 8
             Median :4.000
             Mean   :3.519
             3rd Qu.:4.500
             Max.   :5.000
> |
```

Figure 10-19. *Summary of the data*

As you can see, the Attrition field has 28 "Yes" values, which means that these employees have exited the organization, and 24 "No" values, which means that these employees are still continuing in the organization. You can also see that 28 employees have not been assigned "Challenging Work" and 24 employees have been assigned "Challenging Work." Twenty-eight of the employees are working in teams where the Work Environment is considered as excellent, whereas 24 are working in teams where the work environment is not that great (here marked as Low). Twenty-one of the employees have excellent compensation at or above the market compensation (known here as Excellent), whereas 31 have a compensation that is below the market compensation or low compensation (known here as Low). Of all the employees, 44 have excellent technical expertise, whereas 8 others have Low technical expertise.

Ideally, when the organization is providing challenging work to an employee, work environment within the team is excellent, compensation is excellent, and technical expertise of the employee is high, then there should be low chance for the Attrition.

10.5.1.2 Preprocessing Data

Before we create a model, we should check the data types of the attributes and convert them to proper data types (numerical, logical, categorical, etc.) if required. We use `str()` to check the data type. R reads all the variable types properly. We do not have to convert any data types.

Here is the input:

```
> ##Checking data types
> str(data_df)
```

Figure 10-20 shows the input and the output.

```
> ##Checking data types
> str(data_df)
'data.frame':   52 obs. of  6 variables:
 $ Attrition       : Factor w/ 2 levels "No","Yes": 2 1 1 2 1 2 1 2 2 2 ...
 $ YrsExp          : num  2.5 2 2.5 2 2 2 2 2.5 2 3 ...
 $ WorkChallenging: Factor w/ 2 levels "No","Yes": 1 2 2 1 2 1 1 1 1 2 ...
 $ WorkEnvir       : Factor w/ 2 levels "Excellent","Low": 2 1 1 1 2 2 1 2 1 2 ...
 $ Compensation    : Factor w/ 2 levels "Excellent","Low": 2 1 2 2 2 2 1 1 2 1 ...
 $ TechExper       : Factor w/ 2 levels "Excellent","Low": 1 1 1 1 2 1 2 1 1 1 ...
> |
```

Figure 10-20. *Checking data types*

The other task is to meet the input data requirements specified by the library you are using. In this particular example, we are using the `neuralnet()` R package. This package (in general, any neural network) expects the input to be all numerical values. Thus, as part of preprocessing, the next step of the model-building process is to prepare the input data. We will use `model.matrix()` to convert all our categorical values (Yes/No, Excellent, Low, etc.) to numerical values, as shown in Figure 10-21.

Here is the input:

```
> data_df_mx<-as.data.frame(model.matrix(~WorkChallenging+WorkEnvir+Compensation+TechExp
er+YrsExp+Attrition, data=data_df))


> head(data_df_mx)
```

Figure 10-21 shows the input and the output.

```
> data_df_mx<-model.matrix(~WorkChallenging+WorkEnvir+Compensation+TechExper+YrsExp+Attrition,
+                           data=data_df)
> head(data_df_mx)
  (Intercept) WorkChallengingYes WorkEnvirLow CompensationLow
1           1                  0            1               1
2           1                  1            0               0
3           1                  1            0               1
4           1                  0            0               1
5           1                  1            1               1
6           1                  0            1               1
  TechExperLow YrsExp AttritionYes
1            0    2.5            1
2            0    2.0            0
3            0    2.5            0
4            0    2.0            1
5            1    2.0            0
6            0    2.0            1
> |
```

Figure 10-21. *Data preprocessing, converting text to numerical values*

Just to keep all the values consistent, we will scale the `YrsExp, TechExpereince` variable between 0 and 1 using a custom `scale()` function, as shown in Figure 10-22. Before we do that, the `model.matrix()` function created an additional variable called `intercept`, and this has no significance in our model, and hence we will remove it.

Here is the input:

```
> data_df_2<-as.data.frame(data_df_mx[,c(-1)])
> #Scale YrsExp variable using custom scale function
> scale01<-function(x) {
+    (x-min(x))/(max(x)-min(x))
+ }
> data_df_2$YrsExp<-scale01(data_df_mx$YrsExp)
> head(data_df_2)
```

Figure 10-22 shows the input and the output.

```
> ##Remove intercept from the data_matrix
> # change to data frame
> data_df_2 <- as.data.frame(data_df_mx[,-c(1)])
> ##Scale following variables
> ##YrsExp
> scale01 <- function(x){(x-min(x))/(max(x)-min(x))}
> data_df_2$YrsExp <- scale01(data_df_2$YrsExp)
> head(data_df_2)
  WorkChallengingYes WorkEnvirLow CompensationLow TechExperLow    YrsExp AttritionYes
1                  0            1               1            0 0.1666667            1
2                  1            0               0            0 0.0000000            0
3                  1            0               1            0 0.1666667            0
4                  0            0               1            0 0.0000000            1
5                  1            1               1            1 0.0000000            0
6                  0            1               1            0 0.0000000            1
> 
```

Figure 10-22. *Data preprocessing, scaling variable YrsExp*

10.5.1.3 Preparing the Train and Test Data

Before we create the model, we have to split the data into two parts. One is for training the model, and the other set is for testing the model. There are numerous functions available to split the data. In this example, we will use a library called `caret` and a function called `createDataPartition()`, as shown in Figure 10-23. The training data set is used to build the model, and the test data set is used to test the model.

Here is the input:

```
> set.seed(1234)
> data_partition <- createDataPartition(data_df_2$AttritionYes,
+                                       p=0.8,list=FALSE)
> train<-data_df_2[data_partition,]
> test<-data_df_2[-data_partition,]
> train<-as.data.frame(train)
> head(test)
```

Figure 10-23 shows the input and the output.

```
> ##Split data into two parts - train and test using 'caret' library
> library(caret)
> set.seed(1234)
> data_partition <- createDataPartition(data_df_2$AttritionYes,
+                                       p=0.8,list=FALSE)
> train<-data_df_2[data_partition,]
> test<-data_df_2[-data_partition,]
> head(test)
   WorkChallengingYes WorkEnvirLow CompensationLow TechExperLow    YrsExp
1                   0            1               1            0 0.1666667
7                   0            0               0            1 0.0000000
13                  1            0               1            0 0.0000000
17                  0            1               1            0 0.1666667
19                  0            0               1            0 0.6666667
23                  0            0               1            0 0.6666667
   AttritionYes
1             1
7             0
13            0
17            1
19            1
23            1
```

***Figure 10-23.** Data preparation, dividing data into two parts (test and train)*

10.5.1.4 Creating a Neural Network Model Using the Neuralnet( ) Package

Once we have prepared the training data, we will create a model using the `neuralnet()` package. For this example, the neural network architecture consists of only two hidden layers with five neurons in the first layer and two neurons in the second layer. The `neralnet()` function accepts a number of parameters as input for creating the model. At a minimum, it needs the following parameters (shown in Figure 10-24):

> **Formula**: This is the model target variable and its relationship with other variables in the data.
>
> **Data**: This is the data required to create the model as a dataframe.
>
> **Hidden**: This is a vector of integers specifying the number of hidden neurons (vertices) in each layer.

Learningrate: This is a numeric value specifying the learning rate used by traditional backpropagation. This is used only for traditional backpropagation.

Algorithm: This is a string containing the algorithm type to calculate the neural network. The following types are possible: `backprop`, `rprop+`, `rprop-`, `sag`, or `slr`. `backprop` refers to backpropagation, `rprop+` and `rprop-` refer to the resilient backpropagation with and without weight backtracking, while `sag` and `slr` induce the usage of the modified globally convergent algorithm (grprop).

There are other function parameters that can be used based on the problem you want the neural network to solve. The details of the input parameters are listed in the documentation, and you can refer to the documentation to learn more about the `neuralnet()` function (`https://www.rdocumentation.org/packages/neuralnet/versions/1.44.2/topics/neuralnet`).

Here is the input:

```
> feature_list<-paste(c(colnames(data_df_matrix[,-c(1,7)])), collapse = "+")
> feature_list<-paste(c("AttritionYes~", feature_list), collapse="")
> nn_formula<-formula(feature_list)

> # Build the model
> model_nn<-neuralnet(nn_formula,
+                   hidden=c(5,2),
+                   threshold = 0.01,
+                   learningrate = 0.01,
+                   algorithm="backprop",
+                   data=train,
+                   stepmax = 1e+05,
+                   rep=2,
+                   lifesign = "minimal")
hidden: 5, 2    thresh: 0.01    rep: 1/2    steps:    23290   error: 0.98485 time: 1.6
8 secs
hidden: 5, 2    thresh: 0.01    rep: 2/2    steps:    22947   error: 1.04537 time: 1.5
7 secs
> plot(model_nn)
```

Figure 10-24 shows the input and the output.

```
> library(dplyr)
> # create Neural network with 5 hidden neurons and 2 output neurons
> library(neuralnet)
> nn_formula
AttritionYes ~ WorkChallengingYes + WorkEnvirLow + CompensationLow +
    TechExperLow + YrsExp
> model_nn<-neuralnet(formula=nn_formula,
+                     hidden=c(5,2),
+                     learningrate = 0.01,
+                     algorithm="backprop",
+                     data=train)
> plot(model_nn)
> |
```

Figure 10-24. *Creating a neural network model*

The objective of the neural network model is to predict the outcome of attrition based on the five parameters in the data. In this example, the model is a simple network with two hidden layers and five neurons corresponding to five input variables. This is specified by the `hidden` parameter of the `neuralnet()` function. Since this is a simple binary classification problem, we use the `backprop` algorithm to create the model. Figure 10-25 shows the architecture of the model.

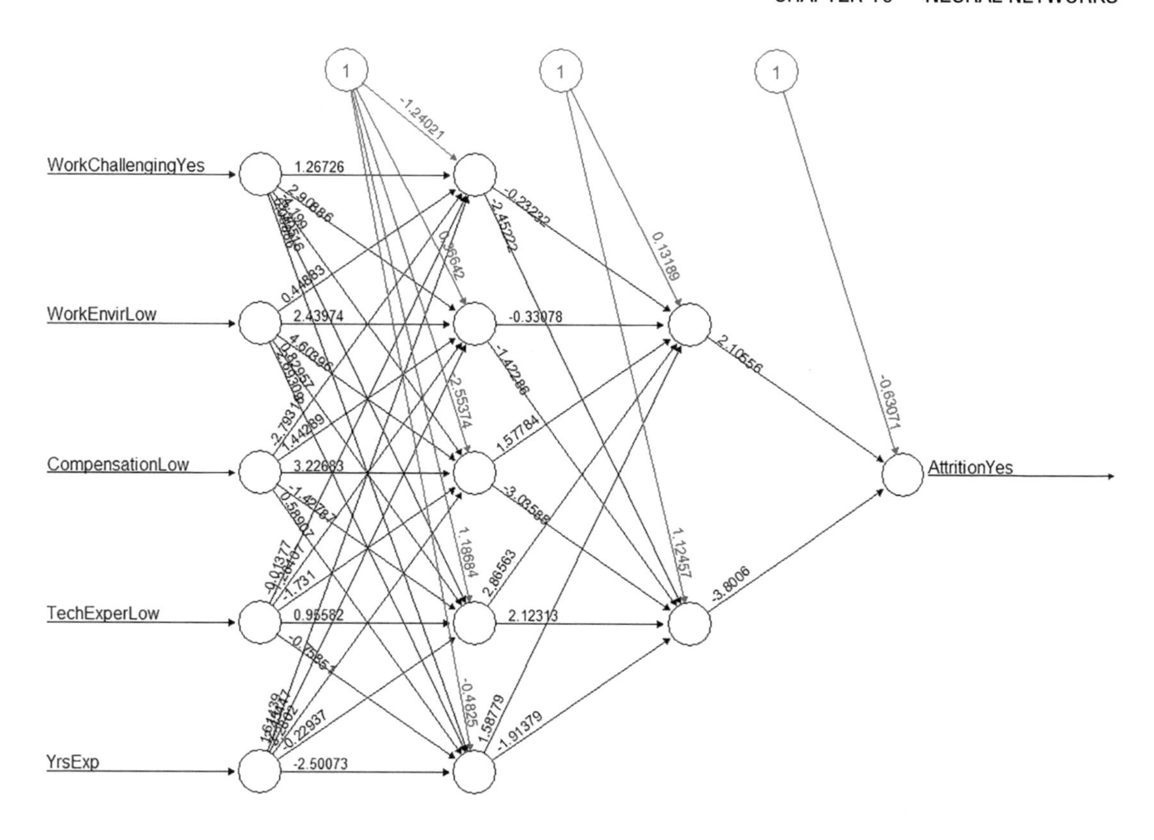

Figure 10-25. *Neural network model architecture*

The `plot()` function provides the model's network architecture and the final weights and biases displayed on each link connected to each neuron, as shown in Figure 10-25. If you want to extract the model's weights and biases, run the following command.

Here is the input:

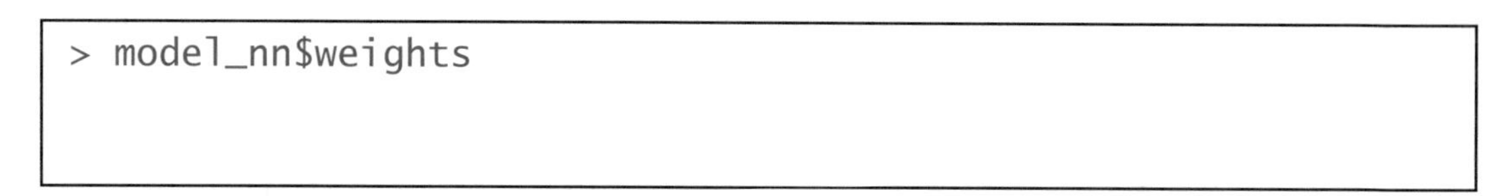

```
> model_nn$weights
```

Figure 10-26 shows the input and the output.

```
> plot(model_nn)
> model_nn$weights
[[1]]
[[1]][[1]]
            [,1]       [,2]      [,3]       [,4]        [,5]
[1,] -1.24020848  0.3664213 -2.553745  1.1868392 -0.48249572
[2,]  1.26726041  2.9088584 -4.199001 -3.3051607  0.05885832
[3,]  0.44882660  2.4397377  4.603956  0.8295744 -2.69308524
[4,] -2.79315565  1.4428889  3.226829 -1.4278665  0.58906931
[5,] -0.01376598 -1.2640717 -1.731003  0.9558220 -0.75850974
[6,]  1.61438754 -2.4544742  3.280200 -0.2293746 -2.50073383

[[1]][[2]]
           [,1]      [,2]
[1,]  0.1318907  1.124573
[2,] -0.2323239 -2.452223
[3,] -0.3307838 -1.422856
[4,]  1.5778397 -3.035879
[5,]  2.8656346  2.123127
[6,]  1.5877865 -1.913793

[[1]][[3]]
           [,1]
[1,] -0.6307147
[2,]  2.1055598
[3,] -3.8005975
```

Figure 10-26. *Final weights and biases of the model*

The model is trained with 80 percent of 52 records, which is only around 42. As it passes through each iteration, weights and biases are adjusted based on the final output-layer error as discussed in our previous sections. Since the training data has 42 records, one epoch consists of 42 iterations. Since we have not specified how many epochs the program should run to optimize, it runs through iterations until it reaches optimum performance conditions and breaks the loop. The resulting error rates, optimal weights, and biases achieved for the last iteration of training the neural net on this data are shown in Figure 10-26. The `neuralnet()` package uses the entire data set to calculate gradients, update weights, and repeat until convergence or `stepmax` steps is reached. The `rep` parameter is the number of *repetition* of training.

10.5.1.5 Predicting Test Data

The output of the model will be a numerical value of probabilities that trigger the response. In binary classification, the response of the output neuron should be either 0 or 1, so we can use just one output node and set a threshold to map a numerical value to one of the two classes. If we have more than two classes, then the neural network output layer will have more than one node at the output layer. Although we typically use 0.5 as a cutoff value with any other classifier, we have a tendency to use 0.5 as a threshold even for the neural network. Using 0.5 as a threshold, we predict the model's output on the test data and calculate the model's accuracy. To calculate the model's accuracy, we use the truth table (confusion matrix) to determine the actual versus predicted values and perform sensitivity analysis just like any other classification method. The following steps demonstrate how to predict the neural network model and perform sensitivity analysis. We are using the `compute` method to predict the test data.

Here is the input:

```
#Test the model using test dataset
## we will use compute() to make predictions
# against our training data and compare the predicted
## outputs against the actual outputs
> pred_results <- neuralnet::compute(model_nn, test[,1:5])
> pred_results$net.result
```

Figure 10-27 shows the input and the output.

```
> #Test the model performance using test dataset
> ## we will use compute() to make predictions and compare the predicted
> ## outputs against the actual values
> pred_results <- neuralnet::compute(model_nn, test[,1:5])
> pred_results$net.result
          [,1]
1   0.98664368
7  -0.15942251
13  0.13334430
17  0.98664368
19  0.96756176
23  0.96756176
25  0.19971185
31  0.99324424
41 -0.11432403
52 -0.01565101
> 
```

Figure 10-27. *Predicting test data*

10.5.1.6 Summary Report

The summary report prints the actual versus predicted values. Since the output of the neural network is probability values ranging between 0 and 1, we have to set a threshold value to conform the output to either 0 or 1, in case of a classification model. In our example, we have set a threshold of 0.8. Any value greater than 0.8 is conformed to '1' and any value less than 0.8 is '0'.

Here is the input:

```
> ###Copying results to a dataframe - Actual Vs Predicted
> comp_results <- data.frame(actual = test$AttritionYes,
                         predicted = pred_results$net.result)
> comp_results

> thr <- function(x) {
+    if (x>0.8) {return(x=1)} else {return (x=0)} }
> comp_results['predicted']<-apply(comp_results['predicted'],1,thr)
> attach(comp_results)
> table(actual,predicted)
```

Figure 10-28 shows the input and the output.

```
> ###Copying results to a dataframe - Actual Vs Predicted
> comp_results <- data.frame(actual = test$AttritionYes,
+                           predicted = pred_results$net.result)
> comp_results
   actual   predicted
1       1  0.98664368
7       0 -0.15942251
13      0  0.13334430
17      1  0.98664368
19      1  0.96756176
23      1  0.96756176
25      0  0.19971185
31      1  0.99324424
41      0 -0.11432403
52      0 -0.01565101
> thr<-function(x){if(x>0.8) {return(x=1)} else {return(x=0)}}
> ##Rouding predicted results to 0 decimal points
> comp_results['predicted']<-apply(comp_results['predicted'],1, thr)
> #attach(comp_results)
> table(actual,predicted)
      predicted
actual 0 1
     0 5 0
     1 0 5
```

Figure 10-28. *Model output, predicted versus actutal*

As you can see from Figure 10-28, the predicted values > 0.8 are considered 1 and <0.8 as 0, and accordingly, the actual versus predicted matrix is displayed.

10.5.1.7 Model Sensitivity Analysis and Performance

There are various measures to calculate the performance of a classification model. In general, we use a confusion matrix to show the predicted values and actual values. From this table we calculate the other measures such as recall, precision, accuracy, and F-score. The following code demonstrates how to calculate these measures. We have shown the actual calculations in R even though there are many libraries available to perform them.

Performance measures such as precision, recall, and F-score are manually calculated and shown in Figure 10-29.

Here is the input:

```
> ## Model performance and accuracy
> ## Confusion matrix, F-Measure and AUC
> truth_table<-table(predicted, actual)
> truth_table
         actual
predicted 0 1
        0 5 0
        1 0 5
> ###Sensitivity Analysis
> # True Positive Rate - RECALL or SENSITIVITY
> # Recall Or TPR = TP/(FN + TP)
> recall<-(truth_table[2 , 2]/sum(truth_table[2,])) * 100
> recall
[1] 100
> # True Negative Rate
> ## TNR = TN /(TN+FP)
> TNR<- (truth_table[1,1] / sum(truth_table[1,])) * 100
> TNR
[1] 100
> ##Accuracy = (TP + TN)/(TN + FP + FN + TP)
> pred_accuracy<-sum(diag(truth_table))/sum(truth_table) * 100
> pred_accuracy
[1] 100
> ## PRECISION
> #  PRECISION = TP/(FP+TP)
> precision <- (truth_table[2,2] / sum(truth_table[,2])) * 100
> precision
[1] 100
> F_Score <- (2 * precision * recall / (precision + recall))/100
> F_Score
[1] 1
```

Figure 10-29 shows the input and the output.

```
> ## Model performance and accuracy
> ## Confusion matrix, F-Measure and AUC
> truth_table<-table(predicted, actual)
> truth_table
         actual
predicted 0 1
        0 5 0
        1 0 5
> ###Sensitivity Analysis
> # True Positive Rate - RECALL or SENSITIVITY
> # Recall or TPR = TP/(FN + TP)
> recall<-(truth_table[2 , 2]/sum(truth_table[2,])) * 100
> recall
[1] 100
> # True Negative Rate
> ## TNR = TN /(TN+FP)
> TNR<- (truth_table[1,1] / sum(truth_table[1,])) * 100
> TNR
[1] 100
> ##Accuracy = (TP + TN)/(TN + FP + FN + TP)
> pred_accuracy<-sum(diag(truth_table))/sum(truth_table) * 100
> pred_accuracy
[1] 100
> ## PRECISION
> #  PRECISION = TP/(FP+TP)
> precision <- (truth_table[2,2] / sum(truth_table[,2])) * 100
> precision
[1] 100
> F_Score <- (2 * precision * recall / (precision + recall))/100
> F_Score
[1] 1
> |
```

Figure 10-29. *Neural network model performance report*

10.5.1.8 ROC and AUC

Another measure we use to check the performance of the classifier model is ROC. Please refer to Chapter 6 to learn about ROC and its significance. To plot the ROC, we use the ROSE library. Our neural network model has a prediction accuracy of 100 percent, and hence the Area Under Curve (AUC) is 1.

Here is the input:

```
> ###ROC curve and Area under the curve
> library(ROSE)
Loaded ROSE 0.0-3

Warning message:
package 'ROSE' was built under R version 3.6.3
> roc.curve(actual, predicted)
Area under the curve (AUC): 1.000
```

Figure 10-30 and Figure 10-31 show the output.

```
> ###ROC curve and Area under the curve
> library(ROSE)
> roc.curve(actual, predicted)
Area under the curve (AUC): 1.000
>
```

Figure 10-30. *Model AUC*

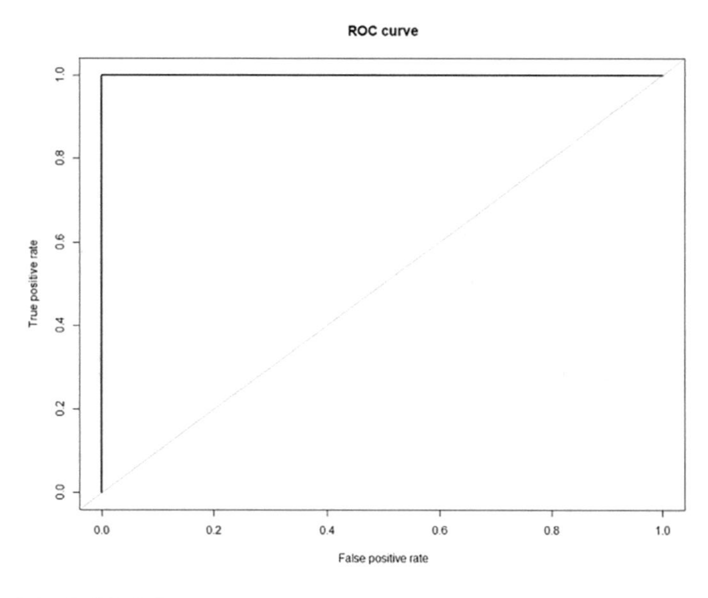

Figure 10-31. *Model ROC*

10.6 Implementation of a Neural Network Model Using Python

This section will demonstrate creating a neural network model in Python. We will be using the same data set and solving the same predicting attrition problem with the same set of attributes. The only difference is that we will implement it using Python's scikit-learn libraries and multilayer perceptron (MLP) algorithm. However, this implementation is not intended for large-scale, practical big data applications. We can also implement using TensorFlow libraries. Though TensorFlow seems easy to implement, it may become challenging unless you have enough knowledge. Also, based on our experience, poor documentation and constant changes to the libraries can be frustrating to users. Our own experience says that the TensorFlow code that worked six months back may not work today. Hence, we will not be discussing the TensorFlow implementation; instead, we will demonstrate our practical learning of creating neural network models using scikit-learn libraries as it has been around a longer time, is more stable, and is reliable.

The process of creating the neural network model is exactly the same as creating a neural network model in R, which is summarized in Figure 10-32. To print the network layers and architecture, Python does not provide an easy library function like in R (we used `plot('model')`). Apart from this, the implementation is similar to R. We will provide the step-by-step process with a brief explanation wherever necessary to avoid repetition.

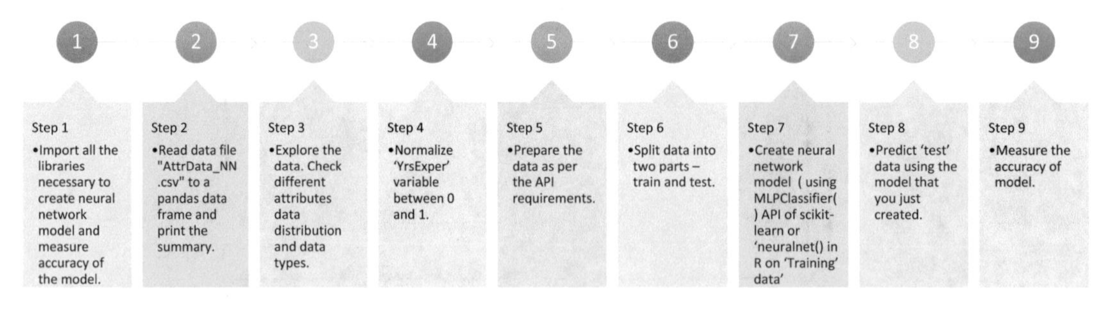

Figure 10-32. *Process of creating a neural network classification model*

Step 1: Import all the libraries necessary to create the neural network model and measure the accuracy of the model. `pandas()` and `numpy()` are required for data ingestion and data manipulation. `Matplotlib()` is for plotting. `Os()` is the basic operating system libraries. We will be importing the sklearn library packages that are essential for building the neural network model and measuring the model performance: `confusion_matrix()`, `roc()`, `classification_report()`, etc.

Here is the input:

```
import sklearn
from sklearn.neural_network import MLPClassifier
from sklearn.neural_network import MLPRegressor

# Import splitting data into train and test
from sklearn.model_selection import train_test_split

## classification accuracy, ROC, AUC score of the model
from sklearn.metrics import classification_report
from sklearn.metrics import confusion_matrix
from sklearn import metrics
from sklearn.metrics import roc_curve
from sklearn.metrics import auc
from sklearn.metrics import roc_auc_score
from sklearn.metrics import roc_curve

##IPYthon display
from IPython.display import display

##Label Encoder for creating dummy variables
from sklearn.preprocessing import LabelEncoder
```

Figure 10-33 shows the input and the output.

```
In [1]: import numpy as np
        import pandas as pd
        import math
        import matplotlib.pyplot as plt
        import random as rd
        import os

In [122]: import sklearn
          from sklearn.neural_network import MLPClassifier
          from sklearn.neural_network import MLPRegressor

          # Import splitting data into train and test
          from sklearn.model_selection import train_test_split

          ## classification accuracy, ROC, AUC score of the model
          from sklearn.metrics import classification_report
          from sklearn.metrics import confusion_matrix
          from sklearn import metrics
          from sklearn.metrics import roc_curve
          from sklearn.metrics import auc
          from sklearn.metrics import roc_auc_score
          from sklearn.metrics import roc_curve

          ##IPYthon display
          from IPython.display import display

          ##Label Encoder for creating dummy variables
          from sklearn.preprocessing import LabelEncoder
```

Figure 10-33. *Results of importing libraries necessary to create our neural network model and measure its accuracy*

Step 2: Read the data file `AttrData_NN.csv` to a Pandas dataframe and print the summary.

Here is the input:

```
data_dir='E:/umesh/Dataset/NN'
filename = "AttrData_NN.csv"
os.chdir(data_dir)
data_df = pd.read_csv(filename)
print(data_df.shape)
```

Figure 10-34 shows the input and the output.

```
In [3]: data_dir='E:/umesh/ResearchAndPublications/WIP/BA-2ndEdition-July2021/BOOK/Book Chapters/Chapter 10 - NeuralNetworks/Dataset/NN'
        filename = "AttrData_NN.csv"
        os.chdir(data_dir)
        data_df = pd.read_csv(filename)
        print(data_df.shape)

        (52, 6)

In [4]: data_df.head(3)

Out[4]:
```

	Attrition	YrsExp	WorkChallenging	WorkEnvir	Compensation	TechExper
0	Yes	2.5	No	Low	Low	Excellent
1	No	2.0	Yes	Excellent	Excellent	Excellent
2	No	2.5	Yes	Excellent	Low	Excellent

***Figure 10-34.** Print and summary of AttrData_NN.csv*

Step 3: Explore the data. Check different attributes' data distribution and data types. You should use all the techniques you have learned in earlier chapters including data normalization, data preprocessing, data type conversion, etc. You can gain insight into the data and the data distribution, distribution of categorical variables, etc., just by exploring the data.

Here is the input:

```
data_df_sub=data_df.select_dtypes(include=['object'])

for c in data_df_sub.columns:
    display(data_df_sub[c].value_counts())
```

Figure 10-35 shows the input and the output.

```
In [10]: data_df_sub=data_df.select_dtypes(include=['object'])

         for c in data_df_sub.columns:
            display(data_df_sub[c].value_counts())

         Yes    28
         No     24
         Name: Attrition, dtype: int64

         No     28
         Yes    24
         Name: WorkChallenging, dtype: int64

         Excellent    28
         Low          24
         Name: WorkEnvir, dtype: int64

         Low          31
         Excellent    21
         Name: Compensation, dtype: int64

         Excellent    44
         Low           8
         Name: TechExper, dtype: int64
```

***Figure 10-35.** Results of data exploration*

Step 4: Normalize the `YrsExper` variable between 0 and 1.

Here is the input:

```
##Normalizing 'YerasOfExp' attribute data between 0 - 1
features = ['YrsExp']
data_df[features] = data_df[features]/data_df[features].max()
```

Figure 10-36 shows the input and the output.

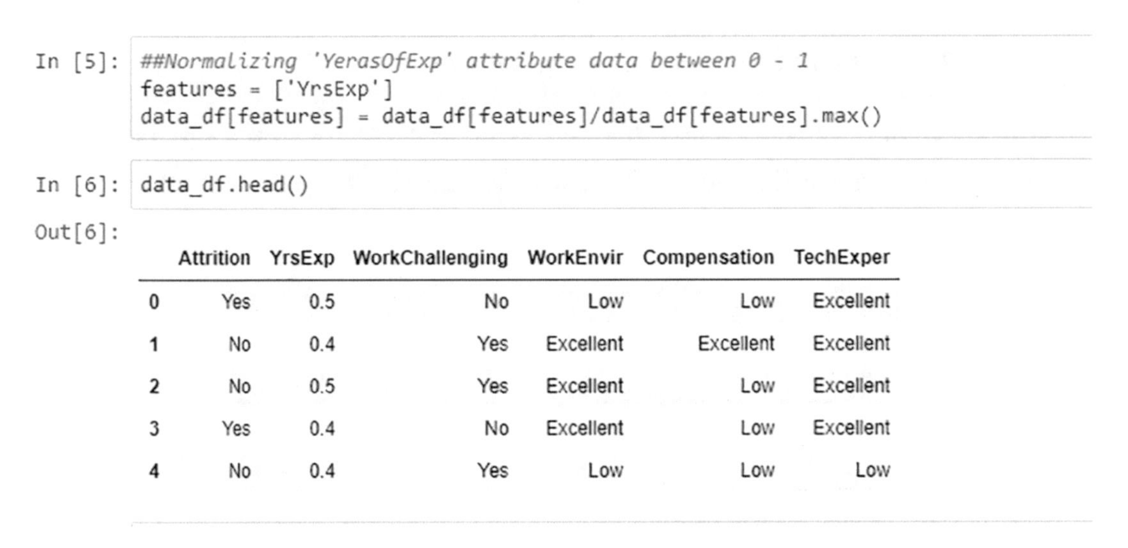

```
In [5]: ##Normalizing 'YerasOfExp' attribute data between 0 - 1
        features = ['YrsExp']
        data_df[features] = data_df[features]/data_df[features].max()

In [6]: data_df.head()

Out[6]:
```

	Attrition	YrsExp	WorkChallenging	WorkEnvir	Compensation	TechExper
0	Yes	0.5	No	Low	Low	Excellent
1	No	0.4	Yes	Excellent	Excellent	Excellent
2	No	0.5	Yes	Excellent	Low	Excellent
3	Yes	0.4	No	Excellent	Low	Excellent
4	No	0.4	Yes	Low	Low	Low

Figure 10-36. *Results of normalizing the YrsExper variable*

Step 5: Prepare the data as per the API requirements. The neural network API does not accept text values. Hence, convert data types with any strings into numerical values using `LabelEncoder()`.

Note If you have more than two levels in your categorical variables, you should use the `dummy_variable()` or `one_hot_encoding()` function and create a dummy variable for all your predictors that has more than two levels. Since in our data we just have only two levels (yes/No, Low and Excellent, etc.), we are using the `LabelEncoder()` function for everything including the `target` variable.

Here is the input:

```
from sklearn.preprocessing import LabelEncoder

le = LabelEncoder()

cat_vars = ['Attrition', 'WorkChallenging','WorkEnvir','Compensation',

        'TechExper']

X[['Attrition','WorkChallenging',

  'WorkEnvir','Compensation',

  'TechExper']] = X[['Attrition','WorkChallenging',

   'WorkEnvir','Compensation','TechExper']].apply(LabelEncoder().fit_transform)

for var in cat_vars:

   X[var] = X[var].astype('category',copy=False)

X.head(3)
```

Figure 10-37 shows the input and the output.

```
In [19]: ##Convert the categorical variables TEXT to numeric values as Neural Network does not understand text strings

In [18]: X = data_df.copy()

In [19]: from sklearn.preprocessing import LabelEncoder
         le = LabelEncoder()
         cat_vars = ['Attrition', 'WorkChallenging','WorkEnvir','Compensation',
                      'TechExper']
         X[['Attrition','WorkChallenging',
            'WorkEnvir','Compensation',
            'TechExper']] = X[['Attrition','WorkChallenging',
             'WorkEnvir','Compensation','TechExper']].apply(LabelEncoder().fit_transform)

         for var in cat_vars:
             X[var] = X[var].astype('category',copy=False)

         X.head(3)
Out[19]:
```

	Attrition	YrsExp	WorkChallenging	WorkEnvir	Compensation	TechExper
0	1	0.5	0	1	1	0
1	0	0.4	1	0	0	0
2	0	0.5	1	0	1	0

```
In [20]: X.dtypes
Out[20]: Attrition          category
         YrsExp              float64
         WorkChallenging    category
         WorkEnvir          category
         Compensation       category
         TechExper          category
         dtype: object
```

Figure 10-37. *Results of preparing the data per API requirements*

Step 6: Split data into two parts, train and test. Train the model using the Training data set and test the model with the Test data set. The sklearn neural `network()` function accepts X parameters and Y parameters separately. Hence, we have to drop `target` variables from the dataframe and create a separate X dataframe and Y dataframe with only the `target` class.

Here is the input:

```
##Data preparation

## For neural network, we have to input X and Y separately

##Y is the response class (target) and X is features Once this is done then split dataset

# into train and test

y = X['Attrition']

X1 = X.drop(columns='Attrition')

X1.head(3)

#Split data into training and test datset

X_train, X_test, y_train, y_test = train_test_split(X1, y, test_size=0.20, random_state=40)

print(X_train.shape); print(X_test.shape);print(y_train.shape);print(y_test.shape)
```

Figure 10-38 shows the input and the output.

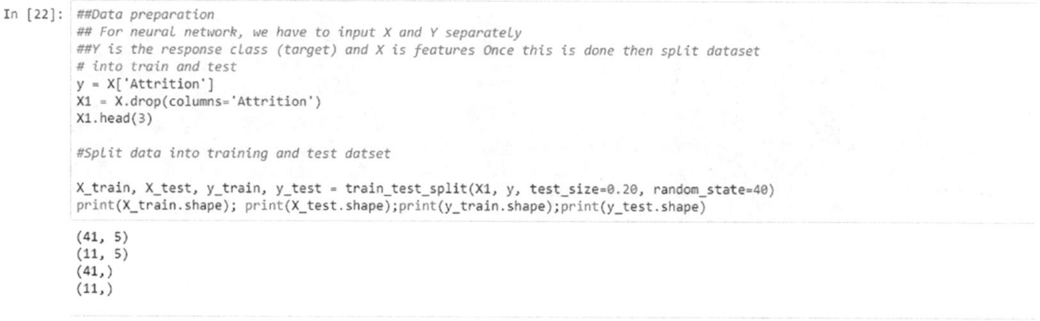

```
In [22]: ##Data preparation
         ## For neural network, we have to input X and Y separately
         ##Y is the response class (target) and X is features Once this is done then split dataset
         # into train and test
         y = X['Attrition']
         X1 = X.drop(columns='Attrition')
         X1.head(3)

         #Split data into training and test datset

         X_train, X_test, y_train, y_test = train_test_split(X1, y, test_size=0.20, random_state=40)
         print(X_train.shape); print(X_test.shape);print(y_train.shape);print(y_test.shape)

         (41, 5)
         (11, 5)
         (41,)
         (11,)
```

Figure 10-38. *Results of splitting the data in two parts*

Step 7: Create a neural network model using an `MLPClassifier()` API of scikit-learn on the training data.

Here is the input:

```
#Generate Neural network Model with (5, 2) 2 hidden layers with 5 neurons

from sklearn.neural_network import MLPClassifier

nn_model = MLPClassifier(hidden_layer_sizes=(5,2), activation='logistic',

            solver='adam',shuffle=True,max_iter=1000)

nn_model.fit(X_train,y_train)

nn_model.hidden_layer_sizes

nn_model.hidden_layer_sizes

nn_model.n_layers_

nn_model.classes_
```

Figure 10-39 shows the input and the output.

```
In [44]:  #Generate Neural network Model with (7, 10, 2) hidden Layers

In [111]: from sklearn.neural_network import MLPClassifier

          nn_model = MLPClassifier(hidden_layer_sizes=(5,2), activation='logistic',
                              solver='adam',shuffle=True,max_iter=1000)
          nn_model.fit(X_train,y_train)

          C:\Users\phdst\AppData\Local\conda\conda\envs\tensorflow\lib\site-packages\sklearn\neural_network\_multilayer_perceptron.py:57
          1: ConvergenceWarning: Stochastic Optimizer: Maximum iterations (1000) reached and the optimization hasn't converged yet.
            % self.max_iter, ConvergenceWarning)

Out[111]: MLPClassifier(activation='logistic', alpha=0.0001, batch_size='auto',
                        beta_1=0.9, beta_2=0.999, early_stopping=False, epsilon=1e-08,
                        hidden_layer_sizes=(5, 2), learning_rate='constant',
                        learning_rate_init=0.001, max_fun=15000, max_iter=1000,
                        momentum=0.9, n_iter_no_change=10, nesterovs_momentum=True,
                        power_t=0.5, random_state=None, shuffle=True, solver='adam',
                        tol=0.0001, validation_fraction=0.1, verbose=False,
                        warm_start=False)

In [112]: nn_model.hidden_layer_sizes

Out[112]: (5, 2)

In [113]: nn_model.n_layers_

Out[113]: 4

In [114]: nn_model.classes_

Out[114]: array([0, 1], dtype=int64)
```

Figure 10-39. *Results of creating the neural network model using the MLPClassifier() API of scikit-learn on the training data*

The multilayer perceptron classifier `MLPClassifier()` API function optimizes the log-loss function using stochastic gradient descent (SGD). SGD is an extension of the gradient descent. It calculates the gradient for only one training example at every iteration. The learning rate is used to calculate the variation or adjustments at every iteration. If the LR is too large, then the variations may be too far past the optimum value. Similarly, if the LR is too small, you may require many iterations to reach a local minimum. A recommended method to vary the LR is to start with learning rate 0.1 and adjust as necessary.

You have to provide several input parameters for the function to work properly. At a minimum, you must specify the number of layers, activation function, learning rate, and maximum number of epochs. We are providing the `MLPClassifier()` API details from the scikit-learn documentation here for reference:

Parameters

`hidden_layer_sizestuple, length = n_layers - 2, default=(100,)`

The ith element represents the number of neurons in the ith hidden layer.

`activation{'identity', 'logistic', 'tanh', 'relu'}, default='relu'`

Activation function for the hidden layer.

- `identity`, no-op activation, useful to implement linear bottleneck, returns f(x) = x
- `logistic`, the logistic sigmoid function, returns f(x) = 1 / (1 + exp(-x))
- `tanh`, the hyperbolic tan function, returns f(x) = tanh(x)
- `relu`, the rectified linear unit function, returns f(x) = max(0, x)

  `solver{'lbfgs', 'sgd', 'adam'}, default='adam'`

  The solver for weight optimization.

- `lbfgs` is an optimizer in the family of quasi-Newton methods.
- `sgd` refers to stochastic gradient descent.
- `adam` refers to a stochastic gradient-based optimizer proposed by Kingma, Diederik, and Jimmy Ba

Note The default solver adam works pretty well on relatively large data sets (with thousands of training samples or more) in terms of both training time and validation score. For small data sets, however, lbfgs can converge faster and perform better.

alphafloat, default=0.0001

L2 penalty (regularization term) parameter.

batch_sizeint, default='auto'

Size of minibatches for stochastic optimizers. If the solver is lbfgs, the classifier will not use minibatch when set to "auto", batch_size=min(200, n_samples).

learning_rate{'constant', 'invscaling', 'adaptive'}, default='constant'

Learning rate schedule for weight updates.

- constant is a constant learning rate given by learning_rate_init.
- invscaling gradually decreases the learning rate at each time step t using an inverse scaling exponent of power_t. effective_learning_rate = learning_rate_init / pow(t, power_t)
- adaptive keeps the learning rate constant to learning_rate_init as long as training loss keeps decreasing. Each time two consecutive epochs fail to decrease training loss by at least tol, or fail to increase the validation score by at least tol if early_stopping is on, the current learning rate is divided by 5.

Only used when solver='sgd'.

learning_rate_initfloat, default=0.001

The initial learning rate used. It controls the step-size in updating the weights. Only used when solver='sgd' or 'adam'.

power_tfloat, default=0.5

The exponent for inverse scaling learning rate. It is used in updating effective learning rate when the `learning_rate` is set to `invscaling`. Only used when `solver='sgd'`.

`max_iterint, default=200`

Maximum number of iterations. The solver iterates until convergence (determined by `tol`) or this number of iterations. For stochastic solvers (`sgd`, `adam`), note that this determines the number of epochs (how many times each data point will be used), not the number of gradient steps.

`shufflebool, default=True`

Whether to shuffle samples in each iteration. Only used when `solver='sgd'` or `adam`.

`random_stateint, RandomState instance, default=None`

Determines random number generation for weights and bias initialization, train-test split if early stopping is used, and batch sampling when `solver='sgd'` or `'adam'`. Passes an int for reproducible results across multiple function calls.

`tolfloat, default=1e-4`

Tolerance for the optimization. When the loss or score is not improving by at least `tol` for `n_iter_no_change` consecutive iterations, unless `learning_rate` is set to `adaptive`, convergence is considered to be reached and training stops.

`verbosebool, default=False`

Whether to print progress messages to stdout.

`warm_startbool, default=False`

When set to True, reuse the solution of the previous call to fit as initialization; otherwise, just erase the previous solution.

`momentumfloat, default=0.9`

Momentum for gradient descent update. Should be between 0 and 1. Only used when `solver='sgd'`.

`nesterovs_momentumbool, default=True`

Whether to use Nesterov's momentum. Only used when `solver='sgd'` and `momentum` > 0.

`early_stoppingbool, default=False`

Whether to use early stopping to terminate training when the validation score is not improving. If set to true, it will automatically set aside 10 percent of the training data as validation and terminate training when the validation score is not improving by at least `tol` for `n_iter_no_change` consecutive epochs. The split is stratified, except in a multilabel setting. If early stopping is False, then the training stops when the training loss does not improve by more than `tol` for `n_iter_no_change` consecutive passes over the training set. Only effective when `solver='sgd'` or `'adam'`.

`validation_fractionfloat, default=0.1`

The proportion of training data to set aside as validation set for early stopping. Must be between 0 and 1. Only used if `early_stopping` is True.

`beta_1float, default=0.9`

Exponential decay rate for estimates of first moment vector in adam; should be in [0, 1). Only used when `solver='adam'`.

`beta_2float, default=0.999`

Exponential decay rate for estimates of second moment vector in adam; should be in [0, 1). Only used when `solver='adam'`.

`epsilonfloat, default=1e-8`

Value for numerical stability in adam. Only used when `solver='adam'`.

`n_iter_no_changeint, default=10`

Maximum number of epochs to not meet `tol` improvement. Only effective when `solver='sgd'` or adam.

New in version 0.20.

> `max_funint, default=15000`
>
> Only used when `solver='lbfgs'`. Maximum number of loss function calls. The solver iterates until convergence (determined by `tol`), number of iterations reaches `max_iter`, or this number of loss function calls. Note that number of loss function calls will be greater than or equal to the number of iterations for the `MLPClassifier`.

Step 8: Predict test data using the model that you just created.

Here is the input:

```
#predict_train = NNCL.predict(X_train)

predict_test = nn_model.predict(X_test)

predict_test
```

Figure 10-40 shows the input and the output.

```
Predict 'Attrition' using Neural Network model ¶

In [115]: #predict_train = NNCL.predict(X_train)
          predict_test = nn_model.predict(X_test)

In [116]: predict_test

Out[116]: array([1, 1, 0, 1, 0, 1, 1, 0, 0, 0, 1], dtype=int64)
```

***Figure 10-40.** Predicting attrition using a neural network model*

Step 9: Measure the accuracy of the model. Use a confusion matrix (a truth table of actual versus predicted). `Sklearn()` provides built-in functions to calculate different performance measures, and we will use the same functions. Also, we will plot the loss curve, which displays the number of iterations it took to adjust the weights and biases to come up with an optimal model.

Here is the input:

```
print(confusion_matrix(y_test,predict_test))

print(classification_report(y_test,predict_test))
```

Figure 10-41 shows the input and the output.

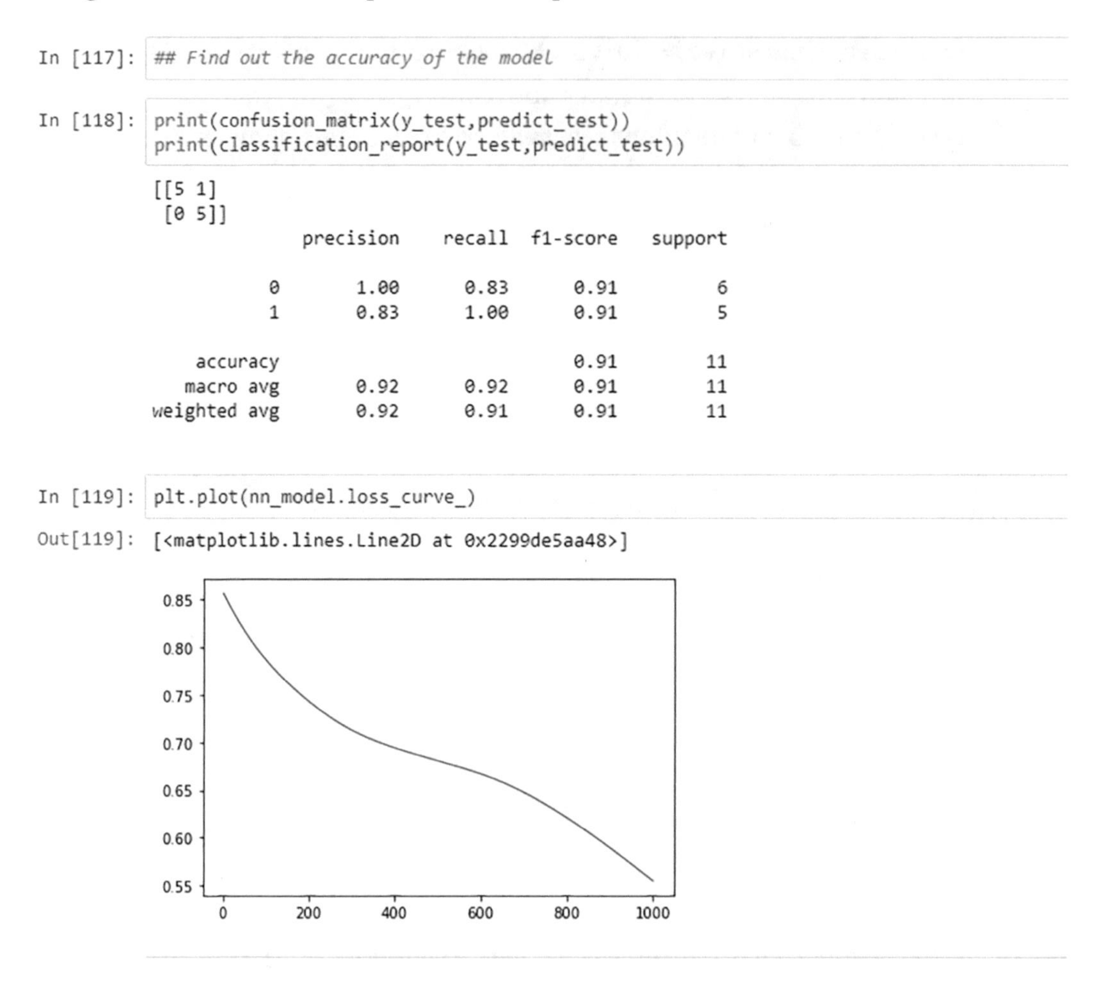

In [117]: ## Find out the accuracy of the model
In [118]: print(confusion_matrix(y_test,predict_test))
print(classification_report(y_test,predict_test))
[[5 1]
[0 5]]
precision recall f1-score support
0 1.00 0.83 0.91 6
1 0.83 1.00 0.91 5
accuracy 0.91 11
macro avg 0.92 0.92 0.91 11
weighted avg 0.92 0.91 0.91 11
In [119]: plt.plot(nn_model.loss_curve_)
Out[119]: [<matplotlib.lines.Line2D at 0x2299de5aa48>]
0.85
0.80
0.75
0.70
0.65
0.60
0.55
0
200
400
600
800
1000

***Figure 10-41.** Measuring the accuracy of the model*

Step 10: Finally, we will print the weights and biases of the optimized model for reference. See Figure 10-42.

```
In [121]: nn_model.coefs_

Out[121]: [array([[ 0.13038589,  0.37105089, -0.2412836 , -0.16881156, -0.48200271],
                  [-1.64279412,  1.87100249, -1.11592665, -1.61885547, -1.2007269 ],
                  [ 1.09244609, -1.10782601,  1.12467809,  0.70618752,  0.94465865],
                  [ 0.58017621, -0.91420091,  0.70659265,  0.63411723,  0.84311529],
                  [-1.04105164,  1.27404746, -1.27456171, -1.77544134, -1.01694948]]),
           array([[-0.90225391,  1.37706551],
                  [ 0.89980021, -1.55736702],
                  [-0.96883396,  1.24830141],
                  [-0.72844235,  1.22966177],
                  [-0.76597002,  0.80152607]]),
           array([[-0.85809687],
                  [ 0.73184114]])]
```

***Figure 10-42.** Printing the weights and biases of the optimized model*

10.7 Strengths and Weaknesses of Neural Network Models

The most prominent advantage of neural networks is their ability to learn complex patterns from the data and provide exemplary performance in predicting. In addition, neural networks have high tolerance to noisy data and can capture patterns and relationships between predictors and a response variable that are complicated.

The model we constructed is based on a small training set and a network consisting of two hidden layers. The model outperformed with 100 percent accuracy, which could be considered a weakness as the model may have been overfitted. One of the weaknesses of the neural network is the model can easily overfit the data unless you control the training epochs. The other problem with neural networks is that they require more training data to properly learn and perform.

The process of identifying the number of epochs, activation functions, learning rate, and a number of network layers required for the optimal performance of the model is called *tuning* the model. Tuning the model for optimal performance requires several iterations. A good indicator for finding optimal epochs and learning rate is finding the minimum error. As the neural network model goes through the iteration process of learning and adjusting weights, during the initial stages of epochs training, the error decreases, but after a while, errors can start to increase, as shown in Figure 10-43 (gradient descent). The error may have several minimal points during the learning, but there is only one lowest point called *global minima*. The point where the function takes the minimum value (error) is called *global minima*. Other points will be called *local minima*. The neural network adopts a gradient descent algorithm to find the minimum error. The learning could also lead to a risk of obtaining local minima weights rather than converging to global minima if the correct number of epochs is not selected during the model creation process.

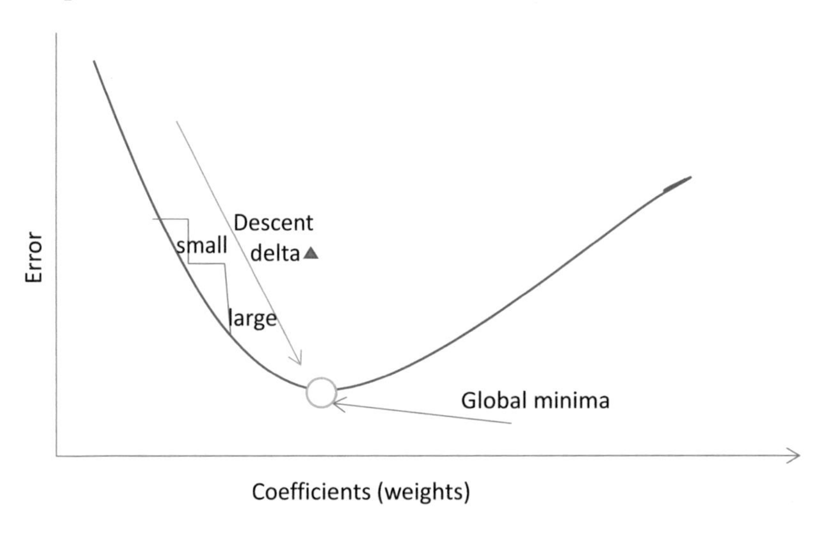

Figure 10-43. *Gradient descent, minimize loss function*

Neural network learning heavily relies on having a sufficient quantity of data for the training process. A neural network may perform poorly with smaller data, as in our example. Similarly, having an imbalanced class of data also leads to poor learning in the minority category.

A practical challenge is computational timeliness. Though neural network concepts have been around for more than 50 years, the computational abilities to process large data was a challenge. This is no longer a problem because of the availability of storage and powerful processors. In general, the neural network is computationally intensive and requires longer runtime than other classifiers. This runtime can grow exponentially with a higher number of variables and more network layers (many more weights to compute). If you have a real-time or near-real-time prediction application, then you should consider runtime measures to make sure it is not causing an unacceptable delay in the decision-making process.

Finally, the challenge is a careful selection of input variables. Since neural networks automatically adjust the weights and biases based on the "gradient descent" errors, we may have no mechanism to remove or add the variable based on output. This can be an advantage or disadvantage depending on the problem.

10.8 Deep Learning and Neural Networks

Our brain is highly complicated and is a structure that does the cognition, connections, and decisions using a network of natural neurons. Computers cannot do this as easily as humans. However, artificial intelligence is currently complemented by the huge data processing capabilities of the cloud and cluster of machines using multiple machines' distributed and collective power. These networks not only learn but can keep improving their performance with the increase of the data and learning cycles.

This has been possible because of deploying a network of artificial neurons. Now, we can deploy a number of hidden layers of various types (depending upon the type of the learning, data, and accuracy of the prediction required), e.g., convolutional layers, recurrent neural layers, max-pooling layers, dropout layers, dense layers, etc., leading to deep learning.

Deep learning is learning through the deep woven architecture of the neural nets. Here, a number of different layers of artificial neurons are typically deployed to ensure effective and high learning from the data.

Convolutional neural networks (CNNs) are a type of neural networks that are highly helpful in image recognition and classification.

Recurrent neural networks (RNNs) are a type of neural networks that are highly helpful in tasks such as speech recognition, handwriting recognition, etc.

There are many other types of neural networks too. Some of them are deep belief network, deep feed-forward network, Jordan network, Elman network, etc.

It is not possible to go into the details of all the network architectures and applications of deep learning in this chapter as deep learning itself is deep. Figure 10-44 shows a typical CNN implementation for deep learning.

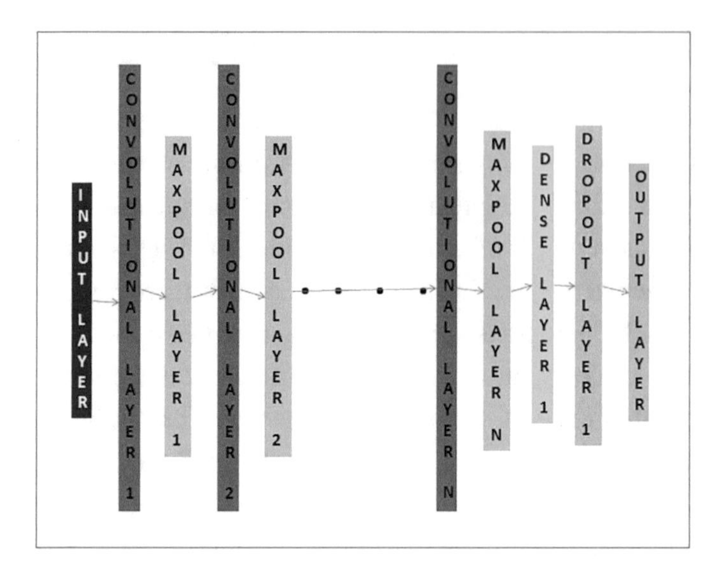

Figure 10-44. *Typical CNN implementation structure*

10.9 Chapter Summary

We started the chapter by providing a background of artificial neurons and how to imitate brain functions using artificial neurons. We explained the fundamentals of perceptrons and building neural networks and how a neural network learns and adjusts the weights and biases to perform better.

We discussed different activation functions such as RELU, sigmoid, and tanh, and you learned about the gradient descent algorithm to find the optimal weights and biases.

You also learned how to create a neural network model using both R and Python with a practical business case.

Finally, we ended the chapter by introducing deep neural networks and their applications.

CHAPTER 11

Logistic Regression

In Chapters 7 and 8, we discussed simple linear regression and multiple linear regression, respectively. In both types of regression, we have a dependent variable or response variable as a continuous variable that is normally distributed. However, this is not always the case. Often, the response variable is not normally distributed. It may be a binary variable following a binomial distribution and taking the values of 0/1 or No/Yes. It may be a categorical or discrete variable taking multiple values that may follow other distributions other than the normal one.

If the response variable is a discrete variable (which can be nominal, ordinal, or binary), you use a different regression method, known as *logistic regression*. If the response variable takes binary values (such as Yes/No or Sick/Healthy) or multiple discrete variables (such as Strongly Agree, Agree, Partially Agree, and Do Not Agree), we can use logistic regression. Logistic regression is still a linear type of regression. Logistic regression with the response variable taking only two values (such as Yes/No or Sick/Healthy) is known as *binomial logistic regression*. A logistic regression with a response variable that can take multiple discrete values is known as *multinomial logistic regression*. (Note that nonlinear regression is outside the scope of this book.)

Consider the following examples:

- You want to decide whether to lend credit to a customer. Your choice may depend on various factors including the customer's earlier record of payment regularity with other companies, the customer's credibility in the banking and financial sector, the profitability of the customer's organization, and the integrity of the promoters and management.
- You are deciding whether to invest in a project. Again, your choice depends on various factors, including the risks involved, the likely profitability, the longevity of the project's life, and the capability of the organization in the particular area of the project.

U. R. Hodeghatta and U. Nayak, *Practical Business Analytics Using R and Python*,
https://doi.org/10.1007/978-1-4842-8754-5_11

- You are deciding whether to hire a particular candidate for a senior management post. This choice may depend on various factors including the candidate's experience, past record, attitude, and suitability to the culture of your organization.
- You are deciding whether a particular employee is likely to stick with the organization. This choice may depend on various factors including the internal working environment of the organization, whether the employee is deployed on a challenging project, whether the compensation to that employee is commensurate with the industry standard, and whether the employee is technically good enough to be lured by other organizations.

Interactions among the predictors as well as multicollinearity may still be issues in logistic regression, and they need to be addressed appropriately. However, the assumptions of linearity, normality, and homoscedasticity that generally apply to regression do not apply to logistic regression.

A starting point for building any logistic regression model is to get your data in a dataframe format, as this is a requirement of the `glm()` function in R. As you saw in previous chapters, the `glm()` function with `family = gaussian(link = "identity")` is equivalent to the `lm()` function; the requirement is that the response or dependent variable is distributed normally. However, when you use `glm()` for logistic regression, the response variable takes the binary values, so we use `family = binomial(link = "logit")`. Why? Because we want the value to be between 0 and 1, which can be interpreted as a probability. We use the natural logarithm (`ln`) of the odds ratio, as shown here:

ln(odds ratio) = ln[probability of success / (1 – probability of success)]

Hence, we use the following bounding function that enables us to determine the probability of success and derive the logistic regression model by combining linear regression with the bounding function:

$$f(a)=1/\left(1+e^{-a}\right)$$

The following is the logistic regression model obtained:

$$P(\text{Response} = \text{Yes, given Independent Variable}_1, \text{Independent Variable}_2, \ldots \text{Independent Variable}_N) =$$

$$1 / (1 + e^{-(\beta_0 + \beta_1 \times \text{Independent Variable}_1 + \beta_2 \times \text{Independent Variable}_2 + \ldots + \beta_N \times \text{Independent Variable}_N)})$$

and

$$P(\text{Response} = \text{No, given Independent Variable}_1, \text{Independent Variable}_2, \ldots \text{Independent Variable}_N) =$$

$$1 - [1 / (1 + e^{-(\beta_0 + \beta_1 \times \text{Independent Variable}_1 + \beta_2 \times \text{Independent Variable}_2 + \ldots + \beta_N \times \text{Independent Variable}_N)})]$$

which is nothing but

$$1 / (1 + e^{(\beta_0 + \beta_1 \times \text{Independent Variable}_1 + \beta_2 \times \text{Independent Variable}_2 + \ldots + \beta_N \times \text{Independent Variable}_N)})$$

From these two equations pertaining to P(Response = Yes) and P(Response = No), we get the odds, as follows:

$$P(\text{Response} = \text{Yes}) / P(\text{Response} = \text{No}) =$$

$$e^{(\beta_0 + \beta_1 \times \text{Independent Variable}_1 + \beta_2 \times \text{Independent Variable}_2 + \ldots + \beta_N \times \text{Independent Variable}_N)}$$

Hence, logit, or log odds, is as follows:

$$\text{logit}(P) = \ln[P(\text{Response} = \text{Yes}) / P(\text{Response} = \text{No})] = \beta_0 + \beta_1 \times \text{Independent Variable}_1 + \beta_2 \times \text{Independent Variable}_2 + \ldots + \beta_N \times \text{Independent Variable}_N$$

We use the maximum likelihood method to generate the logistic regression equation that predicts the natural logarithm of the odds ratio. From this logistic regression equation, we determine the predicted odds ratio and the predicted probability of success, as shown here:

Predicted probability of success = predicted odds ratio / (1 + predicted odds ratio)

Here, we do not use the dependent variable value as is. Instead, we use the natural logarithm of the odds ratio.

11.1 Logistic Regression

In this section, we will demonstrate logistic regression using a data set.

11.1.1 The Data

Let's start by considering the data set we have created, which has six variables.

- `Attrition` represents whether the employee has exited the organization or is still in the organization (Yes for exit, and No for currently working in the organization).

- Yrs_Exp represents the experience of the employee at this time (in years).
- Work_Challenging represents whether the work assigned to the employee is challenging.
- Work_Envir represents whether the work environment is Excellent or Low.
- Compensation represents whether the compensation is Excellent or Low.
- Tech_Exper represents whether the employee is technically expert (Excellent or Low).

The data covers the last six months and pertains only to those employees with 2 to 5 years of experience. The data is extracted from the CSV text file named attr_data.txt by using the read.csv() command.

```
> attrition_data<-read.csv("attr_data.txt")
> summary(attrition_data)
 Attrition     Yrs_Exp        Work_Challenging
 No :24     Min.   :2.000    No :28
 Yes:28     1st Qu.:2.500    Yes:24
            Median :4.000
            Mean   :3.519
            3rd Qu.:4.500
            Max.   :5.000
      Work_Envir     Compensation      Tech_Exper
 Excellent:28    Excellent:21     Excellent:44
 Low      :24    Low      :31     Low      : 8
```

This code also presents a summary of the data. As you can see, the Attrition field has 28 Yes values; this means these employees have exited the organization. The 24 No values represent employees who are still working in the organization. You can also observe from the preceding summary that 28 employees have not been assigned challenging work (Work_Challenging), and 24 employees have been. Furthermore, 28 employees are working in teams where the work environment (Work_Envir) is considered excellent, whereas 24 are working in teams where the work environment is not that great (here marked Low). Finally, 21 employees have excellent compensation, at par or above the market compensation (shown here as Excellent); but 31 have compensation that is below the market compensation or low compensation (shown here as Low). Out of the total employees, 44 have excellent technical expertise (Tech_Exper), whereas 8 others have low technical expertise. The data set contains 52 records.

Ideally, when the organization is providing challenging work to an employee, the work environment within the team is excellent, compensation is excellent, and technical expertise of the employee is low, then the chance for attrition should be low.

Here is a glimpse of the data:

```
> head(attrition_data)
  Attrition Yrs_Exp Work_Challenging Work_Envir
1       Yes     2.5               No        Low
2        No     2.0              Yes  Excellent
3        No     2.5              Yes  Excellent
4       Yes     2.0               No  Excellent
5        No     2.0              Yes        Low
6       Yes     2.0               No        Low
  Compensation Tech_Exper
1          Low  Excellent
2    Excellent  Excellent
3          Low  Excellent
4          Low  Excellent
5          Low        Low
6          Low  Excellent
> tail(attrition_data)
   Attrition Yrs_Exp Work_Challenging Work_Envir
47        No     4.0              Yes  Excellent
48        No     4.5               No  Excellent
49       Yes     5.0               No  Excellent
50        No     5.0               No  Excellent
51       Yes     2.0              Yes  Excellent
52        No     4.0              Yes  Excellent
```

```
   Compensation Tech_Exper
47    Excellent  Excellent
48          Low        Low
49    Excellent  Excellent
50    Excellent  Excellent
51    Excellent  Excellent
52    Excellent  Excellent
```

11.1.2 Creating the Model

Now, let's build a logistic regression model by using the `glm()` function of R. We need to use `family = binomial(link="logit")` along with `glm()` because our response variable is binary. To start, let's generate the model by using all the parameters, since they have `Attrition` as a binary response (or dependent) variable; furthermore, we want to use this model to predict the possibility of any employee resigning later (in tune with this model and the other variables `Yrs_Exp`, `Work_Challenging`, `Work_Envir`, `Compensation`, and `Tech_Exper`). The command in R to generate the logistic regression model is shown here:

```
> attri_logit_model<-glm(
Attrition~Yrs_Exp+Work_Challenging+Work_Envir+Compensation+Tech_Exper,
data=attrition_data,
family =binomial(link="logit"))
> summary(attri_logit_model)
```

The model created by using the `glm()` function is shown here, along with the summary (generated by using `summary(model name)`):

```
> summary(attri_logit_model)

Call:
glm(formula = Attrition ~ Yrs_Exp + Work_Challenging + Work_Envir +
    Compensation + Tech_Exper, family = binomial(link = "logit"),
    data = attrition_data)

Deviance Residuals:
     Min        1Q    Median        3Q       Max
-1.37759  -0.20326   0.04508   0.22389   2.95410

Coefficients:
                    Estimate Std. Error z value Pr(>|z|)
(Intercept)          -1.1964     2.4077  -0.497   0.6193
Yrs_Exp               0.1320     0.5102   0.259   0.7959
Work_ChallengingYes  -3.4180     1.4091  -2.426   0.0153 *
Work_EnvirLow         4.6118     1.6783   2.748   0.0060 **
CompensationLow       2.8160     1.3513   2.084   0.0372 *
Tech_ExperLow        -3.9598     1.7030  -2.325   0.0201 *
---
Signif. codes:  0 '***' 0.001 '**' 0.01 '*' 0.05 '.' 0.1 ' ' 1

(Dispersion parameter for binomial family taken to be 1)

    Null deviance: 71.779  on 51  degrees of freedom
Residual deviance: 26.018  on 46  degrees of freedom
AIC: 38.018

Number of Fisher Scoring iterations: 7
```

Only one value among the categorical variables is shown here. This is because each variable has two levels, and one level is taken as a reference level by the model. An example is the categorical variable `Work_Challenging`, which has two levels: `Work_ChallengingYes` and `Work_ChallengingNo`. Only `Work_ChallengingYes` is shown in the model, as `Work_ChallengingNo` is taken as the reference level.

Note The *weights of evidence* transformation technique can be used if each factor variable has many levels. This ensures that only one combined variable represents multiple levels of each variable, thus reducing the coefficients of the model. This is useful because categorical variables with many levels can create issues in logistic regression.

You can see in the preceding summary of the logistic regression model that except for `Yrs_Exp`, all other variables are significant to the model (as each p-value is less than 0.05). `Work_ChallengingYes`, `Work_EnvirLow`, `CompensationLow`, and `Tech_ExperLow` are the significant variables. `Yrs_Exp` is not a significant variable to the model, as it has a high p-value. It is quite obvious from even a visual examination of the data that `Yrs_Exp` will not be significant to the model, as `attrition` is observed regardless of the number of years of experience. Furthermore, you can see that the model has converged in seven Fisher's scoring iterations, which is good because ideally we expect the model to converge in less than eight iterations.

We can now eliminate Yrs_Exp from the logistic regression model and recast the model. The formula used for recasting the logistic regression model and the summary of the model are provided here:

```
> attri_logit_model_2 <-glm(Attrition~Work_Challenging+Work_
Envir+Compensation+Tech_Exper,
data=attrition_data,
family =binomial(link="logit"))

> summary(attri_logit_model_2)
```

```
> summary(attri_logit_model_2)

Call:
glm(formula = Attrition ~ Work_Challenging + Work_Envir + Compensation +
    Tech_Exper, family = binomial(link = "logit"), data = attrition_data)

Deviance Residuals:
     Min        1Q    Median        3Q       Max
-1.36759  -0.20050   0.05191   0.22273   2.86409

Coefficients:
                    Estimate Std. Error z value Pr(>|z|)
(Intercept)          -0.6216     0.9101  -0.683   0.4946
Work_ChallengingYes  -3.4632     1.4025  -2.469   0.0135 *
Work_EnvirLow         4.5215     1.5995   2.827   0.0047 **
CompensationLow       2.7090     1.2542   2.160   0.0308 *
Tech_ExperLow        -3.8547     1.6065  -2.400   0.0164 *
---
Signif. codes:  0 '***' 0.001 '**' 0.01 '*' 0.05 '.' 0.1 ' ' 1

(Dispersion parameter for binomial family taken to be 1)

    Null deviance: 71.779  on 51  degrees of freedom
Residual deviance: 26.086  on 47  degrees of freedom
AIC: 36.086

Number of Fisher Scoring iterations: 7
```

As you can see, now all the model parameters are significant because the p-values are less than 0.05.

The degrees of freedom for the data are calculated as *n* minus 1 (the number of data points - 1, or 52 - 1 = 51). The degrees of freedom for the model is *n* minus 1 minus the number of coefficients (52 - 1 - 4 = 47). These are shown in the above summary of the model.

`Deviance` is a measure of lack of fit. Null and residual deviance are the most common values used in statistical software to measure the model fit. `Null deviance` is nothing but the deviance of the model with only intercept. The `null deviance` tells us how well the variable can be predicted without other coefficients, i.e., keeping only intercept. The `residual deviance` is the fitness of the model with all the predictors. The lower this value, the better the model and can predict the value more accurately.

Chi-square statistic can be calculated as follows:

$$X^2 = \text{Null deviance} - \text{Residual deviance}$$

Based on the p-value, we can determine whether the model is "fit" enough. The lower the p-value, the better the model compared to the model with only intercept.

In our previous example, we have Null deviance of 71.779 and Residual deviance of 26.086, so the Chi-square statistic would be as follows:

$$X^2 = \text{Null deviance} - \text{Residual deviance}$$

$$X^2 = 71.779 - 26.086$$

$$X^2 = 45.693$$

Here, p = 4, so there are 4 variables degrees of freedom. We use a Chi-Square table to find out the p-value. For X^2 of 45.693 for 4 degrees of freedom, the p-value is 0.00000000285. Since the p-value is less than 0.05, we could conclude that the model is very useful.

Figure 11-1 shows that the `residual deviance` reduces with the addition of each coefficient:

```
> anova(attri_logit_model_2, "PChiSq")
```

```
> anova(attri_logit_model_2, "PChiSq")
Analysis of Deviance Table

Model: binomial, link: logit

Response: Attrition

Terms added sequentially (first to last)


                  Df Deviance Resid. Df Resid. Dev
NULL                                 51     71.779
Work_Challenging  1  15.6908         50     56.089
Work_Envir        1  17.4619         49     38.627
Compensation      1   3.3265         48     35.300
Tech_Exper        1   9.2142         47     26.086
```

Figure 11-1. *Output of ANOVA*

Let's compare both the models—attri_logit_model (with all the predictors) and attri_logit_model_2 (with only significant predictors)—and check how the second model fares with respect to the first one. See Figure 11-2.

```
> anova(attri_logit_model_2, attri_logit_model,
test="Chisq")
```

```
> anova(attri_logit_model_2, attri_logit_model, test = "Chisq")
Analysis of Deviance Table

Model 1: Attrition ~ Work_Challenging + Work_Envir + Compensation + Tech_Exper
Model 2: Attrition ~ Yrs_Exp + Work_Challenging + Work_Envir + Compensation +
    Tech_Exper
  Resid. Df Resid. Dev Df Deviance Pr(>Chi)
1        47     26.086
2        46     26.018  1 0.067459   0.7951
>
```

Figure 11-2. *Model comparision*

This test is carried out by using the anova() function and checking the Chi-square p-value. In this table, the Chi-square p-value is 0.7951. This is not significant. This suggests that the model attri_logit_model_2 without Yrs_Exp works well compared to the model attri_logit_model with all the variables. There is not much difference between the two models. Hence, we can safely use the simpler model without Yrs_Exp (that is, attri_logit_model_2).

From this model (attri_logit_model_2), we can see that the coefficient of Work_ChallengingYes is -3.4632, Work_EnvirLow is 4.5215, CompensationLow is 2.7090, and Tech_ExperLow is -3.8547. These coefficient values are the natural logarithm of the odds ratio. A coefficient with a minus sign indicates that it decreases the potential for Attrition. Similarly, a coefficient with a plus sign indicates that it increases the potential for Attrition. Hence, the odds of having Attrition = Yes is exp(4.5215) = 91.97 times higher than not having any Attrition with all other variables remaining the same (all other things being equal) when Work_Envir = Low compared to when the Work_Envir = Excellent. This means that if Work_Envir = Excellent, the chance of Attrition = Yes is 5 percent, or 0.05, and then the odds for Attrition = No under the same conditions are (0.05 / (1 - 0.05)) × 91.97 = 4.8376. This corresponds to a probability of Attrition = No of 4.8376 / (1 + 4.8376) = 0.8286, or about 82.86 percent. The odds of having Attrition = Yes is exp(2.7090) = 15.02 times higher than not having any Attrition with all other variables remaining the same when Compensation = Low compared to when the Compensation = High. Both Work_Envir = Low and Compensation = Low increase the possibility of Attrition = Yes.

However, other two variables, Work_Challenging = Yes and Tech_Exper = Low, with negative signs to the respective coefficients means that they reduce the possibility of Attrition = Yes. Work_Challenging = Yes with the coefficient value of -3.4632 lowers the possibility of Attrition = Yes by exp(-3.4632) = 0.0313 with all other variables remaining the same compared to when Work_Challenging = No. Similarly, Tech_Exper = Low with the coefficient value of -3.8547 lowers the possibility of Attrition = Yes by exp(-3.8547) = 0.0211 with all other variables remaining the same compared to when Tech_Exper = High.

From this, you can see that the major impact to Attrition is influenced mainly by Work_Envir and Compensation.

11.1.3 Model Fit Verification

It is important for us to verify the model fit. This can be done by computing the pseudo R-squared value. This value is calculated by using the formula 1 - (model$deviance / model$null.dev). This tells us the extent that the model explains the deviance. The following code shows how to compute this value using R:

```
> ##Calculate pseudo-R value
> pseudo_R_Squared<- 1-(attri_logit_model_2$deviance)/(attri_logit_
model_2$null.deviance)
```

```
> pseudo_R_Squared
[1] 0.6365818
```

This calculation shows that the model explains 63.65 percent of the deviance. You can also compute the value of pseudo R-square by using `library(pscl)` and `pR2(model_name)`.

Another way to verify the model fit is by calculating the p-value with the Chi-square method as follows:

p-value <- pchisq[(model_deviance_diff),(df_data – df_model),lower.tail=F]

Here, `df_data` can be calculated by `nrow(data_set) - 1`, or in our case, `nrow(attrition_data) - 1`; `df_model` can be calculated by `model$df.residual`; and `model_deviance_diff` can be calculated as `model$null.dev - model$deviance`. These calculations are shown here:

```
> deviance_diff<-attri_logit_model_2$null.deviance - attri_logit_
model_2$deviance
> deviance_diff
[1] 45.6934
> df_data<-nrow(attrition_data) - 1
> df_data
[1] 51
> df_residual<-attri_logit_model_2$df.residual
> df_residual
[1] 47
> p_value<-pchisq(deviance_diff, (df_data - df_residual),
+                   lower.tail=FALSE)
> p_value
[1] 2.852542e-09
```

Because the p-value is very small, the reduction in deviance cannot be assumed to be by chance. As the p-value is significant, the model is a good fit.

11.1.4 General Words of Caution

The following warning message could be thrown while generating the logistics model:

```
Warning message:
glm.fit: fitted probabilities numerically 0 or 1 occurred
```

This may be due to data or a portion of data predicting the response perfectly. This is known as the *issue of separation* or *quasi-separation*.

Here are some general words of caution with respect to the logistic regression model:

- We have a problem when the null deviance is less than the residual deviance.
- We have a problem when the convergence requires too many Fisher's scoring iterations.
- We have a problem when the coefficients are large in size with significantly large standard errors.

In these cases, we may have to revisit the model and relook again at each coefficient.

11.1.5 Multicollinearity

We talked about multicollinearity in Chapter 8. Multicollinearity can be made out in R easily by using the `vif(model name)` function. VIF stands for *variance inflation factor*. Typically, a rule of thumb for the multicollinearity test to pass is that the VIF value should be greater than 5. The following test shows the calculation of VIF:

```
> ##Multi collinearity
> library(car)
> vif(attri_logit_model_2)
Work_Challenging       Work_Envir
        1.984868         2.461992
    Compensation       Tech_Exper
        1.611694         1.562850
```

As you can see, our model does not suffer from multicollinearity.

11.1.6 Dispersion

Dispersion (variance of the dependent variable) above the value of 1 (as mentioned in the summary of the logistic regression dispersion parameter for the binomial family to be taken as 1) is a potential issue with some of the regression models, including the logistic regression model. This is known as *overdispersion*. Overdispersion occurs when the observed variance of the dependent variable is bigger than the one expected out of the usage of binomial distribution (that is, 1). This leads to issues with the reliability of the significance tests, as this is likely to adversely impact standard errors.

Whether a model suffers from the issue of overdispersion can be easily found using R, as shown here:

```
> # If the ratio of the Residual Deviance of model to
> # its residual degrees of freedom is greater than 1 then
> # the model suffers from the issue of Overdispersion
> # Let us check whether our Logistic Regression Model
> # suffers from this issue
> overdisp_indicator<-attri_logit_model_2$deviance/attri_logit_
model_2$df.residual
> overdisp_indicator
[1] 0.5550193
>
> #As you can see the value is less than 1. Hence, our model
# does not suffer from the issue of overdispersion
```

The model generated by us, `attri_logit_model_2`, does not suffer from the issue of overdispersion. If a logistic regression model does suffer from overdispersion, you need to use quasibinomial distribution in the `glm()` function instead of binomial distribution.

11.1.7 Conclusion for Logistic Regression

For our data, we created a good logistic regression model that we have verified for significance and goodness of fit. The model does not suffer from multicollinearity or overdispersion.

Note Using `family=binomial(link="probit")` in the `glm()` function also works similarly to `family=binomial(link="logit")` in most circumstances. For `logit`, you may use `family=binomial()` directly in `glm()` without using `link="logit"`, as both produce the same result.

11.2 Training and Testing the Model

We can also use separate subsets of the data set for training the model (basically to create the model) and testing the model. Ideally, such data sets have to be generated randomly. Typically, 75 percent to 80 percent of the data set is used to train the model, and another 25 percent to 20 percent of the data set is used to test the model.

Let's use the same data set that we used previously to do this. We have to run `install.packages("caret")` and use `library(caret)`. Here we split the data set into two subsets (training data set and test data set) using R:

```
> library(caret)
> set.seed(1234)
> # setting seed ensures the repeatability of the results on
different trials
> # We are going to partition data into train and test using
> # createDataPartition() function from the caret package
> # we use 80% of the data as train and 20% as test

> Data_Partition<-createDataPartition(attrition_data$Attrition,
p=0.8,list=FALSE)
> Training_Data<-attrition_data[Data_Partition, ]
> Test_Data<-attrition_data[-Data_Partition, ]
> nrow(attrition_data)
```

```
[1] 52
> nrow(Training_Data)
[1] 43
> nrow(Test_Data)
[1] 9
> summary(Training_Data)
 Attrition    Yrs_Exp       Work_Challenging     Work_
Envir    Compensation
 No :20     Min.   :2.000   No :24          Excellent:24   Excellent:18
 Yes:23     1st Qu.:2.500   Yes:19          Low      :19   Low      :25
            Median :4.000
            Mean   :3.547
            3rd Qu.:4.500
            Max.   :5.000
     Tech_Exper
 Excellent:37
 Low      : 6
```

This split of the entire data set into two subsets (Training_Data and Test_Data) has been done randomly. Now, we have 43 records in Training_Data and 9 records in Test_Data.

We now train our model using Training_Data to generate a model, as shown here:

```
## Model 3
## Create a logistic regression model using Training_Data set
#  We will not use Yrs_Exp variable as it is not significant.
#  We already explained what variables to consider in our previous
discussion

train_logit_model<-glm(Attrition~Work_Challenging+Work_
Envir+Compensation+Tech_Exper,
                       data=Training_Data,
                       family =binomial(link="logit"))
summary(train_logit_model)
```

```
> summary(train_logit_model)

Call:
glm(formula = Attrition ~ Work_Challenging + Work_Envir + Compensation +
    Tech_Exper, family = binomial(link = "logit"), data = Training_Data)

Deviance Residuals:
     Min        1Q    Median        3Q       Max
-1.17893  -0.19835   0.05261   0.21902   3.04828

Coefficients:
                    Estimate Std. Error z value Pr(>|z|)
(Intercept)           -1.773      1.294  -1.370   0.1708
Work_ChallengingYes   -2.864      1.464  -1.956   0.0505 .
Work_EnvirLow          4.640      1.821   2.548   0.0108 *
CompensationLow        3.715      1.526   2.435   0.0149 *
Tech_ExperLow         -3.391      1.692  -2.004   0.0450 *
---
Signif. codes:  0 '***' 0.001 '**' 0.01 '*' 0.05 '.' 0.1 ' ' 1

(Dispersion parameter for binomial family taken to be 1)

    Null deviance: 59.401  on 42  degrees of freedom
Residual deviance: 20.394  on 38  degrees of freedom
AIC: 30.394

Number of Fisher Scoring iterations: 7
```

As you can see, the model generated (`train_logit_model`) has taken seven Fisher's scoring iterations to converge.

The next step is to predict the test data and measure the performance of the model as follows:

1. Use the model generated from the training data to predict the response variable for the test data and store the predicted data in the test data set.
2. Compare the values generated from the response variable with the actual values of the response variable in the test data set.
3. Generate the confusion matrix to understand the true positives (TPs), true negatives (TNs), false positives (FPs), and false negatives (FNs).

- True positives are the ones that are actually positives (1) and are also predicted as positives (1).
- True negatives are the ones that are actually negatives (0) but are also predicted as negatives (0).
- False positives are the ones that are predicted as positives (1) but are actually negatives (0).
- False negatives are the ones that are predicted as negatives (0) but are actually positives (1).

4. Check for accuracy, specificity, and sensitivity.
   - Accuracy = (TP + TN) / Total Observations
   - Specificity = True Negative Rate = TN / (FP + TN)
   - Sensitivity = Recall = True Positive Rate = TP / (FN + TP)

In addition, Precision = TP / (FP + TP) and F1 Score = 2TP / (2TP + FP + FN) may be considered.

Higher accuracy, higher sensitivity, and higher specificity are typically expected. Check whether these values are appropriate to the objective of the prediction in mind. If the prediction will affect the safety or health of people, we have to ensure the highest accuracy. In such cases, each predicted value should be determined with caution and further validated through other means, if required.

11.2.1 Example of Prediction

The originally fitted model (`attri_logit_model_2`) not only explains the relationship between the response variable and the independent variables but also provides a mechanism to predict the value of the response variable from the values of the new independent variables. This is done as shown here:

```
> predictor_1<-data.frame(Yrs_Exp=3, Work_Challenging="Yes", Work_
Envir="Excellent",
+                         Compensation="Excellent", Tech_Exper="Low")
> predicted_1<-predict(attri_logit_model_2, newdata=predictor_1,
type="response")
> predicted_1
           1
0.0003562379

##The probability is very low which represents "NO"

#The model has predicted well in this case
```

We take the value of `Attrition` as Yes if the probability returned by the prediction is > 0.5, and we take the same as No if the probability returned by the prediction is not > 0.5. As you can see in the preceding code, the value is far below 0.5, so we can safely assume that `Attrition = No`.

The preceding prediction is determined by using the function `predict(model name, newdata=dataframe_name, type="response")`, where `model name` is the name of the model arrived at from the input data, `newdata` contains the data of independent variables for which the response variable has to be predicted, and `type="response"` is required to ensure that the outcome is not `logit(y)`.

11.2.2 Validating the Logistic Regression Model on Test Data

We could obtain a valid model using the `Train_Data` above. Now we can use the `Test_Data` to predict the values of Attrition on the same and thus validate the model. This is done as follows:

1. We use the model generated to predict the dependent variable values (logit) using the `predicted_value <- predict(model, type="response")` on the test data set that we used to generate the model.

```
> predicted<-predict(train_logit_model,
newdata=Test_Data,
type="response")
```

2. Then we generate a confusion matrix.

```
> ##Set a threshold for anything >0.5 it is 1 and anything
less than 0.5, it is 0
> # 0 means NO and 1 means YES
> table(Test_Data$Attrition, predicted>0.5)

       FALSE TRUE
  No       3    1
  Yes      1    4
```

We can clearly see from the confusion matrix that the model generates very high true positives and true negatives. The accuracy of the model can be calculated using the following formula:

= (TP+TN)/(TP+TN+FP+FN)
=(4+3)/(4+3+1+1)=7/9=0.77

3. Now we use the ROCR package and the prediction() function from it as follows: prediction_object <-prediction(predicted_value, dataset$dependent_variable). With this, we now create a prediction_object.

```
> library(ROCR)
> prediction_object<-prediction(predicted,
                            Test_Data$Attrition)
> prediction_object
```

4. Using the `performance()` function from the ROCR package on the `prediction_object`, we obtain TPR = TP / (FN + TP) = TP / (All Positives) and FPR = FP / (FP + TN) = FP / (All Negatives) and plot TPR against the FPR to obtain a receiver operating characteristic (ROC) curve. This is done using `plot(performance(prediction_object, measure = "tpr", x.measure = "fpr")`. Alternatively, we can use sensitivity/specificity plots by using `plot(performance(prediction_object, measure = "sens", x.measure = "spec")` or precision/recall plots by using `plot(performance(prediction_object, measure = "prec", x.measure = "rec")`. The first of these R commands used to generate the ROC curve is shown here, followed by the curve generated (see Figure 11-3):

```
> # Generate performance measures fro the above
prediction values
> perf<-performance(prediction_object,
measure="tpr",
x.measure="fpr")
> plot(perf)
```

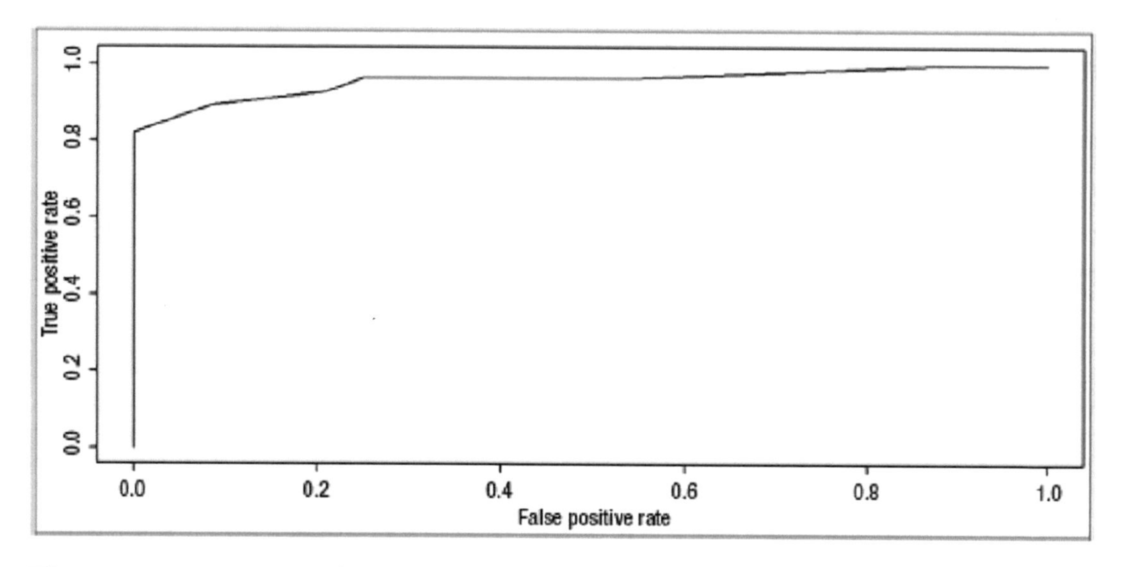

Figure 11-3. *Generated ROC curve*

This ROC curve clearly shows that the model generates almost no false positives and generates high true positives. Hence, we can conclude that the model generated is a good model.

11.3 Multinomial Logistic Regression

Multinomial logistic regression is used with categorical or discrete dependent variables that have more than two levels. The independent variables can be categorical variables or continuous variables. However, the methods to generate and evaluate these models are not straightforward and are tedious. These require lots of understanding of the underlying mathematics and statistics. Hence, we are not covering this in this book.

However, the `mlogit()` function from the `mlogit` package or the `multinom()` function from the `nnet` package is used to generate multinomial logistic regression for dependent variables with more than two unordered levels.

For dependent variables with more than two ordered levels, the `lrm()` function from the `rms` package is used to generate the multinomial logistic regression.

11.4 Regularization

Regularization is a complex subject that we won't discuss thoroughly here. However, we provide an introduction to this concept because it is an important aspect of statistics that you need to understand in the context of statistical models.

Regularization is the method normally used to avoid overfitting. When we keep adding parameters to our model to increase its accuracy and fit, at some point our prediction capability using this model decreases. By taking too many parameters, we are overfitting the model to the data and losing the value of generalization, which could have made the model more useful in prediction.

Using forward and backward model fitting and subset model fitting, we try to avoid overfitting and hence make the model more generalized and useful in predicting future values. This will ensure less bias as well as less variance when relating to the test data.

Regularization is also useful when we have more parameters than the data observations in our data set and the least squares method cannot help because it would lead to many models (not a single unique model) that would fit to the same data. Regularization allows us to find one reasonable solution in such situations.

Shrinkage methods are the most used regularization methods. They add a penalty term to the regression model to carry out the regularization. We penalize the loss function by adding a multiple (λ, also known as the *shrinkage parameter*) of the regularization norm, such as Lasso or Ridge (also known as the *shrinkage penalty*), of the linear regression weights vector. We may use cross validation to get the best multiple (λ value). The more complex the model, the greater the penalty. We use either the L_1 regularizer (Lasso) or the L_2 regularizer (Ridge). Regularization shrinks the coefficient estimates to reduce the variance.

Ridge regression shrinks the estimates of the parameters but not to 0, whereas the Lasso regression shrinks the estimates of some parameters to 0. For Ridge, the fit will increase with the value of λ, and along with that, the value of variance also increases. This can lead to a huge increase in parameter estimates, even for small changes in the training data, and get aggravated with the increase in the number of parameters. Lasso creates less-complicated models, thus making the predictability easier.

Let's explore the concept of regularization on our data set `attrition_data` without `Yrs_Exp`. We don't take `Yrs_Exp` into consideration because we know that it is not significant.

We use the `glmnet()` function from the `glmnet` package to determine the regularized model. We use the `cv.glmnet()` function from the `glmnet` package to determine the best lambda value. We use `alpha=1` for the Lasso and use `alpha=0` for the Ridge. We use `family="binomial"` and `type="class"` because our response variable is binary and we are using the regularization in the context of logistic regression, as required. The `glmnet()` function requires the input to be in the form of a matrix and the response variable to be a numeric vector. This fits a generalized linear model via penalized maximum likelihood. The regularization path is computed for the Lasso or elasticnet penalty at a grid of values for the regularization parameter lambda.

The generic format of this function as defined in the `glmnet` R package is as follows:

```
glmnet(x, y, family=c("gaussian", "binomial", "poisson", "multinomial",
"cox", "mgaussian"),
weights, offset=NULL, alpha = 1, nlambda = 100,
lambda.min.ratio = ifelse(nobs<nvars,0.01,0.0001), lambda=NULL,
standardize = TRUE, intercept=TRUE, thresh = 1e-07, dfmax = nvars + 1,
pmax = min(dfmax × 2+20, nvars), exclude, penalty.factor = rep(1, nvars),
lower.limits=-Inf, upper.limits=Inf, maxit=100000,
type.gaussian=ifelse(nvars<500, "covariance", "naive"),
type.logistic=c("Newton", "modified.Newton"),
standardize.response=FALSE, type.multinomial=c("ungrouped", "grouped"))
```

As usual, we will not be using all the parameters. We will be using only the absolutely required parameters in the interest of simplicity. Please explore the glmnet package guidelines for details of each parameter.

We will first prepare the inputs required. We need the model in the format of a matrix, as the input for the glmnet() function. We also require the response variable as a vector:

```
> ###################
> library(glmnet)
Loading required package: Matrix
Loaded glmnet 4.1-1
Warning message:
package ‘glmnet’ was built under R version 3.6.3
> #converting into a matrix as required for the input.
> x<-model.matrix(Attrition~Work_Challenging+Work_
Envir+Compensation+Tech_Exper,
+                   data = attrition_data)
> y<-attrition_data$Attrition
> glmnet_fit<-glmnet(x,y,
+            family="binomial",
+            alpha=1,
+            nlambda=100)
> summary(glmnet_fit)
           Length Class     Mode
a0          68    -none-    numeric
beta       340    dgCMatrix S4
df          68    -none-    numeric
dim          2    -none-    numeric
lambda      68    -none-    numeric
dev.ratio   68    -none-    numeric
nulldev      1    -none-    numeric
npasses      1    -none-    numeric
jerr         1    -none-    numeric
offset       1    -none-    logical
classnames   2    -none-    character
```

```
call          6     -none-    call
nobs          1     -none-    numeric
```

Explaining the contents of the summary is beyond the scope of this book, but we will show how the regularization is carried out primarily using the graphs. We use the `plot()` function for this purpose. As we are using the binary data and logistic regression, we use `xvar="dev"` (where `dev` stands for *deviance*) and `label = TRUE` to identify the parameters in the plot as inputs to the `plot()` function in Figure 11-4.

```
> ##plot
> plot(glmnet_fit,
xvar="dev",
label=TRUE)
```

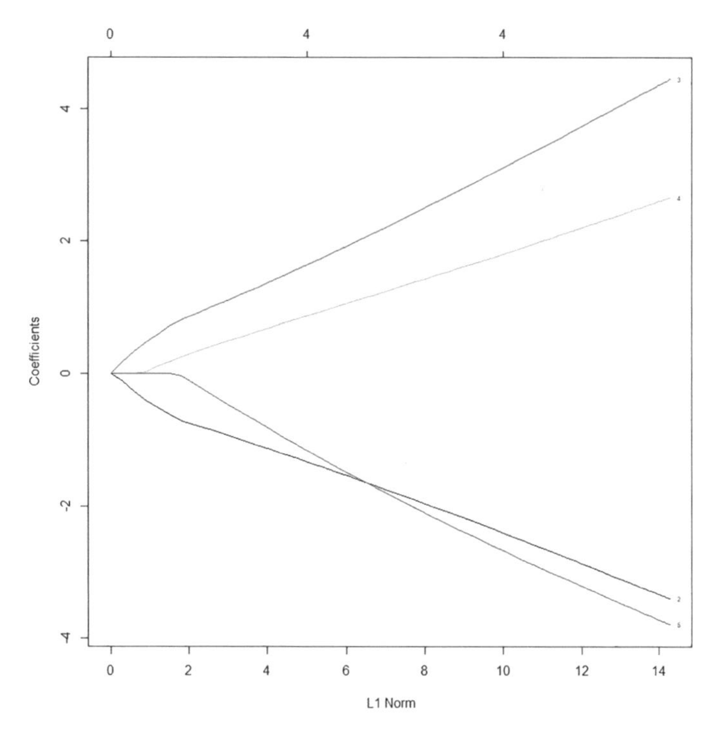

Figure 11-4. *Shows the deviance of each variable: two have + coefficients, and two have – coefficients*

The output of the glmnet_fit using the print() function is shown here:

```
> print(glmnet_fit)
Call:  glmnet(x = x, y = y, family = "binomial", alpha = 1, nlambda = 100)
    Df  %Dev   Lambda
1    0  0.00 0.273000
2    2  5.56 0.248700
3    2 10.91 0.226600
4    3 15.61 0.206500
5    3 19.94 0.188200
....
....
....
```

This primarily shows the degrees of freedom (number of nonzero coefficients), the percentage of null deviance explained by the model, and the lambda value. As you can see, the lambda value keeps on decreasing. As the lambda value decreases, the percent of deviance explained by the model increases, as does the significant number of nonzero coefficients. Even though we supplied nlambda = 100 for the function (this is the default), the lambda value is shown only 68 times. This is because the algorithm ensures that it stops at an optimal time when it sees there is no further significant change in the percent deviation explained by the model.

Now we will make the prediction of the class labels at lambda = 0.05. Here type = "class" refers to the response type:

```
> ##Predict glmnet() model
> predict(glmnet_fit, newx=x[1:4,],
+           type="class",
+           s=.05)
  1
1 "Yes"
2 "No"
3 "No"
4 "Yes"
```

As you can see, all four values are predicted accurately, as they match the first four rows of our data set.

Now we will do the cross validation of the regularized model by using the `cv.glmnet()` function from the `glmnet` package. This function does k-fold cross validation for `glmnet`, produces a plot, and returns a minimum value for lambda. This also returns a lambda value at one standard error. This function by default does a 10-fold cross-validation. We can change the k-folds if required. Here we use `type.measure = "class"` as we are using the binary data and the logistic regression. Here, `class` gives the misclassification error:

```
> ###Cross validation fit
> cv.fit<-cv.glmnet(x,y,
                    family="binomial",
                    type.measure="class")
> summary(cv.fit)
           Length Class  Mode
lambda     68     -none- numeric
cvm        68     -none- numeric
cvsd       68     -none- numeric
cvup       68     -none- numeric
cvlo       68     -none- numeric
nzero      68     -none- numeric
call        5     -none- call
name        1     -none- character
glmnet.fit 13     lognet list
lambda.min  1     -none- numeric
lambda.1se  1     -none- numeric
index       2     -none- numeric
```

We now plot the output of `cv.glmnet()`—that is, `cv.fit`—by using the `plot()` function.

```
> #Plot cv.fit model output
> plot(cv.fit)
```

Figure 11-5 shows the cross-validated curve along with the upper and lower values of the misclassification error against the log(lambda) values. Red dots depict the cross-validated curve.

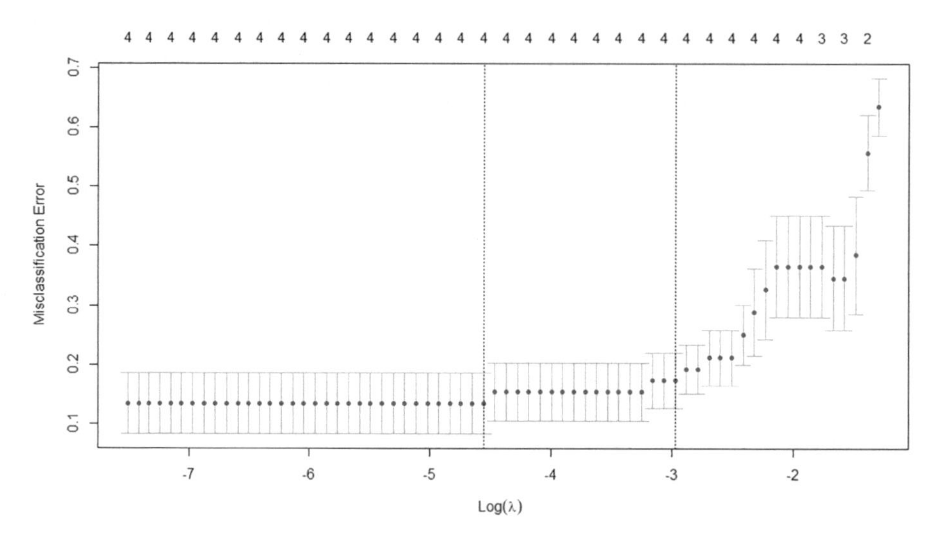

Figure 11-5. *Output of the plot(cv.fit)*

The following shows some of the important output parameters of the `cv.fit` model, including `lambda.min` or `lambda.1se`:

```
> ## We will read some of the important output of cv.fit
> cv.fit$lambda.min
[1] 0.01052007
> cv.fit$lambda.1se
[1] 0.05115488
```

In the previous code, `lambda.min` is the value of lambda that gives minimum `cvm`, and `cvm` is the cross-validation error. Similarly, `lambda.1se` is the largest value of lambda such that the error is within one standard error of the minimum.

We can view the coefficient value at the `lambda.min` value using the `coef(cv.fit, s = "lambda.min")` command in R. The output is a sparse matrix with the second levels shown for each independent factor.

```
> ##Get the coeffiicients of the model
> coef(cv.fit, s="lambda.min")
6 x 1 sparse Matrix of class "dgCMatrix"
                                1
(Intercept)          -0.4770243
(Intercept)           .
Work_ChallengingYes -2.6311567
Work_EnvirLow         3.4122989
CompensationLow       1.9985559
Tech_ExperLow        -2.9425308
```

Let's now see how this regularized model predicts the values by using the `predict()` function and the `s = "lambda.min"` option. We will check this for the first six values of our data set. The results are shown here:

```
> # predict new data
> predict(cv.fit, newx=x[1:6,],s="lambda.min",
type="class")
  1
1 "Yes"
2 "No"
3 "No"
4 "Yes"
5 "No"
6 "Yes"
```

All six values are predicted properly by the `predict()` function. However, please note that we have not validated the results for our entire data set. The accuracy of the model may not be 100 percent, as our objective of regularization was to provide a generalized model for future predictions without worrying about an exact fit (overfit) on the training data.

11.5 Using Python to Generate Logistic Regression

In the previous sections of this chapter, we used the R programming language to generate the model and predict the response variable for the new predictor values.

Here, we are going to use Python and the packages available within the Anaconda framework like Jupyter Notebook, scikit-learn, pandas, NumPy, SciPy, etc. (as relevant), to generate the model and make the predictions using the generated model. To keep it simple and to explain the details in the context of each input and output, the comments are embedded in the code. The extract from the Jupyter Notebook we used to carry out this is embedded in logical pieces in the code in the following sections.

11.5.1 Loading the Required Packages and Importing the Data

Here's the code to load the required packages and import the data:

```
#import necessary base packages
import numpy as np
import pandas as pd
import scipy
import sklearn

attri_data = pd.read_csv('E:/Book_Revision/attri_data_10.txt', sep=',',
header=0)
print(attri_data)
```

The code to read the data from the text file and to print the data is provided below (the code to import data is also provided in the text form in the above text box to facilitate the ease of use by the readers of this book):

```
In [2]: import pandas
        attri_data = pd.read_csv('E:/Book_Revision/attri_data_10.txt', sep=',', header=0)
        print(attri_data)

   Attrition  Yrs_Exp Work_Challenging Work_Envir Compensation Tech_Exper
0        Yes      2.5               No        Low          Low  Excellent
1         No      2.0              Yes  Excellent    Excellent  Excellent
2         No      2.5              Yes  Excellent          Low  Excellent
3        Yes      2.0               No  Excellent          Low  Excellent
4         No      2.0              Yes        Low          Low        Low
5        Yes      2.0               No        Low          Low  Excellent
6         No      2.0               No  Excellent    Excellent        Low
7        Yes      2.5               No        Low    Excellent  Excellent
8        Yes      2.0               No  Excellent          Low  Excellent
9        Yes      3.0              Yes        Low    Excellent  Excellent
10       Yes      3.5               No  Excellent          Low  Excellent
11        No      3.0              Yes  Excellent    Excellent  Excellent
12        No      2.0              Yes  Excellent          Low  Excellent
13        No      2.0              Yes  Excellent    Excellent  Excellent
14        No      2.5              Yes        Low    Excellent        Low
15        No      2.0              Yes  Excellent    Excellent        Low
16       Yes      2.5               No        Low          Low  Excellent
17        No      4.5              Yes  Excellent    Excellent  Excellent
18       Yes      4.0               No  Excellent          Low  Excellent
19       Yes      4.0              Yes        Low    Excellent  Excellent
20       Yes      4.5               No        Low          Low  Excellent
21        No      4.5               No  Excellent    Excellent  Excellent
22       Yes      4.0               No  Excellent          Low  Excellent
```

In the above dataframe, we have 52 records with one response variable (i.e., `Attrition`) and five predictor variables (only a partial view is shown).

11.5.2 Understanding the Dataframe

Here, we will be exploring the dataframe to understand the data types even though from the data it is clear that, except for one of the predictor variables (`i.e., Yrs_Exp`), all the variables are categorical. We also check for any null values as we have to clean up the null values or set them with the appropriate correct values before we generate the model.

Here is the input:

```
attri_data.info()
```

Here is the output:

```
In [3]: #get full information on attri_data dataframe
        attri_data.info()

        <class 'pandas.core.frame.DataFrame'>
        RangeIndex: 52 entries, 0 to 51
        Data columns (total 6 columns):
        Attrition           52 non-null object
        Yrs_Exp             52 non-null float64
        Work_Challenging    52 non-null object
        Work_Envir          52 non-null object
        Compensation        52 non-null object
        Tech_Exper          52 non-null object
        dtypes: float64(1), object(5)
        memory usage: 2.5+ KB
```

11.5.3 Getting the Data Ready for the Generation of the Logistic Regression Model

Here, we first make a copy of the dataframe to further prepare the data in a form to generate the logistic regression model. We need to modify our categorical string data in logical form in terms of numbers for the algorithms to understand and interpret the data. Further, we need to split the data into two separate dataframes: one with the response variable and the other with the predictor variables.

```
#making a copy of the attri_data dataframe
X = attri_data.copy()
```

We use the `labelEncode()` function on categorical variables as sklearn functions need only numerical variables. Strictly speaking, we should convert everything to one-hot variables and perform the model building exercise. However, in our case, we have only two levels of "yes" or "no" or "Low" or "High," and both yield the same results. Though there are several other methods available to perform the same task, we have taken the simplistic approach.

Here is the input:

```
from sklearn.preprocessing import LabelEncoder
le = LabelEncoder()
X[['Attrition','Work_Challenging','Work_Envir',
'Compensation',
   'Tech_Exper']]=X[['Attrition','Work_Challenging','Work_Envir',
   'Compensation','Tech_Exper']].apply(le.fit_transform)
X.head(5)
```

Here is the output:

```
X.head(5)

   Attrition  Yrs_Exp  Work_Challenging  Work_Envir  Compensation  Tech_Exper
0          1      2.5                 0           1             1           0
1          0      2.0                 1           0             0           0
2          0      2.5                 1           0             1           0
3          1      2.0                 0           0             1           0
4          0      2.0                 1           1             1           1
```

11.5.4 Splitting the Data into Training Data and Test Data

In this section, we split the data set into two separate data sets, i.e., the training data set and the test data set. We also generate the logistic regression model using the scikit-learn package. Also, the sklearn `LogisticRegression()` function expects the X and Y factors separately. We have to provide proper input to the API to satisfy the API requirements. These requirements may vary from one API to another. We always suggest referring to the documentation before proceeding to use the functions.

```
#split the dataframe with response variables (y)
#and another with only the predictor variables (X)
y = X.Attrition
X = X[X.columns[1::]]
#split the dataset into two separate sets viz. training set
#and test set manually
#training set for generating the model
#test set for validating the model generated
```

```
train_samp = (40)
from sklearn.model_selection import train_test_split
X_train, X_test, y_train, y_test = train_test_split(X, y,
    train_size=train_samp, test_size=12)
```

11.5.5 Generating the Logistic Regression Model

Once you have all the variables in place and the proper preprocessing is complete, use the sklearn library's LogisticRegression() function to generate a model.

```
from sklearn.linear_model import LogisticRegression
logist_regre = LogisticRegression(random_state = 0, penalty = 'l2',
    solver = 'lbfgs', multi_class = 'multinomial',
    max_iter = 500).fit(X_train, y_train)
```

11.5.6 Predicting the Test Data

After you generate the model, test how well your model is performing.

```
predicted = logist_regre.predict(X_test)
from sklearn.metrics import classification_report
print(classification_report(y_test, predicted))
```

Here is the output:

```
print(classification_report(y_test, predicted))

             precision    recall  f1-score   support

          0       0.75      0.50      0.60         6
          1       0.62      0.83      0.71         6

avg / total       0.69      0.67      0.66        12
```

As you can see, the performance of the model (i.e., the accuracy of prediction on the test data set) is not very high. As we know from the previous discussions on model building in R, Yrs_Exp is not a significant field for this model, so we will now build a model without the variable Yrs_Exp. We will then validate the model using the accuracy of the prediction on the test data.

Note If you want the output to remain the same every time this step is run, then use the `np.random.seed(n)` function. Otherwise, the algorithm may provide different output every time for each specific run.

11.5.7 Fine-Tuning the Logistic Regression Model

Even though we got reasonably good accuracy in our previous model, we will repeat the exercise by removing the `Yrs_Exper` variable to increase the accuracy of the model generated. By repeating this exercise, we may not get the same accuracy. If you look at the significance of each variable, even in Python, the `Yrs_Exp` variable is not significant. So, let's remove `Yrs_Exp` from the dataframe, generate the model, and check for accuracy.

```
Z = X.drop(['Yrs_Exp'], axis=1)
#split the dataset into two separate sets viz. training set
#and test set manually
#training set for generating the model
#test set for validating the model generated
train_samp = (40)
from sklearn.model_selection import train_test_split
Z_train, Z_test, y_train, y_test = train_test_split(Z, y,
    train_size=train_samp, test_size=12)
from sklearn.linear_model import LogisticRegression
logist_regre = LogisticRegression(random_state = 0, penalty = 'l2',
    solver = 'lbfgs', multi_class = 'multinomial',
    max_iter = 500).fit(Z_train, y_train)
```

Predict on the test data and print the confusion matrix score to measure the accuracy of the model. The code in this regard is provided below:

```
predicted = logist_regre.predict(Z_test)
import sklearn.metrics as sm
print(confusion_matrix(y_test, predicted))
```

Output (with partial inputs) is shown below:

```
print(sm.confusion_matrix(y_test, predicted))

[[6 1]
 [0 5]]
```

```
In [18]: #print logistic regression coefficients
         logist_regre.coef_

Out[18]: array([[-0.75862672,  0.96860213,  0.63232032, -0.84379994]])
```

Here, you see higher accuracy. However, depending upon the train-test split carried out by the algorithm, you may get higher accuracy for the initial instance itself.

Note If you want the output to remain same every time this step is run, then use the `np.random.seed(n)` function. Otherwise, the algorithm may provide different output every time for each specific run. Also, please note that the numbers of the Jupyter Notebook cells in the previous example may not be consecutive here as we would carried out some additional steps during our exercise, which are not required to be shown here.

11.5.8 Logistic Regression Model Using the statsmodel() Library

Several methods and related utilities/packages are available to build the logistic regression model in Python and also for various other types of regressions. In the previous sections, we explored one of the ways to generate and validate the logistic regression model using the scikit-learn package. Here, we will use another popular approach using the Statsmodels utility/package. The code and the output are provided below:

```
#Seabold, Skipper, and Josef Perktold. "statsmodels: Econometric and
#statistical modeling with python." Proceedings of the 9th Python in
#Science Conference. 2010. Thanks to Statsmodels
import statsmodels.api as sm
```

```
from statsmodels.formula.api import logit
logit_mod = sm.Logit(y_train, Z_train)
from pandas.core import datetools
logit_resi = logit_mod.fit()
logit_resi.summary()
```

```
In [20]: logit_resi = logit_mod.fit()
         logit_resi.summary()

         Optimization terminated successfully.
                  Current function value: 0.262090
                  Iterations 8

Out[20]:
```

Logit Regression Results

Dep. Variable:	Attrition	**No. Observations:**	40
Model:	Logit	**Df Residuals:**	36
Method:	MLE	**Df Model:**	3
Date:	Sun, 14 Aug 2022	**Pseudo R-squ.:**	0.6156
Time:	11:44:08	**Log-Likelihood:**	-10.484
converged:	True	**LL-Null:**	-27.274
		LLR p-value:	2.428e-07

	coef	**std err**	**z**	**P>|z|**	**[0.025**	**0.975]**
Work_Challenging	-3.0228	1.287	-2.349	0.019	-5.545	-0.501
Work_Envir	4.1816	1.637	2.554	0.011	0.973	7.390
Compensation	2.2552	1.138	1.982	0.047	0.025	4.485
Tech_Exper	-3.9206	1.735	-2.259	0.024	-7.322	-0.519

We can see that the optimization of the previous model was successful as the optimization ended within eight iterations. As you can see from the logit regression results, the model p-value is very small. Hence, the model is significant. We will also check to see the significance of each of the predictors to the model, as follows, which confirms that the selected predictors are significant to the model as the p-value of each of these predictors is less than the selected significance level of 0.05:

```
In [26]: print(logit_resi.pvalues)

         Work_Challenging     0.018816
         Work_Envir           0.010648
         Compensation         0.047485
         Tech_Exper           0.023879
         dtype: float64
```

Once we have the model, the next step is to predict using the model and measure the performance.

Here is the input:

```
#Here, we are applying the model to the full data
#set (excluding Yrs_Exp to predict on the full dataset
predicted_full = logist_regre.predict(Z)
from sklearn.metrics import classification_report
print(classification_report(y, predicted_full))
```

Here is the output:

```
print(classification_report(y, predicted_full))

             precision    recall  f1-score   support

          0       0.88      0.92      0.90        24
          1       0.93      0.89      0.91        28

avg / total       0.90      0.90      0.90        52
```

Print the Confusion Matrix: The code and the corresponding output are shown below:

```
#confusion matrix of the prediction on the
#full data set (except Yrs_Exp)
import sklearn.metrics as sm
from sklearn.metrics import confusion_matrix
print(sm.confusion_matrix(y, predicted_full))
```

```
print(sm.confusion_matrix(y, predicted_full))

[[22  2]
 [ 3 25]]
```

Note You could have also initially generated the model with all the predictors and checked on the significance of each of them. However, in those cases, you may find more predictors as nonsignificant. However, as you reduce one by one of those nonsignificant predictors from the regression equation, you may find in the final model that one or more of these predictors is significant on their own.

11.6 Chapter Summary

In this chapter, you saw that if the response variable is a categorical or discrete variable (which can be nominal, ordinal, or binary), you use a different regression method, called *logistic regression.* If you have a dependent variable with only two values, such as Yes or No, you use *binomial logistic regression.* If the dependent variable takes more than two categorical values, you use multinomial logistic regression.

You looked at a few examples of logistic regression. The assumptions of linearity, normality, and homoscedasticity that generally apply to regressions do not apply to logistic regression. You used the `glm()` function with `"family = binomial(link="logit")` to create a logistic regression model.

You also looked at the underlying statistics and how `logit` (log odds) of the dependent variable is used in the logistic regression equation instead of the actual value of the dependent variable.

You also imported the data set to understand the underlying data. You created the model and verified the significance of the predictor variables to the model by using the p-value. One of the variables (`Yrs_Exp`) was not significant. You reran the model without this predictor variable and arrived at a model in which all the variables were significant.

You explored how to interpret the coefficients and their impact on the dependent variable. You learned about deviance as a measure of lack of fit and saw how to verify the model's goodness of fit by using the p-value of deviance difference using the Chi-square method.

You need to use caution when interpreting the logistic regression model. You checked for multicollinearity and overdispersion.

You then split the data set into training and test sets. You tried to come up with a logistic regression model out of the training data set. Through this process, you learned that a good model generated from such a training set can be used to predict the dependent variable. You can use a classification report to check measures such as accuracy, specificity, and sensitivity.

You also learned by using the `prediction()` and `performance()` functions from the `ROCR` package that you can generate a ROC curve to validate the model, using the same data set as the original.

You learned how to predict the value of a new data set by using the logistic regression model you developed. Then you learned about multinomial logistic regression and the R packages that can be used in this regard.

Also, you learned about regularization, including why it's required and how it's carried out.

Finally, you learned how to generate the logistic regression model using Python and allied utilities like scikit-learn, numpy, Pandas, etc., in Jupyter Notebook using interactive programming. We used the Anaconda framework to do the same thing.

PART III

Time-Series Models

CHAPTER 12

Time Series: Forecasting

12.1 Introduction

Imagine you are an investor in the stock market and want to know how a particular stock is likely to perform in the near future. Or perhaps you are a business owner and have to predict the human resources required for the next year based on your revenue projections. Alternatively, suppose you are a government official who wants to understand how the economy of your country is likely to perform in the next quarter or year. Maybe you are a sales manager and want to make a decision as to the likely demand for a particular item sold by your company. You may already have data related to these various elements from earlier months, quarters, or years. Using this data, you can forecast, or in other words *predict*, all the various outcomes. However, the accuracy of the prediction depends upon the accuracy of the data you have on hand, as well as the capabilities of your algorithms or tools/utilities to forecast the outcome. All businesses, economies, etc., related to the data undergo ups and downs. Many are impacted by the sudden development of situations like war; others may be impacted by natural calamities like floods or earthquakes. In essence there is some or significant impact of these events like war, floods, earthquakes, recession, etc. on various data elements. A particular item sold by the company may be impacted by new competitive products that have just come to market, or the purchasing power of your target consumer may change. There may be some positive impact, there may be some negative impact, or there may be a mix of these. Thus, the data used for the future prediction or forecasting itself may not be clean or good enough to be able to deliver perfect predictions irrespective of the best of the algorithms or tools/utilities used. Thus, expecting perfect predictions in most of the actual scenarios is not advisable. However, forecasting can definitely guide you in the right direction even though most of the forecasting may not be accurate.

U. R. Hodeghatta and U. Nayak, *Practical Business Analytics Using R and Python*,
https://doi.org/10.1007/978-1-4842-8754-5_12

All the previous data we talked about involves lots of time-series data, i.e., data taken over a continuous period of time like minute by minute or hour by hour or day by day or week by week or month by month or quarter by quarter or year by year, etc. Again, this data may be univariate (i.e., data about a single factor or parameter) or multivariate (i.e., data with multiple variables). Examples of univariate data are the closing stock price of a particular stock in a particular stock exchange (e.g., the price of the Walt Disney stock on the New York Stock Exchange), the total revenue of a particular organization (e.g., HP), the sales volume of a particular product of a particular organization (e.g., a Voltas AC 1.5-ton model in California), the price of a particular commodity in a particular market (e.g., the price of gold in New York per 10 grams of 24 carats gold), etc. Examples of multivariate data are open price, close price, volume traded, etc., of a particular stock on a particular stock exchange, revenue, expenses, profit before tax, profit after tax of a company, maximum temperature, minimum temperature, or average rainfall at a particular city across various periods.

If we have a good amount of data, we can forecast the future values of it using the forecasting models. The methodology is simple. First we verify the data for accuracy and clean up the data where relevant, then we generate the model using the clean and accurate data, and finally we use the model generated to predict/forecast.

In this chapter, we will mostly use the functions from the libraries `base, stats, graphics, forecast,` and `tseries` in R. We use pandas, NumPy, scikit-learn, statsmodels, pmdarima, etc., in Python. Please install these packages and load these libraries when you do the hands-on work. If the code does not work, first check if you have failed to load the libraries concerned.

12.2 Characteristics of Time-Series Data

Time-series data may have many characteristics. Some of these are seasonality, trends, or irregular or error components.

Rainfall, for example, has a seasonality; it is high normally during the rainy seasons and is low in other seasons (unless due to climate change, there are nonseasonal rains or because of drought there is less rain during the rainy season, etc.). Similarly, temperatures during summer are high compared to other periods of the year, particularly compared to the temperatures during the winter.

Similarly, stock prices demonstrate an upward trend whenever the stock market is bullish or nearer to the date of the announcement of the results of the company (if the perception or outlook is that the company's results are going to be good) or when the demand for the products of the company is likely to be increasing significantly because of the shortage of the products in the market or the sudden bankruptcy of a major competitor for these products. Data in such cases demonstrates a trend. The trend may be downward also in the case of some stocks due to the performance of the company in a particular quarter or continuously over multiple quarters. The trend of the price may be downward for a particular vegetable (a particularly perishable vegetable) if the crop output significantly increases beyond the demand. Level is the average value, whereas trend is an increasing or decreasing value.

We call the third factor the *irregular or error component*, which typically captures those influences that are not captured by seasonal and trend components and may be pure white noise.

The other way we need to look at these characteristics is whether these are additive or multiplicative in the context of a particular time series. The maximum temperature or minimum temperature over a year is additive as from one period to the next period it cannot increase drastically. In this case, the seasonality, trend, and error components all will be additive. However, the quantum of car sales can demonstrate multiplicative seasonal trend in the sense that the quantum of car sales may increase to, say, 20 to 30 percent just in the month before the financial budget of the country in view of a possibility of additional taxation in the upcoming budget, and in the immediate next month it may come down, say, by 10 to 20 percent.

We consider a time series to be *stationary* if it does not demonstrate either seasonality or trend. However, even a stationary time series will still have random fluctuations. A time series may be demonstrating all three characteristics, i.e., seasonality, trend, and irregular components.

We can also call the characteristics *systematic* or *nonsystematic*. Systematic constituents of the time series can be modeled well and easily, whereas nonsystematic constituents of the time series are difficult to model. For example, the random influence on the prices of a commodity that is not explainable is nonsystematic, whereas normal seasonality that can be easily explained or modeled is systematic. Similarly, a trend is part of the systematic component.

12.3 Decomposition of a Time Series

We have discussed in detail the components of a time series. These are seasonal, trend, and irregular components. The decomposition of a time series can be checked using the `decompose(ts)` or `stl(ts, s.window="periodic")` command.

One of the above two commands are typically used in the R language to decompose a time series. We will be using the `decompose` command. To demonstrate the usage of this, let us first import a data set. Here, we have taken the data set "Monthly, Seasonal and Annual Maximum Temperature Series from 1901 to 2017" from the Government of India website (`https://www.data.gov.in`).

We have downloaded the data from this website to a local machine as a comma-separated Excel file (`.csv`). Using R, first we will load it to our R environment. This is done using the following R code:

```
> ##Get the data into the dataframe from the .csv dataset
> max_temp_imd_2017 <- read.csv("C:/Users/kunku/OneDrive/Documents/Book
  Revision/Max_Temp_IMD_2017.csv")
> ##Get the description of the dataframe in brief
> str(max_temp_imd_2017)
```

```
'data.frame':  117 obs. of  18 variables:
 $ YEAR  : int  1901 1902 1903 1904 1905 1906 1907 1908 1909 1910 ...
 $ JAN   : num  22.4 24.9 23.4 22.5 22 ...
 $ FEB   : num  24.1 26.6 25 24.7 22.8 ...
 $ MAR   : num  29.1 29.8 27.8 28.2 26.7 ...
 $ APR   : num  31.9 31.8 31.4 32 30 ...
 $ MAY   : num  33.4 33.7 32.9 32.6 33.3 ...
 $ JUN   : num  33.2 32.9 33 32.1 33.2 ...
 $ JUL   : num  31.2 30.9 31.3 30.4 31.4 ...
 $ AUG   : num  30.4 30.7 30 30.1 30.7 ...
 $ SEP   : num  30.5 29.8 29.9 30 30.1 ...
 $ OCT   : num  30 29.1 29 29.2 30.7 ...
 $ NOV   : num  27.3 26.3 26.1 26.4 27.5 ...
 $ DEC   : num  24.5 24 23.6 23.6 23.8 ...
 $ ANNUAL : num  29 29.2 28.5 28.5 28.3 ...
 $ JAN.FEB : num  23.3 25.8 24.2 23.6 22.2 ...
 $ MAR.MAY: num  31.5 31.8 30.7 30.9 30 ...
 $ JUN.SEP: num  31.3 31.1 30.9 30.7 31.3 ...
 $ OCT.DEC: num  27.2 26.5 26.3 26.4 26.6 ...
```

Figure 12-1. *Characteristics of the temperature time series imported, output of the R code executed*

Figure 12-1 shows that there are 117 rows of data (i.e., from 1901 to 2017) and that there are 18 variables including the year, maximum temperature data for each of the 12 months of these years, and another five columns, namely, `ANNUAL`, `JAN.FEB`, `MAR.MAY`, `JUN.SEP`, and `OCT.DEC`.

We want to retain only the year and the 12 individual month column data and remove all the other columns. We are doing this using the following code:

```
>new_temp_df <- subset(max_temp_imd_2017, select=-c(ANNUAL, JAN.FEB,MAR.
MAY,JUN.SEP,OCT.DEC))
```

We can test to check whether the intended columns have been dropped from the data set using the `str(new_temp_df)` command from R. We get the output shown in Figure 12-2.

```
'data.frame':  117 obs. of  13 variables:
 $ YEAR: int  1901 1902 1903 1904 1905 1906 1907 1908 1909 1910 ...
 $ JAN : num  22.4 24.9 23.4 22.5 22 ...
 $ FEB : num  24.1 26.6 25 24.7 22.8 ...
 $ MAR : num  29.1 29.8 27.8 28.2 26.7 ...
 $ APR : num  31.9 31.8 31.4 32 30 ...
 $ MAY : num  33.4 33.7 32.9 32.6 33.3 ...
 $ JUN : num  33.2 32.9 33 32.1 33.2 ...
 $ JUL : num  31.2 30.9 31.3 30.4 31.4 ...
 $ AUG : num  30.4 30.7 30 30.1 30.7 ...
 $ SEP : num  30.5 29.8 29.9 30 30.1 ...
 $ OCT : num  30 29.1 29 29.2 30.7 ...
 $ NOV : num  27.3 26.3 26.1 26.4 27.5 ...
 $ DEC : num  24.5 24 23.6 23.6 23.8 ...
```

Figure 12-2. *Characteristics of the temperature time series after removing columns that are not needed, output of the R code executed*

To reduce further complexity, for your ease of understanding, we have created the following small univariate data set with 24 months of data pertaining to the years 2015 and 2016, converting the data to a time series with the start month as Jan 2015 and with a monthly frequency. Further, we have checked if the time series has been created properly by checking the starting month of the time series, the ending month of the time series, and also the frequency. We find that all the data is appropriate with the following code with the embedded output. At the end, we have provided the command to plot the time series using `plot(ts_max_temp)`.

```
> ##In order for you to understand the concepts
> ##clearly, we have created a dataframe with 2015,2016
> ##data manually
> max_temp_df <- c(24.58,26.89,29.07,31.87,34.09,32.48,31.88,31.52,31.5
5,31.04,28.10,25.67,
+26.94,29.72,32.62,35.38,35.72,34.03,31.64,31.79,31.66,31.98,30.11,28.01)
> ts_max_temp <- ts(max_temp_df, start=c(2015,1), frequency=12)
> ##Now let us check if the time series has been set up properly
> start(ts_max_temp)
[1] 2015    1
> end(ts_max_temp)
[1] 2016   12
> frequency(ts_max_temp)
[1] 12
>plot(ts_max_temp)
```

Please Note We have included > on each code line to make it clear to you which is the code and separate it from the embedded output. If you are copying the code from here and inputting it in R, then you should not include the >. The entire R code of this chapter is provided in a separate R script on the accompanying Apress website.

The output of the last code line, `plot(ts_max_temp)`, is provided in Figure 12-3; it shows the fluctuations of the data over these different months over 2 years.

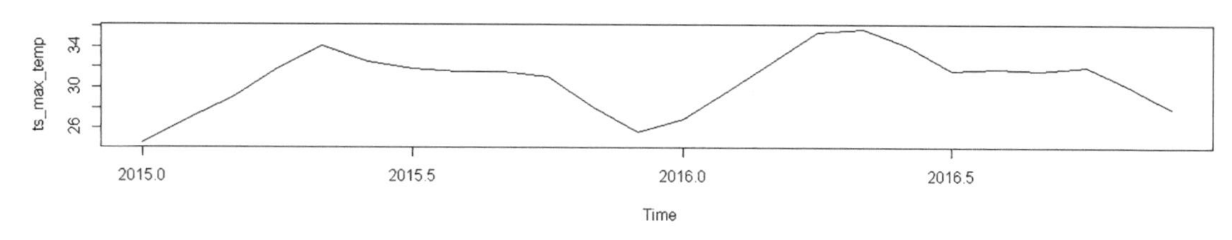

***Figure 12-3.** Plotted time series*

Now we will decompose the time series using the command decompose(ts_max_temp) and look at the seasonal, trend, and irregular components of the time series using the following code. The related outputs are also embedded here.

```
>#Decomposing the timeseries into seasonal, trend and
>#irregular components
> decomp <- decompose(ts_max_temp)
> ##Finding the seasonal, trend, irregular components
> ##of the time series
> decomp$seasonal
              Jan         Feb         Mar         Apr         May
Jun         Jul         Aug         Sep         Oct         Nov         Dec
2015 -4.47826389 -1.69951389 1.18465278 3.90090278 4.11798611
2.24673611 1.63923611 1.06298611 0.82715278 0.02298611                -
3.13118056 -5.69368056
2016 -4.47826389 -1.69951389 1.18465278 3.90090278 4.11798611
2.24673611 1.63923611 1.06298611 0.82715278 0.02298611
-3.13118056 -5.69368056
> decomp$trend
          Jan      Feb      Mar      Apr      May      Jun      Jul
 Aug      Sep      Oct      Nov      Dec
2015       NA       NA       NA       NA       NA       NA
29.99333 30.20958 30.47542 30.76958 30.98375 31.11625
2016 31.17083 31.17208 31.18792 31.23167 31.35458 31.53583
NA       NA       NA       NA       NA       NA
> decomp$irregular
NULL
> ##We can now plot the decomposed ts
> ##to understand the components
> plot(decomp)
```

The output of the plot(decomp) command provided shows clearly the seasonal and trend components graphically in Figure 12-4. The second graph from the bottom of the output shown in Figure 12-4 shows the seasonal trend; i.e., almost the same pattern is repeated with the seasons. As you can see in this graph, nearer to the middle of the year the temperature peaks, and during the end of the year it drops. This happens during both years, i.e., over the entire period of the input data. The trend component is shown in the middle graph. This typically shows if the values have a component that continues to increase or continues to decrease or they are horizontal. In our case, there is a trend component for the partial spectrum of the data, i.e., July of the earlier year to June of the next year (as shown in the output provided along with the code). A random component is an irregular or error component. There is no random component here as the bottommost graph shows a straight line (as you can see from the output shown along with the code, it is NULL).

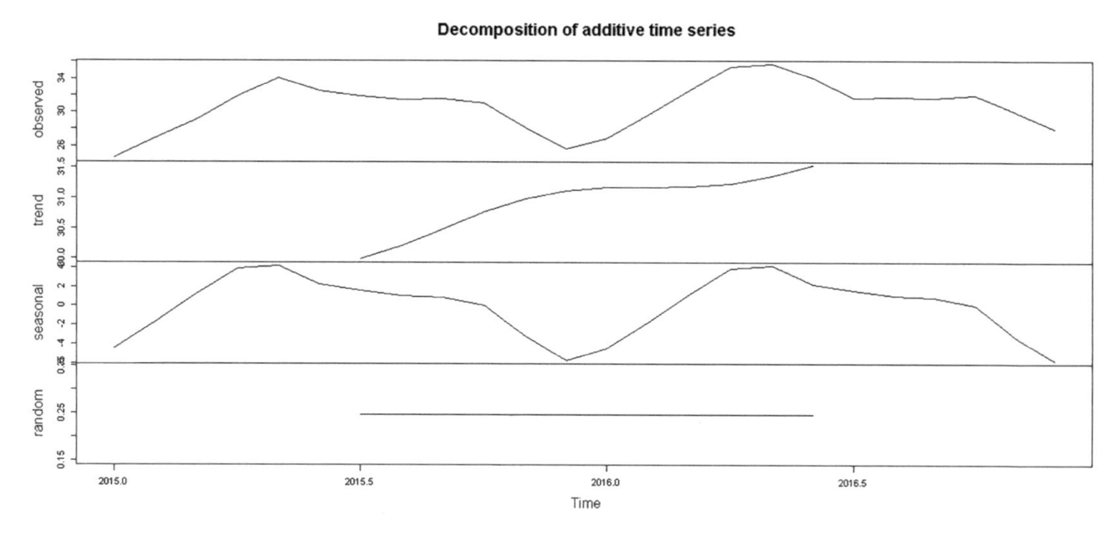

Figure 12-4. *Decomposition plot*

The following command creates a plot clearly showing the patterns of the decomposition:

```
> monthplot(ts_max_temp, xlab="", ylab="")
```

Figure 12-5 shows the output, which clearly shows both the trend and the seasonal patterns.

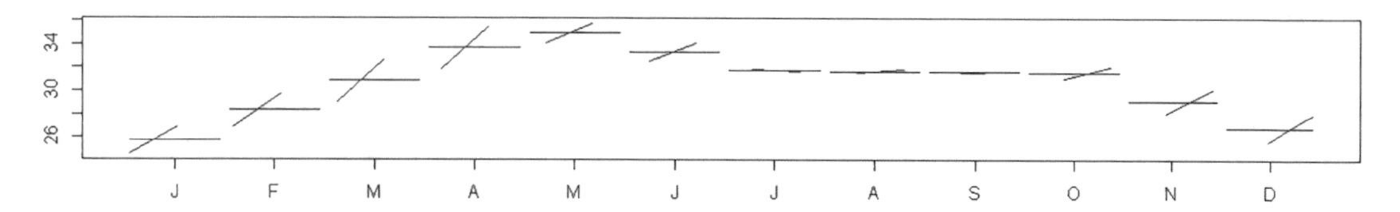

Figure 12-5. *monthplot() showing the patterns*

12.4 Important Forecasting Models

There are many forecasting models available for use. Here, we will look mainly at the exponential forecasting models, called ARMA and ARIMA. We will see how to generate these models and how to predict/forecast using them.

12.4.1 Exponential Forecasting Models

Exponential forecasting models are the easiest to use. There are three types: the single exponential forecasting model, the double exponential forecasting model, and the triple exponential forecasting model. While the single exponential forecasting model uses only the level component, it is used when the time series is not demonstrating the seasonal and the trend components. While the double exponential forecasting model (also known as Holt exponential smoothing) uses both the level and trend, it is used when the time series does not demonstrate the seasonal component. In the case of the triple exponential forecasting model (also known as Holt-Winters exponential smoothing), we take into consideration all three, i.e., level, trend, and seasonal components. Level is the average value.

As mentioned, in the simple exponential forecasting model, we use the level. We weigh each of our time-series values with an exponentially decaying weight as we go back into the past. The recent values get a higher weight compared to the older values. The decay in weight depends upon the smoothing factor alpha (α) for the level. When the alpha is nearer to 1, then the recent observations get relatively higher weights, and when the alpha is nearer to 0, the older observations in the past get relatively higher weights. The exponential smoothing equation gives the level $Z_{(t)}= \alpha Y_{(t)} + (1 - \alpha)Z_{(t-1)}$ where $Z_{(t)}$ is the exponentially smoothed series value being calculated in the time period (t), $Z_{(t-1)}$ is the earlier forecasted value in the time period $Y_{(t-1)}$, $Y_{(t)}$ is the observed value of the time series in the time period t, and α is the smoothing factor (i.e., assigned weight). The forecast equation is Y-hat$_{(t+h/t)} = Z_{(t)}$, i.e., the level.

In the double exponential forecasting model, we use the level and trend components. We will now have both α and β smoothing factors. α is the smoothing factor associated with the level, and β is the smoothing factor associated with the trend. Each of the α and β take a value between 0 to 1; again, the higher this value, the higher the relative weightage to the recent observations in the time series, and the lower this value, the higher the relative weightage to the older observations in the past.

In the triple exponential forecasting model, we use all the three components, i.e., the level, the trend, and the seasonal components. Now, we will add another smoothing factor, i.e., γ. This smoothing factor γ will also take a value between 0 and 1. Again, the higher this value, the higher the relative weightage to the recent observations in the time series, and the lower this value, the higher the relative weightage to the older observations in the past.

To learn more, refer to books on statistics, which cover these aspects and the related formulae in detail.

`ets()` is the most popular function used. You can find it in the `forecast` library of R. `ets(time-series name, model="ZZZ")` will select appropriate type of model without the need for you to specify it. Here, each letter can mean one of the following: A means Additive, M means Multiplicative, and N means None. The first letter stands for the error type, the second letter stands for the trend type, and the third letter stands for the seasonal component type. Let's check how this works with our time-series data, i.e., `ts_max_temp`. The code used for this along with the output is provided here:

```
> #Exponential model creation selecting the parameters automatically
> library(forecast)
> model_1 <- ets(ts_max_temp, model="ZZZ")
> model_1
ETS(A,A,N)

Call:
 ets(y = ts_max_temp, model = "ZZZ")

  Smoothing parameters:
    alpha = 0.9999
    beta  = 0.9999

  Initial states:
    l = 22.2701
    b = 2.3093

  sigma:  1.7486


     AIC     AICc      BIC
108.7209 112.0543 114.6112
```

The previous output shows the additive model using the errors and the trend. This model also shows the AIC, AICc, and BIC values as well as the smoothing parameters alpha and beta used. Alpha and beta are the smoothing parameters for the level and trend, respectively. Both are almost 1. These high values of the alpha and beta suggest that only the recent observations are being taken into account for the forecast.

Note that the values of alpha, beta, and gamma will be between 0 and 1. A value nearer to 1 means the recent observations are given more weight compared to the older observations. A value nearer to 0 means the distant past observations get more relative weights compared to the recent observations.

Let's now forecast the next three max temperatures from the year 2017 using this model and check if this gives near equal predictions. The simple code used along with the corresponding output is shown here:

```
>#Forecasting 3 values using model_1 generated
>forecast(model_1, 3)
    Point            Forecast    Lo 80          Hi 80      Lo 95          Hi 95
Jan 2017        25.90983 23.66890 28.15076 22.48262 29.33704
Feb 2017        23.80963 18.79917 28.82010 16.14679 31.47248
Mar 2017        21.70944 13.32545 30.09343  8.88723 34.53165
```

The output shown also shows the limits for an 80 percent confidence interval and a 95 percent confidence interval. From the raw data for the year 2017 from the government data set, the values for January 2017, February 2017, and March 2017 are, respectively, 26.45, 29.46, and 31.60. As you can see, there is wide variation among the actual values and the predicted values. All the predicted values are on the lower side.

Let's now explore other models we have not explored: AAA and ANA. Among these, we will first explore AAA as we saw a seasonal pattern in our data earlier. The code and the model generated as output are provided here:

```
> model_2 <- ets(ts_max_temp, model="AAA")
> model_2
ETS(A,A,A)

Call:
 ets(y = ts_max_temp, model = "AAA")

  Smoothing parameters:
    alpha = 0.974
```

```
    beta  = 0.1329
    gamma = 0.003

  Initial states:
    l = 28.7322
    b = 0.1554
    s = -4.852 -2.2334 0.4228 0.7817 1.111 1.2383
           2.5547 4.1739 2.903 0.2007 -2.0253 -4.2754

  sigma:  0.8999

      AIC            AICc           BIC
 78.84388 180.84388  98.87079
```

Let's now predict the next three values for the year 2017, i.e., January 2017, February 2017, and March 2017, the actual values of which are 26.45, 29.46, and 31.60.

```
>#Forecasting 3 values using model_2
> forecast(model_2, 3)
    Point          Forecast      Lo 80          Hi 80          Lo 95          Hi 95
Jan 2017       28.81220 27.65893 29.96547 27.04843 30.57597
Feb 2017       31.29460 29.57427 33.01493 28.66359 33.92562
Mar 2017       33.75386 31.51692 35.99081 30.33275 37.17497
```

From the previous, we observe that the values fitted are on the higher side. Let's now explore the model with ANA. The code along with the model and the forecasted three values (pertaining to January 2017, February 2017, and March 2017) are shown here:

```
> #Generating model ANA
> model_3 <- ets(ts_max_temp, model="ANA")
> model_3
ETS(A,N,A)
Call:
 ets(y = ts_max_temp, model = "ANA")
 Smoothing parameters:
    alpha = 0.9998
    gamma = 1e-04
```

```
  Initial states:
    l = 30.6467
    s = -4.8633 -2.1923 0.4212 0.6708 0.9745 1.2463
        2.8205 4.541 3.1301 0.2431 -2.2662 -4.7257
  sigma:  0.9319
      AIC          AICc          BIC
 81.87586 141.87586  99.54667
> forecast(model_3, 3)
         Point     Forecast     Lo 80        Hi 80          Lo 95          Hi 95
Jan 2017       28.14746 26.95320 29.34172 26.32100 29.97393
Feb 2017       30.60699 28.91821 32.29577 28.02423 33.18975
Mar 2017       33.11627 31.04801 35.18453 29.95314 36.27940
```

Still, we observe that the predicted values are differing significantly from the actual values.

The previous may be because of the limited data set we used to create the time series. However, the forecasting of unknown values is not always correct because the actual values may be impacted by many other parameters unknown to us and the past data may not be completely representative of the future data.

12.4.2 ARMA and ARIMA Forecasting Models

The autoregressive model and moving average model are very useful. When we combine these models, we get the autoregressive moving average (ARMA) model. Another variation of these is the autoregressive integrated moving average (ARIMA) model.

There are many fundamental concepts that need to be assimilated if these models are to be completely understood. The fundamental concepts we are referring here are lags, autocorrelation, and partial autocorrelation. We will leave it to the readers to explore these concepts on their own. For the ease of understanding, let's first start with the Automated ARIMA model generation, which is easy to do and which is easy to understand. Automated ARIMA selects the parameters automatically and hence is named automated ARIMA.

We will now use a different time series than we used earlier, i.e., `nhtemp`, which is part of the R time-series data. Let us first check the data in it and plot the data from this time series. Here is the code with the partial output. The output of the plot is given in Figure 12-6.

```
>str(nhtemp)
 Time-Series [1:60] from 1912 to 1971: 49.9 52.3 49.4 51.1 49.4 47.9 49.8 50.9
49.3 51.9 ...
>plot(nhtemp)
```

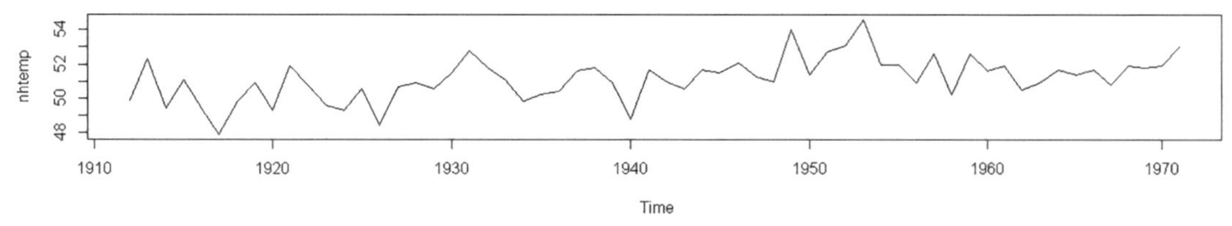

Figure 12-6. *Time-series nhtemp data*

We will now use the `auto.arima` function from `library(forecast)`. The code and the output are given here:

```
> #Finding the hyper parameters automatically using auto_arima
> model_4 <- auto.arima(nhtemp)
> model_4
Series: nhtemp
ARIMA(0,1,1)
Coefficients:
          ma1
      -0.7983
s.e.   0.0956
sigma^2 estimated as 1.313:  log likelihood=-91.76
AIC=187.52   AICc=187.73   BIC=191.67
```

You can see the model's order for values `(p, d, q)` in the previous model is (0, 1, 1). Here, `p` stands for `(p, d, q)` autoregressive model of order `p`, `q` stands for `(p, d, q)` moving average model of order `q`, and `d` stands for the number of times the time series has been differenced.

In the autoregressive model, each value of the time series is predicted using the linear combination of past `p` values. In a moving average model, each value of the time series is predicted using the linear combination of past `q` errors.

We can make the prediction as usual using the `forecast(model_name, n)` command as earlier, where `n` is the number of future values to be predicted. The code and the results for the previous model generated is given here:

```
#Forecasting using model_4
> forecast(model_4, 1)
 Point      Forecast      Lo 80       Hi 80       Lo 95       Hi 95
1972       51.90008 50.43144 53.36871 49.65399 54.14616
```

The accuracy of the model can be tested using accuracy(model_name). The code and output are given here:

```
>#Checking the accuracy of model_4
> accuracy(model_4)
                              ME         RMSE          MAE           MPE
MAPE          MASE            ACF1
Training set 0.1285716 1.126722 0.8952077 0.2080785 1.749865 0.7513123
-0.008258617
```

As you can see, the various accuracy measures of the model are provided including mean error (ME), root mean square error (RMSE), mean absolute error (MAE), mean percentage error (MPE), mean absolute percentage error (MAPE), mean absolute scaled error (MASE), and first order autocorrelation coefficient (ACF1). The errors are very less and almost between 0 to 1. Hence, the accuracy is excellent in our case.

12.4.3 Assumptions for ARMA and ARIMA

The ARMA and ARIMA models require that the time series is stationary; i.e., the statistical properties such as mean, variance, etc., are constant. If not, the time series needs to be made stationary. A stationary time series does not have a trend and seasonal component. Sometimes, we may require transformation of the time series like log transformation or Box_Cox transformation to obtain a stationary time series. Another way to make a nonstationary time series stationary is by differencing. Differencing removes the trend and seasonal components.

The following is the example of a stationary time series:

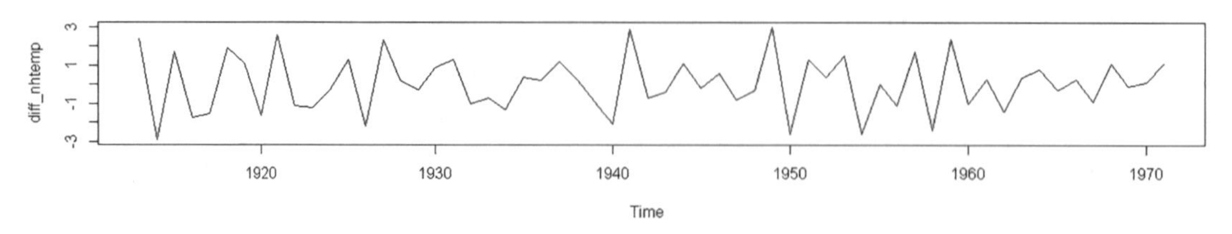

As you can see, timeseries does not have any seasonal component or trend component. Such a time series is known as *stationary*, and such a time series has statistical properties such as mean and variance constant and also a consistent covariance.

The following is an example of a nonstationary time series:

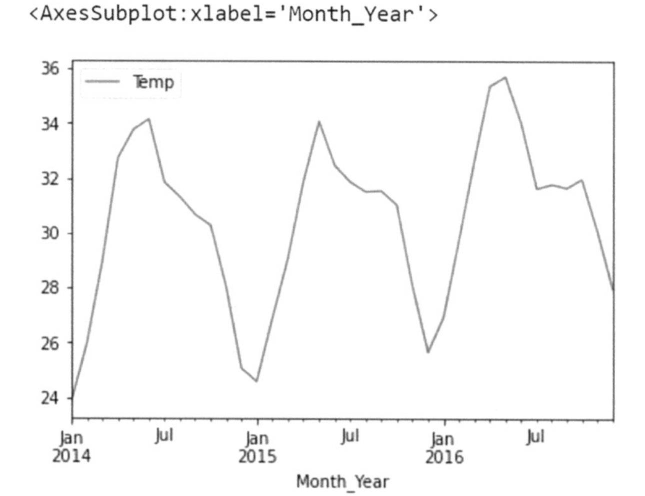

As you can see in the plot, there is at least a seasonal component. Hence, it is a nonstationary time series.

The `ndiffs()` function in R checks and suggests how many differences are required to make the time series stationary. Let's check the time series we have in hand, i.e., `nhtemp`.

```
>#Checking on the number of differences required to
>#make the timeseries stationary
>library(forecast)
> ndiffs(nhtemp)
[1] 1
```

Now we will difference the original time series once (i.e., `differences=d=1`) and then check if the time series is now stationary. Here, the function `diff(time series, differences=d)` is used. By default, this function takes `d` as 1. If it is different than 1, then we need to pass the parameter `differences=d`. In order to check if the time series has become stationary after differencing, we use the augmented Dickey-Fuller test. The code and the output are provided here:

```
> ##Now we will difference the time series data
> ##one time to make it stationary
> diff_nhtemp <- diff(nhtemp)
> ##Now we will use the Augmented Dickey_Fuller Test
> ##to check if the time series is now stationary
> library(tseries)
> adf.test(diff_nhtemp)

        Augmented Dickey-Fuller Test

data:  diff_nhtemp
Dickey-Fuller = -4.6366, Lag order = 3, p-value = 0.01
alternative hypothesis: stationary

Warning message:
In adf.test(diff_nhtemp) : p-value smaller than printed p-value
```

As the p-value is very small (less than the significance level), we reject the null hypothesis that the time series is nonstationary and accept the alternative hypothesis that the time series is stationary.

We will now plot the differenced time series using the following code:

```
>plot(diff_nhtemp)
```

We will get the plot shown in Figure 12-7.

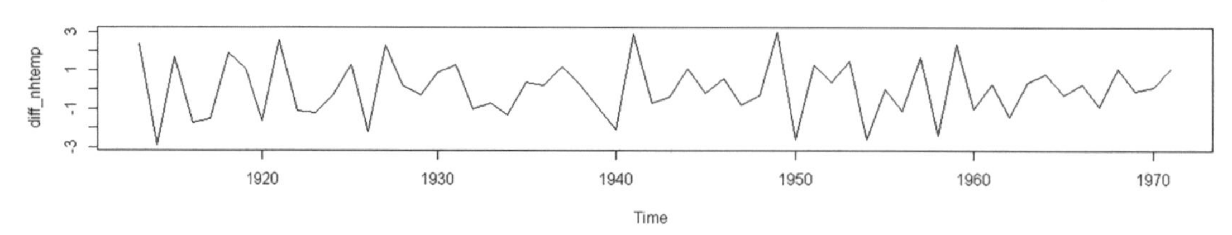

***Figure 12-7.** Differenced and now stationary time series nhtemp*

Now, we can check for the autocorrelation function (ACF) plot and partial autocorrelation function (PACF) using the `Acf(time series)` and `Pacf(time series)` from the library `(forecast)` to check the p and q parameters required for the model building. We already know d=1 from the previous discussions. At lag m, the autocorrelation depicts the relationship between an element of the time series and the value, which is m intervals away from it considering the intervening interval values. Partial autocorrelation depicts the relation between an element of the time series and the value that is m intervals away from it without considering the intervening interval values.

Note that as we have d=1, we will be using the differenced time series in Acf() and in Pacf() instead of the original time series.

Let's plot this and check. The following is the code.

```
> Acf(diff_nhtemp)
> Pacf(diff_nhtemp)
```

Figure 12-8 shows the output.

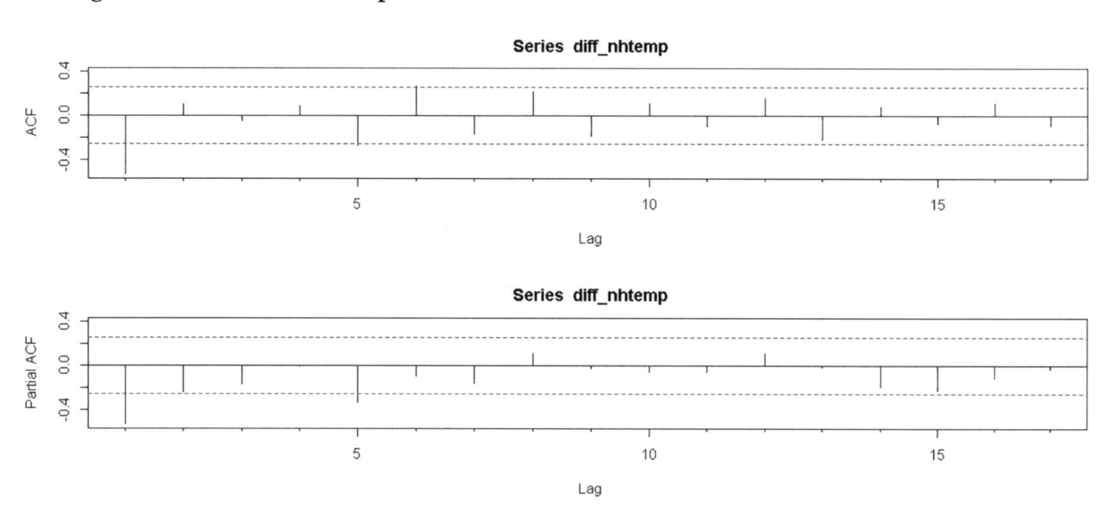

Figure 12-8. *Acf() and Pacf() plots from forecast package*

The following table provides the rules for interpreting the p order for AR() and the q order for MA().

Type of Plot	AR(p)	MA(q)
Acf()	The ACF plot is sinusoidal (i.e., of the form of sine wave) or is gradually or exponentially decaying.	The ACF plot shows significant spike at lag q. The residuals beyond lag q are normally within the confidence interval plotted. Few residuals other than the initial significant ones just outside the confidence interval plotted are OK and may be ignored.
Pacf()	The PACF plot shows significant spike at lag p but no further spikes beyond lag p. The residuals beyond lag p are normally within the confidence interval plotted. Few residuals other than the initial significant ones just outside the confidence interval plotted are OK and may be ignored.	The PACF plot is sinusoidal (i.e., of the form of sine wave) or is gradually or exponentially decaying.

Both the Acf plot and Pacf plots shown in Figure 12-8 have very small residuals. Further, Pacf() shows a significant spike at the first lag, and all the subsequent residuals are within the confidence interval plotted. Hence, we have p=1, and we have AR(p) = AR(1). Acf() shows a significant spike at the first lag, and all the subsequent residuals are within the confidence interval plotted. Hence, we have q=1, and we have MA(q) = MA(1). We already know that d=1.

We now run the arima() model on the original time series using these (p, d, q) values; i.e., the order of the model is (1,1,1). The code and the output are given here:

```
> ##Running arima on the original time series nhtemp
> ##using the d=1 and p, q values selected from the
> ##Acf() and Pacf() plots
> model_6 <- arima(nhtemp, order=c(1,1,1))
> model_6
Call:
arima(x = nhtemp, order = c(1, 1, 1))
Coefficients:
```

```
         ar1      ma1
      0.0073  -0.8019
s.e.  0.1802   0.1285
sigma^2 estimated as 1.291:  log likelihood = -91.76,  aic = 189.52
```

We are now required to evaluate whether the model is fit and the residuals are normal. We use the Box-Ljung test for the model fit verification and use the quantile-to-quantile plots of the residuals to check if the residuals are normal. The code and the output are shown here:

```
> ##Evaluating model fit and the normality of the residuals
> qqnorm(model_6$residuals)
> qqline(model_6$residuals)
> Box.test(model_6$residuals, type="Ljung-Box")
        Box-Ljung test
data:  model_6$residuals
X-squared = 0.010934, df = 1, p-value = 0.9167
```

As the p-value is not significant, we cannot reject the null hypothesis that the model is fit or there is no autocorrelation. Hence, we consider the model to be fit and without autocorrelation.

Further, Figure 12-9 shows the output from the quantile-to-quantile plot, which clearly shows that the residuals are normally distributed, as all the points are on the straight line.

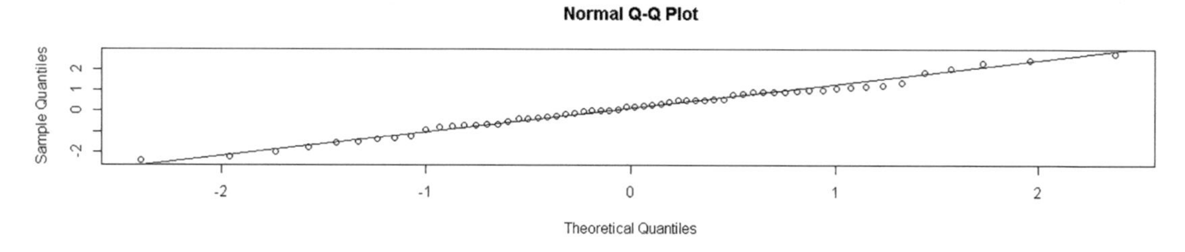

***Figure 12-9.** Quantile-to-quantile plot of the residuals*

From this, you can conclude that both the assumptions are met and the `model_6` generated by us is a very good fit. We can use it to predict the future values as we have done earlier.

As you can see from the previous discussions, the model generated through `auto.arima()` is little different from what we generated here.

Please Note The ARMA model does not have a d component. It will have only (p, q).

We will now plot the forecasted values using model_6 along with the original values of the time series, nhtemp, using the following code:

```
> ##Forecasting using model_6 and then plotting the forecasted values
> forecast(model_6, 3)
      Point  Forecast     Lo 80        Hi 80         Lo 95        Hi 95
1972        51.90458 50.44854 53.36062 49.67776 54.13141
1973        51.89660 50.41015 53.38304 49.62327 54.16992
1974        51.89654 50.38194 53.41114 49.58015 54.21292
> plot(forecast(model_6, 3), xlab="Temperature", ylab="Timeline")
```

Figure 12-10 shows the time series we get with the additional forecasted values.

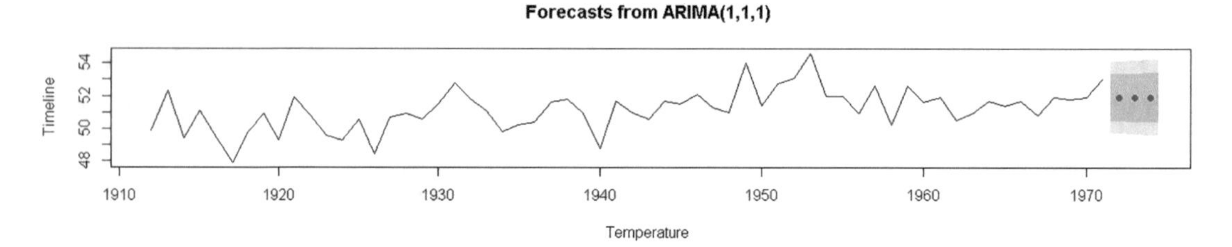

Figure 12-10. *Time series, nhtmp extended with the forecasted values*

12.5 Forecasting in Python

We have so far seen how to import data, convert it to a time series, decompose the time series into various components like seasonal, trend, level, and error; use various methods of generating models; check for the prerequisites (where applicable); validate the model fit and the assumptions; and forecast using the models in R. Now, we will be looking at carrying out the same activities in Python. We are using the Jupyter Notebook from the Anaconda framework here for the coding as it provides you with the output immediately when running the code in each of the cells. This provides you with the capability of interactive programming. Throughout the Jupyter Notebook we have provided "comments" that will help you to understand the reasons behind the specific code or the output.

12.5.1 Loading the Base Packages

First, we will load the base packages required. Here is the code:

```
import pandas as pd
import numpy as np
import sklearn
import statsmodels as sm
```

By using these commands, these models will be loaded into memory and will be available for use.

Please Note We will be loading some of the other packages required subsequently as they become necessary.

12.5.2 Reading the Time-Series Data and Creating a Dataframe

We now get the data required to create the time series from our data source on the dataframe. In this case, the reference is to the local source as we are loading it from our local machine. We have used 36 months' data of maximum temperature, i.e., Jan 2014 to Dec 2016. Here the month is represented by the last date of the month. The data gets uploaded to a dataframe known as `Max_Temp_df`, and we read the first five rows of it to understand what data it contains.

```
Max_Temp_df = pd.read_csv("C:/Users/kunku/OneDrive/Documents/Book
Revision/Max_IMD_Temp_Train.csv", sep=',', header=0)
Max_Temp_df.head(5)
```

The output is given here:

	Date	Temp
0	31-01-2014	23.83
1	28-02-2014	25.97
2	31-03-2014	28.95
3	30-04-2014	32.74
4	31-05-2014	33.77

The output shows that there are two fields in the dataframe, namely, Date and Temp.

12.5.3 Trying to Understand the Data in More Detail

We will now try to understand more about the data, details such as the types of the data elements/fields, the number of data rows, whether any of the data fields have NULL values, etc. The code is as follows:

```
#get full information on Max Temp dataframe
Max_Temp_df.info()
```

We will get the following output in the Jupyter Notebook:

```
<class 'pandas.core.frame.DataFrame'>
RangeIndex: 36 entries, 0 to 35
Data columns (total 2 columns):
 #   Column  Non-Null Count  Dtype
---  ------  --------------  -----
 0   Date    36 non-null     object
 1   Temp    36 non-null     float64
dtypes: float64(1), object(1)
memory usage: 704.0+ bytes
```

As you can see, we have 36 rows of data. The two elements/columns of the data are Date and Temp. We observe that none of the fields has a NULL value. Further, we find that Date is an object type of field, which means it is a text type of field and currently is not recognized as a date when processing it because Date requires us to convert it to a true date format. However, our dates represent the corresponding months of the respective years. Hence, we need to convert them to a Month_Year or Year_Month format to be recognizable for further time-series analysis. The code to convert the Date text data to Month_Year in a way that is recognizable for further time-series analysis is provided here:

```
#Date is currently recognized as text data by
#Python. Hence, will be converted to the
#standard data format recognized by Python
#using the pandas function to_datetime()
#and dt.to_period for converting it to Month_Year
Max_Temp_df['Month_Year'] = pd.to_datetime(Max_Temp_df['Date']).dt.to_
period('M')
Max_Temp_df.info()
```

The dates will be converted to a datetime type and will be converted to a Month_Year format and added as a separate field in the dataframe Max_Temp_df. Further, the following revised information is provided with regard to data types:

```
<class 'pandas.core.frame.DataFrame'>
RangeIndex: 36 entries, 0 to 35
Data columns (total 3 columns):
 #   Column      Non-Null Count  Dtype
---  ------      --------------  -----
 0   Date        36 non-null     object
 1   Temp        36 non-null     float64
 2   Month_Year  36 non-null     period[M]
dtypes: float64(1), object(1), period[M](1)
memory usage: 992.0+ bytes
```

Now, we will drop the Date field, and we will index the dataframe on the Month_Year field. The code is shown here:

```
#As we already have a Month_Year field we do not
#require Date field any longer - Hence, dropping
Max_Temp_df.drop("Date", inplace=True, axis=1)
#indexing the dataframe on Month_Year field so that
#further operations on the data can be done easily
Max_Temp_df.set_index("Month_Year", inplace=True)
Max_Temp_df.info()
```

Now, you can clearly see the dataframe Max_Temp_df as having 36 rows of data, indexed on the Month_Year field.

As the Date field has been dropped and the index has been set on Month_Year, if you check for Max_Temp_df.info() as provided previously, you will get the following information without the Date-related information:

```
<class 'pandas.core.frame.DataFrame'>
PeriodIndex: 36 entries, 2014-01 to 2016-12
Freq: M
Data columns (total 1 columns):
 #   Column  Non-Null Count  Dtype
---  ------  --------------  -----
 0   Temp    36 non-null     float64
dtypes: float64(1)
memory usage: 576.0 bytes
```

However, if you print(Max_Temp_df) or simply execute the command Max_Temp_df, you will get the listing of Temp against each date.

Now, let's plot the time series to check how it looks using the command Max_Temp_df.plot(). The plot generated as output is as follows:

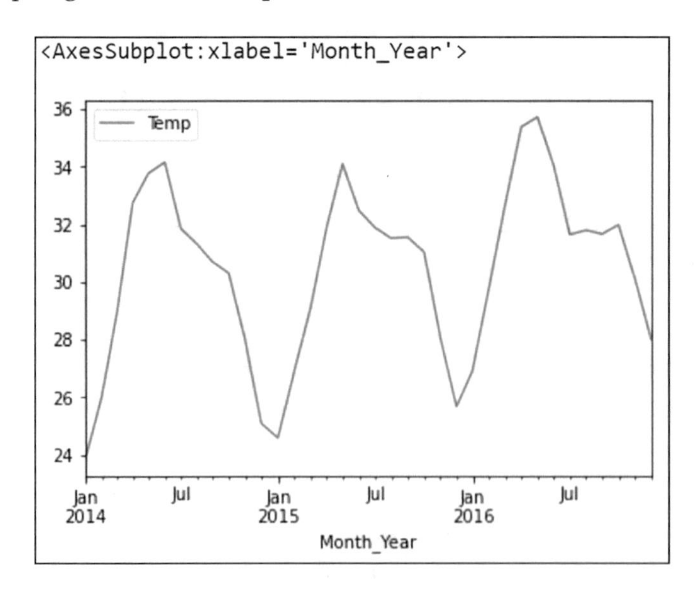

As you can see from the previous plot, there is definitely a seasonal component repeating each year. Let's now decompose the time series to understand its composition better.

12.5.4 Decomposition of the Time Series

Now we will decompose the time-series data into seasonal, trend, and residual (also known as irregular or error) components. We will also load the relevant packages required for this purpose. We will use the popular statsmodels tool/utility to decompose the time series. Here is the relevant code:

```
#Citation: Seabold, Skipper, and Josef Perktold. "statsmodels:
#Econometric and statistical modeling with python." Proceedings of the
#9th Python in Science Conference. 2010. Thanks to them for the
#excellent contribution to the society
#Now, we will use the statsmodels tool to carryout the
#time series decomposition into seasonal, trend, residual/
#irregular/noise/error components
#We have used the "additive" model as we see seasonal pattern
#but it is varying by small quantity
from statsmodels.tsa.api import seasonal_decompose
from statsmodels.tools.eval_measures import rmspe, rmse
decomp = seasonal_decompose(Max_Temp_df, model="additive", period=12)
#Let us look at the seasonal plot
decomp.seasonal.plot()
```

In this code, after decomposition of the time series, we are plotting the seasonal component. Figure 12-11 shows the output.

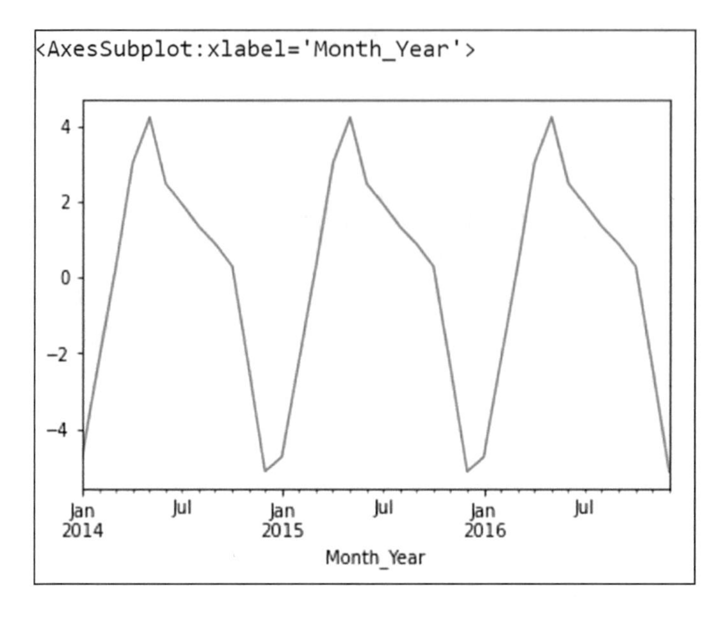

Figure 12-11. *Plot of the seasonal component*

You can see that there is a seasonal impact on the time series. Now, we will plot the other two components. The code in this regard is provided here:

```
#let us now look at the trend and residual plot
decomp.trend.plot()
decomp.resid.plot()
```

Figure 12-12 shows the plot of the trend and residual components.

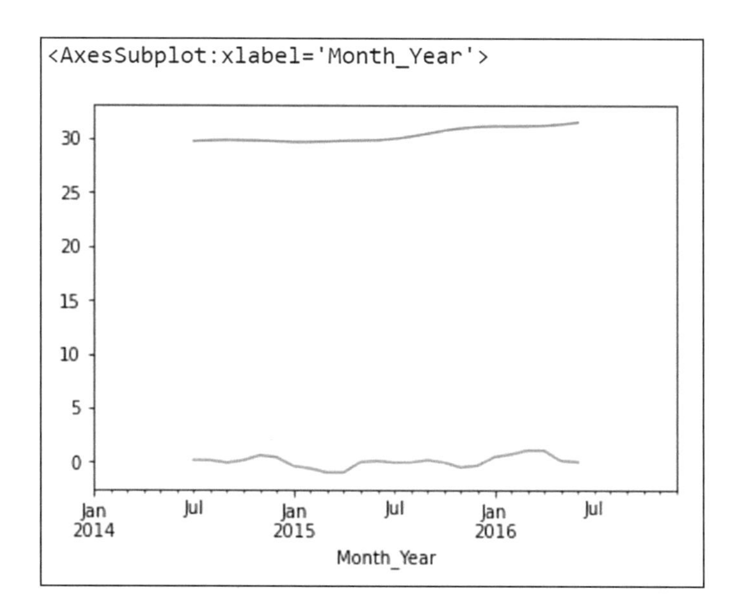

Figure 12-12. *Plot of the trend and residual components*

As you can see from the previous plots, we do have a small upward trend component and some residual/irregular components. Hence, all three components are relevant to the time series `Max_Temp_df`.

12.5.5 Test Whether the Time Series Is "Stationary"

To proceed with the model generation activity, we need to first make sure that the time series is stationary, as mentioned earlier in this chapter. We will be using the augmented Dickey-Fuller test to check this. The code is provided next. We are using here the `pmdarima` package.

```
#Check if the time series is stationary
#otherwise we have to make it stationary
#To understand it we run Augmented Dickey-Fuller
#Test
import pmdarima
from pmdarima.arima import ADFTest
ADF_Test_Res = ADFTest(alpha=0.05)
p_value, should_diff = ADF_Test_Res.should_diff(Max_Temp_df)
print(p_value)
print(should_diff)
```

```
Output:
0.2964816976648261
True
```

From this output, we can see that the p-value is not significant. Hence, we will not reject the null hypothesis that the time series is not stationary. The "True" in the output means that the time series is currently nonstationary, and we need to make the time series stationary before we proceed for model generation, by using *differencing*.

12.5.6 The Process of "Differencing"

We need to carry out differencing to make the time series stationary. We use the `pandas.diff()` command for this purpose. We need to first check how many times the differencing has to be carried out using the `ndiffs()` command before commencing the actual differencing activity.

```
#The above shows the time series is not stationary.  We
#also saw from the above plots that there is a seasonal component
#as well as trend component
#The way to find out how many times differencing is required
from pmdarima.arima.utils import ndiffs
n_adf = ndiffs(Max_Temp_df, test='adf')
n_adf
Output:
2
```

The `ndiffs()` command shows the need for differencing twice. We also know from the decomposition done earlier that we have to differentiate twice, i.e., for trend and for seasonal component. Now, we will carry out differencing twice and create a new dataframe, i.e., `Max_Temp_diff`. When this is done, some of the beginning values become NaN. We need to drop them. This is done by using the following code:

```
#As the time series is not stationary
#We need to difference the time series 2 times as shown
#above to make it stationary.
#dropna() is added as some of the values will become
```

```
#NaN after differencing
Max_Temp_diff = Max_Temp_df.diff().diff().dropna()
Max_Temp_diff
```

The output of the differencing and some of the values (partial view of the output) of the diffs are provided in Figure 12-13.

Month_Year	Temp
2014-03	0.84
2014-04	0.81
2014-05	-2.76
2014-06	-0.65
2014-07	-2.68
2014-08	1.77
2014-09	-0.11
2014-10	0.25
2014-11	-1.85
2014-12	-0.73
2015-01	2.47
2015-02	2.81

Figure 12-13. *Differenced fields (partial listing of the output)*

The NaN values will be dropped from the time series after applying the previous code.

Now, we will check whether the differenced time series is now stationary using the same method, i.e., the augmented Dickey-Fuller test. If the p-value is significant (less than or equal to the level of significance, then we reject the null hypothesis that the time series is not stationary and hold the alternative hypothesis; i.e., the time series is stationary. Otherwise, we cannot reject the null hypothesis that the time series is not stationary. In our case, it turns out that the differenced time series is now stationary. Figure 12-14 shows the code and output.

```
#Check if the time series is stationary

#otherwise we have to make it stationary

#To understand it we run Augmented Dickey-Fuller

#Test

import pmdarima

from pmdarima.arima import ADFTest

ADF_Test_Res = ADFTest(alpha=0.05)

p_value, should_diff = ADF_Test_Res.should_diff(Max_Temp_diff)

print(p_value)

print(should_diff)

Output:

0.0183348400728848
False
```

Figure 12-14. *Augmented Dickey_Fuller test to check if the time series is stationary*

From this, we can conclude that `d=2`.

12.5.7 Model Generation

We will now generate the model using the `pmdarima` package. We will be using the `auto_arima` utility within the `pmdarima` package for this purpose. This `auto_arima` function does the heavy lifting for us and suggests the best possible model after iterating on the various options. We will use `d=differencing=2` in the input as we know the `d` is 2. The code is shown here:

```
#Now that the time series is stationary we will
#check for the best model using auto_arima from
#pmdarima.arima which does the heavy lifting for
#We know the d value as 2 from the differencing
#exercise we carried out earier
import pmdarima as pm
from pmdarima.arima import auto_arima
Max_Temp_Best_fit = pm.auto_arima(Max_Temp_df, start_p=0, start_q=0,
max_p=3, max_q=3, m=12, start_P=0, seasonal=True, d=2, start_D=0,
trace=True, error_action='ignore',                    suppress_
warnings=True,  stepwise=True)
Max_Temp_Best_fit.summary()
```

The model is generated. The output of the previous `Max_Temp_Best_fit.summary()` is shown in Figure 12-15.

```
Performing stepwise search to minimize aic
 ARIMA(0,2,0)(0,1,1)[12]             : AIC=inf, Time=0.06 sec
 ARIMA(0,2,0)(0,1,0)[12]             : AIC=84.697, Time=0.01 sec
 ARIMA(1,2,0)(1,1,0)[12]             : AIC=69.977, Time=0.05 sec
 ARIMA(0,2,1)(0,1,1)[12]             : AIC=inf, Time=0.13 sec
 ARIMA(1,2,0)(0,1,0)[12]             : AIC=73.100, Time=0.02 sec
 ARIMA(1,2,0)(2,1,0)[12]             : AIC=inf, Time=0.29 sec
 ARIMA(1,2,0)(1,1,1)[12]             : AIC=71.968, Time=0.21 sec
 ARIMA(1,2,0)(0,1,1)[12]             : AIC=inf, Time=0.11 sec
 ARIMA(1,2,0)(2,1,1)[12]             : AIC=73.945, Time=0.40 sec
 ARIMA(0,2,0)(1,1,0)[12]             : AIC=74.525, Time=0.05 sec
 ARIMA(2,2,0)(1,1,0)[12]             : AIC=71.863, Time=0.05 sec
 ARIMA(1,2,1)(1,1,0)[12]             : AIC=71.878, Time=0.06 sec
 ARIMA(0,2,1)(1,1,0)[12]             : AIC=inf, Time=0.12 sec
 ARIMA(2,2,1)(1,1,0)[12]             : AIC=inf, Time=0.19 sec
 ARIMA(1,2,0)(1,1,0)[12] intercept   : AIC=71.921, Time=0.06 sec

Best model:  ARIMA(1,2,0)(1,1,0)[12]
Total fit time: 1.827 seconds
```

SARIMAX Results

Dep. Variable:	y	**No. Observations:**	36
Model:	SARIMAX(1, 2, 0)x(1, 1, 0, 12)	**Log Likelihood**	-31.988
Date:	Sat, 10 Sep 2022	**AIC**	69.977
Time:	18:02:02	**BIC**	73.250
Sample:	01-31-2014	**HQIC**	70.748
	- 12-31-2016		
Covariance Type:	opg		

	coef	**std err**	**z**	**P>|z|**	**[0.025**	**0.975]**
ar.L1	-0.5416	0.179	-3.030	0.002	-0.892	-0.191
ar.S.L12	-0.6835	0.206	-3.325	0.001	-1.086	-0.281
sigma2	0.7490	0.333	2.249	0.025	0.096	1.402

Ljung-Box (L1) (Q):	0.01	**Jarque-Bera (JB):**	1.15
Prob(Q):	0.92	**Prob(JB):**	0.56
Heteroskedasticity (H):	0.52	**Skew:**	0.40
Prob(H) (two-sided):	0.40	**Kurtosis:**	2.22

Figure 12-15. *ARIMA model summary*

Now, we will interpret the model summary. The model clearly shows that the best possible model is the seasonal ARIMA model (i.e., SARIMAX) with (p,d,q)x(P,D,Q,S) as (1,2,0)x(1,1,0,12). We clearly know that this model has a seasonal component. We observe that the d value that we arrived at earlier and the d value suggested by the auto_arima are same, i.e., 2. The S clearly shows 12 as we have 12 months of data for every year and depicts the seasonal component. The ar.L1 has a significant p-value of 0.002, and the Sigma2, which represents the error, also has a significant p-value. The Prob(Q), i.e., the p-value of the Ljung-Box test, is insignificant. Hence, we cannot reject the null hypothesis that the model does not have the autocorrelation and is fit with only the white noise. Also, the p-value of the Jarque-Bera test, i.e., Prob(JB), is 0.56, which is insignificant. Hence, we cannot reject the null hypothesis that the data is normally distributed. Further, Prob(H) of the "heteroskedasticity" is 0.40, which is insignificant. Hence, we cannot reject the null hypothesis that the error residuals have the same variance. Hence, the model generated is validated from all significant perspectives.

12.5.8 ACF and PACF Plots to Check the Model Hyperparameters and the Residuals

One of the excellent packages now available for business analytics is the statsmodels tool. Thanks to statsmodels, many of the analyses have become easy. We will use the plots of Acf() and Pacf() to check the model hyperparameters already generated and also check the residuals of the model. Here, we are discussing the best-fit model, i.e., the Max_Temp_Best_fit model generated by auto_arima(). Please note that we provide the differenced dataframe, i.e., Max_Temp_diff, as the input to these commands.

The code for generating the ACF plot is provided here:

```
#Thanks to the statsmodels.  We acknowledge with hearty
#thanks excellent design and development of statsmodels
#and making it available to the community
import statsmodels
from statsmodels import graphics
from statsmodels.graphics.tsaplots import plot_acf
plot_acf(Max_Temp_diff, alpha=0.05, use_vlines=True, title="ACF Plot",
auto_ylims=True, zero=False)
```

Figure 12-16 shows the output.

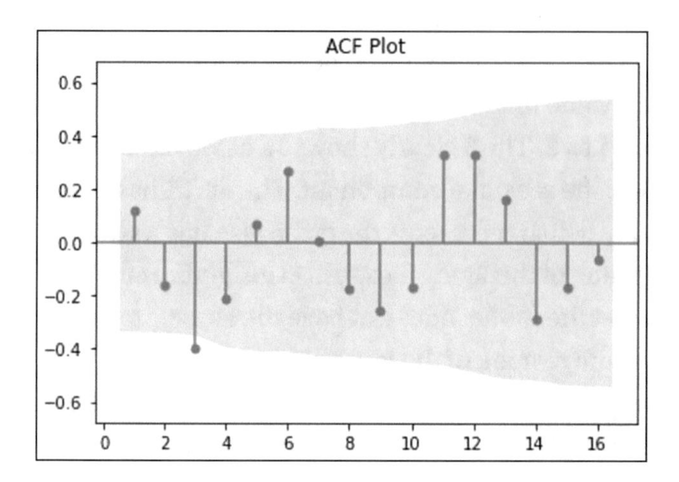

***Figure 12-16.** The Acf() plot from statsmodels on the Max_Temp_diff dataframe*

The code for generating the `Pacf()` plot is provided here:

```
#Thanks to the statsmodels.  We acknowledge with hearty
#thanks excellent design and development of statsmodels
#and making it available to the community
import statsmodels
from statsmodels import graphics
from statsmodels.graphics.tsaplots import plot_pacf
plot_pacf(Max_Temp_diff, alpha=0.05, use_vlines=True, method="ywm",
title="PACF Plot", auto_ylims=True, zero=False)
```

Figure 12-17 shows the output.

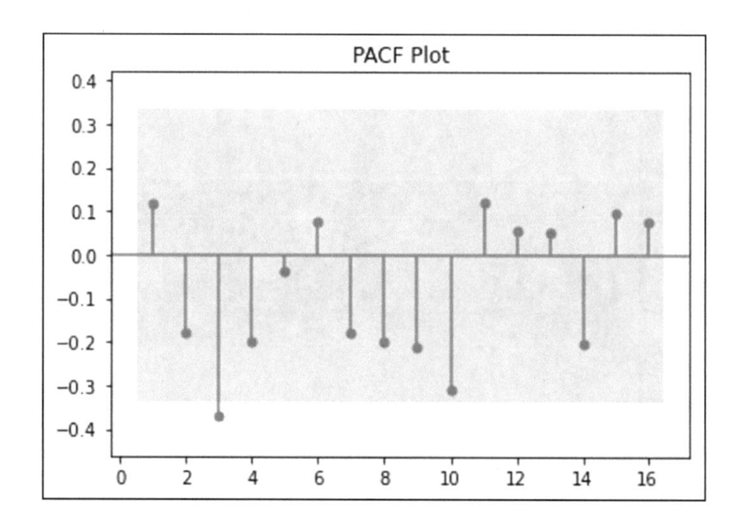

***Figure 12-17.** The Pacf() plot from statsmodels on the Max_Temp_diff dataframe*

As you can see from the ACF plot, it is sinusoidal in form; i.e., it follows a sine-wave pattern. This means the order of `MA()`, i.e., `q`, is 0. If we observe the PACF plot, we find that none of the residuals is significant (just one point is beyond the confidence interval, which we can ignore). However, the first lag is showing a positive value before it turns into negative for the further lags. Hence, we can conclude that the order of `AR()`, i.e., `p`, is 1. This is also in tune with the `Max_Temp_Best_fit` model hyperparameters. We already know that `d=2`.

Now, we will use the residuals of the model generated, i.e., `Max_Temp_Best_fit`, and generate the ACF and PACF plots.

The code to generate the ACF plot on the residuals of the model is given here:

```
#plot residuals of the above model generated
#i.e., Max_Temp_Best_fit
plot_acf(Max_Temp_Best_fit.resid(), zero=False)
```

Figure 12-18 shows the output ACF plot.

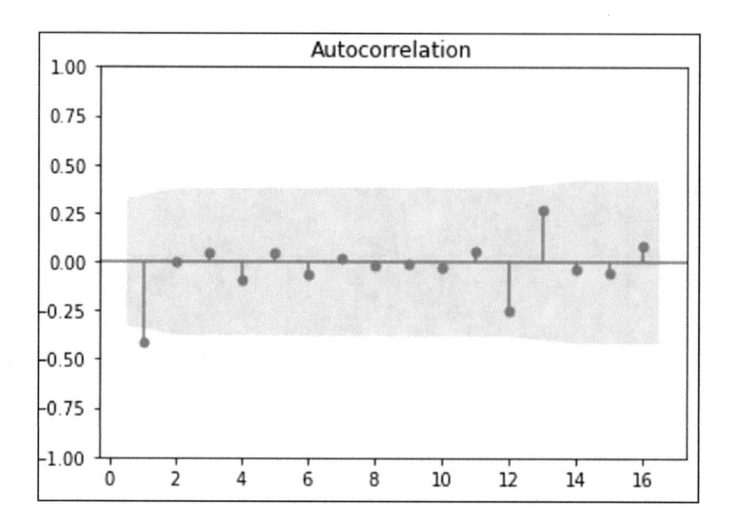

***Figure 12-18.** Acf() plot on the residuals of the model Max_Temp_Best_fit*

The code for generating the PACF plot on the residuals of the model is given here:

```
#plot residuals of the above model generated
#i.e., Max_Temp_Best_fit
plot_pacf(Max_Temp_Best_fit.resid(), method="ywm", zero=False)
```

Figure 12-19 shows the output PACF plot.

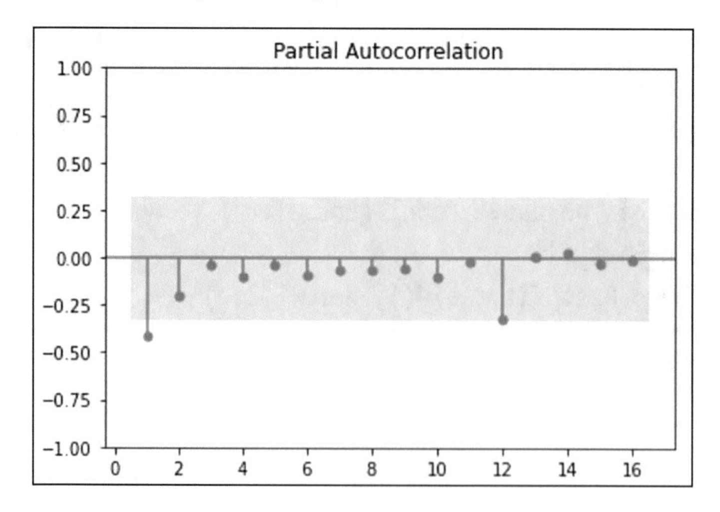

***Figure 12-19.** Pacf() plot on the residuals of the model Max_Temp_Best_fit*

We can see that almost all the residuals are very small or near zero. Hence, we can conclude that the model is fit to be used.

12.5.9 Forecasting

Forecasting is normally made on the test data kept out of the training data to ensure that our testing of the prediction is appropriate. In our case, we have the `Max_Temp_IMD_Test.csv` data for the further dates, i.e., from Jan 2017 to Dec 2017, which we kept separately for the purpose of testing.

We will import the test data to our working environment and then use the `Date` field to get the `Month_Year` field as we did in the case of our original training data. We will then drop the `Date` field and index the `Max_Temp_Test_df` on `Month_Year`. The code is as follows:

```
#Importing the test time series which has 12 months data post the Train #data
used above
Max_Temp_Test_df = pd.read_csv("C:/Users/kunku/OneDrive/Documents/Book
Revision/Max_Temp_IMD_Test.csv", sep=",", header=0)
Max_Temp_Test_df['Month_Year'] = pd.to_datetime(Max_Temp_Test_df['Date']).
dt.to_period('M')
Max_Temp_Test_df.drop("Date", inplace=True, axis=1)
#indexing the dataframe on Date field so that
#further operations on the data can be done easily
Max_Temp_Test_df.set_index("Month_Year", inplace=True)
Max_Temp_Test_df.head(5)
```

Figure 12-20 shows the output.

	Date	Temp
0	31-01-2017	26.45
1	28-02-2017	29.46
2	31-03-2017	31.60
3	30-04-2017	34.95
4	31-05-2017	35.84

***Figure 12-20.** The first five records of the Max_Temp_Test_df used for testing of the model*

Now, we will forecast using the model Max_Temp_Best_fit (without using any data input) for the next six months beyond the training data using the following code:

```
#forecasting further 6 values based on the model generated
forecasts = Max_Temp_Best_fit.predict(n_periods=6)
forecasts
```

Figure 12-21 shows the output.

```
2017-01     28.711060
2017-02     31.842207
2017-03     34.905397
2017-04     38.357296
2017-05     40.641819
2017-06     39.668761
Freq: M, dtype: float64
```

Figure 12-21. *The predicted or forecasted results using the model Max_Temp_Best_fit*

The model has forecasted the output. If we compare these values with the 2017 month values in our Max_Temp_Test_df, we may find that the forecasted values may not be a near match in many of the cases. This means that the model built is still not able to explain fully all the components of the time series. You can further appreciate the fact that in many cases it may be difficult to forecast accurately based on the past data as there may be specific aspects that have impacted the further values that the model has not seen and the past values may not be representative of the future values for these reasons.

We will now predict the values using the Max_Temp_Best_fit model on our current data used for the training, i.e., Max_Temp_df, which is known as *in-sample* or *in-series* prediction. The code used for the same is provided here:

```
#You can also predict on the sample used for the model generation
#This may help for close comparision with the actual values
#and the predicted values
predictions_in_sample = Max_Temp_Best_fit.predict_in_sample(alpha=0.05)
predictions_in_sample
```

The following is the output:

```
Month_Year
2014-01      0.000000
2014-02     39.716663
2014-03     28.110012
2014-04     31.929993
2014-05     36.530003
2014-06     34.800001
2014-07     34.530000
2014-08     29.550003
2014-09     30.789995
2014-10     30.040004
2014-11     29.899995
2014-12     25.810010
2015-01     27.304002
2015-02     16.740004
2015-03     30.039972
2015-04     32.584781
2015-05     32.013921
2015-06     34.477478
2015-07     29.915671
2015-08     31.045280
2015-09     31.890271
2015-10     31.537564
2015-11     29.147156
2015-12     24.666804
2016-01     25.196811
(partial output)
```

12.6 Chapter Summary

In this chapter, you learned what a time series is and the uses/benefits of time series in the real world.

You learned about the components of a time series such as seasonal, trend, and irregular/error components.

You learned practically, through the help of examples, how to carry out exponential smoothing modeling in R.

You learned, through the help of examples, how to carry out ARIMA and ARMA modeling in R. In the process, you also learned the prerequisites that need to be met before such modeling is carried out (e.g., the time series needs to be stationary) and what assumptions your models need to fulfil to be of use.

You also learned how to use the model for forecasting.

You then experimented with decomposition, model generation, its validation, and forecasting using Python instead of R.

PART IV

Unsupervised Models and Text Mining

CHAPTER 13

Cluster Analysis

Clustering is an unsupervised learning technique to categorize data in the absence of defined categories in the sample data set. In this chapter, we will explore different techniques and how to perform clustering analysis.

13.1 Overview of Clustering

Clustering analysis is an unsupervised technique. Unlike supervised learning, in unsupervised learning, the data has no class labels for the machines to learn and predict the class. Instead, the machine decides how to group the data into different categories. The objective of the clusters is to enable the business to make meaningful analysis. Clustering analysis can uncover previously undetected relationships in a data set. For example, cluster analysis can be applied in marketing for customer segmentation based on demographics to identify groups of people who purchase similar products. Similarly, identify clusters based on consumer spending to estimate the potential demand for products and services. These kind of analysis help businesses to formulate marketing strategies.

Nielsen (and earlier, Claritas) were pioneers in cluster analysis. Through its segmentation solution, Nielsen helped customize demographic data to understand geography based on region, state, ZIP code, neighborhood, and block. This has helped the company to come up with effective naming and differentiation of groups such as movers and shakers, fast-track families, football-watching beer aficionados, and casual, *and* sweet palate drinkers.

In a human resources (HR) department, cluster analysis can help to identify employee skills and performance. Furthermore, you can cluster based on interests, demographics, gender, and salary to help a business act on HR-related issues such as relocating, improving performance, or hiring an appropriately skilled labor force for forthcoming projects.

U. R. Hodeghatta and U. Nayak, *Practical Business Analytics Using R and Python*,
https://doi.org/10.1007/978-1-4842-8754-5_13

In finance, cluster analysis can help create risk-based portfolios based on various characteristics such as returns, volatility, and P/E ratio. Similarly, clusters can be created based on revenues and growth, market capital, products and solutions, and global presence. These clusters can help a business position itself in the market. Other applications of clustering include grouping newspaper articles based on topics such as sports, science, or politics; grouping the effectiveness of the software development process based on defects and processes; and grouping various species based on classes and subclasses.

The clustering algorithm takes the raw data as input, runs the clustering algorithm, and segregates the data into different groups. In this example, shown in Figure 13-1, based on the size, bags are clustered together by the clustering algorithm. The purpose of cluster analysis is to segregate data into groups. The idea of clustering is not new and has been applied in many areas, including archaeology, astronomy, science, education, medicine, psychology, and sociology.

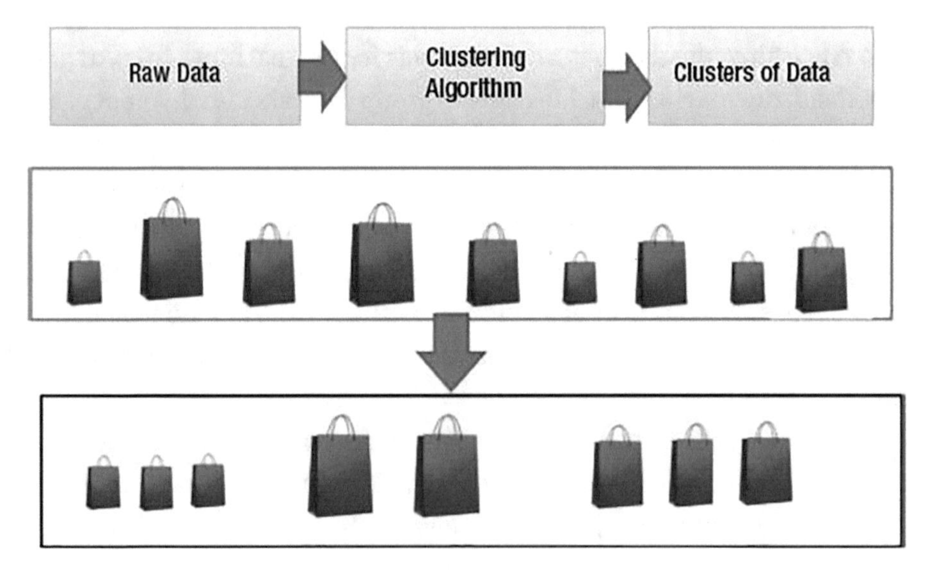

***Figure 13-1.** Example of cluster analysis*

There are different clustering algorithms to perform this task, and in the next section, we will discuss various clustering techniques and how to perform the clustering technique with an example.

13.1.1 Distance Measure

To understand the cluster techniques, we have to first understand the distance measured between the sample records and how to group records into two or more different clusters. There are several different metrics to measure the distance between the two records. The most common measures are Euclidian distance, Manhattan distance, and Minkowski.

13.1.2 Euclidean Distance

Euclidean distance is the simplest and most common measure used. The Euclidean distance, E_{ij}, between the two records, i and j, are the variables defined as follows:

$$E_{ij} = \sqrt{(X_{i1} \quad X_{j1})^2 + (X_{i2} \quad X_{j2})^2 + (X_{i3} \quad X_{j3})^2 + \ldots\ldots\ldots(X_{ip} \quad X_{jp})^2} \qquad (1)$$

As an extension to the equation, you can also assign weights to each variable based on its importance. A weighted Euclidean distance equation is as follows:

$$E_{ij} = \sqrt{W_1(X_{i1} \quad X_{j1})^2 + W_2(X_{i2} \quad X_{j2})^2 + W_3(X_{i3} \quad X_{j3})^2 + \ldots\ldots\ldots W_p(X_{ip} \quad X_{jp})^2} \qquad (2)$$

where X_{ij} are the p samples from different variables.

13.1.3 Manhattan Distance

Another well-known measure is *Manhattan* (or *city block*) *distance*, which is defined as follows:

$$M_{ij} = |x_{i1} - x_{j1}| + |x_{i2} - x_{j2}| + |x_{i3} - x_{j3}| + \ldots + |x_{ip} - x_{jp}| \qquad (3)$$

Here, X_{ij} is the p sample from different variables. Both the Euclidian distance and the Manhattan distance should also satisfy the following conditions:

$E_{ij} \geq 0$ and $M_{ij} \geq 0$: The distance is a non-negative number.

$E_{ii} = 0$ and $M_{ii} = 0$: The distance from an object to itself is 0.

$E_{ij} = E_{ji}$ and $M_{ij} = M_{ji}$: The distance is a symmetric function.

13.1.4 Distance Measures for Categorical Variables

A *categorical variable* can take two or more states. Therefore, we cannot use the previous distance measure to calculate the distance between the two categorical variables. One approach is to compute the distance using a contingency table, as shown in Table 13-1. In this example, we compute the distance between two variables with two categories, 0 and 1. The simple measure to calculate the distance between the two variables is defined here:

(p+s)/(p+q+r+s)

Here, p is the total number of samples with both "yes" categories, and s is the total number of samples with both "no" categories. Similarly, r and q are the total number of samples with the yes/no and no/yes combination.

Table 13-1. *Contingency Table*

Variable 2 \ Variable 1	Yes	NO	Total
Yes	p	q	p+q
No	r	s	r+s
Total	p+r	q+s	p+q+r+s

Usually, data sets have a combination of categorical and continuous variables. The Gower similarity coefficient is a measure to find the distance between quantitative (such as income or salary) variables and categorical data variables (such as spam/not spam).

13.2 Distance Between Two Clusters

The clustering algorithm groups sample records into different clusters based on similarity and the distance between different clusters. The similarity between the two variables is based on the distance measure that we discussed earlier, and to put them in different cluster boundaries, it takes the distance between the clusters. There are several measures to measure the distance between the clusters. The commonly used measures are single linkage, complete linkage, centroid, and average linkage.

For a given cluster B, with B_1, B_2, B_3 … B_m samples and cluster C with C_1, C_2, C_3 … $C_{m,}$ records, *single linkage* is defined as the shortest distance between the two pairs of records in C_i and B_j. This is represented as Min(distance(C_i, B_j)) for i = 1, 2, 3, 4 … *m*; j = 1, 2, 3 … *n*. The two samples that are on the edge make the shortest distance between the two clusters. This defines the cluster boundary. The boundaries form the clusters, *as* shown in Figure 13-2.

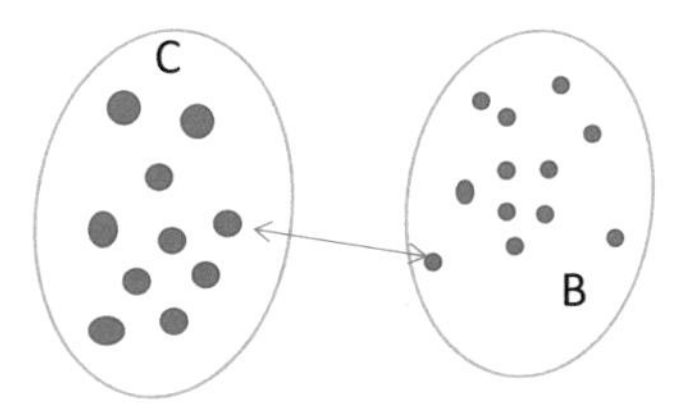

Figure 13-2. *Single linkage*

Complete linkage is the distance between two clusters and is defined as the *longest* distance between two points in the clusters, as shown in Figure 13-3. The farthest distance between two records in clusters C_i and B_j is represented as Max (distance (C_i, B_j)) for I = 1, 2, … *m*; j = 1, 2, 3 … *n*. The farthest samples make up the edge of the cluster.

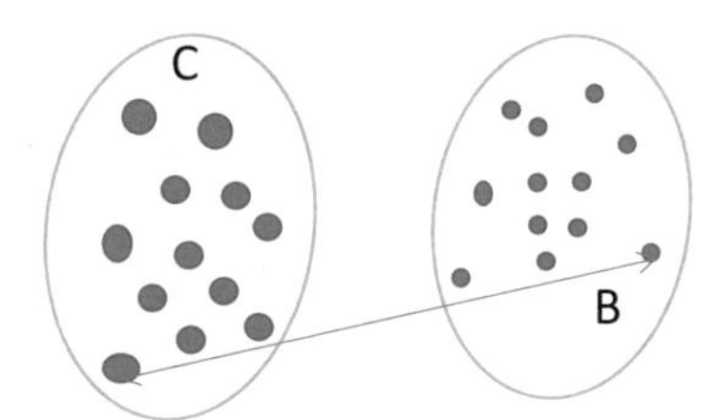

Figure 13-3. *Complete linkage*

The *average linkage* is a measure that indicates the average distance between each point in the clusters, as shown in Figure 13-4. The average distance between records in one cluster and records in the other cluster is calculated as Average (distance(C_i, B_j)) for I =1, 2, 3 … *m*; j = 1, 2, 3 … *n*.

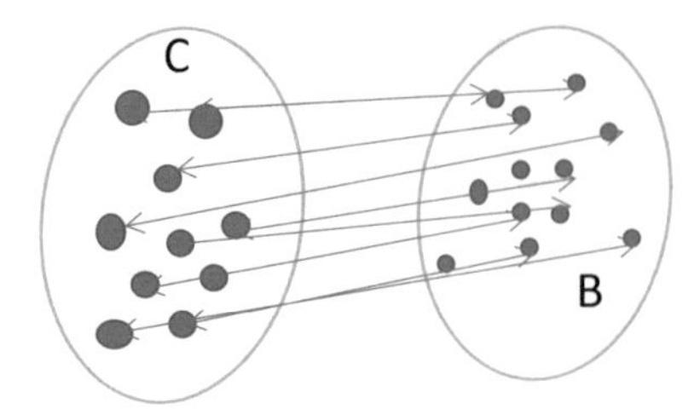

Figure 13-4. *Average linkage*

The *centroid* is the center of the cluster. It is calculated by taking the average of all the records in that cluster. The centroid distance between the two clusters A and B is simply the distance between centroid(A) and centroid(B).

Selecting the clustering distance measure depends on the data, and it also requires some amount of domain knowledge. When the data is well spread, a single linkage may be a good choice. On the other hand, the average or complete linkage may be a better choice if the data is somewhat in sequence and the data appears to be a spherical shape. Selecting a specific cluster method is always a challenge, and also, we are unsure how many clusters will be formed by the algorithm. Based on the domain knowledge, we need to check each element in a cluster and decide whether to keep two or three clusters. Our research and experience have shown that the unsupervised learning methods are not intuitive or easy to comprehend. It takes effort to understand and label the clusters properly.

13.3 Types of Clustering

Clustering analysis is performed on data to gain insights that help you understand the characteristics and distribution of data. The process involves grouping the data into similar groups based on similar characteristics, and different groups are as dissimilar as possible from one cluster to another. If the two samples from two or more variables have close similarity measures, they are grouped under the same cluster. Unlike in classification, clustering algorithms do not rely on predefined class labels in the sample data. Instead, the algorithms are based on the similarity measures discussed earlier between the different variables and different clusters.

There are several clustering techniques based on the procedure used in similarity measures, the thresholds in constructing the clusters, and the flexibility of cluster objects to move around different clusters. Irrespective of the procedure used, the resulting cluster must be reviewed by the user. The clustering algorithms fall under one of the following: hierarchical and nonhierarchical clustering.

13.3.1 Hierarchical Clustering

Hierarchical clustering constructs the clusters by dividing the data set into similar records by constructing a hierarchy of predetermined order from top to bottom. For example, all files and folders on the hard disk are organized in a hierarchy. In hierarchical clustering, as the algorithm steps through the samples, it creates a hierarchy based on distance. It starts with one large cluster and slowly moves the data to different clusters. There are two types: the agglomerative method and the divisive method. Both methods are based on the same concept. In agglomerative, the algorithm starts with ***n*** clusters and then merges similar clusters until a single cluster is formed. In divisive, the algorithm starts with one cluster and then moves the elements into multiple clusters based on dissimilarities, as shown in Figure 13-5.

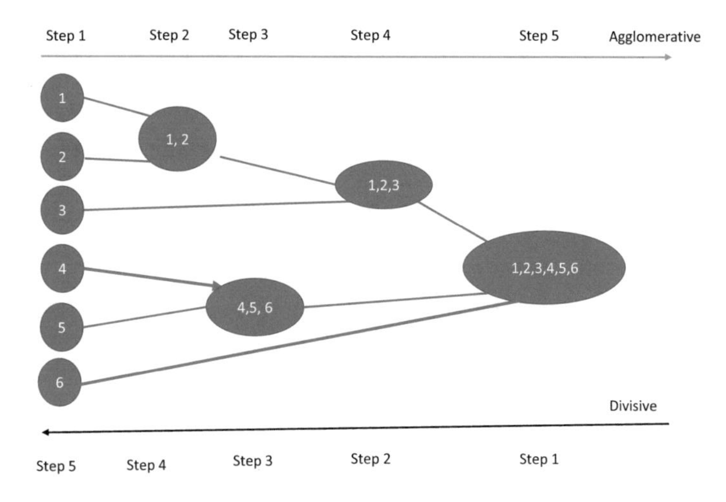

Figure 13-5. *Hierarchical clustering*

The hierarchical agglomerative clustering algorithm is as follows:

1. Start with *n* clusters. Each record in the data set can be a cluster by itself.
2. The two similar cluster observations are merged in the next step.
3. The process repeats until a single cluster is formed. At every step, the clusters with the smallest distance measure are merged together until all the clusters are combined to form a single cluster. This creates a hierarchy of clusters.

The hierarchical divisive clustering algorithm is just the opposite of the previous steps:

1. Start with one single cluster where all the samples in the data set are in one cluster.
2. The two dissimilar cluster observations are separated in the next step.
3. The process repeats until all the cluster elements are separated. At every step, the clusters with the largest distance measure are separated. This also creates a hierarchy of clusters.

Several tools support several hierarchical clustering algorithm implementations with little variations in creating clusters. Some of the common ones are as follows:

- BIRCH (1996): Uses CF-tree and incrementally adjusts the quality of subclusters
- ROCK (1999): Clustering categorical data by neighbor and link analysis
- CHAMELEON (1999): Hierarchical clustering using dynamic modeling

13.3.2 Dendrograms

A *dendrogram* is the representation of the hierarchy. A dendrogram is a tree-like structure that summarizes the clustering process and the hierarchy pictorially, as shown in Figure 13-6.

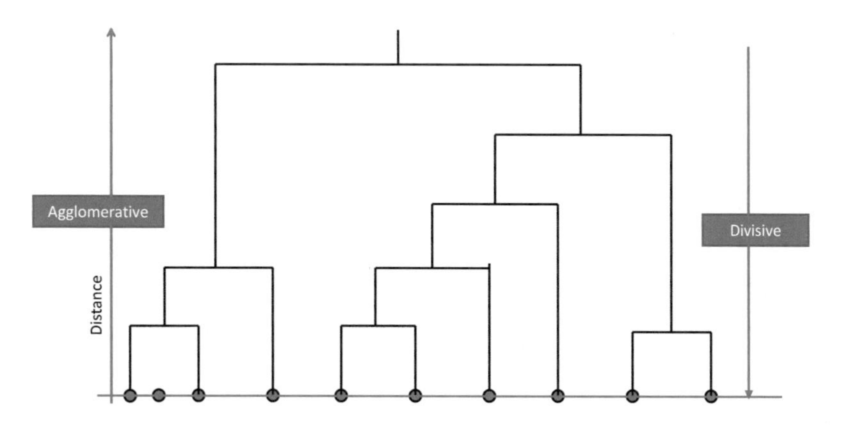

***Figure 13-6.** A dendrogram representation of hierarchical clustering*

13.3.3 Nonhierarchical Method

In nonhierarchical clustering, no hierarchy of clusters is formed; instead, the number of clusters is prespecified with k partitions. The partitions are formed by minimizing the error. The objective is to minimize total intracluster variance, using the squared error function.

$$E = \sum_{i=1}^{n} \sum_{i=1}^{k} \left| \left(x_i - m_j \right) \right|^2$$

Here, *k* is the number of partitions, *x* is the sample, and *m* is the mean distance of the cluster. The algorithm intends to partition *n* objects into *k* clusters with the nearest mean. The goal is to divide the samples into the k-clusters so that clusters are as homogeneous as possible. The end result is to produce k different clusters with clear distinctions. There are many methods of partitioning clusters implemented by different tools. The common ones are k-means, probabilistic clustering, k-medoids method, and partitioning around mediods (PAM). All these algorithms are optimization problems that try to minimize the error function. The most common is the k-means algorithm.

13.3.4 K-Means Algorithm

The objective of k-means clustering is to minimize total intracluster variance.

The k-means clustering algorithm starts with K centroids. The initial values of the cluster centroids are selected randomly or from a prior information. Then it assigns objects to the closest cluster center based on the distance measure. It recalculates the centroid after each assignment. After recalculating the centroid, it checks the data point's distance to the centroid of its own cluster. If it is closest, then it is left as is. If not, it moves to the next closest cluster. This process is repeated until all the data points are covered and no data point is moving from one cluster to another cluster. The following example demonstrates the k-means algorithm.

We want to group the visitors to a website using just their age (a one-dimensional space) using the k-means clustering algorithm. Here is the one-dimensional age vector:

[15,15,16,19,19,20,20,21,22,28,35,40,41,42,43,44,60,61,65]

Iteration 1:

Let k = 2. Let's choose 2 centroids randomly, 16 and 22. Move all the points to 16 or 22 based on distance. Now the two clusters are as follows:

C1 = [15, 15, 16,19]

C2 = [20,21,22,28,35,40,41,42,43,44,60,61,65]

Iteration 2:

Updated centroid 1 : 16.25 (average of centroid 1 from previous)

Updated centroid 2: 36.85

Move elements closer to the new updated centroid.

New C1 elements = [15,15,16,19, 20, 21, 22, 28]

New C2 elements = [35,40,41,42,43,44,60,61,65]

Iteration 3:

Updated centroid 1:19.5

Updated centroid 2: 47.88

Move the elements to the updated centroid point.

New C1 elements = [15,15,16,19,20,21,22,28]

New C2 elements = [35,40,41,42,43,44,60,61,65]

There is no change in the centroid from iteration 3 and iteration 4; thus, the algorithm would stop here. Using k-means clustering, two groups have been identified: C1[15–28] and C2[35–65]. However, the initial random selection of cluster centroids can affect the cluster iterations and elements. To overcome this, run the algorithm multiple times with different starting conditions to get a fair view of what the clusters should be.

13.3.5 Other Clustering Methods

In addition to the hierarchical and nonhierarchical clustering, several other clustering methods have been developed to overcome some of the limitations of clustering.

- *Birch*: This builds a tree called the *clustering feature tree* (CFT) for the given data. It typically works well when the number of variables is less than 20.
- *MeanShift*: This is similar to k-means method. It works by updating centroids to be the mean within a given cluster region. Unlike in k-means, the algorithm automatically sets the number of clusters. The algorithm is not highly scalable due to the search method used.
- *Spectral clustering*: This is most ideal for image clustering. It performs a low-dimension embedding between sample data followed by clustering.
- *Affinity propagation*: This method creates clusters by measuring the distance between the pair of samples and adjusting the cluster group until the convergence. This is ideal for small data sets.
- *DBSCAN*: It looks at clusters as areas of high density separated by areas of low density. It is an expensive memory algorithm.

13.3.6 Evaluating Clustering

Unlike supervised machine learning algorithms like classifications, or regression, it is hard to evaluate the performance of clustering algorithms and compare one with the other. However, researchers have developed common techniques based on cluster homogeneity. Homogeneity is how the samples are distributed within the clusters that contain members from a proper class. Completeness is all samples in the data set are assigned suitable cluster classes. Intraclustering distance is another measure used to find out how well clusters are formed. The objective of the clustering algorithm is to develop more distinct clusters that are well separated from each other.

13.4 Limitations of Clustering

In general, clustering is unsupervised learning, and there are no predefined class labels in the sample data set. The algorithm reads all the data, and based on different measures, it tries to group the data into different clusters. Hierarchical clustering is simple to understand and interpret. It does not require you to specify the number of clusters to form. It has the following limitations:

- For large data sets, computing and storing the n × n matrix may be expensive and slow and have low stability.
- The results may vary when the metric is changed from one measure to another.

K-means is a simple and relatively easy and efficient method. The problem with this method is that a different k can vary the results and cluster formation. A practical approach is to compare the outcomes of multiple runs with different k values and choose the best one based on a predefined criterion. Selecting the initial k is driven by external factors such as previous knowledge, practical constraints, or requirements. If the selection of k is random and not based on prior knowledge, then you have to try a few different values and compare the resulting clusters.

Clustering higher-dimensional data is a major challenge. Many text documents and pharmaceutical applications have higher dimensions. The distance measures may become meaningless to these applications because of the equidistance problem. Several techniques have been developed to address this problem, and some of the newer methods include CLIQUE, ProClus, and frequent pattern-based clustering.

13.5 Clustering Using R

In this section, we will create a k-means clustering model using R. The process of creating a clustering model is the same as for any other supervised model. The steps are to read the data set, then explore the data set, prepare the data set for the clustering function, and finally create the model. Since there are no proper metrics to measure the performance of the cluster models, after the clusters are formed, the clusters have to be examined manually before finalizing the cluster names and the cluster elements.

Step 1: Load the essential libraries to the development environment and read data from the source. This k-means clustering model aims to create clusters based on students' performance on assignments and exams. The data contains `StudentID`, `Quiz1`,

Quiz2, Quiz3, Quiz4, and Oquiz5 variables. The goal is to group students into different clusters based on their performance and assign grades to each student.

```
> # Read the Training data
> grades_df<-read.csv("grades.csv",header=TRUE, sep=",")
> ##Read the Test data
> head(grades_df)

  StudentID Quiz1 Quiz2 Quiz3 Quiz4 Quiz5
1  20000001    10  10.0  27.0    95     8
2  20000002     8   3.0  23.5    75     9
3  20000003    14  15.0  28.0    70     8
4  20000004    12  16.5  28.0   100     9
5  20000005    13  11.5  21.5    95     9
6  20000006     8   0.0  25.0   100     9
```

Step 2: Check the data types and remove the StudentID column from the data.

```
> ##Removing 'StudentID' column, that is not required for the
clustering analysis
> grades_df2<-grades_df[-c(1)]
> head(grades_df2)

  Quiz1 Quiz2 Quiz3 Quiz4 Quiz5
1    10  10.0  27.0    95     8
2     8   3.0  23.5    75     9
3    14  15.0  28.0    70     8
4    12  16.5  28.0   100     9
5    13  11.5  21.5    95     9
6     8   0.0  25.0   100     9
```

Step 3: Standardize data to a single scale using the scale() function. Since all the variables are on different scales, it is always recommended to scale the data for better performance.

```
> ##Scale the data
> # Standardize the Data
> grades_df2<-scale(grades_df2)
> head(grades_df2)
```

```
          Quiz1      Quiz2       Quiz3      Quiz4      Quiz5
[1,] -0.2291681 -0.6336322  0.57674037  0.4707916 -0.6022429
[2,] -0.8481581 -1.4423179 -0.55332570 -1.2543467  1.0763489
[3,]  1.0088120 -0.0559995  0.89961639 -1.6856313 -0.6022429
[4,]  0.3898220  0.1172903  0.89961639  0.9020761  1.0763489
[5,]  0.6993170 -0.4603424 -1.19907774  0.4707916  1.0763489
[6,] -0.8481581 -1.7888975 -0.06901167  0.9020761  1.0763489
```

Step 4: Build the clustering model using the k-means function in R. Initially choose the k-value as 3. The `kmeans()` function has several input parameters that are listed next; we use `nstart = 25` to generate 25 initial configurations. For all other parameters, we use default values.

Kmeans( ) Function Arguments (from the R documentation)

X	This is a numeric matrix of data, or an object that can be coerced to such a matrix (such as a numeric vector or a dataframe with all numeric columns).
`centers`	This is either the number of clusters, say *k*, or a set of initial (distinct) cluster centers. If a number, a random set of (distinct) rows in x is chosen as the initial centers.
`iter.max`	This is the maximum number of iterations allowed.
`nstart`	If `centers` is a number, how many random sets should be chosen?
`Algorithm`	This is a character that may be abbreviated. Note that `"Lloyd"` and `"Forgy"` are alternative names for one algorithm.
`object`	This is an R object of class `kmeans`, typically the result ob of ob `<- kmeans(..)`.
`method`	Th is a character, which may be abbreviated. `centers` causes `fitted` to return cluster centers (one for each input point), and `classes` causes `fitted` to return a vector of class assignments.
`trace`	This is a logical or integer number, currently only used in the default method (`Hartigan-Wong`). If positive (or true), tracing information on the progress of the algorithm is produced. Higher values may produce more tracing information.
`...`	This is not used.

```
> km_model<-kmeans(grades_df2, centers=3, nstart=25)
> #Model summary
> summary(km_model)
             Length Class  Mode
cluster      131    -none- numeric
centers       15    -none- numeric
totss          1    -none- numeric
withinss       3    -none- numeric
tot.withinss   1    -none- numeric
betweenss      1    -none- numeric
size           3    -none- numeric
iter           1    -none- numeric
ifault         1    -none- numeric
```

kmeans returns an object of class kmeans. It lists the following components:

Cluster	This is a vector of integers (from 1:k) indicating the cluster to which each point is allocated.
centers	This is a matrix of cluster centers.
totss	This is the total sum of squares.
withinss	This is a vector of within-cluster sum of squares, one component per cluster.
tot.withinss	This is a total within-cluster sum of squares, i.e., sum(withinss).
betweenss	This is a between-cluster sum of squares, i.e., totss-tot.withinss.
size	This is the number of points in each cluster.
iter	This is the number of (outer) iterations.
ifault	This is an integer that is an indicator of a possible algorithm problem; this is for experts.

Step 5: Summarize the model by printing the number of clusters, cluster distribution, assignment, and cluster centers using the following functions:

```
> # Cluster details of the data points
> km_model$cluster

  [1] 3 2 3 1 1 1 1 2 3 3 1 3 1 2 3 3 3 2 3 3 3 3 2 2 2 1 2 2 3 3
 [31] 3 2 3 3 3 1 3 3 1 1 3 3 1 2 3 1 3 3 3 3 2 3 3 3 3 3 3 3 3 1
 [61] 2 1 3 1 1 2 2 1 3 2 2 3 2 3 1 3 2 2 3 2 2 3 2 1 1 1 3 1 3 1
 [91] 2 1 3 3 3 3 1 3 1 3 3 1 1 3 3 1 3 2 2 3 1 1 3 3 3 3 3 3 3 3
[121] 3 3 3 1 3 3 1 3 3 2 1
> # Centers of each cluster for each variables
> km_model$centers
        Quiz1      Quiz2      Quiz3      Quiz4      Quiz5
1  0.08032695  0.1920428  0.2063826  0.2678341  1.4713117
2 -0.80230698 -0.6550260 -1.0854732 -1.1265587 -0.2913925
3  0.27044532  0.1593750  0.3184396  0.3044389 -0.6022429

>
```

Step 6: Plot the clusters using the `fvizcluster()` function. Since there are more than two dimensions in the data, `fvizcluster()` uses principal component analysis (PCA) to plot the first two principal components, which explains the majority of the variance in the data. Figure 13-7 shows the output.

```
> ##Plotting cluster
> library(factoextra)
Loading required package: ggplot2
Warning message:
package 'factoextra' was built under R version 3.6.3
> fviz_cluster(km_model, data = grades_df2)
```

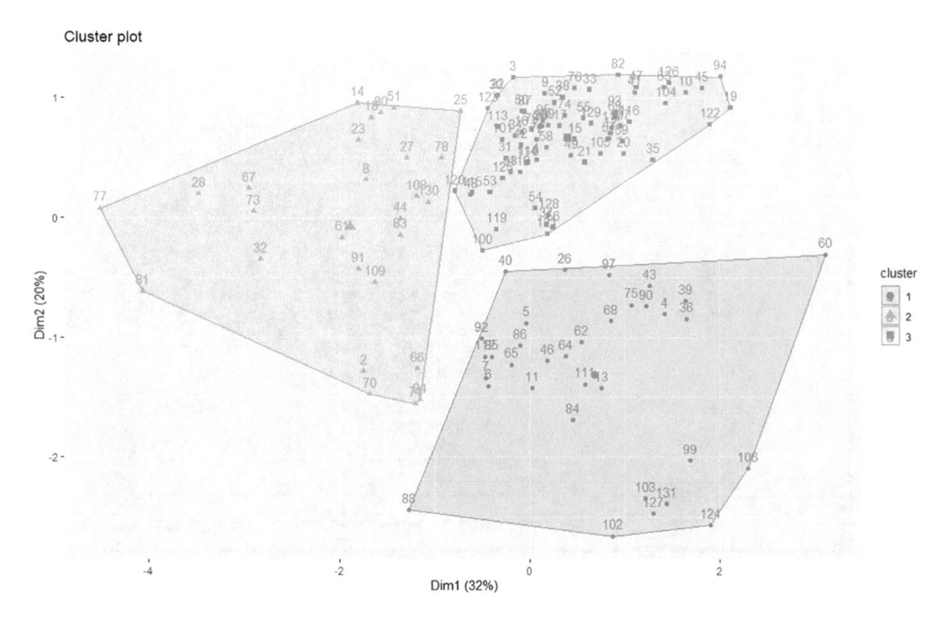

***Figure 13-7.** K-means clustering example*

Step 7: Find the optimal value of k by using the elbow method. Fortunately, `fviz_nbclust()` supports this.

```
> # Fortunately, the "Elbow method" is supported by
> # a single function (fviz_nbclust):
> set.seed(123)
> fviz_nbclust(grades_df, kmeans, method = "wss")
```

The challenge of the unsupervised clustering method is determining the optimal number of clusters. The elbow and silhouette are the two most common methods used to determine the optimal value of k. The elbow plot shows the cluster errors on the y-axis to the number of k values. In the elbow plot, you want to see a sharp decline of the error from one k value to another rather than a more gradual decrease in slope. The "good" k value would be the last value before the slope of the plot levels off.

Similarly, the silhouette measure provides how distinct the two clusters are, that is, how well the two clusters are separated. It has an arrangement of [-1, 1]. The y-axis on the plot provides the silhouette width for different k values (x-axis). Silhouette coefficients close to +1 indicate that the sample is far away from the neighboring clusters (this is what we want), and a 0 value or near 0 value indicates that the sample is close to the decision boundary between two neighboring clusters. A silhouette of negative values indicates that the data samples have been in the wrong cluster. See Figure 13-8.

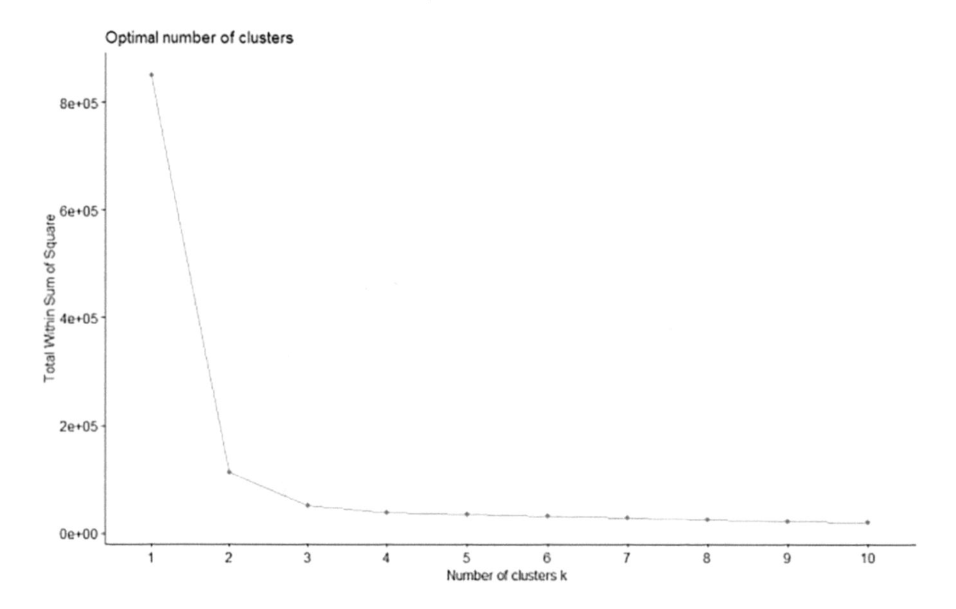

Figure 13-8. *Finding optimal value of k using the elbow method*

Step 8: We use the silhouette method as well. See Figure 13-9.

```
##Using "silhouette" method
> fviz_nbclust(df, kmeans, method = "silhouette")
```

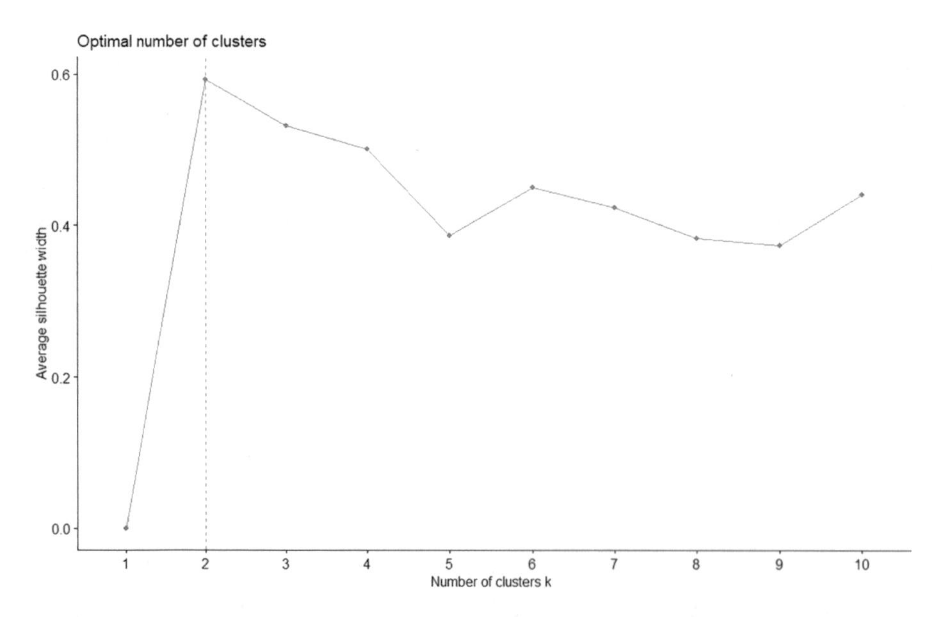

Figure 13-9. *Optimal value of k using the silhouette method*

Step 9: From both methods, the optimal value is found to be 2. We would use 2 and re-create the model. See Figure 13-10.

```
> ## The results shows 2 or 3 is the most optimal k-value
> # We will computer using k=2
> km_model_k2<-kmeans(grades_df2, centers=2, nstart=25)
> fviz_cluster(km_model_k5, data = grades_df2)
> # Cluster details of the data points
> km_model_k2$cluster
  [1] 2 1 2 2 2 2 1 1 2 2 2 2 2 1 2 2 2 1 2 2 2 1 1 1 1 2 1 1 2 1
 [31] 2 1 2 2 2 2 2 2 2 1 2 2 2 1 2 2 2 2 2 2 1 2 2 2 2 2 2 2 2 2
 [61] 1 2 2 2 2 1 1 2 2 1 1 2 1 2 2 2 1 1 2 1 1 2 1 2 2 2 2 1 2 2
 [91] 1 2 2 2 2 2 2 2 2 2 2 2 2 2 2 2 2 1 1 2 2 1 2 2 1 2 2 2 2 2
[121] 2 2 2 2 2 2 2 2 2 1 2
```

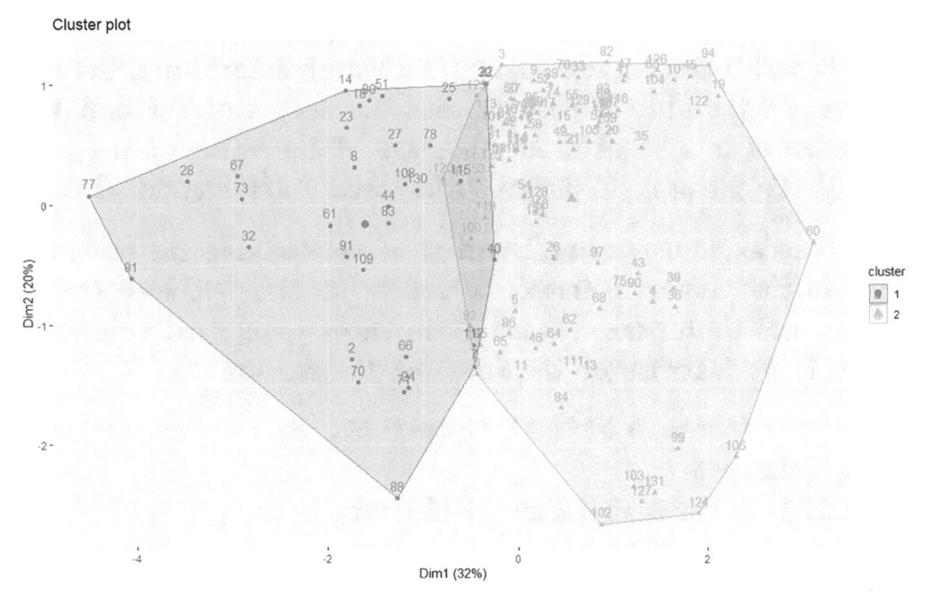

Figure 13-10. K-means cluster for optimal value (k=2)

13.5.1 Hierarchical Clustering Using R

In this section, we briefly discuss the hierarchical clustering approach. R supports the `hclust()` function, and the `hclust()` method has different options to calculate the distances between the clusters and between the variables. There are five distance methods available: single linkage, average linkage, complete linkage, centroid, and ward. The average and centroid methods are the most popular. We will use Euclidian for observations and average linkage for between clusters methods.

```
> ##### Hierachical clustering
> dist_obs_grades<-dist(grades_df2, method="euclidian")
> cluster_hier<-hclust(dist_obs_grades,method="average")
```

Optimize the cluster using the `NbCLust()` function. To learn more about `NbClust()`, please read the documentation. See Figure 13-11.

```
> optimal_cluster<-NbClust(data=grades_df2, distance="euclidean",
+                          min.nc=3, max.nc=15,
+                          method="average")
*** : The Hubert index is a graphical method of determining the number
of clusters. In the plot of Hubert index, we seek a significant knee
that corresponds to a significant increase of the value of the measure
i.e the significant peak in Hubert index second differences plot.

*** : The D index is a graphical method of determining the number of
clusters. In the plot of D index, we seek a significant knee (the
significant peak in Dindex  second differences plot) that corresponds
to a significant increase of the value of the measure.

*******************************************************************
* Among all indices:
* 5 proposed 3 as the best number of clusters
* 3 proposed 4 as the best number of clusters
* 6 proposed 5 as the best number of clusters
* 1 proposed 8 as the best number of clusters
* 1 proposed 9 as the best number of clusters
* 3 proposed 10 as the best number of clusters
* 1 proposed 13 as the best number of clusters
```

```
* 1 proposed 14 as the best number of clusters
* 3 proposed 15 as the best number of clusters

                   ***** Conclusion *****

* According to the majority rule, the best number of clusters is  5

*******************************************************************
```

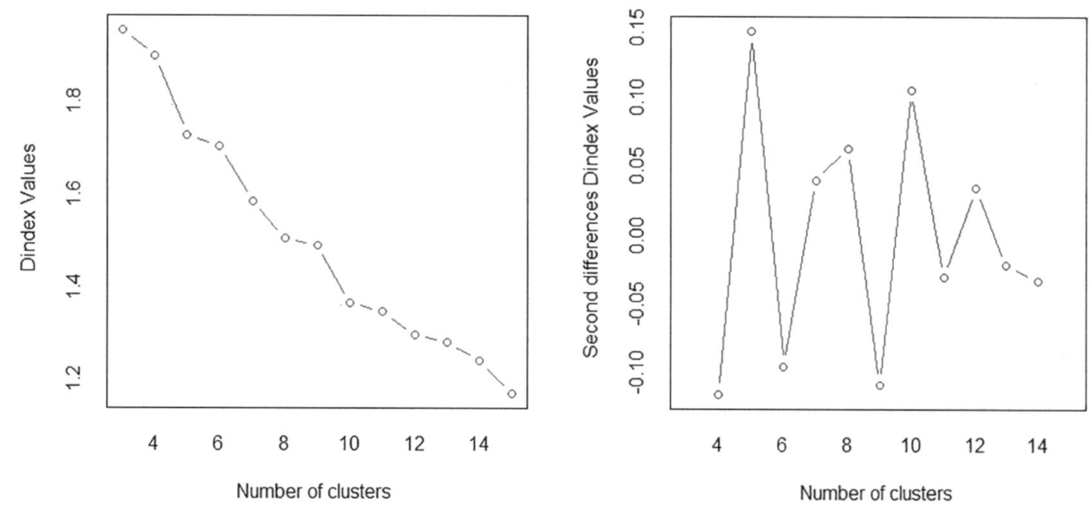

Figure 13-11. *Finding optimal clusters in the hierarchical clustering method*

Finally, we plot the dendrograms to represent the clusters, as shown here:

```
> require(factoextra)
> fviz_dend(x = cluster_hier,
+           rect = TRUE,
+           cex = 0.5, lwd = 0.6,
+           k = 5,
+           k_colors = c("purple","red",
+           "green3", "blue", "magenta"),
+           rect_border = "gray",
+           rect_fill = FALSE)
Warning message:
`guides(<scale> = FALSE)` is deprecated. Please use `guides(<scale> =
"none")` instead.
```

The dendrogram in Figure 13-12 shows how the clusters have grouped students' based on their performance in various assignment components. The graph is hard to read and interpret because of too many observations on the x-axis. The bigger the area of the plot, the better the visual representation. For example, if we display this on a 55- or 65-inch TV screen, the graph may look bigger, and it would be easier to read some of the x-axis values. If we display it on a 10ft by 10ft area, the area of plot is even bigger, and we would probably be able to read all the x-axis values. But, practically, this is not possible. Hence, we develop a better visualization program to zoom the plot and understand the details. This includes adding a horizontal and vertical scroll bar, adding a zoom mechanism, or simply plotting the limited values on the x- and y-axis.

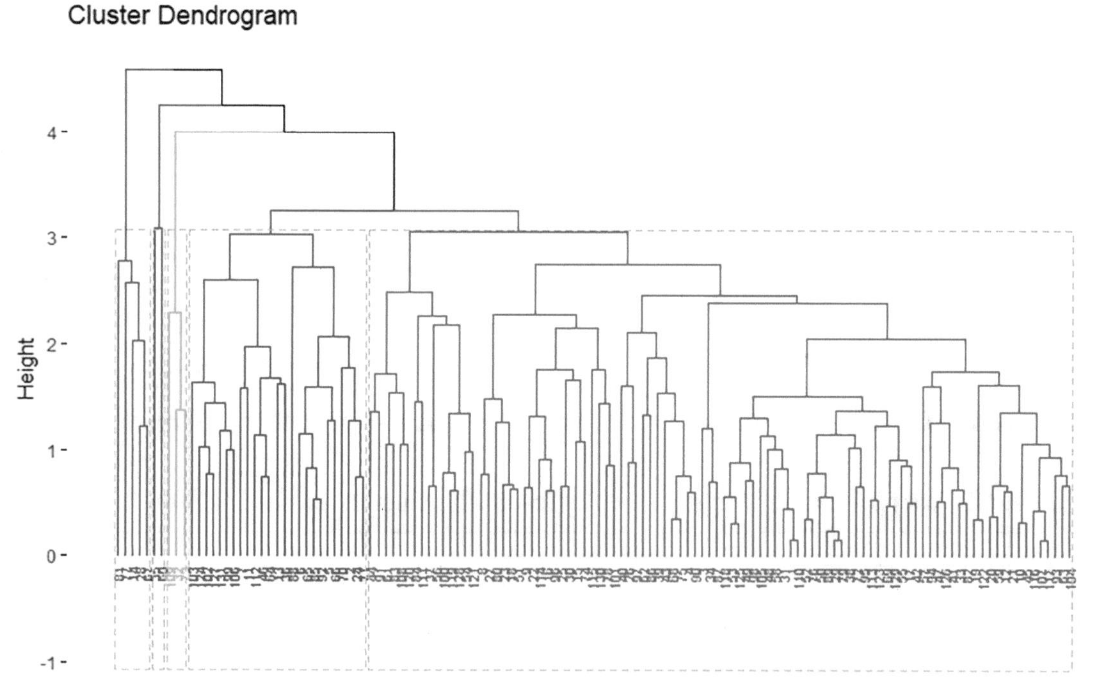

***Figure 13-12.** Dendrogram showing the number of hierarchical clusters*

13.6 Clustering Using Python sklearn()

We will repeat the same exercise using Python and the `sklearn()` libraries. We can create the clustering models using Python in exactly the same way we did for R. The steps and the process for creating the model remain the same. First we create the k-means clustering, and then we create the hierarchical clustering. In both cases, we have to find the optimal value of the k-number of clusters using the elbow method and the silhouette method, which were explained earlier.

Step 1: Import all the necessary libraries to the development environment and read the data. The k-means clustering model aims to create clusters based on students' performance in the different components of assignments and exams. The data contains `StudentID`, `Quiz1`, `Quiz2`, `Quiz3`, `Quiz4`, and `Qquiz5` variables. The goal is to group students into different clusters based on their performance and assign grades to each student.

```
#Step 1: Import all the libraries required to the jupyter environment
from __future__ import absolute_import, division, print_function
import codecs
import glob
import logging
import multiprocessing
import os
import pprint
import re

##Load pandas and numpy library
import pandas as pd
import numpy as np

## Loading plot libraries
%matplotlib inline
import matplotlib.pyplot as plt
from pandas.plotting import scatter_matrix
from sklearn.metrics.pairwise import cosine_similarity
from scipy.spatial.distance import cosine
import sklearn as sk
import matplotlib.pyplot as plt
import seaborn as sns

##sklearn() clustering libraires
from sklearn import cluster
from scipy.cluster import hierarchy
```

```
# sklearn() scaler() function to scale data
from sklearn.preprocessing import MinMaxScaler
from sklearn.preprocessing import StandardScaler
# TO learn more about the function, please refer to sklearn()
documentation
#https://scikit-learn.org/stable/modules/generated/
sklearn.preprocessing.MinMaxScaler.html
#https://scikit-learn.org/stable/modules/generated/
sklearn.preprocessing.StandardScaler.html
#https://scikit-learn.org/stable/modules/clustering.html
```

Step 2: Read the data.

```
#Step 2: Reading Data
#Set the working directory where the data set is stored before reading
the data set
data_dir = 'E:/Umesh-MAY2022/Personal-May2022/BA2ndEdition/2ndEdition/
Book Chapters/Chapter 13 - Clustering/scripts'
os.chdir(data_dir)
grades_df = pd.read_csv("grades.csv")
grades_df.head()
   StudentID  Quiz1  Quiz2  Quiz3  Quiz4  Quiz5
0   20000001     10   10.0   27.0     95      8
1   20000002      8    3.0   23.5     75      9
2   20000003     14   15.0   28.0     70      8
3   20000004     12   16.5   28.0    100      9
4   20000005     13   11.5   21.5     95      9
```

Step 3: For the data exploration and data preparation, we remove `StudentID` since it is not required for the clustering analysis. Then we scale all the other variables.

```
#Step 3: Data exploration, data preparation
##Removing 'StudentID' column, which is not required for the clustering
analysis
# Scale all the variables
grades_df2 = grades_df.drop('StudentID',axis=1)
scaler = MinMaxScaler()
```

```
grades_scaled = pd.DataFrame(scaler.fit_transform(grades_df2),
                             columns=grades_df2.columns)
grades_scaled.head()

     Quiz1  Quiz2     Quiz3  Quiz4  Quiz5
0 0.666667   0.20 0.727273    0.9    0.0
1 0.533333   0.06 0.515152    0.5    0.5
2 0.933333   0.30 0.787879    0.4    0.0
3 0.800000   0.33 0.787879    1.0    0.5
4 0.866667   0.23 0.393939    0.9    0.5
```

Step 4: Create the k-means clustering model. Choose K randomly and then optimize the k-means. Please read the documentation to understand the input parameters. Here, we are setting only the `n_cluster`, `random_state`, and `n_init` parameters, and the rest of the values are kept as the default.

Parameters (from sklearn() libraries)

n_clusters*int, default=8*

The number of clusters to form as well as the number of centroids to generate.

init*{'k-means++', 'random'}, callable or array-like of shape (n_clusters, n_features), default='k-means++'*

Method for initialization:

`k-means++`: Selects the initial cluster centers for k-mean clustering in a smart way to speed up convergence. See "Notes in k_init" for more details.

`random`: Choose `n_clusters` observations (rows) at random from data for the initial centroids.

If an array is passed, it should be of shape (n_clusters, n_features) and gives the initial centers.

If a callable is passed, it should take arguments X, n_clusters, and a random state and return an initialization.

n_init*int, default=10*

Number of times the k-means algorithm will be run with different centroid seeds. The final results will be the best output of `n_init` consecutive runs in terms of inertia.

max_iter*int, default=300*

Maximum number of iterations of the k-means algorithm for a single run.

tol*float, default=1e-4*

Relative tolerance with regard to Frobenius norm of the difference in the cluster centers of two consecutive iterations to declare convergence.

verbose*int, default=0*

Verbosity mode.

random_state*int, RandomState instance or None, default=None*

Determines random number generation for centroid initialization. Use an int to make the randomness deterministic.

copy_x*bool, default=True*

When precomputing distances, it is more numerically accurate to center the data first. If `copy_x` is `True` (the default), then the original data is not modified. If `False`, the original data is modified and put back before the function returns, but small numerical differences may be introduced by subtracting and then adding the data mean. Note that if the original data is not C-contiguous, a copy will be made even if `copy_x` is `False`. If the original data is sparse but not in CSR format, a copy will be made even if `copy_x` is `False`.

algorithm*{"lloyd", "elkan", "auto", "full"}, default="lloyd"*

K-means algorithm to use. The classical EM-style algorithm is `lloyd`. The `elkan` variation can be more efficient on some data sets with well-defined clusters, by using the triangle inequality. However, it's more memory intensive due to the allocation of an extra array of shape `(n_samples, n_clusters)`.

auto and full are deprecated, and they will be removed in Scikit-Learn 1.3. They are both aliases for lloyd.

Changed in version 0.18: Added Elkan algorithm.

Changed in version 1.1: Renamed full to lloyd and deprecated auto and full. Changed auto to use lloyd instead of elkan.

```
# Create the K-Means cluster model
# From the model, identify which studentID is in which cluster
kmm_model = cluster.KMeans(n_clusters=5, max_iter=50, n_init=5, random_
state = 10)
kmm_model.fit(grades_scaled)
grades_clusters = pd.DataFrame(kmm_model.labels_ ,
                               columns=['Cluster ID'],
                               index = grades_df.StudentID)
```

Step 5: Check the cluster model output and display the cluster allocation to the StudentIDs.

```
#Print the cluster centroids
pd.DataFrame(kmm_model.cluster_centers_,
             columns=grades_df2.columns)

      Quiz1     Quiz2     Quiz3     Quiz4         Quiz5
0  0.608696  0.264348  0.565217  0.752174  5.217391e-01
1  0.837500  0.393750  0.734848  0.931250  7.187500e-01
2  0.852482  0.306596  0.710509  0.885106  1.110223e-16
3  0.685714  0.283333  0.474747  0.395238 -2.775558e-17
4  0.497222  0.326250  0.540404  0.895833 -2.775558e-17

grades_clusters = pd.DataFrame(kmm_model.labels_ ,
                               columns=['Cluster ID'],
                               index = grades_df.StudentID)
grades_clusters.head(10)
```

```
          Cluster ID
StudentID
20000001           2
20000002           0
20000003           3
20000004           1
20000005           0
20000006           0
20000007           0
20000008           3
20000009           2
20000010           2
```

Step 6: Find the optimal value of k using the elbow method, as shown in Figure 13-13.

```
# To determine the value of K, number of clusters in the data,
# We can use elbow method or "silhouette" method
# We will use the "Elbow" method
# we use k-values from 1 to 10 and compute their
# corresponding sum-of-squared errors (SSE)
# as shown in the example below.
# Find the "elbow" from the plot of SSE versus
# number of clusters
def k_elbowFunc(ks, data_df):
    error = []
    for k in ks:
        kmm_mods = cluster.KMeans(n_clusters=k, max_iter=50, n_init=5,
                   random_state = 10)
        kmm_mods.fit(data_df)
        error.append(kmm_mods.inertia_)
    plt.plot(ks, error)
    plt.xlabel('Number of Clusters')
    plt.ylabel('error SSE')

kmm_values = [1,2,3,4,5,6,7,8,9,10]
k_elbowFunc(kmm_values, grades_scaled)
```

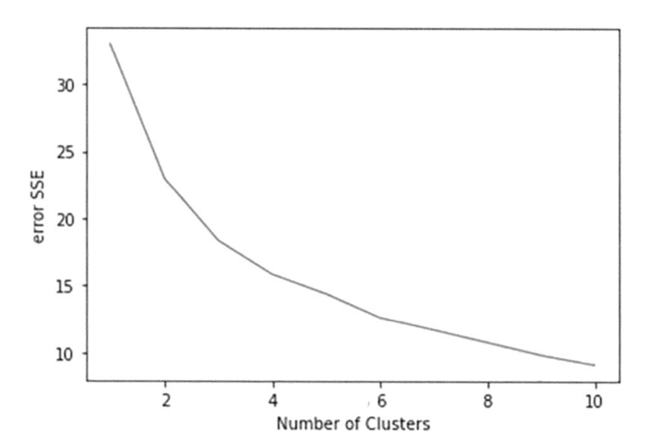

***Figure 13-13.** Finding the optimal value of k using the elbow method*

Note The challenge of the unsupervised clustering method is determining the optimal number of clusters. The elbow and silhouette are the two most common methods used to determine the optimal value of k. The elbow plot shows the cluster errors on the y-axis for the number of k values. In the elbow plot, you want to see a sharp decline of errors from one k value to another rather than a more gradual decrease in slope. The "good" k value would be the last value before the slope of the plot levels off.

Similarly, the silhouette measure provides how distinct the two clusters are, that is, how well the two clusters are separated. It has an arrangement of [-1, 1]. The y-axis on the plot provides the silhouette width for different k values (x-axis). Silhouette coefficients close to +1 indicate that the sample is far away from the neighboring clusters (this is what we want), and a 0 value or near 0 value indicates that the sample is very close to the decision boundary between the two neighboring clusters. A silhouette of negative values indicates that the data samples have been in the wrong cluster.

Step 7: From Figure 13-13, the optimal value of k is 2 or even 4, so we can consider and rebuild the model.

```
# Create the K-Means cluster model with optimized K =3
# From the model, identify which studentID is in which cluster
kmm_model_k3 = cluster.KMeans(n_clusters=3, max_iter=50, n_init=5,
                              random_state = 10)
kmm_model_k3.fit(grades_scaled)
```

```
grades_clusters = pd.DataFrame(kmm_model_k3.labels_ ,
                               columns=['Cluster ID'],
                               index = grades_df.StudentID)
grades_clusters.head(10)
          Cluster ID
StudentID
20000001           1
20000002           3
20000003           0
20000004           3
20000005           3
20000006           3
20000007           3
20000008           0
20000009           1
20000010           1
```

Plotting anything in Python is always challenging. Unlike in R, you have to write explicit code to plot the desired graphs. We know only how to plot two-dimensional graphs; hence, one way is to reduce the number of components in the data using PCA (Principal Component Analysis) and fit the two PCA components to the clusters, as shown in Figure 13-14. As you can see from the legend, the data is clustered into three different clusters: 0, 1, and 2. After a manual examination of the cluster contents, you should label the clusters with more meaningful names.

```
from sklearn.manifold import MDS
import seaborn as sns
pca_embed = MDS(n_components=2)
pca_mds = pd.DataFrame(pca_embed.fit_transform(grades_scaled),
            columns = ['pca_1','pca_2'])
pca_mds['K_cluster'] = kmm_model_k3.predict(grades_scaled)
sns.scatterplot(data=pca_mds,x = "pca_1",y="pca_2",hue="K_cluster")
<AxesSubplot:xlabel='pca_1', ylabel='pca_2'>
```

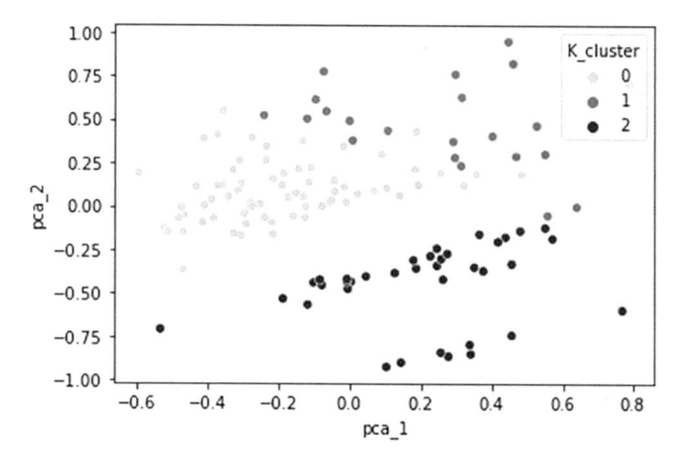

***Figure 13-14.** Optimal k-means clustering graphical representation*

The next section explains the hierarchical clustering models and how to create them using the sklearn `hierarchy()` function. It is pretty straightforward. Since we already explained earlier in theory how hierarchical clustering works, we will just present the code here. As you learned earlier, two measures are important. The first measure is the distance among different observations in every variable, and the other is the distance between the clusters. Dendrograms show the cluster formation for a specific distance measure. See Figure 13-15.

```
#Hierarchical Clustering
# Create clustering using Single linkage
cluster_hier  = hierarchy.linkage(grades_scaled, 'single')
cluster_dendo = hierarchy.dendrogram(cluster_hier, orientation='right',
                                     labels = grades_df['StudentID'].to_list())
```

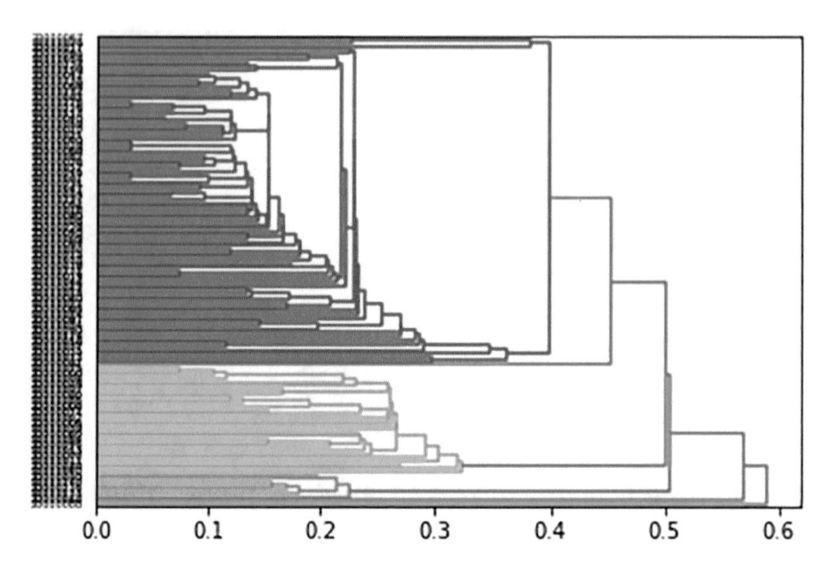

Figure 13-15. *Hierarchical clustering Dendrograms representation using single linkage*

This is hierarchical clustering using average linkage.

```
#Hierarchical Clustering
# Create clustering using "Group Average linkage"
cluster_hier  = hierarchy.linkage(grades_scaled, 'average')
cluster_dendo = hierarchy.dendrogram(cluster_hier, orientation='left',
           labels = grades_df['StudentID'].to_list())
```

The dendrogram in Figure 13-16 shows how the clusters have grouped students based on their performance in various assignment components. The graph is hard to read and interpret as the number of observations becomes large. The bigger the area of the plot, the better the visual representation. If we display the same graph on a 55- or 65-inch TV screen, the graph may look bigger and more readable for some of the x-axis values. If we display the same on a 10ft by 10ft area, it is even bigger, and all the x-axis values would be able to be read. But, practically, this is not possible. Hence, we have to develop visualization program to zoom the plot to understand the details. This includes adding a horizontal and vertical scroll bar, adding a zoom mechanism, or simply plotting the limited values on the x- and y-axes.

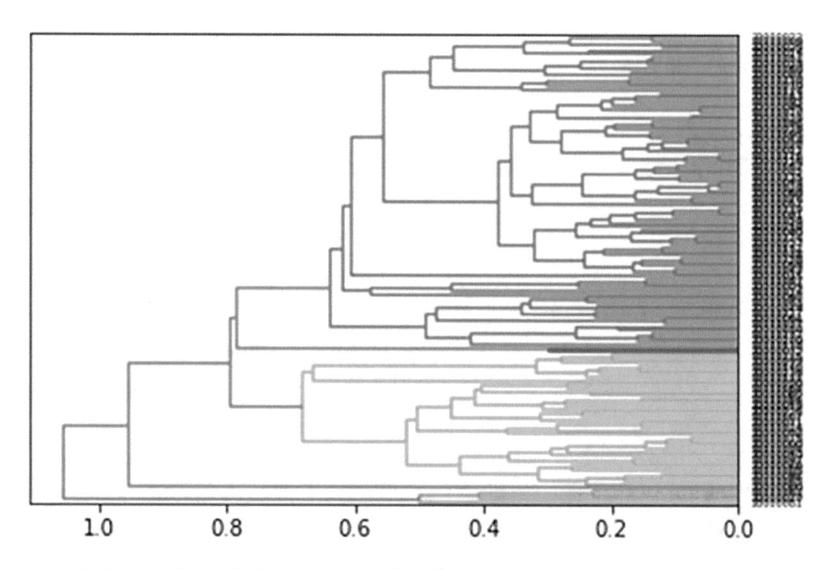

Figure 13-16. *Hierarchical clustering dendrogram representation using average linkage*

Various other clustering methods are in use depending on the use case and applications. The following table lists the clustering methods supported by the `sklearn()` library:

Method Name	Parameters	Scalability	Use Case	Geometry (Metric Used)
`K-means`	Number of clusters	Very large n_samples, medium n_clusters with `MiniBatch code`	General-purpose, even cluster size, flat geometry, not too many clusters, inductive	Distances between points
`Affinity propagation`	Damping, sample preference	Not scalable with n_samples	Many clusters, uneven cluster size, nonflat geometry, inductive	Graph distance (e.g., nearest-neighbor graph)

(continued)

Method Name	Parameters	Scalability	Use Case	Geometry (Metric Used)
Mean-shift	Bandwidth	Not scalable with n_samples	Many clusters, uneven cluster size, nonflat geometry, inductive	Distances between points
Spectral clustering	Number of clusters	Medium n_samples, small n_clusters	Few clusters, even cluster size, nonflat geometry, transductive	Graph distance (e.g., nearest-neighbor graph)
Ward hierarchical clustering	Number of clusters or distance threshold	Large n_samples and n_clusters	Many clusters, possibly connectivity constraints, transductive	Distances between points
Agglomerative clustering	Number of clusters or distance threshold, linkage type, distance	Large n_samples and n_clusters	Many clusters, possibly connectivity constraints, non Euclidean distances, transductive	Any pairwise distance
DBSCAN	Neighborhood size	Very large n_samples, medium n_clusters	Nonflat geometry, uneven cluster sizes, outlier removal, transductive	Distances between nearest points
OPTICS	Minimum cluster membership	Very large n_samples, large n_clusters	Nonflat geometry, uneven cluster sizes, variable cluster density, outlier removal, transductive	Distances between points
Gaussian mixtures	Many	Not scalable	Flat geometry, good for density estimation, inductive	Mahalanobis distances to centers

(*continued*)

Method Name	Parameters	Scalability	Use Case	Geometry (Metric Used)
BIRCH	Branching factor, threshold, optional global cluster	Large n_clusters and n_samples	Large data set, outlier removal, data reduction, inductive	Euclidean distance between points
Bisecting k-means	Number of clusters	Very large n_samples, medium n_clusters	General-purpose, even cluster size, flat geometry, no empty clusters, inductive, hierarchical	Distances between points

13.7 Chapter Summary

In this chapter, you learned what unsupervised learning is and what clustering analysis is.

You also looked at various clustering analysis techniques, including hierarchical clustering and nonhierarchical clustering.

You learned the various distance measures used for creating clusters including Euclidian distance, Manhattan distance, single linkage, average linkage, etc.

You also learned how to create k-means and `hclust()` models using both R and Python and also selecting optimal value of K for the right number of clusters.

CHAPTER 14

Relationship Data Mining

The association rule, also called *market basket analysis*, is the most popular model to find the relationship among the transactional data. In this chapter, we will discuss the association rule and modeling technique to find the buying relationship patterns.

14.1 Introduction

The growth of e-commerce, digital transactions, and the retail industry has led to the generation of a humongous amount of data and an increase in database size. The customer transactional information and the relationships between the transactions and customer buying patterns are hidden in the data. Traditional learning and data mining algorithms that exist may not be able to determine such relationships. This has created an opportunity to find new and faster ways to mine the data to find meaningful hidden relationships in the transactional data. To find such associations, the association rule algorithm was developed. Though many other algorithms have been developed, the apriori algorithm introduced in 1993 by Srikanth and Agarwal (AIS93) is the most prominent one. The apriori algorithm mines data to find association rules in a large real-world transactional database. This method is also referred to as *association-rule* analysis, *affinity analysis*, *market-basket analysis* (MBA), or *relationship data mining*.

The association rule analysis is used to find out "which item goes with what item." This association is used in the study of customer transaction databases, as shown in Figure 14-1. The association rules provide a simple analysis indicating that when an event occurs, another event occurs with a certain probability. Knowing such probability and discovering such relationships from a huge transactional database can help companies manage inventory, product promotions, product discounts, the launch of new products, and other business decisions. Examples are finding the relationship between phones and phone cases, determining whether customers who purchase a mobile phone also purchase a screen guard, or seeing whether a customer buys milk

U. R. Hodeghatta and U. Nayak, *Practical Business Analytics Using R and Python*,
https://doi.org/10.1007/978-1-4842-8754-5_14

and pastries together. Based on such association probability, stores can promote the new product, sell extra services, or sell additional products at a promotional price. Such analysis might encourage customers to buy a new product at a reduced price and might encourage companies to increase sales and revenues. Association rules are probabilistic relationships using simple if-then rules computed from the data.

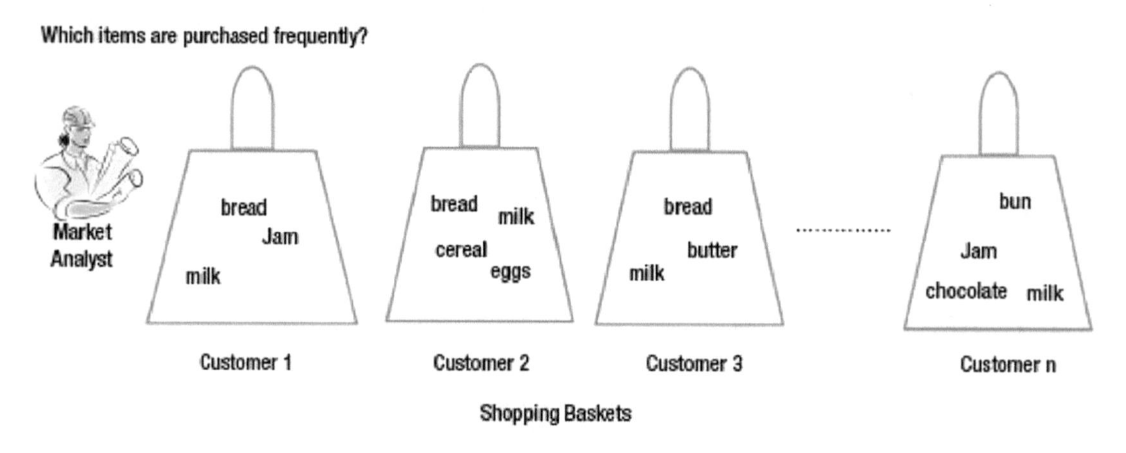

Figure 14-1. *Transactional data and relationship mining*

Association rules fall under unsupervised learning and are used for the discovery of patterns rather than the prediction of an outcome. Though the association rules are applied for transactional data, the rules can also be applied in other areas such as biomedical research to find patterns of DNA, find insurance fraud patterns, find credit card transaction frauds, etc.

The apriori algorithm (Agrawal and Srikant, 1995) generates frequent-item sets. The algorithm begins with just one item and then generates a two-item set with two items frequently purchased together, and then it moves on to three-item sets with three items frequently purchased together, and so on, until all the frequent-item sets are generated. The key idea is to generate frequent-item sets with one item and then generate two-item sets, three-item sets, and so on, until all the items are covered in the transactional database. Once the list of all frequent-item sets is generated, you can find out how many of those frequent-item sets are in the database. In general, generating n-item sets uses the frequent $n - 1$ item and a complete run through the database once. The apriori algorithm is faster and more efficient even for a large database with many items.

The apriori rules are derived based on the association of frequent items in the data. Transaction data provides the knowledge of frequent items in the data. By having such knowledge, we can generate rules and efficiently reduce the number of frequent items of interest. The apriori algorithm is based on the assumption that "the entire subset of a frequent item set must also be frequent." The item set is frequent only if all of its subsets, pairs, triples, and singles occur frequently and are considered "interesting" and "valuable." Before discussing how to generate the association rules, to discover the most frequent patterns, we will discuss the support, confidence, and lift metrics used to deduce the associations and rules.

14.2 Metrics to Measure Association: Support, Confidence, and Lift

Let "T" be a database of all the transactions.

Let "I" be a set of all items in a store, I = { i_1, i_2, i_3,....i_m}.

Let "D" be the data set of all the item sets.

All the items in "I" are from a set of all transactional databases, T.

Each t_i is a set of items t such that t ε I. For each transaction t_i, a transaction ID is assigned.

$I_A \rightarrow I_B$, (Item B --> Item A) means Item B is purchased after Item A, or I_B follows I_A, where A and B are item sets.

The first part is called the *antecedent*, and the second part is called the *consequent*.

For example, in the following example, one item set, I{chocolate), is purchased after a two-item set, I{Milk, Jam}, is purchased:

I{Milk, Jam} ➤ I{chocolate}

14.2.1 Support

Support (S) is the fraction of transactions that contain both A and B (antecedent and consequent). The *support* is defined as the number of transactions that include both the antecedent and the consequent item sets. It is expressed as a percentage of the total number of records in the database.

Support(A & B) = Freq(A & B) / N (where N is the total number of transactions in database)

For example, if the two-item set {Milk, Jam} in the data set is 5 out of a total of 10 items, then *Support* = S = 5/10 = 50%.

14.2.2 Confidence

Confidence (A --> B) is a ratio of *support for* A and B to the *support for* A. It is expressed as a ratio of the number of transactions that contain A and B together to the number of transactions that contain A.

$$\begin{aligned}\text{Conf}(\text{A} -> \text{B}) &= \frac{\text{Trans}(\text{A,B})}{\text{Trans}(\text{A})} \\ &= \text{P}(\text{B} \mid \text{A})\end{aligned}$$

Though support and confidence are good measures to show the strength of the association rule, sometimes it can be deceptive. For example, if the antecedent or the consequent have high support, it can have high confidence even though both are independent.

14.2.3 Lift

Lift is a measure when the occurrence of the consequent item in a transaction is independent of the occurrence of the antecedents. It gives a better measure to compare the strength of the association. Mathematically, *lift* is defined as follows:

$$lift(A \rightarrow B) = \frac{conf(A \rightarrow B)}{p(B)} = \frac{\frac{p(A \cap B)}{p(A)}}{p(B)} = \frac{p(A \cap B)}{p(A)p(B)}$$

In other words, Lift(A --> B) = Support(A & B) / [Support(A) × Support(B)]

We will calculate the support, confidence, and lift ratios for the following example:

Transaction 1: Shirt, pant, tie, belt

Transaction 2: Shirt, belt, tie, shoe

Transaction 3: Socks, tie, shirt, jacket

Transaction 4: Pant, tie, belt, blazer

Transaction 5: Pant, tie, hat, sweater

For the first case, calculate support, confidence, and lift for Shirt-> tie; the antecedent is shirt, and the consequent is tie. Whenever someone buys shirt, they may also buy a tie.

There are five transactions in total. Out of five transactions, there are three transactions that have shirt -> tie.

Support(A & B) = Freq(A & B) / N (where N is the total number of transactions in database)

Support (shirt->tie) = 3/5 = 0.6

Confidence(A -->B) = Support(A & B) / Support(A) = Freq(A & B) / Freq(A)

Confidence (shirt->tie) = 3/3 = 1

Lift(A -->B) = Support(A & B) / [Support(A) × Support (B)]

Lift = (3/5) / (3/5 * 5/5) = 1

Figure 14-2 lists the other combinations' support, confidence, and lift.

Rule	Support(A&B)	Confidence(A→B)	Lift(A→B)
shirt → tie	3/5 = 0.6	3/3 = 1	(3/5)/[(3/5)*(5/5)]=1
socks → shirt	1/5 = 0.2	1/1 = 1	(1/5)/[(1/5)*(3/5)]=5/3= 1.67
pant & tie → belt	2/5 = 0.4	2/3 = 0.67	(2/5)/[(3/5)*(3/5)]=1.11

Figure 14-2. *Calculation of support, confidence, and lift example*

Go ahead and calculate the support, confidence and lift measures for the other examples, socks -> shirt (pant, tie) -> belt.

14.3 Generating Association Rules

The apriori algorithm generates rules for *k* products as follows:

1. The user sets a minimum support criterion.
2. Generate a list of one-item sets that meets the support criterion.
3. Use the list of one-item sets to generate a list of two-item sets that meets the support criterion.
4. Use the two-item list to generate a three item-list, and so on.

5. Continue the process through *k*-item sets.
6. Decide the final rule based on support, confidence, and lift.

For example, there are four transactions in database D with Transaction ID 200, 201, 202, 203 with items pencil, pen, eraser, scale and notebook. For simplicity, we will give numbers to these items as 1, 2, 3, and 4 in sequence.

The first step is to set a support value. We will set it to 50%, so support S = 50%.

The next step is to generate a one-item set, two-item set, three-item set, etc., based on the support.

There are five items; hence, we have five one-item sets and seven two-item sets, as shown in Figure 14-3.

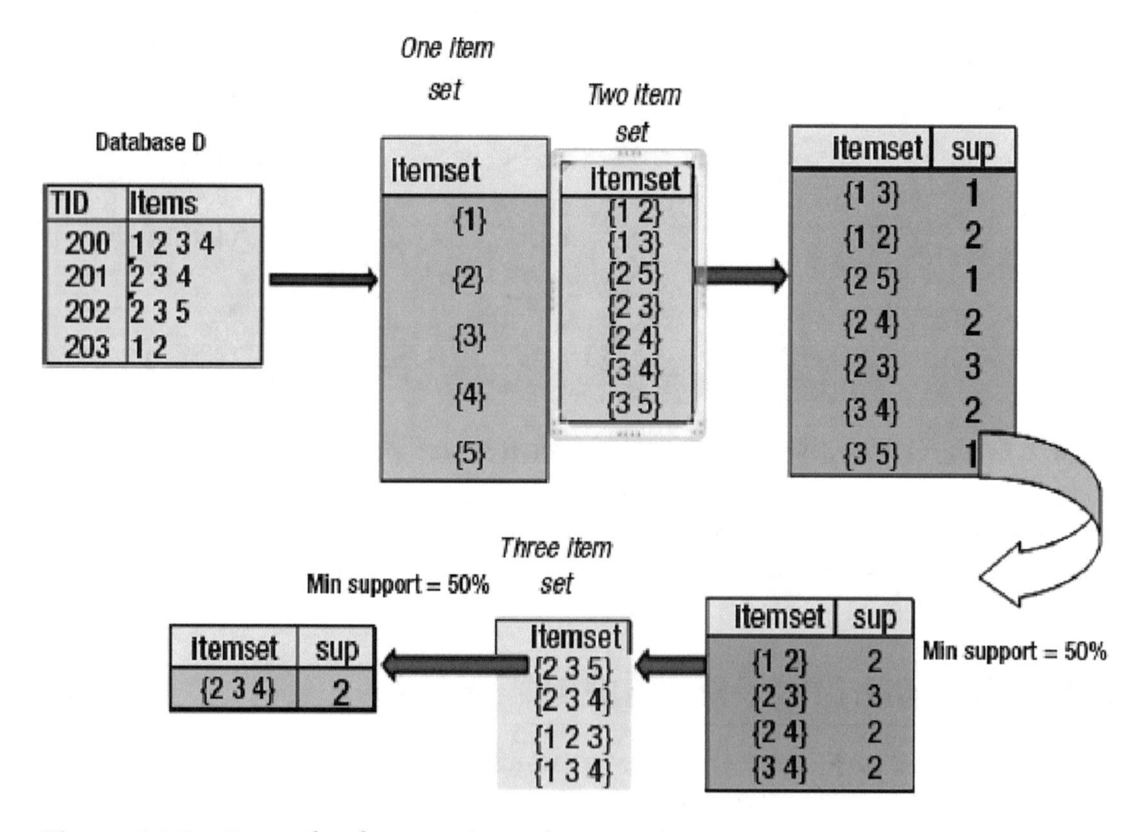

***Figure 14-3.** Example of generating a frequent-item set*

Out of seven two item sets, there are only four items that meet the support criterion (50 percent support). They are {1,2}, {2,3}, {2,4}, and {3,4}. The next step is to generate {3} item sets from this item set. {2,3,5}, {2,3,4}, {1,2,3}, and {1,3,4} are four possible {3} item sets. Out of these four {3 item sets}, only one meets the support criterion: item set {2,3,4}.

Once we generate the rules, the goal is to find the rules that indicate a strong association between the items and indicate dependencies between the antecedent (previous item) and the consequent (next item) in the set.

The support gives you an indication of overall transactions and how they affect the item sets. If you have only a small number of transactions with minimum support, the rule may be ignored. The lift ratio provides the strength of the consequent in a random selection. But the confidence gives the rate at which a consequent can be found in the database. Low confidence indicates a low consequent rate, and deciding whether to promote the consequent is a worthwhile exercise. The more records, the better the conclusion. Finally, the more distinct the rules that are considered, the better the interpretation and outcome. We recommend looking at the rules from a top-down approach rather than automating the decision by searching thousands of rules.

A high value of confidence suggests a strong association rule. But when B is independent of A—that is, $p(B) = p(B \mid A)$—and $p(B)$ is high, then we'll have a rule with high confidence. For example, if p("buy pen") = 85 percent and is independent of "buy pencil," then the rule "buy pen" ⇒ "buy pencil" will have a confidence of 85 percent. If nearly all customers buy *pen* and nearly all customers buy *pencil*, then the confidence level will be high regardless of whether there is an association between the items. Similarly, if *support* is very low, it is not worth examining.

Note that association rules do not represent causality or correlation between the two items. A --> B does not mean B causes A or, no causality, and A --> B can be different from B --> A, unlike correlation.

14.4 Association Rule (Market Basket Analysis) Using R

In this example, we use a transactional database that consists of customer transactional data. The database consists of customer transactions of 14 different items purchased by customers, as shown in Table 14-1. Each transaction has a transactional ID. If a customer has purchased a certain item, it is recorded as 1; otherwise, it is recorded as 0. There are 1,000 total transactions in the database that will be used to find the association rule using the apriori algorithm. First, we demonstrate how to perform apriori using R; then we use the same database to do the same thing using Python.

***Table 14-1.** Customer Transactional Database*

A	B	C	D	E	F	G	H	I	J	K	L	M	N	O	P
Trans. Id	Belt	Shoe	Perfume	Dress	Shirt	Jackets	Trouser	Tie	Wallet	TravelBag	NailPolish	Socks	Hats	Fitbit	
1	0	1	1	1	1	0	1	1	1	0	0	0	0	1	
2	0	0	1	0	1	0	1	1	0	0	1	1	0	0	
3	0	1	0	0	1	1	1	1	1	1	1	1	1	0	
4	0	0	1	1	1	0	1	0	0	0	1	0	0	1	
5	0	1	0	0	1	0	1	1	1	1	0	1	1	0	
6	0	0	0	0	1	1	0	0	0	0	0	0	0	1	
7	0	1	1	1	1	0	1	1	1	1	1	1	0	0	
8	0	0	1	1	0	0	1	0	1	1	0	1	0	0	
9	0	0	0	0	1	0	0	0	0	0	1	0	1	0	
10	1	1	1	1	0	0	0	0	1	1	0	0	0	0	
11	0	0	1	0	0	0	1	0	0	0	0	1	1	1	
12	0	0	1	1	1	0	1	0	1	1	1	0	0	0	
13	0	1	0	0	1	0	0	1	1	1	0	1	0	0	
14	0	1	1	1	1	0	1	1	1	1	1	1	1	0	
15	0	1	0	0	1	0	1	1	1	1	1	1	1	1	
16	0	0	1	1	1	0	1	1	1	1	1	1	1	1	
17	1	1	1	1	1	0	1	1	1	1	0	0	0	1	
18	0	0	0	0	1	0	0	0	0	0	1	1	1	1	

Step 1: Read the database files.

The first step is to read the database files. In this case, we are reading the CSV file type. We read it as an R data frame and view the results using the `head()` function.

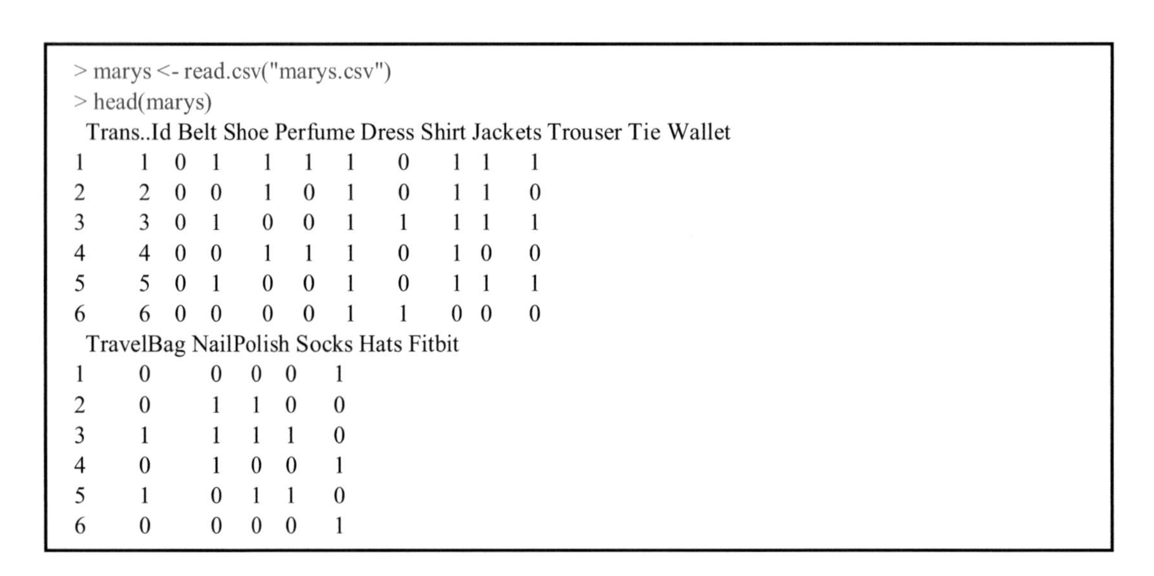

```
> marys <- read.csv("marys.csv")
> head(marys)
  Trans..Id Belt Shoe Perfume Dress Shirt Jackets Trouser Tie Wallet
1         1    0    1       1     1     1       0       1   1      1
2         2    0    0       1     0     1       0       1   1      0
3         3    0    1       0     0     1       1       1   1      1
4         4    0    0       1     1     1       0       1   0      0
5         5    0    1       0     0     1       0       1   1      1
6         6    0    0       0     0     1       1       0   0      0
  TravelBag NailPolish Socks Hats Fitbit
1         0          0     0    0      1
2         0          1     1    0      0
3         1          1     1    1      0
4         0          1     0    0      1
5         1          0     1    1      0
6         0          0     0    0      1
```

***Figure 14-4.** Step 1, reading data*

Step 2: Preprocess the data.

The second step in the process of data mining is to check the data types, check for any missing values, and prepare the data for the input library function. However, we have to drop the `Trans.ID` column as it does not add any value to our analysis.

```
> str(marys)
'data.frame':           100 obs. of 15 variables:
 $ Trans..Id  : int 1 2 3 4 5 6 7 8 9 10 ...
 $ Belt       : int 0 0 0 0 0 0 0 0 0 1 ...
 $ Shoe       : int 1 0 1 0 1 0 1 0 0 1 ...
 $ Perfume    : int 1 1 0 1 0 0 1 1 0 1 ...
 $ Dress      : int 1 0 0 1 0 0 1 1 0 1 ...
 $ Shirt      : int 1 1 1 1 1 1 1 0 1 0 ...
 $ Jackets    : int 0 0 1 0 0 1 0 0 0 0 ...
 $ Trouser    : int 1 1 1 1 1 0 1 1 0 0 ...
 $ Tie        : int 1 1 1 0 1 0 1 0 0 0 ...
 $ Wallet     : int 1 0 1 0 1 0 1 1 0 1 ...
 $ TravelBag  : int 0 0 1 0 1 0 1 1 0 1 ...
 $ NailPolish : int 0 1 1 1 0 0 1 0 1 0 ...
 $ Socks      : int 0 1 1 0 1 0 1 1 0 0 ...
 $ Hats       : int 0 0 1 0 1 0 0 0 1 0 ...
 $ Fitbit     : int 1 0 0 1 0 1 0 0 0 0 ...
> sum(is.na(marys))
[1] 0
```

***Figure 14-5.** Step 2, checking data types*

Step 3: The `Apriori()` function accepts only `True` or `False` as inputs. Hence, convert all the data to logical data types. In this case, the data set is clean, and there is no need for any further data cleaning.

```
> # apriori() function accepts logical values and hence convert data to Logical
> marys_1 <- marys %>% mutate_if(is.numeric,as.logical)
> marys_2<-subset(marys_1, select = -c(Trans..Id))
> #str(marys_2)
> head(marys_2)
   Belt  Shoe Perfume Dress Shirt Jackets Trouser   Tie Wallet TravelBag NailPolish
1 FALSE  TRUE    TRUE  TRUE  TRUE   FALSE    TRUE  TRUE   TRUE     FALSE      FALSE
2 FALSE FALSE    TRUE FALSE  TRUE   FALSE    TRUE  TRUE  FALSE     FALSE       TRUE
3 FALSE  TRUE   FALSE FALSE  TRUE    TRUE    TRUE  TRUE   TRUE      TRUE       TRUE
4 FALSE FALSE    TRUE  TRUE  TRUE   FALSE    TRUE FALSE  FALSE     FALSE       TRUE
5 FALSE  TRUE   FALSE FALSE  TRUE   FALSE    TRUE  TRUE   TRUE      TRUE      FALSE
6 FALSE FALSE   FALSE FALSE  TRUE    TRUE   FALSE FALSE  FALSE     FALSE      FALSE
  Socks  Hats Fitbit
1 FALSE FALSE   TRUE
2  TRUE FALSE  FALSE
3  TRUE  TRUE  FALSE
4 FALSE FALSE   TRUE
5  TRUE  TRUE  FALSE
6 FALSE FALSE   TRUE
```

***Figure 14-6.** Step 3, converting to True or False*

Step 4: Find the frequent item sets and association rules using the `apriori()` algorithm with support, confidence, and lift. In this case, we have set support to 0.5 and confidence to 0.7.

```
> #Find frequent itemsets and association rules by applying apripori() algorithm
> #by setting support and confidence limits
> rules<-apriori(marys_2,
+             parameter = list(minlen=3, support=0.5, conf=0.7))
Apriori

Parameter specification:
 confidence minval smax arem  aval originalSupport maxtime support minlen maxlen
        0.7    0.1    1 none FALSE            TRUE       5     0.5      3     10
 target  ext
  rules TRUE

Algorithmic control:
 filter tree heap memopt load sort verbose
    0.1 TRUE TRUE  FALSE TRUE    2    TRUE

Absolute minimum support count: 50

set item appearances ...[0 item(s)] done [0.00s].
set transactions ...[14 item(s), 100 transaction(s)] done [0.00s].
sorting and recoding items ... [10 item(s)] done [0.00s].
creating transaction tree ... done [0.00s].
checking subsets of size 1 2 3 done [0.00s].
writing ... [9 rule(s)] done [0.00s].
creating S4 object  ... done [0.00s].
```

***Figure 14-7.** Step 4, Apriori() rules to find frequent patterns rules*

Step 5: List the set of association rules and explore the output rules.

```
> #Inspect the top 10 rules
> rules.sorted <- sort(rules, by="lift")
> inspect(head(rules.sorted, n=10, by="lift"))
```

Figure 14-8 shows both Input and Output of the association rules.

```
> inspect(head(rules.sorted, n=10, by="lift"))
    lhs                     rhs         support confidence coverage lift      count
[1] {Shirt,Wallet}       => {Shoe}      0.5     0.8333333  0.60     1.5151515 50
[2] {Shoe,Shirt}         => {Wallet}    0.5     1.0000000  0.50     1.4285714 50
[3] {Shirt,TravelBag}    => {Wallet}    0.5     1.0000000  0.50     1.4285714 50
[4] {Shirt,Wallet}       => {TravelBag} 0.5     0.8333333  0.60     1.3888889 50
[5] {Shirt,Trouser}      => {Wallet}    0.5     0.8333333  0.60     1.1904762 50
[6] {Shirt,Wallet}       => {Trouser}   0.5     0.8333333  0.60     1.1904762 50
[7] {Shoe,Wallet}        => {Shirt}     0.5     0.9090909  0.55     1.0695187 50
[8] {Trouser,Wallet}     => {Shirt}     0.5     0.9090909  0.55     1.0695187 50
[9] {Wallet,TravelBag}   => {Shirt}     0.5     0.8333333  0.60     0.9803922 50
> |
```

Figure 14-8. *Step 5, inspecting the rules()*

The `apriori()` algorithm created a set of frequent items and corresponding association rules for the given transactional data set. This set of association rules is based on the confidence of 0.7 and support of 0.5. The rules are sorted by "lift" values. From the previous rules, we can say that {Shirt, wallet} and Shoe have strong associations and are sold together. Similarly, the next association is {Shirt, Travelbag} and {wallet} and are sold together. This association helps to make certain business decisions such as inventory management or promoting new products {Shoe} or {Wallet} and thus managing stocks appropriately.

14.5 Association Rule (Market Basket Analysis) Using Python

In this example, we use a transactional database that consists of customer transactional data. The database consists of customer transactions of 14 different items purchased by customers, as shown in Table 14-2. Each transaction has a transactional ID. If a customer has purchased a certain item, it is recorded as 1; otherwise, it is recorded as 0. There are 1,000 total transactions in the database that will be used to find the association rule using the apriori algorithm. here, we are repeating the same exercise as earlier but using Python.

Table 14-2. *Customer Transactional Database*

Trans. Id	Belt	Shoe	Perfume	Dress	Shirt	Jackets	Trouser	Tie	Wallet	TravelBag	NailPolish	Socks	Hats	Fitbit
1	0	1	1	1	1	0	1	1	1	0	0	0	0	1
2	0	0	1	0	1	0	1	1	0	0	1	1	0	0
3	0	1	0	0	1	1	1	1	1	1	1	1	1	0
4	0	0	1	1	1	0	1	0	0	0	1	0	0	1
5	0	1	0	0	1	0	1	1	1	1	0	1	1	0
6	0	0	0	0	1	1	0	0	0	0	0	0	0	1
7	0	1	1	1	1	0	1	1	1	1	1	1	0	0
8	0	0	1	1	0	0	1	0	1	1	0	1	0	0
9	0	0	0	0	1	0	0	0	0	0	1	0	1	0
10	1	1	1	1	0	0	0	0	1	1	0	0	0	0
11	0	0	1	0	0	0	1	0	0	0	0	1	1	1
12	0	0	1	1	1	0	1	0	1	1	1	0	0	0
13	0	1	0	0	1	0	0	1	1	1	0	1	0	0
14	0	1	1	1	1	0	1	1	1	1	1	1	1	0
15	0	1	0	0	1	0	1	1	1	1	1	1	1	1
16	0	0	1	1	1	0	1	1	1	1	1	1	1	1
17	1	1	1	1	1	0	1	1	1	1	0	0	0	1
18	0	0	0	0	1	0	0	0	0	0	1	1	1	1

Step 1: Read the database files.

The first step is to read the database files. In this case, we are reading the CSV file type. We read it as a Pandas dataframe and view the results using the `head()` function. We import several libraries including Pandas, `apriori()`, and `association_rules()` to the development environment.

```
import os
import pandas as pd
import numpy as nd
import matplotlib.pyplot as plt
from mlxtend.frequent_patterns import apriori, association_rules
data_dir = 'C:/Personal/dataset'
os.chdir(data_dir)
marys_df = pd.read_csv('marys.csv')
marys_df.head()
  Trans. Id Belt Shoe Perfume Dress Shirt Jackets Trouser Tie \
0     1  0  1    1  1  1   0   1  1
1     2  0  0    1  0  1   0   1  1
2     3  0  1    0  0  1   1   1  1
3     4  0  0    1  1  1   0   1  0
4     5  0  1    0  0  1   0   1  1


  Wallet TravelBag NailPolish Socks Hats Fitbit
0  1     0     0  0  0   1
1  0     0     1  1  0   0
2  1     1     1  1  1   0
3  0     0     1  0  0   1
4  1     1     0  1  1   0
print(marys_df.shape)
(100, 15)
```

Figure 14-9. *Step 1, reading data*

Step 2: Preprocess the data.

The second step in data mining is to check the data types, check for any missing values, and prepare the data for the input library function. In this case, the data set is clean, and there is no need for any data cleaning. However, we have to drop the `Trans. ID` column as it does not add any value to our analysis.

```
marys_df2=marys_df.drop(columns = {'Trans. Id'})

marys_df2.head()

   Belt  Shoe  Perfume  Dress  Shirt  Jackets  Trouser  Tie  Wallet  \
0     0     1        1      1      1        0        1    1       1
1     0     0        1      0      1        0        1    1       0
2     0     1        0      0      1        1        1    1       1
3     0     0        1      1      1        0        1    0       0
4     0     1        0      0      1        0        1    1       1

   TravelBag  NailPolish  Socks  Hats  Fitbit
0          0           0      0     0       1
1          0           1      1     0       0
2          1           1      1     1       0
3          0           1      0     0       1
4          1           0      1     1       0
```

Figure 14-10. *Step 2, data preprocessing*

Step 3: Find the frequent item sets using the `apriori()` algorithm with a support of 0.5.

```
item_sets = apriori(marys_df2, min_support=0.5, use_colnames=True,
low_memory=False, verbose=1)

 Processing 90 combinations | Sampling itemset size 2 Processing 72
combinations | Sampling itemset size 3 Processing 8 combinations |
Sampling itemset size 4

item_sets.head(6)

   support   itemsets
0     0.55     (Shoe)
1     0.60  (Perfume)
2     0.85    (Shirt)
3     0.70  (Trouser)
4     0.50      (Tie)
5     0.70   (Wallet)
```

***Figure 14-11.** Step 3, executing the aprori() algorithm to find frequent item sets*

Step 4: Find the association rules from the frequent data set. We use the `association_rules()` function.

```
mba_rules.head(10)

           antecedents          consequents  antecedent support  \
25              (Shoe)      (Shirt, Wallet)                0.55
20     (Shirt, Wallet)               (Shoe)                0.60
19            (Wallet)          (TravelBag)                0.70
18         (TravelBag)             (Wallet)                0.60
3               (Shoe)             (Wallet)                0.55
21       (Shirt, Shoe)             (Wallet)                0.50
32  (Shirt, TravelBag)             (Wallet)                0.50
2             (Wallet)               (Shoe)                0.70
24            (Wallet)        (Shirt, Shoe)                0.70
37            (Wallet)   (Shirt, TravelBag)                0.70

    consequent support  support  confidence      lift  leverage
conviction
25                0.60     0.50    0.909091  1.515152     0.170
4.40
20                0.55     0.50    0.833333  1.515152     0.170
2.70
19                0.60     0.60    0.857143  1.428571     0.180
2.80
18                0.70     0.60    1.000000  1.428571     0.180
inf
3                 0.70     0.55    1.000000  1.428571     0.165
inf
21                0.70     0.50    1.000000  1.428571     0.150
inf
32                0.70     0.50    1.000000  1.428571     0.150
inf
2                 0.55     0.55    0.785714  1.428571     0.165
2.10
24                0.50     0.50    0.714286  1.428571     0.150
1.75
37                0.50     0.50    0.714286  1.428571     0.150
1.75

mba_rules[mba_rules['antecedents']==frozenset({'Shirt'})].sort_values(by
= 'lift', ascending = False)
```

Figure 14-12. *Step 4, association rules*

The `apriori()` algorithm created a set of frequent items and corresponding association rules for the given transactional data set. This set of association rules is based on the confidence of 0.7 and the support of 0.5.

Figure 14-13 shows both Input and Output of the association rules of one item set.

```
In [31]: mba_rules[mba_rules['antecedents']==frozenset({'Shirt'})].sort_values(by = 'lift', ascending = Fals
```

Out[31]:

	antecedents	consequents	antecedent support	consequent support	support	confidence	lift	leverage	conviction
15	(Shirt)	(NailPolish)	0.85	0.55	0.55	0.647059	1.176471	0.0825	1.275000
8	(Shirt)	(Tie)	0.85	0.50	0.50	0.588235	1.176471	0.0750	1.214286
0	(Shirt)	(Shoe)	0.85	0.55	0.50	0.588235	1.069519	0.0325	1.092857
23	(Shirt)	(Wallet, Shoe)	0.85	0.55	0.50	0.588235	1.069519	0.0325	1.092857
29	(Shirt)	(Trouser, Wallet)	0.85	0.55	0.50	0.588235	1.069519	0.0325	1.092857
6	(Shirt)	(Trouser)	0.85	0.70	0.60	0.705882	1.008403	0.0050	1.020000
10	(Shirt)	(Wallet)	0.85	0.70	0.60	0.705882	1.008403	0.0050	1.020000
12	(Shirt)	(TravelBag)	0.85	0.60	0.50	0.588235	0.980392	-0.0100	0.971429
35	(Shirt)	(TravelBag, Wallet)	0.85	0.60	0.50	0.588235	0.980392	-0.0100	0.971429

***Figure 14-13.** Printing rules with support 0.7, one item set*

We use `frozenset()` to build a Python function to find out the set of rules for promoting any item set. The frozenset() function is commonly used to remove duplicates from a sequence, computing mathematical operations such as intersection, union, and symmetric differences. In this example, we use it to find {shirt}-only rules and then rules for a {shirt, wallet} combination.

Figure 14-14 shows both Input and Output of the association rules of two item set.

```
In [32]: mba_rules[mba_rules['antecedents']==frozenset({'Shirt','Wallet'})].sort_values(by = 'lift', ascending = False)
```

Out[32]:

	antecedents	consequents	antecedent support	consequent support	support	confidence	lift	leverage	conviction
20	(Shirt, Wallet)	(Shoe)	0.6	0.55	0.5	0.833333	1.515152	0.17	2.7
33	(Shirt, Wallet)	(TravelBag)	0.6	0.60	0.5	0.833333	1.388889	0.14	2.4
27	(Shirt, Wallet)	(Trouser)	0.6	0.70	0.5	0.833333	1.190476	0.08	1.8

***Figure 14-14.** Printing rules with support 0.7, two-item set*

The rules are sorted by "lift" values. From the previous rules, we can say that {Shirt, wallet} and Shoe have strong associations and are sold together. Similarly, the next association is {Shirt, Travelbag} and {wallet} and are sold together. This association helps to make certain business decisions such as inventory management or promoting new {Shoe} or {Wallet} and thus managing stocks appropriately.

Note Using the `frozenset()` Python function because of immutable association rules.

14.6 Chapter Summary

In this chapter, we explained another unsupervised learning technique called relationship mining. Relationship mining is also referred to as association rules mining or market basket analysis.

Association rules find interesting associations among large transactional item sets in the database. You learned the basic concepts of association rule analysis, how to perform such analysis using both Python and R, and what metrics are used to measure the strength of the association with a case study.

CHAPTER 15

Introduction to Natural Language Processing

> *A language is not just words. It's a culture, a tradition, a unification of a community, a whole history that creates what a community is. It's all embodied in a language.*
>
> Noam Chomsky

Natural language processing (NLP) aims to make computers process human languages. NLP has a wide variety of applications today. You may already be using them when you are buying a book on Amazon, talking to Google Assistant or Siri when checking the weather, or talking to an automated chat agent when seeking customer service. In this chapter, we start with an overview of NLP and discuss NLP applications, key concepts, various NLP tasks, how to create models, the Python and R NLP libraries, and case studies.

15.1 Overview

Today technology has become ubiquitous. Without smartphones, it's hard to conduct daily business. Our morning routine often starts by talking to Siri or Google Assistant on your phone or similar AI bots asking about weather or traffic reports. We talk to these voice assistants in our natural language, not computer programming languages. But, computers know how to interpret and process only binary data. How can you make computers understand human language?

U. R. Hodeghatta and U. Nayak, *Practical Business Analytics Using R and Python*,
https://doi.org/10.1007/978-1-4842-8754-5_15

In computer science, NLP is an area that deals with methods, processes, and algorithms to process languages. There are various steps to perform NLP tasks. This chapter will explain the various NLP tasks and commonly used methods, models, and techniques to process language. This chapter will help you solve NLP problems with various techniques and methods and suggest the best method to choose based on the type of problem you are solving.

We start with an overview of numerous applications of NLP in real-world scenarios, then cover a basic understanding of language and what makes NLP complex, and next discuss various NLP tasks involved in building NLP applications. We will also be discussing machine learning and deep learning methods to solve NLP.

15.2 Applications of NLP

There are several applications of NLP, including Amazon Alexa, Google Assistant, and Siri. Some other applications of NLP include sentiment analysis, question answering, text summarization, machine translation, voice recognition and analysis, and user recommendation systems. Grammar checkers, spelling checkers, Google language translators, and voice authentication are some other applications of NLP. Other examples of NLP are your smartphone messaging application's next word prediction functionality and the recent Microsoft Outlook feature that suggests automatic email responses.

15.2.1 Chatbots

Chatbots can converse with humans using natural languages. AI chatbots have recently become popular and are mainly used as the first customer support answer to the point simple user queries and collect customer information for further assistance. AI chatbots have several benefits; they enhance efficiency, cut operational time and cost, have zero customer bias, and offer no emotions. According to Salesforce, 69 percent of consumers are satisfied with the use of chatbots to resolve their issues and the speed at which they can communicate.

Figure 15-1 is an example of a chatbot from the CITI Bank account website. When you go online and contact customer support, an AI chatbot triggers the conversation.

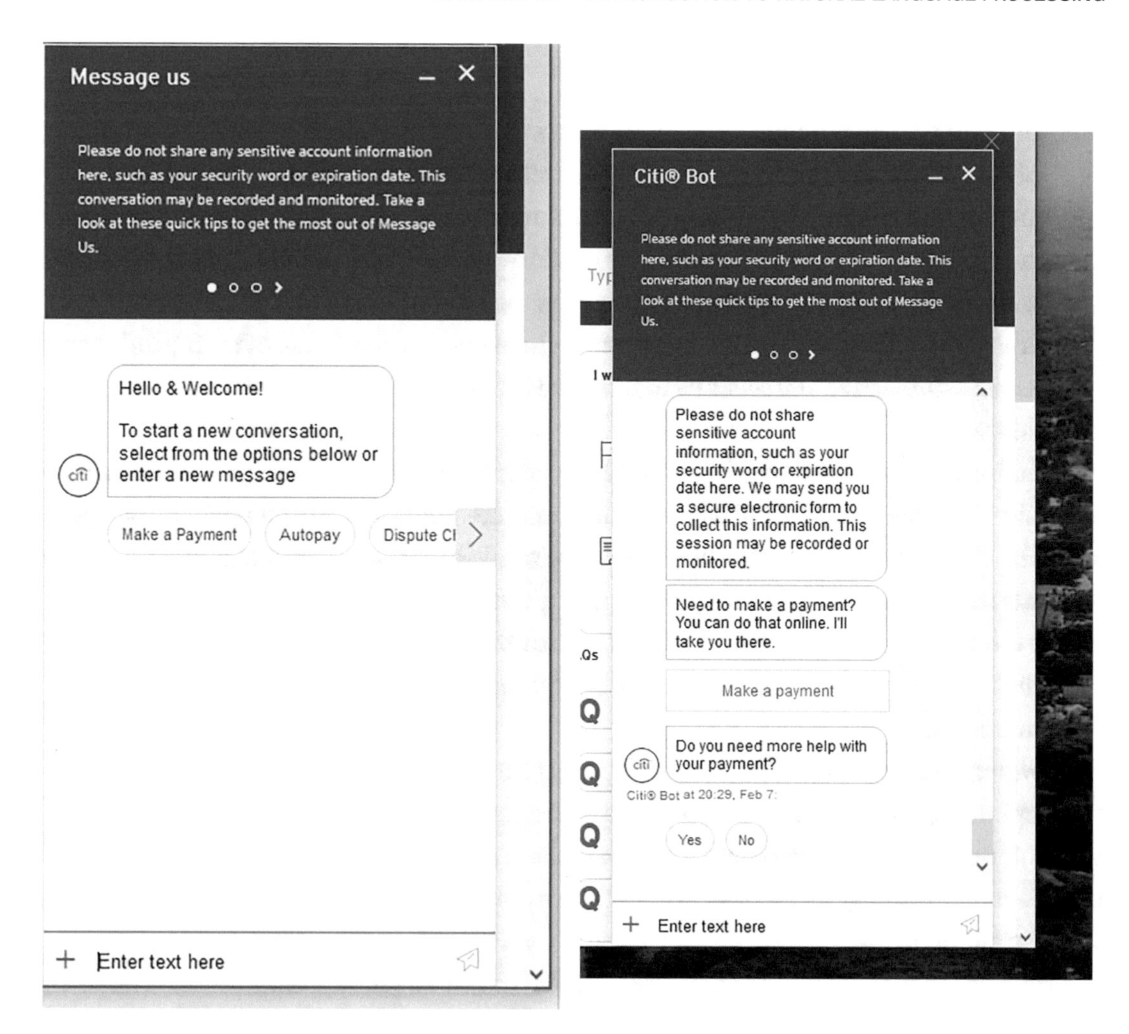

Figure 15-1. *Example of AI chatbot*

According to a Mordor Intelligence report, the chatbot market in 2020 was valued at USD $17 billion and is projected to reach USD $102 billion by 2026.

Voice-assisted intelligent devices such as Alexa, Google Home, etc., are gaining tremendous popularity due to convenience. In addition, many recent advancements in deep learning neural networks and machine learning techniques have made the technology more efficient. Though the technology is heavily adopted in many sectors, the banking sector has been a significant adopter due to its speed, efficiency, and quicker response time. In addition, chatbots in the banking sector are also assisting in capturing critical behavior data and performing cognitive analytics. These analytics help businesses to build a better customer experience. For the complete market analysis, read the full report at `https://www.mordorintelligence.com/industry-reports/chatbot-market`.

15.2.2 Sentiment Analysis

It is common to read feedback and reviews on social media before buying any product or service. There has been a rapid growth of online forums to post product-related information and customers' opinions. This information is useful for both companies and customers. Customers read these reviews and can assess the product's quality, customer experience, price, and satisfaction level in purchasing decisions. At the same time, companies are interested in such information to assess the feedback of their products and services and arrive at decisions to improve the product quality and customer experience.

Sentiment analysis has been studied by academic and enterprise applications and is gaining importance, particularly to understand customers' emotions such as happy, satisfied, dissatisfied, angry, or simply positive or negative. Sentiment classification is valuable in many business intelligence applications and recommendation systems, where thousands of feedback and ratings are summarized quickly to create a snapshot. Sentiment classification is also useful for filtering messages, such as spam (Tatemura, 2000).

Twitter, Facebook, Tumblr, and other social media encourage users to freely express their opinions, thoughts, and feelings and share them with their friends. As a result, a vast amount of unstructured data on social media may contain useful data to marketers and researchers. Most messages contain little information, but millions of such messages would yield better analysis.

Here are some examples of user sentiments that can be analyzed:

- A stock investor conversation on recent NASDAQ market sentiment
- An individual's opinion on a recent Hollywood movie
- An opinion of a new iPhone and Samsung phone

A sentiment analysis system can be a simple lexicon-based dictionary lookup or ML/DL-based classifier. The choice of method is dictated by the business requirements, the complexity of the data, and other development constraints. There are several online tools available to check the sentiment classification of your text. One such popular tool available online is Brand24 (`https://brand24.com/ai-driven-sentiment-analysis/`).

As shown in Figure 15-2, you can use these services on a free trial basis. Then all you have to do is copy and paste a review, and the tool provides the sentiment using the pretrained NLP sentiment analyzer.

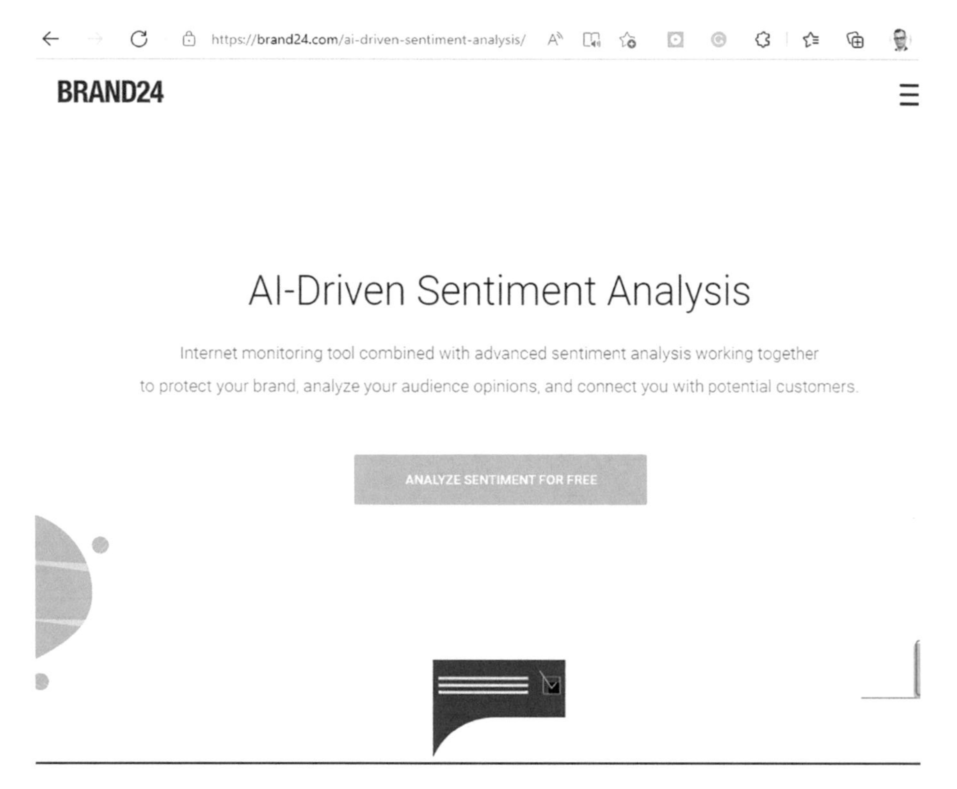

***Figure 15-2.** AI-driven sentiment analyzer*

15.2.3 Machine Translation

There are so many languages spoken globally that language translation is always needed to communicate with people of different languages. Even today, we have human translators who have learned multiple languages to help others by translating from one language to another. Therefore, there was always a need for language translation using the machine. The early language translation methods were developed during Cold War times to translate some of the Russian languages to English using NLP. In 1964, the U.S. government created a committee called the Automatic Language Processing Advisory Committee (ALPAC) to explore machine translation techniques. Although the ALPAC could not come up with promising results during its time, today the advancements in computational technology can create models that can translate languages with high accuracy.

In today's highly interconnected global business communities, machine translation (MT) has been one of the most critical business applications. Although businesses often rely on human translators to translate conversations in hospitals, legal firms, courts, etc., the use of NLP applications has been increasing.

Several translators are available online, from Google, Microsoft, Apple, etc., translating languages with significantly high accuracy.

Google Translate (`https://translate.google.com/`), which is based on AI/deep learning NLP, translates multiple languages. Figure 15-3 is an example of translating using Google Translate.

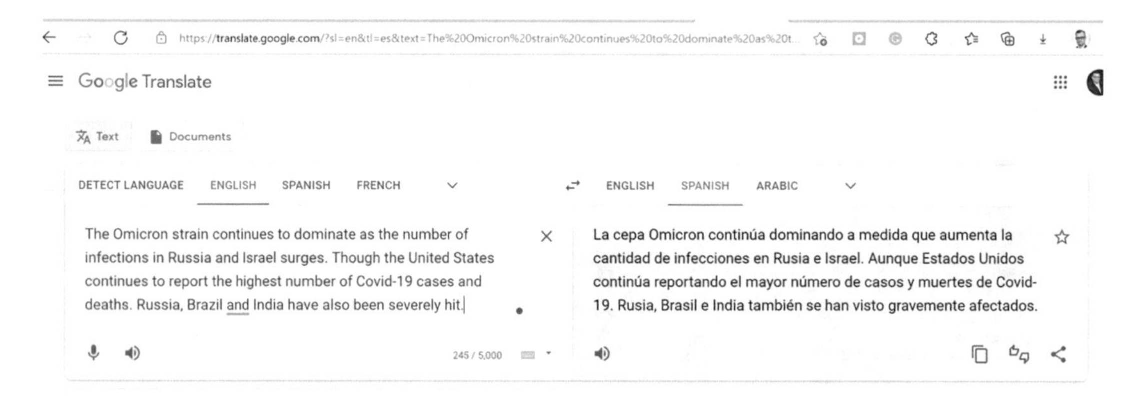

Figure 15-3. *Translating from English to Spanish using Google Translate*

Google supports many languages including Russian, Albanian, Danish, Korean, Hebrew, Latin, Polish, etc. Figure 15-4 lists the languages you can translate using Google Translate.

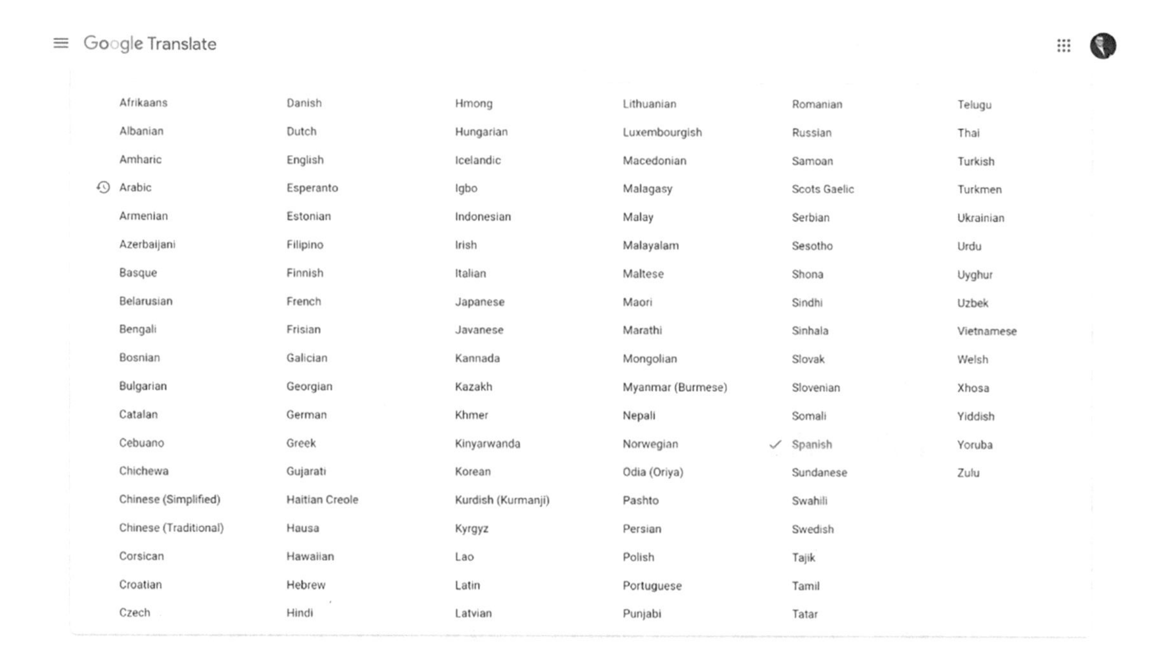

Figure 15-4. *Languages supported by Google Translate*

Microsoft also has a machine translator. Figure 15-5 is the translator output for the exact text used earlier (`https://www.bing.com/translator`).

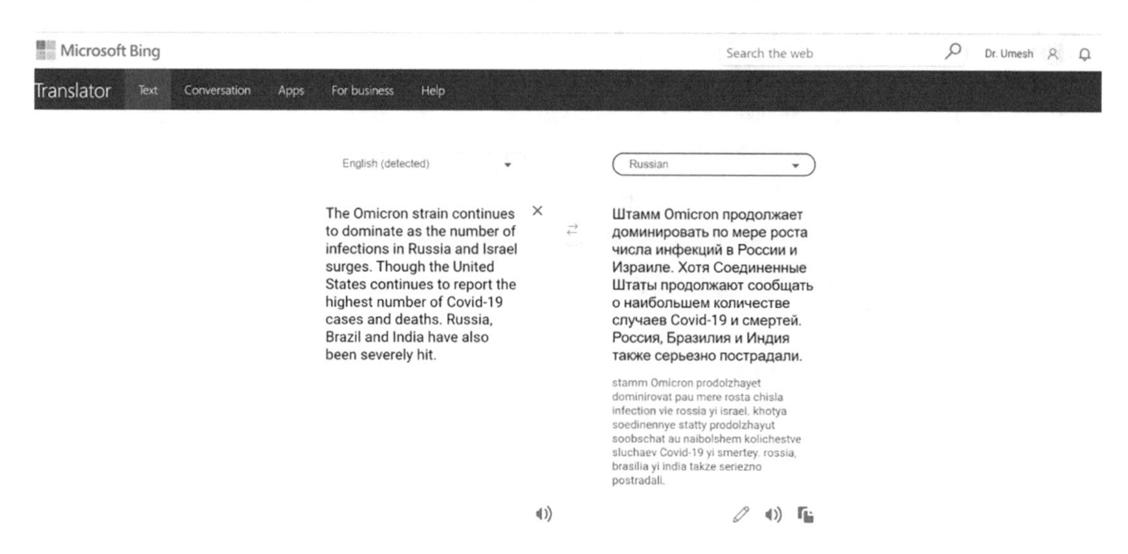

Figure 15-5. *Microsoft Bing Translator*

It should be noted that the Microsoft translator time was better than Google, and the translation time was almost instantaneous. So, there are significant improvements from where we were to where we are today, even though there is still scope for improvements and research in this area.

Chatbots, machine translators, and sentiment analyzers are only a few applications of NLP. There are many applications of NLP, and they are text summarization, information extraction, named entity recognition (NER), automatic text indexing, recommendation systems, and trend analysis.

Text summarization is the process of generating a short summary of a paragraph or a text document. Information retrieval is the retrieval of specific information related to a selected topic from a body of text or document. Named entity recognition (NER) is a subtask of information retrieval that locates and classifies the named entities, such as organization, person, or animal, in unstructured text. Based on the previous customer behavior of the customers, a recommendation system recommends similar material to the customer such as Netflix movie recommendations or Amazon purchase recommendations.

Let's begin our NLP discussion with a shared understanding of language and how computers process language.

15.3 What Is Language?

Before the existence of languages, humans used to communicate using signs or drawings. Though they were simple and easy to understand, not all emotions were easy to express. Language is a mode of communication between two parties that involves a complex combination of its constituent elements such as scripts, characters, and letters. Language also comprises words, structure, tone, and grammar. It has both syntax and semantics. It is easy to learn all the aspects of a native language as humans. However, suppose one has to learn another language; that requires a lot more effort to learn all aspects such as the structure, the grammar, and the ambiguities that exist in that language.

Learning any natural language, whether it is English, French, or German, requires years of learning and practice. To speak a language means understanding many aspects of the language such as concepts of words, phrases, spellings, grammar, and structure in a meaningful way. Though it may be easier for humans to learn, making a computer learn and master a language is challenging.

It has already been proven that computers can solve complex mathematical functions, but they have yet to master all the components of a spoken or written language.

Pronunciation and sounds are also an essential part of the language. Similarly, the roots of the words, how they are derived, and the context when used in a phrase are all critical aspects of any language. In formal language definitions, a language consists of phonemes (sounds and pronunciations), morphemes and lexemes (the roots of the words), syntax, semantics, and context. It is vital to understand all aspects of a language and how a language is structured in order for the computer to process the language. Figure 15-6 shows the different components of language and some of the associated NLP tasks and applications.

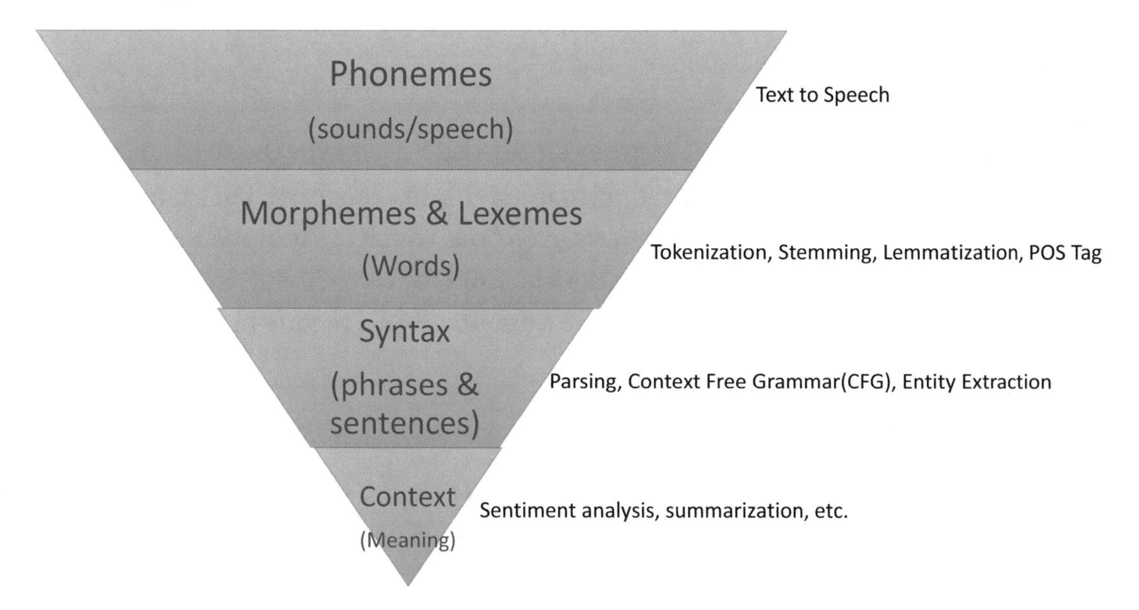

Figure 15-6. *Building blocks of a language and use case*

Let's quickly look at the building blocks of languages.

15.3.1 Phonemes

Phonemes are the various sounds for different letters in a language such as the way you pronounce *a* or *b* or the words *apple* and *ability*. When uttered with a combination of other letters, phonemes can produce a different sound, and they are essential to the meaning of the words. For example, the standard English language has around

44 phonemes. These phonemes are a combination of single letters or a combination of letters. Phonemes play an important role in speech recognition systems such as Alexa or Siri, speech-to-text transcriptions, and text-to-speech conversions. We will not go into the details of phonemes as this is beyond the scope of this book.

15.3.2 Lexeme

A *lexeme* is an abstract unit of a word from various inflectional endings. For example, *take, taking,* and *taken* are all inflected forms of the same lexeme *take*. Similarly, *dogs* and *dog* and *cat* and *cats* are inflected forms of the same lexeme *dog* or *cat*. A lexeme is the smallest unit of a word without the end prefixes.

15.3.3 Morpheme

Morphemes are similar to lexemes, but they are formed by removing prefixes and suffixes. Prefixes and suffixes add meaning to the words in a language. For example, by removing *d* in the English word *followed*, we get *follow*, which is a morpheme. Similarly, the words *reliable, reliability,* and *rely* are derived from the same morpheme, *rel*. Not all morphemes are words. Morphemes can have a grammatical structure in a text.

15.3.4 Syntax

Syntax is a set of grammar rules to construct sentences from a set of words in a language. In NLP, syntactic structure is represented in many ways. A common approach is a tree structure, as shown in Figure 15-7. A constituent is a group of words that appears as a single unit in a phrase. A sentence is a hierarchy of constituents. The sentence's syntactic structure is guided by a set of grammar rules for the language. The earliest known grammar structure is in the Sanskrit language and was defined by Panini in the 4th century BC.

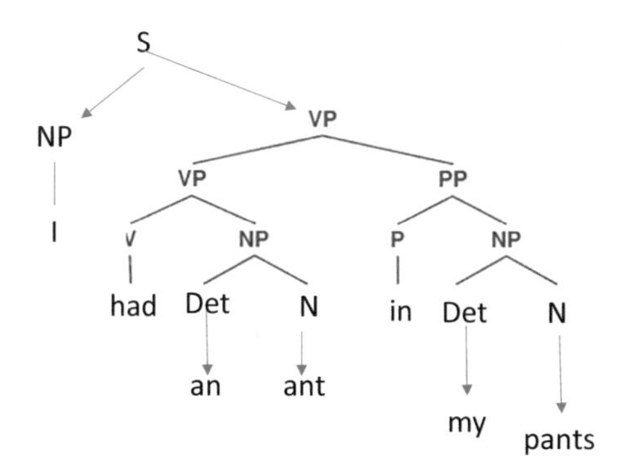

Figure 15-7. *Syntactic structure of a sentence: "I had an ant in my pants"*

In this example, a sentence "I had an ant in my pants" is parsed syntactically, as shown in Figure 15-7.

The sentence has a hierarchical structure of language. The words represent the lowest level, followed by the association of words by their part of speech, whether the word is a noun or verb or adjective, the phrase structure, whether it is a noun phrase or verb phrase, and completing the entire sentence at the highest level. Parsing a sentence is one of the fundamental NLP tasks. This example was created using the Python library; N stands for noun, V for verb, NP for noun phrase, and VP for verb phrase. It should be noted that this parsing structure created is specific to the English language. The syntax and grammar structure may vary from one language to another language, and the language processing approach and methods can also vary.

15.3.5 Context

The words can have different meanings depending on the context. Context sets the situation. For example, the word *fly* can have two different meanings. Suppose it is associated with the word *aeroplane*; in that case, it can mean travel, whereas *fly* also can mean an insect. The context of the word *fly* depends on its association. Languages can have multiple meanings for the words; thus, the meaning of a sentence can change depending on the context and how the words are used.

15.4 What Is Natural Language Processing?

NLP is a linguistic analysis of a language to achieve human-to-machine communication. It is an intelligent system that extracts text, sentences, and compound phrases to perform valuable tasks enabling human-to-machine communication.

In the early days, NLP was referred to as *natural language understanding* (NLU). Ideally, an NLP system would be able to do the following:

1. Paraphrase an input text.
2. Answer questions about the contents of the text.
3. Draw inferences from the text.
4. Translate one language to another language.

15.4.1 Why Is NLP Challenging?

Natural language is vague and has a lot of ambiguity. Although humans are trained to understand languages, a proper understanding of a language requires extensive knowledge of grammar structure, vocabulary, and context. Humans tend to omit a lot of commonsense knowledge and assume that listeners possess common knowledge. Humans also assume that the listener will resolve any ambiguity and abbreviations, slang, etc. For example, as an English native speaker, you can easily interpret a sentence like "She is jumping from a window." Should the word *jumping* be analyzed as a verb or a noun? Depending on the context, a window could also refer to a house panel, operating system, or time frame. Though it is easy for humans to distinguish the meaning of the sentence, it is difficult for a computer to interpret the context. Will computers ever understand natural language the same way we do? Computers have to be trained to parse sentences and understand grammatical structures and semantics to understand a language and identify different words.

Suppose the computer can be built with enough intelligence to understand and resolve enough of the structural ambiguities of the language. Then, it may be possible for a computer to understand and interpret the language. We have to build an "intelligent" logic to make the computer understand letters, words, sentences, and grammar. This specialized software field of computer science is what NLP is.

Common knowledge: Human language assumes certain aspects as "common knowledge." In any conversation, we assume some facts are known, and we do not explicitly specify them, but they do have meaning in the sentences. For example, in the sentence "The cat is chasing a boy," we assume that the pet cat and boy are playing and there is no danger. Further, we know cats usually do not chase anyone, and cats are pet animals. This common knowledge is required to come to the conclusion that the boy is not in danger. However, if a computer processes this, it may find it difficult to come to any conclusion and interpret it differently because it lacks common knowledge. The key challenge in NLP is to encompass common knowledge in a computational model.

Ambiguity: Most languages are inherently ambiguous. Ambiguity in NLP refers to sentences and phrases that potentially have two or more possible interpretations. Three types of ambiguity are lexical ambiguity, semantic ambiguity, and syntactic ambiguity. *Lexical ambiguity* refers to a word used as a verb, noun, or adjective. *Semantic ambiguity* refers to the interpretation of a sentence in context. *Syntactic ambiguity* refers to the confusion of meaning and understanding differently. The meaning of a sentence can differ depending on the words and its association. The same sentence may have multiple meanings depending on the word's use, and its association with the other words makes a phrase ambiguous. Other times, a word may have multiple meanings, and more than one of those meanings could apply in the phrase. Often, context helps humans understand the meaning of the phrase, even if the phrase on its own is ambiguous.

Lexical ambiguity is when the word has two or more possible meanings. Syntactic ambiguity is the expression or a sentence having two or more possible meanings within a sequence of words or a sentence. When an expression is semantically ambiguous, it can have multiple meanings.

Here's an example:

"I heard the Billy Joel tracks last night."

This phrase could be ambiguous when spoken out loud. Songs are referred to as *tracks*, so one could interpret this phrase as someone listened to songs by Billy Joel last night. Tracks are also what path one travels on or how a train travels, so this sentence could mean someone heard a Billy Joel passing nearby last night.

Similarly, in "Anita is looking for a match" and "The boy's shirt and wall color match," the word *match* can have multiple meanings, and depending on the context, the interpretation of the sentence can change. Similarly, think about "Joe could not lift the bag." Is the bag so heavy that Joe could not lift it, or was Joe restricted from lifting the bag?

Though humans can resolve ambiguity, it is difficult for the machines to resolve and thus makes NLP challenging.

Portability of NLP solution: Since every language is different, an NLP solution developed in one language may not work with another language. This makes it hard to develop one NLP solution that models all the languages. This means you should build a language-agnostic solution or build separate solutions for each language. Both are challenging and hard.

15.5 Approaches to NLP

We can solve NLP problems using the traditional approach, parsing words, sentences, grammar; using a heuristic approach; or using probabilistic and machine learning approaches. We will briefly describe all the methods with examples using Python libraries. Of course, you can do all the NLP tasks using R as well.

The heuristic approach is the early attempt of building an NLP system. It is based on creating a set of rules and developing a program to map the rules. Such systems required a corpus containing word dictionaries and their meaning compiled over time. Besides dictionaries and meaning of words, a more elaborate knowledge base has been built over a period of time to support a rule-based heuristic NLP system, including semantic relationships, synonyms, and hyponyms. One such example is the WordNet corpus (`https://wordnet.princeton.edu/`).

15.5.1 WordNet Corpus

WordNet is an open-source public corpus with a large lexical database of nouns, verbs, adjectives, and adverbs grouped into sets of synonyms. They are referred to as *synsets* and are interlinked using conceptual-semantic and lexical relations. Figure 15-8 demonstrates how to access WordNet using the NLTK library and the word association of *sweet*.

Here is the input:

```
#import NLTK library
import nltk
#Load WordNet corpus
from nltk.corpus import wordnet

synnames = wordnet.synsets("walk")

##Definition of walk according to wordNet
synnames[0].definition()

##Lemmas of walk
synnames[0].lemma_names()

##All the synonyms associated with word "SWEET"
for names in wordnet.synsets('sweet'):
    print(names.lemma_names())
```

Figure 15-8 shows the input and output.

```
In [1]:  #import NLTK Library
         import nltk
         #Load WordNet corpus
         from nltk.corpus import wordnet

In [16]: synnames = wordnet.synsets("walk")

In [32]: ##Definition of walk according to wordNet
         synnames[0].definition()

Out[32]: 'the act of traveling by foot'

In [31]: ##Lemmas of walk
         synnames[0].lemma_names()

Out[31]: ['walk', 'walking']

In [33]: ##All the synonyms associated with word "SWEET"
         for names in wordnet.synsets('sweet'):
             print(names.lemma_names())

         ['Sweet', 'Henry_Sweet']
         ['dessert', 'sweet', 'afters']
         ['sweet', 'confection']
         ['sweet', 'sweetness', 'sugariness']
         ['sweetness', 'sweet']
         ['sweet']
         ['angelic', 'angelical', 'cherubic', 'seraphic', 'sweet']
         ['dulcet', 'honeyed', 'mellifluous', 'mellisonant', 'sweet']
         ['sweet']
         ['gratifying', 'sweet']
         ['odoriferous', 'odorous', 'perfumed', 'scented', 'sweet', 'sweet-scented',
         'sweet-smelling']
         ['sweet']
         ['fresh', 'sweet']
         ['fresh', 'sweet', 'unfermented']
         ['sugared', 'sweetened', 'sweet', 'sweet-flavored']
         ['sweetly', 'sweet']
```

Figure 15-8. *Accessing the WordNet corpus using the NLTK library*

15.5.2 Brown Corpus

Brown University created the Brown corpus in 1961. This corpus has more than one million words in the English vocabulary. This corpus contains text from nearly 500 sources and has been categorized by genre, such as news, editorial, and so on. Figure 15-9 shows the example of accessing the Brown corpus using the NLTK library.

Here is the input:

```
from nltk.corpus import brown

brown.categories()

brown.words(categories='news')
```

Figure 15-9 shows the input and output.

```
In [40]:  from nltk.corpus import brown

In [41]:  brown.categories()

  Out[41]: ['adventure',
            'belles_lettres',
            'editorial',
            'fiction',
            'government',
            'hobbies',
            'humor',
            'learned',
            'lore',
            'mystery',
            'news',
            'religion',
            'reviews',
            'romance',
            'science_fiction']

In [42]:  brown.words(categories='news')

  Out[42]: ['The', 'Fulton', 'County', 'Grand', 'Jury', 'said', ...]
```

Figure 15-9. *Accessing the Brown corpus*

15.5.3 Reuters Corpus

The Reuters corpus contains nearly 1.3 million words from 10,000 news articles. The documents have been classified into 90 different topics. Figure 15-10 shows an example of accessing the Reuters corpus.

Here's the input:

```
from nltk.corpus import reuters

reuters.categories()
```

Figure 15-10 shows the input and output.

```
In [43]:  ##Reuters Corpus

In [46]:  from nltk.corpus import reuters

In [47]:  reuters.categories()

Out[47]: ['acq',
           'alum',
           'barley',
           'bop',
           'carcass',
           'castor-oil',
           'cocoa',
           'coconut',
           'coconut-oil',
           'coffee',
           'copper',
           'copra-cake',
           'corn',
           'cotton',
           'cotton-oil',
           'cpi',
           'cpu',
           'crude',
```

Figure 15-10. *Accessing the Reuters corpus*

15.5.4 Processing Text Using Regular Expressions

Regular expressions (*regex*) have been around for a long time to process text, look for patterns in text, and build rule-based NLP systems. Both the Python and R libraries support regular expressions. There are several methods associated with regexes. We will not be able to cover every aspect of regex as it is not in the scope of the book. The following examples are some of the most common methods used in regex.

15.5.4.1 re.search() Method

The `re.search()` method looks for a specific pattern. In the example shown in Figure 15-11, `re.search()` searches all the strings that match *RE*. The method found *RE* at the 27th character in a string. In this example, the regex is looking for the pattern *ex* and *or*. `re.search()` and `re.findall()` are two different methods. You can examine the output to understand the difference between the two.

Here is the input:

```
import re

text = 'Regular expressions called REs, or regexes, or regex patterns are es
sentially a tiny, highly specialized programming language embedded inside Py
thon and made available through the re module.'

text

re.search(r'RE',text)
re.search(r'ex',text)
re.search(r'or',text)
```

Figure 15-11 shows the input and output.

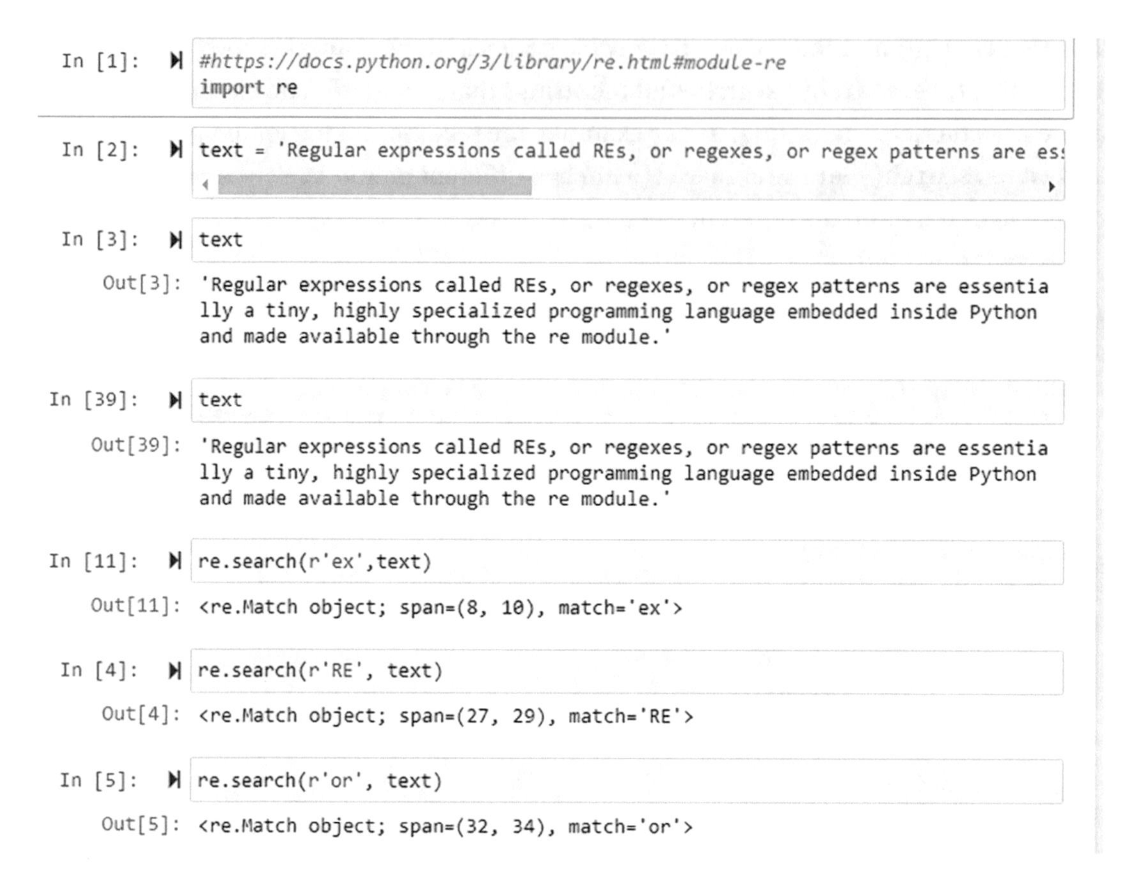

***Figure 15-11.** Demonstration of the re.search() method*

15.5.4.2 re.findall()

This method is similar to `re.search()`, but it finds all non-overlapping matches of a pattern in a string, as shown in Figure 15-12.

Here's the input:

```
re.findall(r'or',text)
re.findall(r'regex',text)
re.findall(r'[RE]',text.lower())
```

Figure 15-12 shows the input and output.

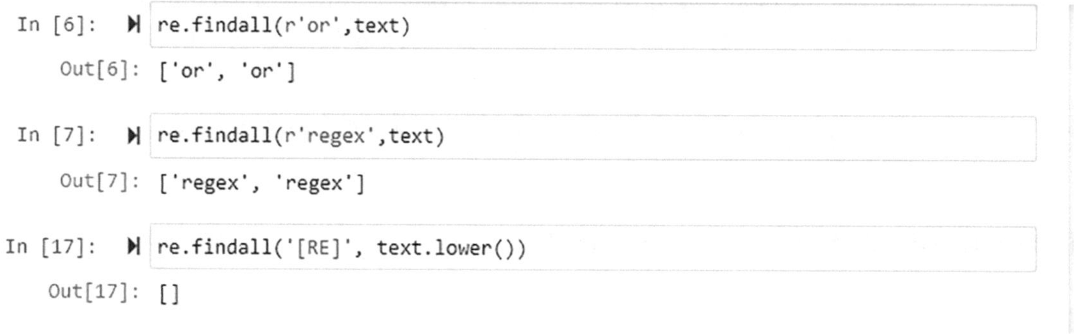

```
In [6]:  re.findall(r'or',text)
Out[6]: ['or', 'or']

In [7]:  re.findall(r'regex',text)
Out[7]: ['regex', 'regex']

In [17]:  re.findall('[RE]', text.lower())
Out[17]: []
```

Figure 15-12. *Demonstration of the re.findall() method*

15.5.4.3 re.sub()

The substring function `sub` substitutes a search string with a new string. It is similar to Find and Replace in Microsoft Word. In Figure 15-13, `re.sub()` searches for the pattern *RE* in the string, which is the third parameter passed in the function, and substitutes the new string REGEXPR.

Here is the input:

```
re.sub(r'RE','REGEXPR',text)
```

Figure 15-13 shows the input and output.

```
In [61]:  re.sub(r'RE','REGEXPR',text)

Out[61]: 'Regular expressions called REGEXPRs, or regexes, or regex patterns are ess
          entially a tiny, highly specialized programming language embedded inside Py
          thon and made available through the re module.'

In [72]:  text

Out[72]: 'Regular expressions called REs, or regexes, or regex patterns are essentia
          lly a tiny, highly specialized programming language embedded inside Python
          and made available through the re module.'
```

Figure 15-13. *Demonstration of re.sub()*

The purpose of this section was to introduce the regular expression tool library. The syntax and different pattern types for a regular expression can be found in the documentation. For the Python documentation, refer to: `https://docs.python.org/3/library/re.html`. For the R regular expression documentation, please refer to `https://www.rdocumentation.org/packages/base/versions/3.6.2/topics/regex`.

15.6 Important NLP Python Libraries

Though several Python libraries support NLP, we are listing the most popular and widely used libraries.

- The Natural Language Tool Kit (NLTK) is one of the most popular Python libraries for NLP. NLTK was developed to support research and teaching in NLP by Steven Bird and Edward Loper in the Department of Computer and Information. NLTK supports many of the NLP tasks such as tokenization, stemming, tagging, parsing, semantic reasoning functionalities, and text classification and clustering. A free online NLTK book is also available.

- The Scikit-learn library is another popular library in Python that supports various NLP tasks. Scikit-learn's `CounterVectorizer()` function supports text preprocessing, tokenization, stop word removal, text vectorization, and other NLP functions. For more information, refer to the Scikit-learn documentation.
- TextBlob supports APIs that support natural language processing tasks such as tokenization, parsing, word and phrase frequencies, n-gram, part-of-speech tagging, noun phrase extraction, sentiment analysis, classification, translation, and more.
- Gensim is considered the fastest library for training vector embeddings such as Word2Vec and Glove. It is an open-source library, and all the source code is available on GitHub under the GNU LGPL license agreement.
- Spacy is another library that is mainly helpful for extracting large-scale text extraction. If your application has to process entire web dumps, you can consider using Spacy. Spacy supports NLP for 64+ languages and has pretrained word vectors, transformers, and BERT. It also provides named entity recognition, part-of-speech tagging, dependency parsing, sentence segmentation, text classification, lemmatization, morphological analysis, entity linking, and more.

15.7 Important NLP R Libraries

There are several packages in R that support NLP. Here is the list of commonly used R packages:

Text2Vec: This is a `memory0friendly` library for word embedding, text vectorization, finding similarities of two words, etc. This package allows processing of text documents that are larger than the available memory on your machine. Reference: `https://cran.r-project.org/web/packages/text2vec/index.html`

TidyText: This is another popular library for text mining and text processing. This is based on the same principles of `tidyr` and `dplyr`. Reference: `https://cran.r-project.org/web/packages/tidytext/tidytext.pdf`

Stringr: Another simple, easy-to-use set of libraries around the `stringi` package. Reference: `https://cran.r-project.org/web/packages/stringr/index.html`

SpacyR: SpacyR is an R wrapper around the Python Spacy package for R programmers. It supports all the functionalities supported by Spacy for text processing and NLP including tokenization, lemmatization, extracting token sequences, entities, phrases, etc. Reference: `https://spacy.io/universe/project/spacyrTM`

tm: This is the most popular package in R for creating a corpus object and processing text within a dataframe. It supports a number of NLP functions. Reference: `https://cran.r-project.org/web/packages/tm/tm.pdf`

15.8 NLP Tasks Using Python

Both Python and R support NLP libraries. Both are open-source programming languages. Python is a high level programing language that is more popular in the NLP research community. Python is easy to use, is easy to debug, and is an object-oriented modular language. Though both R and Python support several NLP libraries, all our examples shown here use Python and Jupyter Notebook as a user interface. Those who are new to Python may consider installing the Anaconda distribution package, which bundles all the libraries required for machine learning and analytics. You can download the Anaconda Python package (`https://www.anaconda.com`) and other interactive development environments (IDEs) such as Spyder and Jupyter Notebook.

15.8.1 Text Normalization

In the earlier sections, we talked about the properties of language such as phonemes, morphemes, lexemes, etc. These are important steps in building the vocabulary, and sometimes this is called *normalizing* the text. In this section, we will discuss the steps to normalize text for further processing and prepare it for various NLP tasks. Some tasks include normalizing text to all lowercase letters and removing unnecessary punctuation, decimal points, etc., to have all the terms in the same form. For example, you would remove the periods in U.S.A. and N.Y.C. Similarly, you would keep the root of the word and remove its prefixes and suffixes. For example, just keep the word *leader* when *leaders* and *leader* appear in the text.

15.8.2 Tokenization

The first step in NLP is to break the larger chunks of documents into smaller sentences and the sentences into smaller chunks of words known as *tokens*. Each token is significant as it has a meaning associated with it. Tokenization is also the fundamental task in text analytics. Tokens can be words, numbers, punctuation marks, symbols, and even emoticons in the case of social media texts.

Let's go through a few examples of tokenization techniques using the NLTK library.

NLTK supports the `word_tokenize()` function, which tokenizes words based on space. The first step is to invoke the libraries using the `import` function and then call the `word_tokenize()` function with the appropriate parameters. In this example, we use the text and tokenize the complete text string using the `word_tokenize()` method.

Here is the input:

```
##Import NLTK Library
import re
import nltk
from nltk.tokenize import word_tokenize

## Sample Text
my_text = "The Omicron surge of coronavirus cases had not yet peaked in the US
and the country is expected to see an increase in hospitalisations and deaths
in the next few weeks"
```

Tokenization

```
from nltk.tokenize import word_tokenize
mytext_tokens = word_tokenize(my_text)
print(mytext_tokens)
```

Figure 15-14 shows the input and output.

```
In [4]:  ##Import NLTK Library
         import re
         import nltk
         from nltk.tokenize import word_tokenize

In [5]:  my_text

Out[5]:  'The Omicron surge of coronavirus cases had not yet peaked in the US and th
         e country is expected to see an increase in hospitalisations and deaths in
         the next few weeks'
```

Tokenization

```
In [7]:  from nltk.tokenize import word_tokenize
         mytext_tokens = word_tokenize(my_text)
         print(mytext_tokens)

         ['The', 'Omicron', 'surge', 'of', 'coronavirus', 'cases', 'had', 'not', 'ye
         t', 'peaked', 'in', 'the', 'US', 'and', 'the', 'country', 'is', 'expected',
         'to', 'see', 'an', 'increase', 'in', 'hospitalisations', 'and', 'deaths',
         'in', 'the', 'next', 'few', 'weeks']
```

Figure 15-14. *Tokenization using NLTK*

This method does not deal appropriately with apostrophes. For example, *I'm* and *we'll* should be ideally split into "I am" and "we will." But, this method does not do it.

Let's see how our library handles this situation with another example.

Here is the input:

```
sent = "When I'm hungry, I do not eat"

sent

sent.split()

sent2="I love Apple's iPhone"

sent2

word_tokenize(sent2)
```

Figure 15-15 shows the input and output.

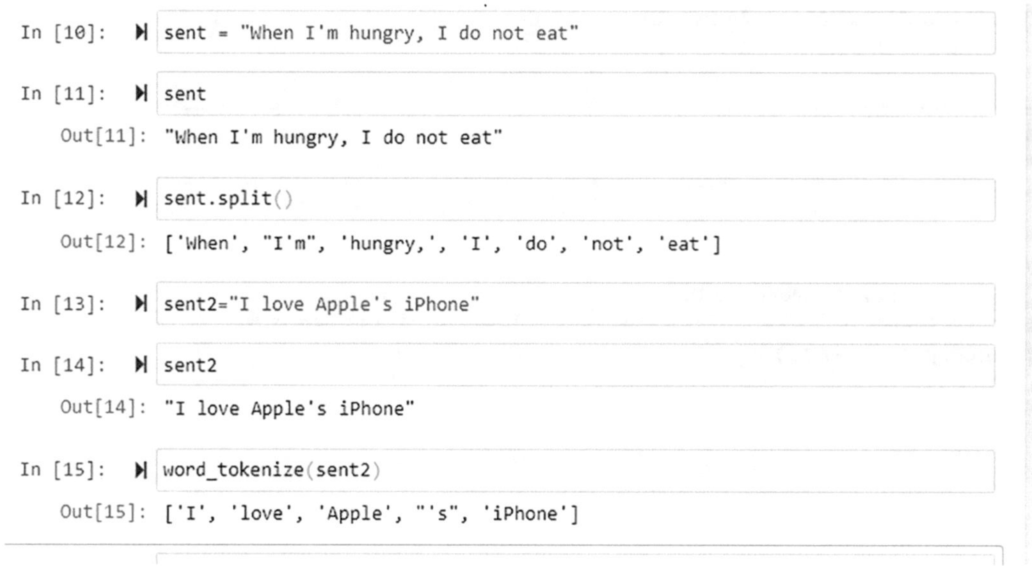

```
In [10]: sent = "When I'm hungry, I do not eat"

In [11]: sent
Out[11]: "When I'm hungry, I do not eat"

In [12]: sent.split()
Out[12]: ['When', "I'm", 'hungry,', 'I', 'do', 'not', 'eat']

In [13]: sent2="I love Apple's iPhone"

In [14]: sent2
Out[14]: "I love Apple's iPhone"

In [15]: word_tokenize(sent2)
Out[15]: ['I', 'love', 'Apple', "'s", 'iPhone']
```

***Figure 15-15.** Tokenizing apostrophes*

As you can see from the previous example, the tokenizer did not do a good job with apostrophes. Nevertheless, such situations should be handled appropriately using regex.

There are other tokenizer libraries available that do the same job. Some other tokenizers include Treebank Tokenizer, WordPunct Tokenizer, and TweetsTokenizer. The TreebankTokenizer does a better job splitting words such as *I'm* into *I* and *m*.

15.8.3 Lemmatization

Lemmatization is the process of removing ending letters to create a base form of the word. For example, the word *car* may appear in different forms in the whole document such as *car, cars, car's, cars'*, etc., and the lemmatization process brings these to a single base form of *car*. The base form of the dictionary word is called the *lemma*. Several lemmatizer libraries are available such as WordNet, Spacy, TextBlob, and Gensim. The most commonly used lemmatizer is the WordNet lemmatizer. In our example, we will explore the WordNet and Spacy lemmatizers.

Here is the input:

```
from nltk.stem import WordNetLemmatizer
my_text

##Lemmatization
from nltk.stem import WordNetLemmatizer
lemma = WordNetLemmatizer()
mytext_lemma = ' '.join([lemma.lemmatize(word) for word in mytext_tokens])
print(mytext_lemma)
```

Figure 15-16 shows the input and output.

NLTK WordNet Lemmatizer

```
In [1]:  from nltk.stem import WordNetLemmatizer
         from nltk.stem import PorterStemmer

In [8]:  my_text

Out[8]:  'The Omicron surge of coronavirus cases had not yet peaked in the US and th
         e country is expected to see an increase in hospitalisations and deaths in
         the next few weeks'

In [10]: ##Lemmatization
         from nltk.stem import WordNetLemmatizer
         lemma = WordNetLemmatizer()
         mytext_lemma = ' '.join([lemma.lemmatize(word) for word in mytext_tokens])
         print(mytext_lemma)

         The Omicron surge of coronavirus case had not yet peaked in the US and the
         country is expected to see an increase in hospitalisation and death in the
         next few week
```

Figure 15-16. *Lemmatization using WordNetLemmatizer()*

The first example, shown in Figure 15-16, is using `WordNetLemmatizer()`, and the second example, as shown in Figure 15-17, is using the Spacy lemmatizer. As we can see from the example, the Spacy lemmatizer has done a better job than the NLTK WordNet lemmatizer. The Spacy lemmatizer was able to remove *ed* from the two words *peaked* and *expected,* which WordNet missed.

Here is the input:

```
#Spacy Lemmatizer
import spacy
load_spacy = spacy.load('en_core_web_sm')
spacy_lemma = load_spacy(my_text)
print(" ".join([word.lemma_ for word in spacy_lemma]))
```

Figure 15-17 shows the input and output.

```
Spacy Lemmatizer

In [11]:  import spacy
          load_spacy = spacy.load('en_core_web_sm')
          spacy_lemma = load_spacy(my_text)
          print(" ".join([word.lemma_ for word in spacy_lemma]))

          the Omicron surge of coronavirus case have not yet peak in the US and the c
          ountry be expect to see an increase in hospitalisation and death in the nex
          t few week
```

Figure 15-17. *Lemmatization using Spacy()*

15.8.4 Stemming

Stemming is a process where words are reduced to their root form. As part of stemming, the inflationary form of the word is changed to the base form called the *stem*. The affixes are chopped off during the stemming process. For example, *digit, digitization,* and *digital* will be stemmed to get *digit*. The stem may not be always a valid dictionary word.

The two most common stemmers used are the Porter stemmer and the Snowball stemmer. The Porter stemmer was the earliest stemmer developed. The Snowball stemmer supports multiple languages, whereas the Porter stemmer supports only the English language. The following examples demonstrate the stemming process using the Porter stemmer and the Snowball stemmer.

Here's the input:

```
my_text

from nltk.stem import PorterStemmer
ps = PorterStemmer()
myporter_stem = ' '.join([ps.stem(word) for word in mytext_tokens])
print(myporter_stem)
```

Figure 15-18 shows the input and output.

```
[12]: my_text

[12]: 'The Omicron surge of coronavirus cases had not yet peaked in the US and the country is expected to see an in
      crease in hospitalisations and deaths in the next few weeks'

[13]: from nltk.stem import PorterStemmer
      ps = PorterStemmer()
      myporter_stem = ' '.join([ps.stem(word) for word in mytext_tokens])
      print(myporter_stem)

      the omicron surg of coronaviru case had not yet peak in the US and the countri is expect to see an increas in
      hospitalis and death in the next few week
```

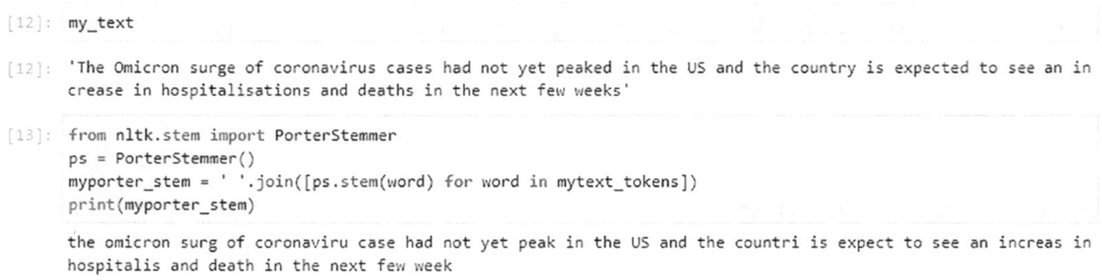

Figure 15-18. *Stemming process using the Porter stemmer*

Another example is using the Snowball `stemmer()` in Figure 15-19.

Here's the input:

```
#SnowballStemmer()
from nltk.stem.snowball import SnowballStemmer
print(SnowballStemmer.languages)


snow_stem= SnowballStemmer(language='english')

snowballstem_words = ' '.join([snow_stem.stem(word) for word in mytext_tokens])
print(snowballstem_words)
```

Figure 15-19 shows the input and output.

```
[14]: from nltk.stem.snowball import SnowballStemmer
      print(SnowballStemmer.languages)

      ('arabic', 'danish', 'dutch', 'english', 'finnish', 'french', 'german', 'hungarian', 'italian', 'norwegian',
      'porter', 'portuguese', 'romanian', 'russian', 'spanish', 'swedish')

[15]: snow_stem= SnowballStemmer(language='english')

[16]: snowballstem_words = ' '.join([snow_stem.stem(word) for word in mytext_tokens])
      print(snowballstem_words)

      the omicron surg of coronavirus case had not yet peak in the us and the countri is expect to see an increas i
      n hospitalis and death in the next few week

[ ]:
```

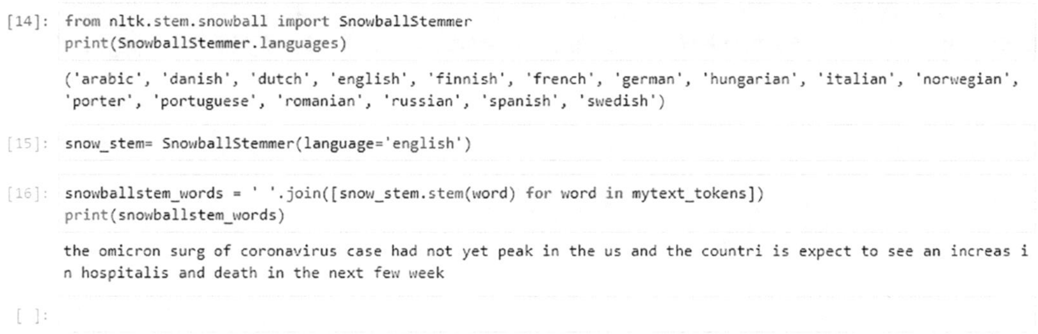

Figure 15-19. *Stemming process using the Snowball stemmer*

As you can see from the previous example, the Porter stemmer stemmed the words *surge, cases, peak, expect, increase, hospitalization*, and *weeks*, and the Snowball stemmer also stemmed the words *surge, coronavirus, cases, peak, expect, increase, hospitalization*, and *weeks*. Both stemmers have done a similar job. Practically you could use either stemmer.

15.8.5 Stop Word Removal

Words such as *a, an, in*, and *it* do not carry much information and occur frequently in the text. Without these words, the contextual meaning of the text does not change. These words are required only to make the sentence grammatically correct and complete the sentence. In NLP, these words are referred to as *stop words*, and in most cases they are filtered out. This also reduces the vocabulary size and improves computational efficiency. Though there are libraries that support stop words, they may not be comprehensive and specific to the use cases. You can add your own list of words as per your requirements based on the problem you are working on.

Here is the input:

#NLTK Stopwords

```
from nltk.corpus import stopwords

nltk.download('stopwords')
stop_words = nltk.corpus.stopwords.words('english')
print(stop_words)

mytext_nostop = ' '.join([word for word in mytext_tokens if word not in stop_words])
print(mytext_nostop)
```

Figure 15-20 shows the input and output.

```
[34]: from nltk.corpus import stopwords

[37]: nltk.download('stopwords')
      stop_words = nltk.corpus.stopwords.words('english')
      print(stop_words)

      ['i', 'me', 'my', 'myself', 'we', 'our', 'ours', 'ourselves', 'you', "you're", "you've", "you'll", "you'd",
      'your', 'yours', 'yourself', 'yourselves', 'he', 'him', 'his', 'himself', 'she', "she's", 'her', 'hers', 'her
      self', 'it', "it's", 'its', 'itself', 'they', 'them', 'their', 'theirs', 'themselves', 'what', 'which', 'wh
      o', 'whom', 'this', 'that', "that'll", 'these', 'those', 'am', 'is', 'are', 'was', 'were', 'be', 'been', 'bei
      ng', 'have', 'has', 'had', 'having', 'do', 'does', 'did', 'doing', 'a', 'an', 'the', 'and', 'but', 'if', 'o
      r', 'because', 'as', 'until', 'while', 'of', 'at', 'by', 'for', 'with', 'about', 'against', 'between', 'int
      o', 'through', 'during', 'before', 'after', 'above', 'below', 'to', 'from', 'up', 'down', 'in', 'out', 'on',
      'off', 'over', 'under', 'again', 'further', 'then', 'once', 'here', 'there', 'when', 'where', 'why', 'how',
      'all', 'any', 'both', 'each', 'few', 'more', 'most', 'other', 'some', 'such', 'no', 'nor', 'not', 'only', 'ow
      n', 'same', 'so', 'than', 'too', 'very', 's', 't', 'can', 'will', 'just', 'don', "don't", 'should', "should'v
      e", 'now', 'd', 'll', 'm', 'o', 're', 've', 'y', 'ain', 'aren', "aren't", 'couldn', "couldn't", 'didn', "did
      n't", 'doesn', "doesn't", 'hadn', "hadn't", 'hasn', "hasn't", 'haven', "haven't", 'isn', "isn't", 'ma', 'migh
      tn', "mightn't", 'mustn', "mustn't", 'needn', "needn't", 'shan', "shan't", 'shouldn', "shouldn't", 'wasn', "w
      asn't", 'weren', "weren't", 'won', "won't", 'wouldn', "wouldn't"]

      [nltk_data] Downloading package stopwords to
      [nltk_data]     C:\Users\phdst\AppData\Roaming\nltk_data...
      [nltk_data]   Package stopwords is already up-to-date!

[40]: mytext_nostop = ' '.join([word for word in mytext_tokens if word not in stop_words])
      print(mytext_nostop)

      The Omicron surge coronavirus cases yet peaked US country expected see increase hospitalisations deaths next
      weeks
```

Figure 15-20. *Removing stop words using the NLTK library*

In the previous example, as shown in Figure 15-20, NLTK removed the common stop words from the text; however, *the* was not removed as it is not in the NLTK stop words corpus. If you want to remove such common and frequently appearing words that may not add much value to the sentence, you can create your own dictionary of such words and apply it to the text before further processing.

15.8.6 Part-of-Speech Tagging

Part-of-speech (POS) tagging is a process where the algorithm reads the text and assigns parts of speech associated to each word. By the end of this process, each word is annotated and assigned with its POS tag such as verb, noun, adverb, preposition, etc. The POS tag aims to help parse text and resolve word sense disambiguation. Sometimes POS tagging is also used for identifying named entities, coreference resolution, and speech recognition. It is impossible to manually annotate each word and assign POS tags in a corpus. Fortunately, many libraries and corpus are available with POS tags assigned for the words. The most popular POS tag corpus are the following:

- *Brown corpus*: Developed by Francis in 1979 at Brown University; has 87 tags
- *Penn Treebank*: Created by Marcus et al. in 1993 at the University of Pennsylvania; has 45 tags
- *British National Corpus (BNC)*: Developed by Garside et al. in 1997; has 61 tags in the C5 tagset, known as the basic tags version, and 146 tags in the C7 tagset, also known as the enriched tag version

Figure 15-21 shows a sample POS tagset from the Penn Treebank corpus; for example, tag JJ means it is an adjective, VB is for verb, and RB is for adverb.

Tag	Description	Example	Tag	Description	Example
CC	coordin. conjunction	*and, but, or*	SYM	symbol	+,%, &
CD	cardinal number	*one, two, three*	TO	"to"	*to*
DT	determiner	*a, the*	UH	interjection	*ah, oops*
EX	existential 'there'	*there*	VB	verb, base form	*eat*
FW	foreign word	*mea culpa*	VBD	verb, past tense	*ate*
IN	preposition/sub-conj	*of, in, by*	VBG	verb, gerund	*eating*
JJ	adjective	*yellow*	VBN	verb, past participle	*eaten*
JJR	adj., comparative	*bigger*	VBP	verb, non-3sg pres	*eat*
JJS	adj., superlative	*wildest*	VBZ	verb, 3sg pres	*eats*
LS	list item marker	*1, 2, One*	WDT	wh-determiner	*which, that*
MD	modal	*can, should*	WP	wh-pronoun	*what, who*
NN	noun, sing. or mass	*llama*	WP$	possessive wh-	*whose*
NNS	noun, plural	*llamas*	WRB	wh-adverb	*how, where*
NNP	proper noun, singular	*IBM*	$	dollar sign	$
NNPS	proper noun, plural	*Carolinas*	#	pound sign	#
PDT	predeterminer	*all, both*	"	left quote	' or "
POS	possessive ending	*'s*	"	right quote	' or "
PRP	personal pronoun	*I, you, he*	(	left parenthesis	[, (, {, <
PRP$	possessive pronoun	*your, one's*	)	right parenthesis	],), }, >
RB	adverb	*quickly, never*	,	comma	,
RBR	adverb, comparative	*faster*	.	sentence-final punc	. ! ?
RBS	adverb, superlative	*fastest*	:	mid-sentence punc	: ; ... – -
RP	particle	*up, off*			

Figure 15-21. *Penn Treebank P.O.S. tags (upenn.edu)*

The manual design of POS tags by human experts requires linguistic knowledge. It is laborious to manually annotate and assign POS tags to each word and each sentence of the text in a corpus. Instead, many computer-based algorithms are used, and one such popular method is a sequence model using hidden Markov models (HMMs). Maximum entropy models (MaxEnt) and sequential conditional random fields (CRFs) can also be used to assign POS tags. We will not discuss HMM, MaxEnt, or CRF POS tags models here. You can refer to the published papers by the inventors given in the reference section to learn more about the models for POS tags.

As shown in Figure 15-22, we tag single words using the NLTK library and Penn Treebank POS. The NLTK POS tagger has tagged both *dog* and *beautiful* as nouns.

Here's the input:

```
#POS TAGGING
nltk.pos_tag(['beautiful'])

nltk.pos_tag(['dog'])
```

Figure 15-22 shows the input and output.

```
[44]: nltk.pos_tag(['beautiful'])
[44]: [('beautiful', 'NN')]
[45]: nltk.pos_tag(['dog'])
[45]: [('dog', 'NN')]
```

Figure 15-22. *POS Tag using Penn Treebank*

In the following example, as shown in Figure 15-23, we tag each word in a sentence using the NLTK library.

Here's the input:

```
mynew_text = 'Novak Djokovic lost 2021 US Open to Daniil Medvedev in the final'
[nltk.pos_tag([word]) for word in mynew_text.split(' ')]
```

Figure 15-23 shows the input and output.

```
[47]: mynew_text = 'Novak Djokovic lost 2021 US Open to Daniil Medvedev in the final'
      [nltk.pos_tag([word]) for word in mynew_text.split(' ')]

[47]: [[('Novak', 'NN')],
       [('Djokovic', 'NN')],
       [('lost', 'VBN')],
       [('2021', 'CD')],
       [('US', 'NN')],
       [('Open', 'VB')],
       [('to', 'TO')],
       [('Daniil', 'NN')],
       [('Medvedev', 'NN')],
       [('in', 'IN')],
       [('the', 'DT')],
       [('final', 'JJ')]]
```

Figure 15-23. *POS tagging of a sentence using NLTK*

The tagger parsed the sentences and tagged each word. It has properly identified and tagged them, for example, *Novak* as Noun, *Open* as Verb, etc.

You do not have to memorize the tags and their association. You can call `help` to learn more about the meaning of different POS tags used by the library.

Here's the input:

```
import nltk
nltk.download('tagsets') ##First time only
nltk.help.upenn_tagset()
```

Figure 15-24 shows the input and output.

```
import nltk
nltk.download('tagsets') ##First time only
nltk.help.upenn_tagset()
```

```
$: dollar
    $ -$ --$ A$ C$ HK$ M$ NZ$ S$ U.S.$ US$
'': closing quotation mark
    ' ''
(: opening parenthesis
    ( [ {
): closing parenthesis
    ) ] }
,: comma
    ,
--: dash
    --
.: sentence terminator
    . ! ?
:: colon or ellipsis
    : ; ...
CC: conjunction, coordinating
    & 'n and both but either et for less minus neither nor or plus so
    therefore times v. versus vs. whether yet
CD: numeral, cardinal
    mid-1890 nine-thirty forty-two one-tenth ten million 0.5 one forty-
    seven 1987 twenty '79 zero two 78-degrees eighty-four IX '60s .025
    fifteen 271,124 dozen quintillion DM2,000 ...
DT: determiner
    all an another any both del each either every half la many much nary
    neither no some such that the them these this those
EX: existential there
    there
FW: foreign word
    gemeinschaft hund ich jeux habeas Haementeria Herr K'ang-si vous
    lutihaw alai je jour objets salutaris fille quibusdam pas trop Monte
    terram fiche oui corporis ...
IN: preposition or conjunction, subordinating
    astride among uppon whether out inside pro despite on by throughout
    below within for towards near behind atop around if like until below
    next into if beside ...
JJ: adjective or numeral, ordinal
    third ill-mannered pre-war regrettable oiled calamitous first separable
    ectoplasmic battery-powered participatory fourth still-to-be-named
    multilingual multi-disciplinary ...
```

Figure 15-24. *Penn treebank POS tags list*

15.8.7 Probabilistic Language Model

A sentence is nothing but a sequence of words in a specific order. In speech recognition or machine translation, applications often have to predict the next word that occurs in a sentence. For example, what is the next word in this sentence?

It will also cut down the _______.

The next word depends on the previous word. The next word could be a verb or it could be noun. It depends on the probability of association of the previous word. We can build such models if we know the probability of each word and its occurrence in a corpus. This approach in NLP is the *language model* (LM). In the LM, the probability of the words depends on the training set. Though linguists prefer grammar models, the language model has given good results and has become the standard in NLP.

15.8.8 N-gram Language Model

In a language model, we can calculate the probability of occurrence for a one-word sequence, two-word sequence, three-word sequence, four-word sequence, or N-word, also called N-gram. N can be 1, 2, or 3. If N is 1, then it is unigram; if N is 2, then it is bigram; and so on.

For instance, in the previous example, the unigram sequence would be simply *It, will, also, cut, down, the*. Similarly, the bigram model would be a two-word sequence of words, for example, *it will, will also, also cut*, and so on.

To predict the probability of the next word, we have to estimate the probability of other words in the sequence. The simplest form of the language model is the *unigram language model,* which estimates each term independently and deletes all the conditions.

To predict the next word $\mathbf{w}_k$, given $w_1, w_2, w_3, \ldots.w_{n-1} = P(w_k|w_1,w_2,\ldots w_{n-1})$, for a unigram:

$P_{uni}(w_1,w_2w_3w_4) = P(w_1)P(w_2)P(w_3)P(w_4)$

Similarly, for the bigram language model:

$P_{bigram}(w_1,w_2w_3w_4) = P(w_1)P(w_2 \mid w_1)P(w_3 \mid w_2)P(w_4 \mid w_3)$

For example, we can compute the probability of a sentence using the chain rule of probability:

$P(w_1w_2w_3..w_n) = p(w_1)p(w_2|w_1)p(w_3|w_1w_2)\ldots\ldots.p(w_n|w_1w_2w_3\ldots w_{n-1})$

We can compute the probability of a sentence. For example, the probability of the sentence "hard to find a true friend" is as follows:

P("hard to find a true friend") = P(hard) X P(to|hard) X P(find|hard to) X P(true|hard to find a) X P(friend|its hard to find a)

Using Markov model assumptions, we can approximate the model to the last word instead of all the words of the context. Simplifying assumptions using 1-gram:

P("hard | to find a true friend") = P(hard| friend)

Or using bigram:

P("hard | to find a true friend") = P(hard| true friend)

In summary, for predicting the next word w* given the history w_1, w_2, w_3,....w_{n-1}, we have this:

Unigram model: $P(w_n | w_{n-0})$

Bigram model: $P(w_n|w_{n-0},w_{n-1})$

Trigram model: $P(w_n|w_{n-0},w_{n-1},w_{n-2})$

Language models have applications used in speech recognition, part-of-speech tagging, and optical character recognition. They are vital in spelling correction and machine translations. Though this model is a good approximation for many NLP applications, it is insufficient because English and any natural language are not always simple as a set of sequence of structured words. Language has what linguists call *long-distance dependencies*. There are many more complex models such as trigram and probabilistic context-free grammars (CFGs).

Several language modeling toolkits are publicly available. The SRI Language Modeling Toolkit and Google N-Gram model are popular.

15.9 Representing Words as Vectors

So far we have looked at processing text using the heuristic approach, probability, and language modeling techniques. In this section, we will discuss how text data can be represented as vectors and matrices. This representation helps in performing vector algebra on the text data. Once text data is represented in vector form, it is easy to find synonyms, antonyms, similar documents, etc.

A vector is a quantity that has both a magnitude and a direction, whereas a quantity with a magnitude but no direction is referred to as a *scalar quantity*. Examples of vector quantities are acceleration, force, weight, and thrust. Examples of scalar quantities are time, temperature, speed, etc.

For example, a document vector can be represented as a matrix, as in the following example. Each sentence is tokenized and represented as single words in a row, thus forming a matrix of words, as shown in Figure 15-25. We can also represent the documents in a similar context.

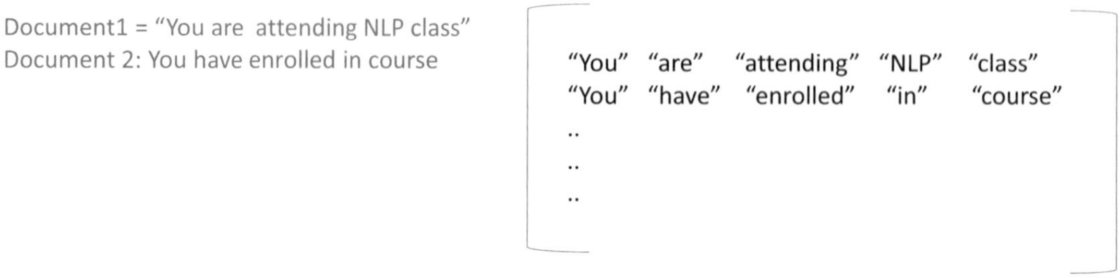

Document1 = "You are attending NLP class"
Document 2: You have enrolled in course

"You"	"are"	"attending"	"NLP"	"class"
"You"	"have"	"enrolled"	"in"	"course"
..				
..				
..				

Figure 15-25. *Matrix representation of documents (sentences)*

15.9.1 Bag-of-Words Modeling

The machine learning models work with numerical data rather than textual data. Using the bag-of-words (BoW) techniques, we can convert text into equivalent fixed-length vectors. The BoW model provides a mechanism for representing textual data as vectors. The model relies on the count of actual terms present in a document. Each entry in the vector corresponds to a term in the vocabulary, and the number in that particular entry indicates the frequency of the term that appeared in the sentence under consideration. Once the vocabulary is available, each sentence can be represented as a vector using the BoW model. This technique is also popularly known as *one-hot vectorization*.

Assume that a document has two sentences, shown here:

Sentence 1: This is course on NLP.

Sentence 2: Welcome to NLP class. Let's start learning NLP.

Using the BoW technique, we can represent both sentences as vectors after obtaining the list of vocabulary words. In this document, the vocabulary size is 11. Figure 15-25 represents the vector representation of two sentences in the document. If the word appeared in the sentence, it would be counted as 1; otherwise, it is 0, as shown in Figure 15-26. For example, in sentence 2, NLP has appeared twice, and hence the NLP word has 2 counts, and the rest have only 1.

	Welcome	To	NLP	Class	Let's	Start	Learning	This	Is	Course	on
Sentence 1	0	0	1	0	0	0	0	1	1	1	1
Sentence 2	1	1	2	1	1	1	1	0	0	0	0

Figure 15-26. *Vector representation of text*

The sentence 1 and sentence 2 vector representation is as follows:

Sentence 1 = [0 0 1 0 0 0 0 1 1 1 1]

Sentence 2 = [1 1 2 1 1 1 1 0 0 0 0]

This model does not take semantics and grammar structure into account. Hence, it does not capture the word's context in a phrase. Also, the BoW model suffers from poor performance and scalability for a large vocabulary.

We use the sklearn library's `Vectorization()` function to achieve the vector representation of a document. The `CountVectorizer()` module helps us vector each sentence in a document and then combines the vectors into a document vector to create the matrix.

Here's the input:

```
document

from sklearn.feature_extraction.text import CountVectorizer
my_vectorizer = CountVectorizer(stop_words = 'english') #Removing stop words
document_vector = my_vectorizer.fit_transform(document) #creating to BoW-model
print(document_vector.todense())

#[[1 0 0 0 1 1 0 1 1 0 0]
# [0 1 0 0 2 0 1 1 1 0 1]
# [0 0 1 1 1 0 0 1 1 1 0]]

##print the vocabulary words
print(my_vectorizer.get_feature_names())
```

Figure 15-27 shows the input and output.

```
[4]: document

[4]: ['We are learning Natural Language Processing in this class',
      'Natural Language Processing is all about how to make computers to understand language',
      'Natural Language Processing techniques are evolving everyday']

[5]: from sklearn.feature_extraction.text import CountVectorizer
     my_vectorizer = CountVectorizer(stop_words = 'english') #Removing stop words
     document_vector = my_vectorizer.fit_transform(document) #creating to BoW-model
     print(document_vector.todense())

     [[1 0 0 0 1 1 0 1 1 0 0]
      [0 1 0 0 2 0 1 1 1 0 1]
      [0 0 1 1 1 0 0 1 1 1 0]]

[7]: ##print the vocabulary words
     print(my_vectorizer.get_feature_names())

     ['class', 'computers', 'everyday', 'evolving', 'language', 'learning', 'make', 'natural', 'processing', 'techniques', 'understan
     d']
```

Figure 15-27. *BoW model using the CounterVectorizer() module*

15.9.2 TF-IDF Vectors

As we noticed, the BoW model considers the frequency of words that appear across a document. If a word is not present, it is ignored or given low weightage. This type of vector representation does not consider the importance of the word in the overall context. Some words may occur in only one document but carry important information, whereas other words that appear more frequently in a document may not carry that much information, and hence the pattern that can be found across similar documents may be lost. This weighting scheme is called TF-IDF. The TF-IDF vector representation mitigates such issues in the text corpus.

15.9.3 Term Frequency

Term frequency (TF) is how frequently a term occurs in a document. In a given a corpus, each document size can vary; it is more likely that a term could appear more frequently in a bigger document than smaller ones. Normalizing the vectors makes the vector size the same throughout the corpus. One representation of TF is to normalize vectors by dividing the frequency of the term by the total terms in the document. The representation to calculate TF is as follows:

$$tf_{w,d} = \text{Count of word w in a document} / \text{Count of total words in the document} \quad (1)$$

Thus, with this normalized representation, two documents of different length, "John is driving faster" and "Jack is driving faster than John," seem to be identical.

15.9.4 Inverse Document Frequency

In TF, all terms have equal weights. Some terms occur in every document, and some occur less frequently. This is a problem in TF. To overcome this problem, consider the number of documents in the corpus. This helps in assigning proper weights to the words based on how frequently a word occurs in different documents of the corpus. This method is referred to as *inverse document frequency* (IDF). The common representation of calculating IDF is as follows:

$$\text{idf}_w = \log(\text{total number of documents} / \text{number of documents containing word w}) \quad (2)$$

In Figure 15-28, IDF values for different words are shown for the documents in the corpus. In this example, there are four words (w1, w2, w3, and w4), and the table shows the frequency of these terms appearing in different documents (d1, d2, d3, and d4). As you can see from the example, even though w3 and w4 appear less frequently, they have higher IDF values.

	w1	w2	w3	w4
d1	16	10	2	9
d2	15	4	1	6
d3	17	6	1	5
d4	18	7	2	3
idf	0.266816	0.654996	1.308209	0.724632

Figure 15-28. *Example of calculating IDF*

15.9.5 TF-IDF

By combining the two, i.e., TF and IDF, you obtain a composite weight called TF-IDF.

Multiplying both TF and IDF, the term frequency and document frequency of the word w is as follows:

$$\text{tfidf}(w,d) = \text{tf}(w,d) \ X \ \text{idf}_w \tag{3}$$

The TF-IDF of the word in a document is the product of TF of word w in document times the IDF of word w across the entire corpus.

TF-IDF for term w and document d assigns the following:

- A higher score when the term w occurs more frequently in a small number of documents
- A lower score when the term w occurs fewer times in a document or occurs in a document
- A lower score when the term w occurs in many documents
- The lowest score when the term w occurs in every document or almost all documents

TF-IDF is also a vector representation of each document with each element corresponding to each word in the dictionary.

Most packages, including the sklearn package, provide a function to create a TFIDF vector representation. In this example, we demonstrate TF-IDF using the library `TfidfVectorizer()` function supported by the sklearn package. The first step is to initiate the `TfidfVectorizer` and then call the `fit_transform()` function to create the vector.

Here is the input:

```
#TF-IDF - Term Frequency (TF) and Inverse Document Frequency (IDF)¶
my_corpus  # Corpus containing three documents d1, d2, d3

from sklearn.feature_extraction.text import TfidfVectorizer
tfidf = TfidfVectorizer()  # Initialize my tfidf function
tfidf_matrix = tfidf.fit_transform(my_corpus) # transform my_corpus to TFIDF vector
print(tfidf_matrix.toarray())  # Print TF-IDF vector
print(tfidf.get_feature_names()) # Print the vocabulary
```

Figure 15-29 shows the input and output.

```
[41]: my_corpus  # Corpus containing three documents d1, d2, d3

[41]: ['We are learning Natural Language Processing in this class',
       'Natural Language Processing is all about how to make computers to understand language',
       'Natural Language Processing techniques are evolving everyday']

[42]: from sklearn.feature_extraction.text import TfidfVectorizer
      tfidf = TfidfVectorizer()  # Initialize my tfidf function
      tfidf_matrix = tfidf.fit_transform(my_corpus) # transform my_corpus to TFIDF vector
      print(tfidf_matrix.toarray())  # Print TF-IDF vector
      print(tfidf.get_feature_names()) # Print the vocabulary

      [[0.         0.         0.29547781 0.38851782 0.         0.
        0.         0.         0.38851782 0.         0.22946488 0.38851782
        0.         0.22946488 0.22946488 0.         0.38851782 0.
        0.         0.38851782]
       [0.27636371 0.27636371 0.         0.         0.27636371 0.
        0.         0.27636371 0.         0.27636371 0.32644971 0.
        0.27636371 0.16322486 0.16322486 0.         0.         0.55272741
        0.27636371 0.        ]
       [0.         0.         0.35364183 0.         0.         0.46499651
        0.46499651 0.         0.         0.         0.27463443 0.
        0.         0.27463443 0.27463443 0.46499651 0.         0.
        0.         0.        ]]
      ['about', 'all', 'are', 'class', 'computers', 'everyday', 'evolving', 'how', 'in', 'is', 'language', 'learning',
      'make', 'natural', 'processing', 'techniques', 'this', 'to', 'understand', 'we']
```

***Figure 15-29.** TF-IDF representation using sklearn*

The function `TfidfVectorizer()` transforms the document into a TF-IDF vector form. The output of the function can be verified by printing both vectors and the TF-IDF score of each word that appeared in the document.

15.10 Text Classifications

In the case of library management, we want to classify library books into different categories. In the case of email, it is required to separate spam emails. Similarly, if you have a lot of customer feedback and reviews, you want to identify the behavioral pattern and classify them into different behaviors. While understanding the grammar and semantics of a language is still a distant goal, researchers have taken a divide-and-conquer approach. According to Manning (2009), sentences can be clearly grammatical or ungrammatical. Most of the time words are used as a single part of speech, but not always. Our discussions have shown how words in a language can be explained using probability theory. Further, we have also shown how we can represent text in a mathematical structure as vectors. Since we have learned how to convert words to vectors and apply probability theory, can we use advanced machine learning techniques to solve this problem?

Given a set of documents $D = \{d_1,d_2,d_3....d_n\}$, each with a label (category), we can train a classifier using supervised machine learning. For any unlabeled document, our machine learning classifier model should be able to *predict* the class of the document. If we have document *d*, from a fixed set of categories $C = \{c1,c2,...cn\}$, then after the model is trained, the classifier should be able to predict the class of a new document for $\boldsymbol{d} \in \boldsymbol{C}$, as shown in Figure 15-30.

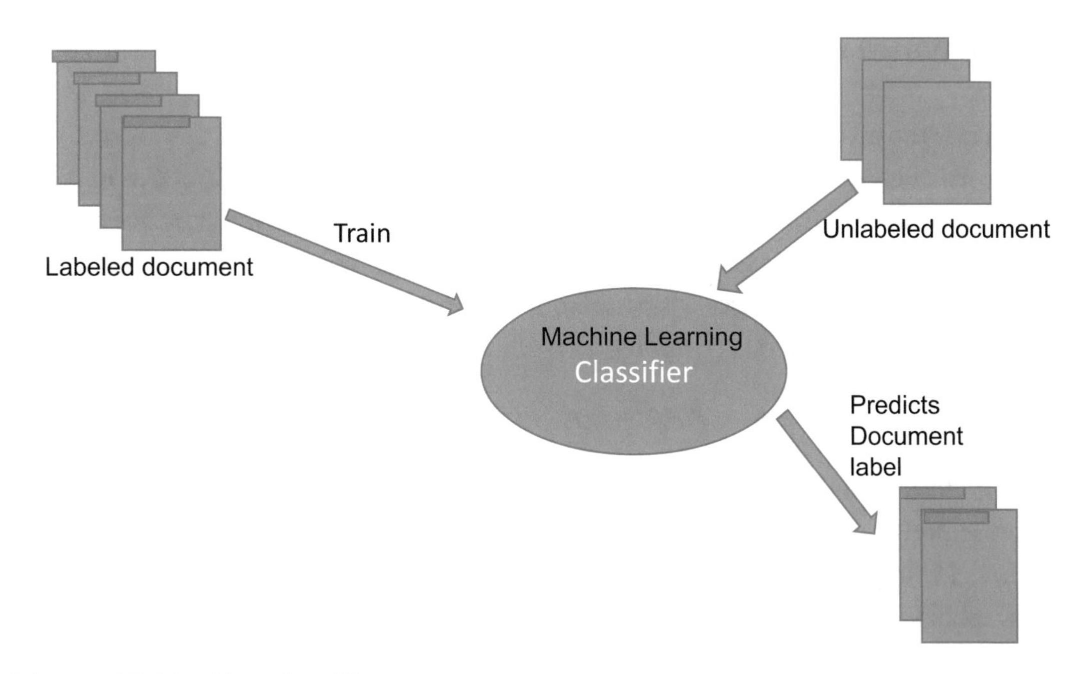

Figure 15-30. *Text classifier*

This task in NLP is referred to as *text classification*. We can use any supervised machine learning classification technique to achieve text classification. The most commonly used classifiers are naïve Bayes, SVM, and neural nets. The naïve Bayes classifier is a simple probabilistic classifier based on the Bayes theorem with an assumption that each word in a sentence is independent of the others. It is one of the most basic classification techniques used in many applications such as email spam detection, email sorting, and sentiment analysis. Even though naïve Bayes is a simple technique, research has shown it gives good performance in many complex real-world problems. Although conditional independence does not hold in real-world situations, naïve Bayes performs well.

The naïve Bayes (NB) classifier is derived based on the simple Bayes rule. Classification is done based on the highest probability score calculated using the following formula:

$$P\left(C_i|,X_1|,X_2|,X_3|,\ldots X_p\right)=\frac{P\left(X_1,|X_2,|X_3,|\ldots X_p,||C_i\right)P\left(C_i\right)}{P\left(X_1,|X_2,|\ldots X_p,|C_1\right)+P\left(X_1,|X_2,|\ldots X_p,|C_2\right)\ldots\ldots P\left(X_1,|X_2,|\ldots X_p,|C_m\right)}$$

Here, $P(C_i)$ is the prior probability of belonging to class C_i in the absence of any other attributes.

$(C_i |X_i)$ is the posterior probability of X_i belonging to class C_i. To classify a record using Bayes' theorem, first compute its chance belonging to each class C_i. Naïve Bayes assumes that the predictor attributes are all mutually independent within each class; that is, the order of the word does not matter.

Applying the Bayes rule to classify document d and a class C, the probability of a document d being in class C is computed as follows:

$$P(c|d) = P(c) . P(d|c) / P(d)$$

Here, *Document* $d \in D$, where D denotes the training document set. Document d can be represented as a bag of words (the order and position of words does not matter). Each word $w \in d$ comes from a set W of all the feature words (vocabulary set).

To classify a new document, the product of the probability of each word of the document given a particular class (likelihood) is estimated. Then, this product is multiplied by the probability of each class (prior). Probabilities are calculated using the previous equation for each class. The one with the highest probability decides the final class of the new document.

When the previous calculations are done using computers, sometimes it may lead to floating-point errors. It is, therefore, better to perform computation by adding algorithms of probabilities. The class with the highest log probability is still the most significant class.

Table 15-1 illustrates an example of sample documents containing documents with known and unknown classes. The training set contains four documents with a known class and one document with an unknown class.

Table 15-1. *Example of Probabilistic Classification Model (Naïve Bayes)*

	DocID	**Document Words**	**Positive Class or Negative Class (c1 and c2)**
Training set	1	Very good movie	C1
	2	Enjoyed movie	C1
	3	Excellent acting	C1
	4	Bad movie	C2
	5	Worst movie	C2
Test data (unknown class)	6	Bad movie.	?

We will illustrate how to find the class of unknown document using the naive Bayes algorithm.

Step 1: Calculate the prior probability of C P(c) :

P(Positive) = N_{pos} / (N_{pos} + N_{neg}) = 3/(2+3) = 3/5 = 0.6 ....P (c1)

P(Negative) = N_{neg} / (N_{pos}+N_{neg}) = 2/(3+2) = 2/5 = 0.4 ....P(c2)

Step 2: Using add-one or laplace smoothing, calculate the probability of words as shown:

P(very|c1) = 1+count(very,positive) /|vocab| + count(pos) (using laplace smoothing)

P(very|c1) = 1+1 / (7+9) = 2/16 = 0.125

(There are a total of nine words including seven positive words.)

Step 3: Similarly, calculate the probability of other words.

P(good|positive) = (1+1)/(7+9) = 2/16 = 0.125

P(movie|positive) = (2+1)/(7+9) = 3/16 = 0.1875

P(bad|positive) = (0+1)/(7+9) = 1/16 = 0.0625

P(very|Negative) = count(very,negative)+1 /|vocab| + count(negative) (using laplace smoothing)

P(very|negative) = (0+1) / (4+9) = 1/13 = 0.0769

p(movie|negative) = (2+1)/(4+9) = 3/13 = 0.2307

p(bad|negative) = (1+1) /(4+9) = 2/13= 0.1584

Step 4: Let's classify document d6 applying naïve Bayes.

Document d6 = "bad movie".

We first calculate the probability of each word for the positive class and then for the negative class.

P(d6|positive = P(bad|positive) * P(movie|positive)

Substituting the number from earlier, we can calculate P(d6|positive) = 0.0625 * 0.1875 = 0.01171.

Similarly, calculate P(D6|negative).

P(d6|negative) = P(bad|negative) * P(movie|negative)

Substituting the number from earlier, we can calculate P(d6|negative) = 0.1584 * 0.2307 = 0.0365.

Now, P(positive|d6) = P(d6|positive) * P(positive) = 0.01171 * 0.6 = 0.007

P(negative|d6) = P(d6|negative) *P(negative) = 0.0365 * 0.4 = 0.014

As you can see, using the NB classifier, d6 is classified as negative because (P(negative|d6) > P(positive|d6).

15.11 Word2vec Models

Languages are ambiguous. For example, *small* and *tiny* are similar words but can be used in different contexts. But, these ambiguities make it hard for computers to understand and interpret, thus making NLP tasks complex.

In our earlier sections, we discussed representing words as vectors. We started with one-hot vector representation, and then we looked at TF, IDF, and TF-IDF. Though these representations are a good start for solving many problems, they take away some of the information that appears within a sentence. For example:

- I am *cycling* to the shopping mall now
- I am going to the shopping mall by *bicycle*

Both sentences have similar meanings except for the words *cycling* and *bicycle*. Humans can interpret them easily, whereas for a computer to find the similarity, we have to represent the sentences as vectors. We can represent both words using one-hot representation: Cycle = [0 0 1 0 0 0 0 0] and Bicycle = [0 0 0 0 0 0 0 1]. Both word vectors are orthogonal to each other. By using similarity measures, such as cosine similarity or the Jaccard similarity measure, it gives 0; i.e, there is no similarity. This is the problem.

In 2013, Tomas Mikolov et al. developed an algorithm called Word2vec, a new model that minimizes computational complexity. This model uses a neural network to learn word vectors from a given corpus. The resulting vectors seem to have semantic and syntactic meaning. Empirical results have shown that this method has given successful results in learning the meaning of words. In addition, these representations have shown better results in identifying synonyms, antonyms, and other relationships between words.

There are two methods for this language model:

- Continuous bag of words (CBOW)
- Skip-gram

For example, given a sentence of five words, w_{t-1}, w_{t-2}, < w_t > . w_{t+1}, w_{t+2}, w_t is the center word(target word), and w_{t-1}, w_{t-2}, w_{t+1}, w_{t+2} are the surrounding words (context words).

In the CBOW, the method predicts w_t , whereas in Skip-gram, the method predicts the surrounding words, w_{t-1}, w_{t-2}, w_{t+1}, w_{t+12}, etc. Research and studies have shown Skip-gram tends to produce better word embedding than CBOW. Figure 15-31 shows the architecture of both CBOW and Skip-gram. If you want to learn the architecture and algorithms, we suggest reading the paper "Distributed representations of words and phrases and their compositionality."

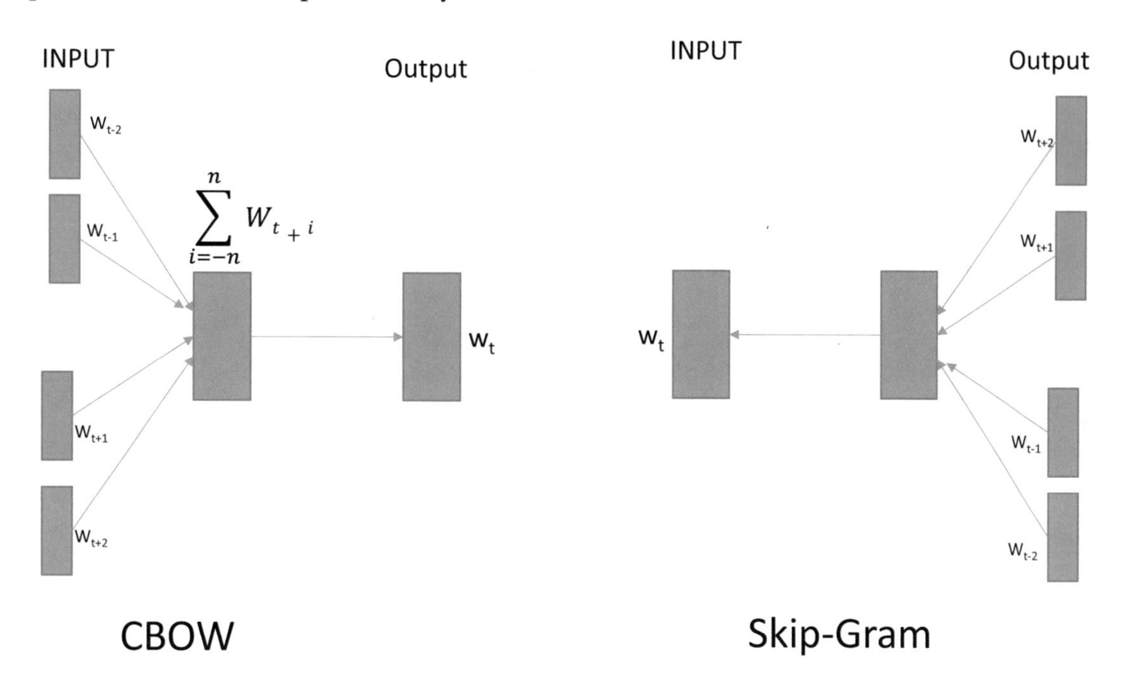

Figure 15-31. *Word2vec embedding architecture of CBOW and Skip-gram*

The output of the Word2vec algorithm is a |v| X d matrix, where |v| is the size of the vocabulary, and **d** is the number of dimensions used to represent each word vector. Google has already trained the Word2vec model on the Google News data set. Google Word2vec has a vocabulary size of 3 million words, with each vector having a dimension of 300.

This model can be downloaded from Google at `https://code.google.com/archive/p/word2vec/`. The download size is around 1.5 GB. Also, there are many libraries that have implemented Word2vec. We will use the Gensim package in this example. Python's Gensim library provides various methods to use the pretrained model directly. The Gensim library function also allows creating a Word2vec model from scratch based on your corpus.

Step 1: Install Genism in your environment.

Step 2: Load the Google pretrained Word2vec model file.

Here is the input:

```
import gensim
from gensim.models import KeyedVectors

#Load the pertained vectors from the pertained Word2Vec model file
model=KeyedVectors.load_word2vec_format('C:/Users/phdst/gensim-data/word2vec-goo
gle-news-300/word2vec-google-news-300.gz', binary=True, limit=300000)
```

Figure 15-32 shows the input and output.

```
[1]: import gensim
     from gensim.models import KeyedVectors

[2]: #Load the pertained vectors from the pertained Word2Vec model file
     model=KeyedVectors.load_word2vec_format('C:/Users/phdst/gensim-data/word2vec-google-news-300/word2vec-google-news-
```

Figure 15-32. *Loading the Word2vec model*

Step 3: Extract Word2vec for any word in the database.

In this example, we will extract the word vector of *girl* and *Boston*. The word vector is based on the principles of Word2vec trained on Google's news documents. The vector limit has been set to 300 by Google.

Here is the input:

```
model['girl']

model['Boston']
```

Figure 15-33 and Figure 15-34 show both the input and output.

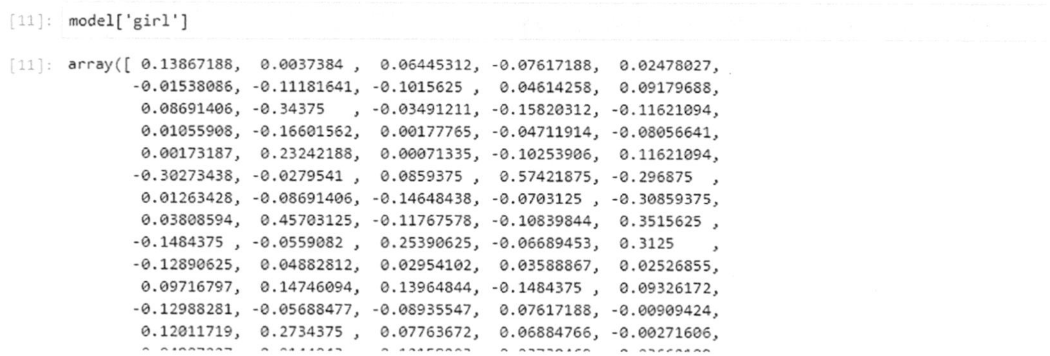

```
[11]: model['girl']

[11]: array([ 0.13867188,  0.0037384 ,  0.06445312, -0.07617188,  0.02478027,
             -0.01538086, -0.11181641, -0.1015625 ,  0.04614258,  0.09179688,
              0.08691406, -0.34375   , -0.03491211, -0.15820312, -0.11621094,
              0.01055908, -0.16601562,  0.00177765, -0.04711914, -0.08056641,
              0.00173187,  0.23242188,  0.00071335, -0.10253906,  0.11621094,
             -0.30273438, -0.0279541 ,  0.0859375 ,  0.57421875, -0.296875  ,
              0.01263428, -0.08691406, -0.14648438, -0.0703125 , -0.30859375,
              0.03808594,  0.45703125, -0.11767578, -0.10839844,  0.3515625 ,
             -0.1484375 , -0.0559082 ,  0.25390625, -0.06689453,  0.3125    ,
             -0.12890625,  0.04882812,  0.02954102,  0.03588867,  0.02526855,
              0.09716797,  0.14746094,  0.13964844, -0.1484375 ,  0.09326172,
             -0.12988281, -0.05688477, -0.08935547,  0.07617188, -0.00909424,
              0.12011719,  0.2734375 ,  0.07763672,  0.06884766, -0.00271606,
```

Figure 15-33. *Word2vec representation of the word girl*

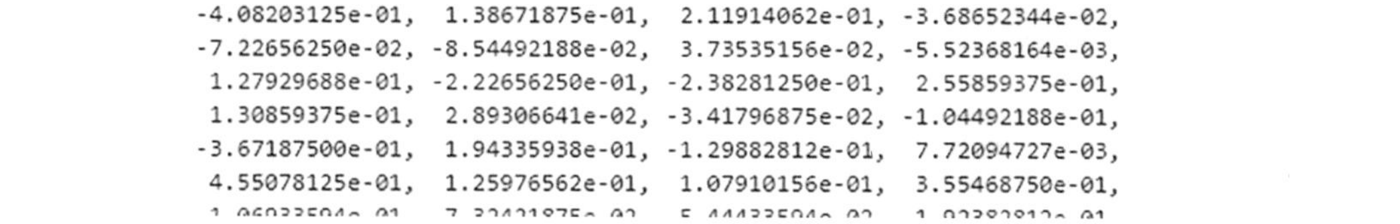

```
[13]: model['Boston']

[13]: array([-1.75781250e-01,  1.90429688e-01,  3.19824219e-02,  2.75390625e-01,
              1.04980469e-01,  4.71191406e-02, -2.83203125e-01, -1.47460938e-01,
              6.17675781e-02, -1.53320312e-01, -2.26562500e-01,  1.54418945e-02,
              7.86781311e-05,  2.03125000e-01,  2.53906250e-01,  7.37304688e-02,
             -2.05078125e-02, -1.42578125e-01,  1.84570312e-01, -6.29882812e-02,
             -9.17968750e-02, -1.79687500e-01, -2.99072266e-02, -1.70898438e-01,
             -7.66601562e-02,  3.82812500e-01,  7.71484375e-02,  1.41601562e-01,
             -4.08203125e-01,  1.38671875e-01,  2.11914062e-01, -3.68652344e-02,
             -7.22656250e-02, -8.54492188e-02,  3.73535156e-02, -5.52368164e-03,
              1.27929688e-01, -2.22656250e-01, -2.38281250e-01,  2.55859375e-01,
              1.30859375e-01,  2.89306641e-02, -3.41796875e-02, -1.04492188e-01,
             -3.67187500e-01,  1.94335938e-01, -1.29882812e-01,  7.72094727e-03,
              4.55078125e-01,  1.25976562e-01,  1.07910156e-01,  3.55468750e-01,
```

Figure 15-34. *Word2vec representation of the word Boston*

Step 4: Find the similarity measure of two words.

This example shows the similar words of *excellent* and *fraud*. The similarity of words just means they are synonyms. By knowing the word vector of the word *excellent*, we can run the similarity (cosine similarity or Jaccard similarity) function to find the similarity of vectors and print the words with the similarity measure. If they are very similar, then the measure is close to 1. If the two words are dissimilar, the measure will be close to zero.

Here is the input:

```
model.most_similar('excellent')

model.most_similar(['fraud','cheat'])
```

Figure 15-35 shows the output.

```
[15]: model.most_similar('excellent')

[15]: [('terrific', 0.7409726977348328),
       ('superb', 0.7062715888023376),
       ('exceptional', 0.681470513343811),
       ('fantastic', 0.6802847981452942),
       ('good', 0.6442928910255432),
       ('great', 0.6124600172042847),
       ('Excellent', 0.6091997623443604),
       ('impeccable', 0.5980967283248901),
       ('exemplary', 0.5959650278091431),
       ('marvelous', 0.582928478717804)]

[19]: model.most_similar(['fraud','cheat'])

[19]: [('cheating', 0.7557769417762756),
       ('frauds', 0.6911643147468567),
       ('fraudulent', 0.6668995022773743),
       ('cheats', 0.6456533074378967),
       ('cheated', 0.6222769618034363),
       ('swindle', 0.6081023812294006),
       ('Fraud', 0.6024792790412903),
       ('scam', 0.6021971106529236),
       ('defraud', 0.5997282862663269),
       ('fraudsters', 0.5786300301551819)]
```

Figure 15-35. *Word similarity using Word2vec*

Step 5: If you want to get the Word2vec representation for your own corpus.

In this example, we have a corpus with three documents. We will limit the vector size to 100 and the frequency of words occurring in the document to 1. Once we import the Word2vec model, you will have a vector representation for your words in the corpus. In the previous example, we have three sentences in the document. We will import the vectors for our words from the Word2vec model we created earlier. After that, we can check the word vector of our corpus.

Here is the input:

```
# Defining corpus. This cirpus has 3 documents
my_corpus = [["I", "am", "trying", "to", "learn", "how",
"to", "fly", "an", "aeroplane"],
["Flying", "is", "fun", "and", "adventurous"],
["Landing", "aeroplane", "requires", "skill", "and", "practice"]]

# Importing Word2Vec vectors
# This will import vectors from the model and assign to all the words in my corpus
model_2 = Word2Vec(my_corpus, min_count = 1)

# dimention of my vector limited to 100, you can go maximum to 300
model_2.vector_size

#100

# Word vector of 'fly'
model_2.wv.get_vector('fly')
```

Figure 15-36 shows the input and output.

```
[1]:  from gensim.models import Word2Vec # Import word2Vec model from Gensim library package

[26]: # Defining corpus. This cirpus has 3 documents
      my_corpus = [["I", "am", "trying", "to", "learn", "how",
      "to", "fly", "aeroplane"],
      ["Flying", "is", "fun", "and", "adventurous"],
      ["Landing", "aeroplane", "requires", "skill", "and", "practice"]]

[61]: # Importing Word2Vec vectors
      # This will import vectors from the model and assign to all the words in my corpus
      model_2 = Word2Vec(my_corpus, min_count = 1)

[62]: # dimention of my vector limited to 100, you can go maximum to 300
      model_2.vector_size

[62]: 100

[64]: # Word vector of 'fly'
      model_2.wv.get_vector('fly')

[64]: array([-9.5813982e-03,  8.9456905e-03,  4.1652643e-03,  9.2360079e-03,
              6.6455095e-03,  2.9253769e-03,  9.8057324e-03, -4.4252356e-03,
             -6.8043000e-03,  4.2288131e-03,  3.7297788e-03, -5.6660306e-03,
              9.7066183e-03, -3.5592420e-03,  9.5516816e-03,  8.3475752e-04,
             -6.3402876e-03, -1.9776404e-03, -7.3783039e-03, -2.9793626e-03,
              1.0419025e-03,  9.4843479e-03,  9.3575213e-03, -6.5968377e-03,
              3.4764954e-03,  2.2755123e-03, -2.4894474e-03, -9.2315283e-03,
              1.0278671e-03, -8.1681050e-03,  6.3218698e-03, -5.8019590e-03,
              5.5366694e-03,  9.8362258e-03, -1.5980411e-04,  4.5294156e-03,
             -1.8091050e-03,  7.3628318e-03,  3.9417055e-03, -9.0122130e-03,
             -2.3998821e-03,  3.6299105e-03, -9.9658420e-05, -1.2014904e-03,
             -1.0563055e-03, -1.6716262e-03,  6.0576608e-04,  4.1664303e-03,
             -4.2543593e-03, -3.8344301e-03, -5.2106283e-05,  2.6977799e-04,
             -1.6937690e-04, -4.7862623e-03,  4.3148198e-03, -2.1730361e-03,
              2.1048370e-03,  6.6660641e-04,  5.9710480e-03, -6.8441834e-03,
```

Figure 15-36. *Using Word2vec for your corpus*

15.12 Text Analytics and NLP

Text analytics is the process of extracting large volumes of unstructured text data into quantitative data to find meaningful patterns and insights. Text mining, text analytics, and text analysis are sometimes used interchangeably. Text mining helps to understand the structure of the textual contents, what words are being spoken frequently, which words are repeated, etc. This analysis enables a business to understand better what is happening, who is the influencer, or which words are making noise, thus making informed decisions. Text analytics provides meaningful information from various different types of unstructured data such as social media posts, blogs, emails, chats, and surveys to provide quantitative analysis using data visualization techniques.

On the other hand, NLP is related to making computers understand the text and spoken words for effective communication between humans and computers. NLP combines linguistics with statistical machine learning and deep learning models to enable the computers to process language and understand its meaning, sentiment, and intent.

The following example demonstrates how to visualize the words that occur more frequently using the `wordcloud()` library. We have created our corpus (`my_text`) and plot the words that frequently occur in the text. In our text, the word *language* seems to appear more frequently than other words; hence, it has been highlighted in a large font, as shown in Figure 15-37.

Here is the input:

```
my_text =

'We are learning Natural Language Processing in this class. Natural Langua
ge Processing is all about how to make computers to understand language.
Natural Language Processing techniques are evolving everyday.'

# importing the necessary modules:
import numpy as np
import pandas as pd
from wordcloud import WordCloud, STOPWORDS, ImageColorGenerator

import matplotlib.pyplot as plt
from PIL import Image

myword_cloud = WordCloud().generate(my_text)
# Display the generated image:
plt.imshow(myword_cloud)
plt.axis("off")
plt.show()
```

Figure 15-37 shows the output.

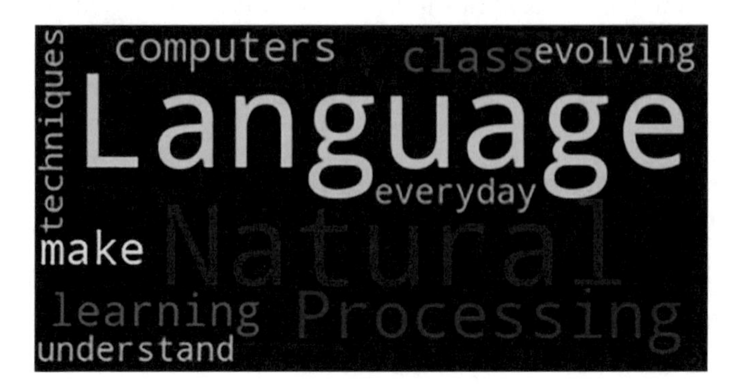

Figure 15-37. *Plotting important terms using the Wordcloud() function*

15.13 Deep Learning and NLP

With recent advancements and research into neural and deep neural networks, people can solve many complex and unstructured data problems. Language is also complex and unstructured. Neural networks have shown good results in modeling complex language tasks. Several deep neural networks have become increasingly popular to solve some of the NLP problems. Recurrent neural networks (RNNs), long-short-memory (LSTM), and convolution neural networks (CNNs) are some of the popular deep learning models applied to NLP tasks. CNNs have performed better in text classification tasks.

Similarly, transformers and autoencoders have also found applications in solving some of the NLP tasks. We will not be discussing CNN and RNNs in this chapter.

15.14 Case Study: Building a Chatbot

Amazon has provided a dataset of its customers' queries and the corresponding responses to the public.

There are many different data sets available in different categories. We will be using only the electronics appliance data to train our chatbot. We train the chatbot on electronics appliance Q&A data; once it is trained, our chatbot can be deployed as automated Q&A support under the appliance items section. The data set is in JavaScript Object Notation (JSON) format.

The design of the chatbot, is as follows:

Step 1: Load the questions from the corpus to a list.

Step 2: Load the corresponding answers from the corpus in another list.

Step 3: Preprocess the question data.

Step 4: Vectorize (TF-IDF) question data.

Step 5: Similarly, preprocess and vectorize (TF-IDF) the answers.

Step 6: Collect the user's query and preprocess the user's query.

Step 7: Vectorize (TF-IDF) the user's query.

Step 8: Use cosine similarity to find the most similar question to the user's query.

Step 9: Return the corresponding answer to the most similar question as a chat response.

The full code can be downloaded from the GitHub repository.

15.15 Chapter Summary

In this chapter, we discussed what NLP is and what text analytics are and how they are interrelated. We also discussed the various applications of NLP.

We talked about how a computer can process the natural language and the basics of language processing, including tokenization, stemming, lemmatization, POS tagging, etc.

We also discussed language modeling, representing words as vectors, term frequency, and inverse document frequency (TF-IDF), Word2vec, and other models.

We demonstrated the basic concepts with examples using various libraries and also mentioned various libraries for NLP in R and Python.

We ended the chapter by discussing how deep learning can be applied to solve some of the NLP problems and applications.

CHAPTER 16

Big Data Analytics and Future Trends

Data is growing at the speed of light, making the management of data an increasingly huge challenge. This chapter will discuss various technologies available to store, process, and analyze such Big Data.

16.1 Introduction

Data is power and is going to be another dimension of value in any enterprise. Data is and will continue to be the major decision driver going forward. All organizations and institutions have woken up to the value of data and are trying to collate data from various sources and mine it for its value. Businesses are trying to understand consumer/ market behavior in order to get the maximum out of each consumer with the minimum effort possible. Fortunately, these organizations and institutions have been supported by the evolution of technology in the form of increasing storage power and computing power, as well as the power of the cloud, to provide infrastructure as well as tools. This has driven the growth of data analytical fields, including descriptive analytics, predictive analytics, machine learning, deep learning, artificial intelligence, and the Internet of Things.

This chapter does not delve into the various definitions of Big Data. Many pundits have offered many divergent definitions of Big Data, confusing people more than clarifying the issue. However, in general terms, Big Data means a huge amount of data that cannot be easily understood or analyzed manually or with limited computing power or limited computer resources. Analyzing Big Data requires the capability to crunch data

U. R. Hodeghatta and U. Nayak, *Practical Business Analytics Using R and Python*,
https://doi.org/10.1007/978-1-4842-8754-5_16

of a diverse nature (from structured data to unstructured data) from various sources (such as social media, structured databases, unstructured databases, and the Internet of Things).

In general, when people refer to Big Data, they are referring to data with three characteristics: variety, volume, and velocity, as shown in Figure 16-1.

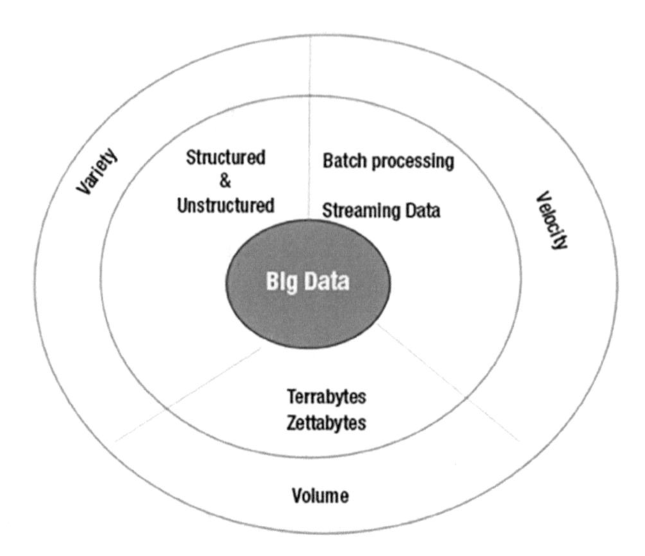

Figure 16-1. *Characteristics of Big Data*

Variety refers to the different types of data that are available on the Web, the Internet, in various databases, etc. This data can be structured or unstructured and can be from various social media and/or from other sources. *Volume* refers to the size of the data that is available for you to process. Its size is *big*—terabytes and petabytes. *Velocity* refers to how fast you can process and analyze data, determine its meaning, arrive at the models, and use the models that can help business.

The following have aided the effective use of Big Data:

- Huge computing power of clusters of computing machines extending the processing power and the memory power by distributing the load over several machines
- Huge storage power distributing the data over various storage resources
- Significant development of algorithms and packages for machine learning

- Developments in the fields of artificial intelligence, natural language processing, and others
- Development of tools for data visualization, data integration, and data analysis

16.2 Big Data Ecosystem

Over the past 10 years or so, organizations such as Google, Yahoo, IBM, SAS, SAP, and Cloudera have made significant efforts to design and develop a substantial system of tools to facilitate the collating and integrating of huge amounts of data from diverse sources, mine the data to reveal patterns or truth hidden in the data, visualize the data, present the learning from the data, and suggest proactive actions.

The Apache Hadoop ecosystem—with its Hadoop Distributed File System coupled with various tools/mechanisms such as MapReduce, YARN, Pig, Mahout, Hive, HBase, Storm, and Oozie—has strengthened the drive of Big Data analysis, along with commercial tools including SAP and SAS. Figure 16-2 shows the Hadoop framework.

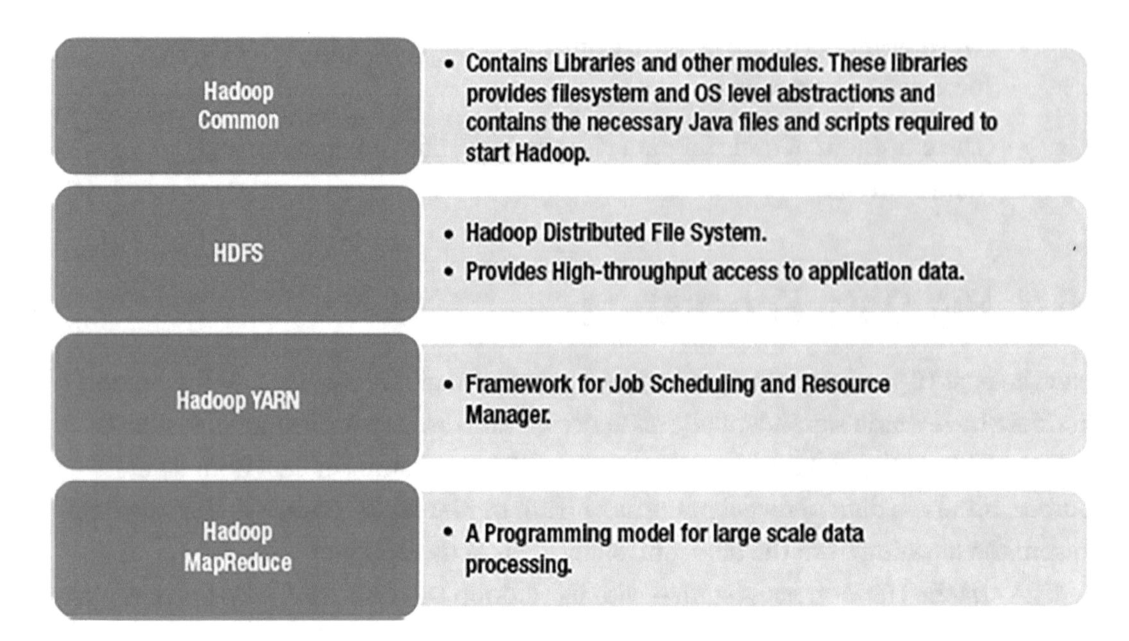

Figure 16-2. *Hadoop ecosystem*

The Hadoop Distributed File System (HDFS) allows data to be distributed and stored among many computers. Further, it allows the use of the increased processing power and memory of multiple clustered systems. This has overcome the obstacle of not being able to store huge amounts of data in a single system and not being able to analyze that data because of a lack of required processing power and memory. The Hadoop ecosystem consists of modules that enable us to process the big data and perform the analysis.

A user application can submit a job to Hadoop. Once data is loaded onto the system, it is divided into multiple blocks, typically 64 MB or 128 MB. Then the Hadoop Job client submits the job to the JobTracker. The JobTracker distributes and schedules the individual tasks to different machines in a distributed system; many machines are clustered together to form one entity. The tasks are divided into two phases: Map tasks are done on small portions of data where the data is stored, and Reduce tasks combine data to produce the final output. The TaskTrackers on different nodes execute the tasks as per MapReduce implementation, and the reduce function is stored in the output files on the file system. The entire process is controlled by various smaller tasks and functions. Figure 16-3 shows the full Hadoop ecosystem and framework.

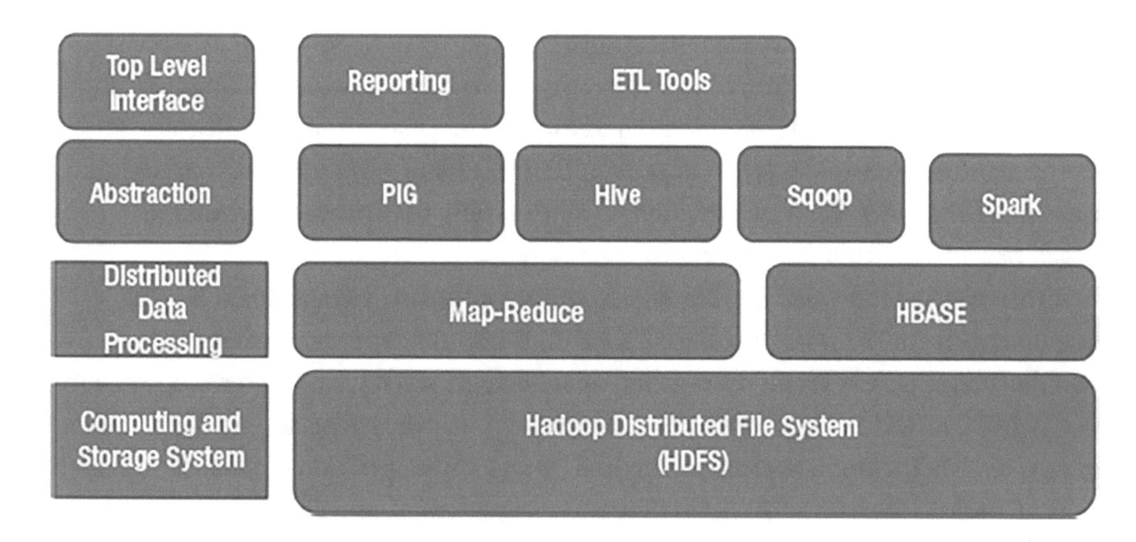

Figure 16-3. *Hadoop framework and ecosystem*

Apache Hadoop YARN is a tool used for resource provisioning as well as management and job scheduling/monitoring activities.

Apache Pig can analyze a huge amount of data. This platform uses a simple scripting language that allows users to create their own programs to complement the capability of predefined functions of Pig. Pig's compiler produces sequences of MapReduce programs. Hence, this eliminates the need for writing lengthy and cumbersome MapReduce programs. Pig also has the capability to carry out parallel processing.

Apache Mahout provides the environment and programming capability to write scalable machine-learning algorithms and applications. Today, Mahout includes Samsara, a vector math experimentation environment.

Apache Hive provides the capability to read, write, and manage data by providing SQL capability.

Apache HBase is the Hadoop distributed database, with high scalability and a huge storage capability. It provides real-time access to data. HBase is a NoSQL database for Hadoop.

Apache Storm enables distributed real-time processing of streams of data in the same way that Hadoop MapReduce does batch processing. Storm is a computation engine that performs continuous computation, real-time analytics, online machine learning, and many other related capabilities such as distributed RPC.

Apache Oozie enables effective workflow scheduling to manage Apache Hadoop jobs. It is extremely well integrated with other tools in the Apache Hadoop ecosystem.

In addition to these tools, NoSQL (originally referring to *not only SQL*) databases such as Cassandra, ArangoDB, MarkLogic, OrientDB, Apache Giraph, MongoDB, and Dynamo have supported or complemented the Big Data ecosystem significantly. These NoSQL databases can store and analyze multidimensional, structured or unstructured, huge data effectively. This has provided significant fillip to the consolidation and integration of data from diverse sources for analysis.

Currently Apache Spark is gaining momentum in usage. Apache Spark is a fast, general engine for Big Data processing, with built-in modules for streaming, SQL, machine learning, and graph processing. Apache Spark, an interesting development in recent years, provides an extremely fast engine for data processing and analysis. It allows an easy interface to applications written in R, Java, Python, or Scala. Apache Spark has a stack of libraries such as MLib, Spark Streaming, Spark SQL, and GraphX. It can run in stand-alone mode as well as on Hadoop. Similarly, it can access various sources from HBase to Cassandra to HDFS. Many users and organizations have shown interest in this tool and have started using it, resulting in it becoming very popular in a short period of time. This tool provides significant hope to organizations and users of Big Data.

Tools such as Microsoft Business Intelligence and Tableau provide dashboards and the visualization of data. These have enabled organizations to learn from the data and leverage this learning to formulate strategies or improve the way they conduct their operations or their processes.

The following are some of the advantages of using Hadoop for Big Data processing:

- Simple parallel architecture on any cheap commodity hardware
- Built-in fault-tolerance system and application-level failure detection and high availability
- Dynamic cluster modification without interruption
- Easy addition or removal of machines to/from the cluster
- Java based and platform independent

Microsoft Azure, Amazon, and Cloudera are some of the big providers of cloud facilities and services for effective Big Data analysis.

16.3 Future Trends in Big Data Analytics

The applications of Big Data and areas in which Big Data analytics are used is increasing year by year. This section provides some of the trends that are going to be accelerating in the years to come.

16.3.1 Growth of Social Media

Over a short period of time, social media has created a significantly large amount of data, and the trend is growing almost exponentially. This data, related to people across the globe, is bringing an understanding to organizations of hitherto unknown facts and figures. This is driving those organizations to think differently and formulate their strategies differently.

16.3.2 Creation of Data Lakes

Organizations are getting data from diverse sources. But often they don't know the value of the data. They dump the data into data *lakes* without even looking at the value it provides. Once the data is inside the data lake, organizations look for patterns and/or allow data to "discover itself." This attitude is of immense value going ahead. Data lakes are storage repositories, similar to databases and data warehouses, but can store large amounts of structured, unstructured, and semistructured data.

16.3.3 Visualization Tools at the Hands of Business Users

The delay that has occurred between the analysis of data by data analysts and the presentation of that data to the business is no longer acceptable in the context of increasing competitiveness. Businesses are required to take quick action to use the knowledge gained from the data effectively. Tools such as Microsoft Business Intelligence, Tableau, and SAS Visual Analytics have provided significant utility to business users. More visualization helps business users quickly understand the patterns and trends and make effective and timely decisions. The tool capabilities are trending toward being dynamic in order to depict additional aspects evidenced by the data in such visualization charts or dashboards.

16.3.4 Prescriptive Analytics

Data analysis is no longer focused only on understanding the patterns or value hidden in the data. The future trend is to prescribe the actions to be taken, based on the past and depending on present circumstances, without the need for human intervention. This is going to be of immense value in fields such as healthcare, aeronautics, automotive and other fields.

16.3.5 Internet of Things

The *Internet of Things* is a driving force for the future. It has the capability to bring data from diverse sources such as home appliances, industrial machines, weather equipment, and sensors from self-driving vehicles or even people. This has the potential to create a huge amount of data that can be analyzed and used to provide proactive solutions to potential future and current problems. This can also lead to significant innovations and improvements.

16.3.6 Artificial Intelligence

Neural networks can drive artificial intelligence—in particular, making huge data learn from itself without any human intervention, specific programming, or the need for specific models. *Deep learning* is one such area that is acting as a driver in the field of Big Data. This may throw up many of the "whats" that we are not aware of. We may not understand the "whys" for some of them, but the "whats" of those may be very useful. Hence, we may move away from the perspective of always looking for cause and effect. The speed at which the machine learning field is being developed and used, drives significant emphasis in this area. Further, natural language processing (NLP), and property graphs (PGs) are also likely to drive new application design and development, putting the capabilities of these technologies in the hands of organizations and users.

16.3.7 Whole Data Processing

With the current infrastructure and the current level of tool support available, particularly in the cloud, it is possible in most cases to analyze a complete set of data rather than sample the data for analysis. This avoids the downsides of sampling. Using huge amounts of data drives us in the right direction, even though the data may not be complete or perfect. We do not need to always be looking for perfect data.

16.3.8 Vertical and Horizontal Applications

Analytics is moving in the direction of consolidating vertical moves specific to industries such as healthcare, banking, and automotive. On the other hand, horizontal moves cut across industry spheres including consumer purchasing behavior, transaction monitoring for prevention of fraud and money laundering, and understanding the exits of the customers from a particular brand or a particular organization.

16.3.9 Real-Time Analytics

Organizations are hungry to understand the opportunities available to them. They want to understand in real time what is happening—for example, what a particular person is purchasing or what a person is planning for—and use the opportunity appropriately to offer the best possible solutions or discounts or cross-sell related products or services. Organizations are no longer satisfied with a delayed analysis of data. which results in missed business opportunities because they were not aware of what was happening in real time.

16.4 Putting the Analytics in the Hands of Business Users

Delays occur before data analysts can bring the results of their data analysis to the attention of business users. Because of this time loss, the analysis, even though wonderful, may not be useful and may not lead organizations to create effective strategies. Instead, making tools more useful and directly usable by business users who can then make decisions immediately will put business users in an advantageous position. The world is moving toward making this a reality to a large extent.

16.5 Migration of Solutions from One Tool to Another

The diversity of tools means that solutions provided for one tool are not easily migrated to other systems. This is primarily because of the different programming languages and environments used. The world is moving toward making the solutions from one tool easily implemented or migrated to other tools. Efforts in Predictive Model Markup Language (PMML) are already moving in this direction. The Data Mining Group (DMG) is working on this.

16.6 Cloud Analytics

Both data and tools in the cloud have provided a significant boost to Big Data analysis. More and more organizations and institutions are using cloud facilities for their data storage and analysis. Organizations are moving from the private cloud to the public or hybrid cloud for data storage and data analysis. This has provided the organization with cost-effectiveness, scalability, and availability. Of course, security may be a concern, and significant efforts to increase security in the cloud are in progress.

16.7 In-Database Analytics

In-database analytics have increased security and reduced privacy concerns, in part by addressing governance. Organizations, if required, can do away with intermediate requirements for data analysis such as data warehouses. Organizations that are more conscious about governance and security concerns will provide significant fillips to in-database analytics. Lots of vendor organizations have already made their presence felt in this space.

16.8 In-Memory Analytics

The value of in-memory analytics is driving transactional processing and analytical processing side-by-side. This may be very helpful in fields where immediate intervention based on the results of analysis is essential. Systems with hybrid transactional/analytical processing (HTAP) are already being used by some organizations. However, using HTAP for the sake of using it may not be of much use, even when the rate of data change is slow and you still need to bring in data from various diverse systems to carry out effective analysis. Instead, it may be overkill, leading to higher costs to the organization.

16.9 Autonomous Services for Machine Learning

More and more organizations may reach out to autonomous services for machine learning (such as Microsoft Azure Machine Learning Studio, Amazon Machine Learning, or Google Prediction API) if they can't get reliable and experienced data scientists who can help them out effectively. An example of autonomous service is handling customer support using AI chatbots or providing recommendations to the customers on products and services available using AI.

16.10 Addressing Security and Compliance

There is an increased awareness of information security and compliance requirements with consumers as well as organizations. This is leading to significant efforts to increase security in data analytical environments. This trend is going to continue to make the tools and systems more secure. Diverse requirements of different countries may put certain restrictions on how the data will be used and protected. Compliance is likely to be a strain on the organizations using data for analysis.

16.11 Big data Applications

Big Data can play a major role in healthcare, finance, manufacturing, logistics, and many other areas. Big healthcare data can predict epidemics, prevent diseases, and improve value-based healthcare and quality of living. An abundance of data is generated from various modern devices such as smartphones, Fitbit products, and pedometers that measure, for example, how far you walk in a day or the number of calories you burn. This data can be used to create diet plans or prescribe medicines. Other unstructured data—such as medical device logs, doctor's notes, lab results, X-ray reports, and clinical and lab data—can be big enough data to analyze and improve patient care and thus increase efficiency. Other data that can be generated and processed for Big Data analytics includes claims data, electronic health/medical record data (EHR or EMR), pharmaceutical R&D data, clinical trials data, genomic data, patient behavior and sentiment data, and medical device data.

Similarly, the finance industry generates a lot of data, both structured and unstructured. There is even more unstructured data generated through discussion boards, social media, and other areas related to finance. This huge amount of data can provide real-time market insights, fraud detection and prevention, risk analysis, better financial advice to customers, improved portfolio analysis, and many more applications.

In manufacturing, the data generated by machinery and processes can be used for predictive maintenance, quality control, process optimization, machinery downtime optimization, anomaly detection, production forecasting, product lifecycle management (PLM), etc.

Big volumes of data are being generated every day by people who use social media. Many researchers are interested in analyzing such data and finding out useful patterns. Though the data can be useful from a business perspective, it can provide many perspectives on different aspects of society. It can also lead society on a negative path because of malicious intentions by some. This discussion is beyond the scope of this book.

16.12 Chapter Summary

In this chapter, we briefly introduced Big Data and Big Data analytics. We also discussed various terms and technologies to manage Big Data and analyze it.

We also touched upon some of the challenges of AI, machine learning, and the future trends.

We hope that this introduction to Big Data helps you to explore these and related areas based on your needs and interest.

PART V

Business Analytics Tools

CHAPTER 17

R for Analytics

This chapter introduces the R tool. We'll discuss the fundamentals of R required to perform business analytics, data mining, and machine learning. This chapter provides enough basics to start R programming for data analysis. This chapter also introduces the data types, variables, and data manipulations in R and explores some of the packages of R and how they can be used for data analytics.

17.1 Data Analytics Tools

Many commercial and free tools are available to perform data analytics functions. This book focuses on R and Python which are most popular statistical/analytical open-source tools available today. They are free and widely used by academia and the research community. Table 17-1 lists some of the most popular statistical/data analytics tools available.

U. R. Hodeghatta and U. Nayak, *Practical Business Analytics Using R and Python*,
https://doi.org/10.1007/978-1-4842-8754-5_17

Table 17-1. Business Analytics and Statistical Tools

Software Package	Functionality Supported	URL
Microsoft Excel	Descriptive statistics. Hypothesis testing, F-tests, Chi-squared tests, t-tests. Analysis of variance (ANOVA). Bar graphs, pie charts, linear regression.	`https://www.microsoft.com/en-us/microsoft-365/excel`
gretl (open source)	Regression and time-series analysis. Least squares, maximum likelihood, GMM; single-equation and system methods. Time-series methods: ARIMA, a wide variety of univariate models.	`http://gretl.sourceforge.net/`
MathWorks MATLAB	Full set of statistics and machine learning functionality. Nonlinear optimization, system identification, and financial modeling. MapReduce functionality for Hadoop and by connecting interfaces to ODBC/JDBC databases.	`http://uk.mathworks.com/products/matlab/`
PSPP (open-source alternative for IBM SPSS Statistics)	Comparison of means (t-tests and one-way ANOVA); linear regression, logistic regression, reliability (Cronbach's alpha, not failure or Weibull), and reordering data, nonparametric tests, factor analysis, cluster analysis, principal component analysis, Chi-square analysis, and more.	`www.gnu.org/software/pspp/`
OpenStat (open source)	OpenStat contains a large variety of parametric, nonparametric, multivariate, measurement, statistical process control, financial, and other procedures.	`https://www.openstat.info/OpenStatMain.htm`
Salstat (open source)	Descriptive statistics, inferential statistics, parametric and nonparametric analysis, bar charts, box plots, histograms, and more.	`www.salstat.com`

(continued)

Table 17-1. *(continued)*

Software Package	Functionality Supported	URL
IBM SPSS	Full set of statistical analysis, parametrics, nonparametric analysis, classification, regression, clustering analysis. Bar charts, histograms, box plots. Social media analysis, text analysis, and so forth.	`https://www.ibm.com/spss`
Stata by StataCorp	Descriptive statistics, ARIMA, ANOVA, MANOVA, linear regression, time-series smoothers, generalized linear models (GLMs), cluster analysis.	`www.stata.com`
Statistica	Statistical analysis, graphs, plots, data mining, data visualization, and so forth.	`https://www.tibco.com`
SciPy (pronounced "sigh *pie*") (open source)	Python library used by scientists and analysts doing scientific computing and technical computing. SciPy contains modules for optimization, linear algebra, interpolation, digital signal and image processing, and machine learning techniques.	`www.scipy.org`
Weka, or Waikato Environment for Knowledge Analysis (open source)	Contains a collection of visualization tools and algorithms for data analysis and predictive modeling, together with graphical user interfaces for easy access to these functions.	`www.cs.waikato.ac.nz/ml/weka/`
RapidMiner (open source)	Integrated environment for machine learning, data mining, text mining, predictive analytics, and business analytics.	`https://rapidminer.com/`
R (open source)	Full set of functions to support statistical analysis, histograms, box plots, hypothesis testing, inferential statistics, t-tests, ANOVA, machine learning, clustering, and so forth.	`www.r-project.org`
Minitab by Minitab Statistical Software	Descriptive statistical analysis, hypothesis testing, data visualization, t-tests, ANOVA, regression analysis, reliability, and survival analysis.	`www.minitab.com`

(continued)

Table 17-1. *(continued)*

Software Package	Functionality Supported	URL
Tableau Desktop by Tableau Software	Statistical summaries of your data, experiment with trend analyses, regressions, correlations. Connect directly to your data for live, up-to-date data analysis that taps into the power of your data warehouse.	`www.tableau.com/products/desktop`
TIBCO by Cloud Software Group	Statistical and full predictive analytics. Integration of R, S+, SAS and MATLAB into Spotfire and custom applications.	`https://www.tibco.com`
SAS by SAS	Advanced statistical and machine learning functions and much more.	`www.sas.com`

R is an integrated suite of software packages for data handling, data manipulation, statistical analysis, graphical analysis, and developing learning models. R is an extension of the S software, initially developed by Bell Labs. It is open-source, free software, licensed under the GNU General Public License (`www.gnu.org/licenses/gpl-2.0.html`) and supported by large research communities spread all over the world.

R has the following advantages:

- Flexible, easy, and friendly graphical capabilities that can be displayed on your computer or stored in different file formats
- Availability of a large number of free libraries to perform various analytics
- Supports all the capabilities of a programming language
- Supports importing data from a wide variety of sources, including text files, database management systems, web XML files, and other repositories
- Supports multiple operating system or platforms (Windows, Unix, and macOS)

Several free books are available to learn R. In this section, instead of focusing on R basics, we demonstrate some of the operations performed on a data set with an example. The examples include checking for nulls and NAs, handling missing values, cleaning and exploring data, plotting various graphs, and using some of the loop functions, including `apply()`, `cut()`, and `paste()`. These examples are only to use as a reference in case you need them. We will not claim this is the only or the best solution.

17.2 Data Wrangling and Data Preprocessing Using R

Normally, tools provide an interface for performing data cleaning and data preprocessing before the analysis. This section describes how this can be achieved using R. Multiple solutions are available to perform the same tasks, so this example is just one solution. The data preprocessing methods discussed are as follows:

- Understanding the variable types and checking whether variables are categorical, numerical, or logical.
- Changing the variable types. Sometimes in order to categorize the data properly and/or to make them amenable for the analysis needed by us, we may have to change one type of data into another data type e.g., categorical data imported as numerical data.

Finding missing values: We usually see `NA` and `NaN` in R. `NA` is generally interpreted as a missing value. `NaN` indicates a mathematical error (for example, dividing by 0). However, `is.na()` returns `TRUE` for both `NA` and `NaN`. The following are some of the solutions in this regard:

- Replacing missing values and NAs with appropriate values e.g., median, average or representative value
- Cleaning missing values with appropriate methods

All variables in R are represented as vectors. The following are the basic data types in R:

Numeric: Real numbers.

Integer: Whole numbers.

Factor: Categorical data to define various categories. In R, categorical variables are represented as factors.

Character: Data strings of characters defining, for example, the name of a person, animal, or place.

We use the Summer Olympics medals data set, which lists gold, silver, and bronze medals won by different countries from 1896 to 2008. We have added some dummy countries with some dummy values to demonstrate some of the concepts.

1. Read the data set and create sample variables. The `dplyr()` package supports a function called `sample_frac()`, which takes a random sample from the data set.

`dplyr()` is a library that helps you to address some of the common data manipulation challenges. It supports several functions. Here are some examples:

- `mutate()` adds new variables that are functions of existing variables.
- `select()` picks variables based on their names.
- `filter()` picks cases based on their values.
- `summarise()` reduces multiple values to a single summary.
- `arrange()` changes the ordering of the rows.

The code and the corresponding output are provided below:

```
> ##Read data as a R dataframe
> medal_df<-read.csv("Summer_Olympic_medallists_1896-2008.csv")
> tail(medal_df,10)

                 Country NOC.CODE Total Golds Silvers
131                 Togo      TOG     1     0       0
132                Tonga      TGA     1     0       1
133 United Arab Emirates      UAE     1     1       0
134       Virgin Islands      ISV     1     0       1
135          Country XYZ      XYZ    NA    NA    NULL
136             Countr B        A    NA     1    NULL
137            Country C        C    NA    NA    NULL
138            Country D        D    NA    NA    NULL
```

2. Check the size of the data, the number of columns, and the rows (records). The code and the corresponding output are provided below:

```
> ##Check the data size (number of rows and columns)
> nrow(medal_df)
[1] 140
```

```
> ncol(medal_df)
[1] 6
```

3. Select random data from the sample and copy it to another dataframe. We are using the `dplyr()` library for some of the following operations. The code and the corresponding output are provided below:

```
> ##Selecting Random N samples (90%) from data
> library(dplyr)
> sample_data<-sample_frac(medal_df, 0.90)
> sample_data<-as.data.frame(sample_data)
> nrow(sample_data)
[1] 126

> str(sample_data)
'data.frame':    126 obs. of  6 variables:
 $ Country : Factor w/ 140 levels "Afghanistan",..: 13 90 10 70 22 26 61
103 114 112 ...
 $ NOC.CODE: Factor w/ 128 levels "","A","AFG","ALG",..: 14 86 11 68 21 2
54 99 104 36 ...
 $ Total   : int  1 10 1 17 385 NA 23 1 1 113 ...
 $ Golds   : int  0 3 0 2 163 1 8 0 0 34 ...
 $ Silvers : Factor w/ 51 levels "0","1","11","112",..: 1 25 1 3 5 51 39 2
2 33 ...
 $ Bronzes : int  1 4 1 4 105 NA 8 0 0 30 ...
```

4. Check the data type of each column and convert it to the appropriate data type. In this example, the column `Silvers` is recognized as a factor instead of an integer by R. And this has to be converted to the proper data type, as shown next.

We will use the as.integer() function to convert the factor to the integer. Similarly, the as.factor(), as.numeric(), as.character(), and as.date() functions are predefined for coercing one data type to other data type. Similarly, other functions like as.vector(), as.matrix(), as.data.frame() are used to convert one object type to the other object type.

Select only a set of variables from the dataframe. This is often the case in data processing tasks. In the following code we explore a first case where we select the two columns required, and explore the second case where we get all the columns except the two columns mentioned. We use the head() function to display the first six rows and tail() to display the last six rows of the dataframe. The code and the corresponding outputs are provided below:

```
> sample_data$Silvers<-as.integer(sample_data$Silvers)
> str(sample_data)

'data.frame':    126 obs. Of  6 variables:
 $ Country : Factor w/ 140 levels "Afghanistan",..: 13 90 10 70 22 26 61
103 114 112 ...
 $ NOC.CODE: Factor w/ 128 leve"" """""",""F"",""LG",..: 14 86 11 68 21 2
54 99 104 36 ...
 $ Total   : int  1 10 1 17 385 NA 23 1 1 113 ...
 $ Golds   : int  0 3 0 2 163 1 8 0 0 34 ...
 $ Silvers : int  1 25 1 3 5 51 39 2 2 33 ...
 $ Bronzes : int  1 4 1 4 105 NA 8 0 0 30 ...
```

Note str() gives the summary of the object in active memory, and head() enables you to view the first few (six) lines of data.

```
> ###Selecting set of variables only
> head(sample_data)

   Country NOC.CODE Total Golds Silvers Bronzes
1  Bermuda      BER     1     0       1       1
2 Pakistan      PAK    10     3      25       4
3 Barbados      BAR     1     0       1       1
4   Latvia      LAT    17     2       3       4
5    China      CHN   385   163       5     105
6 Countr B        A    NA     1      51      NA
> X3 = select(sample_data, Total, Golds)
> head(X3)

  Total Golds
1     1     0
2    10     3
3     1     0
4    17     2
5   385   163
6    NA     1
> X4 = select(sample_data, -NOC.CODE, - Total)
> head(X4)

   Country Golds Silvers Bronzes
1  Bermuda     0       1       1
2 Pakistan     3      25       4
3 Barbados     0       1       1
4   Latvia     2       3       4
5    China   163       5     105
6 Countr B     1      51      NA
```

5. Sometimes, the data set column name may require corrections; we can do this using the `rename()` function as shown. The code and the corresponding output are provided below:

```
> X5 = rename(sample_data, GoldMedal=Golds)
> colnames(sample_data)
[1] "Country"  "NOC.CODE" "Total"    "Golds"    "Silvers"
[6] "Bronzes"
> colnames(X5)
[1] "Country"   "NOC.CODE"  "Total"     "GoldMedal" "Silvers"
[6] "Bronzes"
```

6. Often, we have to get a specific column name from the data. Here is an example of how to get the specific column name from the data demonstrated using the code and the corresponding outputs:

```
> #Get the column name of selected columns
> colnames(sample_data)[6]
[1] "Bronzes"
> colnames(sample_data)[3]
[1] "Total"
```

7. Filter and search for specific information in the data set. For example, the following function demonstrates how to search and filter data with `NOC.CODE = "PRK"` or when `Gold medal` is greater than 10. The code and the corresponding output are provided below:

```
> ## Filter data with a certain 'string' name
> head(sample_data[sample_data$NOC.CODE == 'PRK', ])
       Country NOC.CODE Total Golds Silvers Bronzes
13 North Korea      PRK    41    10       6      19
> ## get data with certain 'condition'
> head(subset(sample_data, Golds > 10 ))
          Country NOC.CODE Total Golds Silvers Bronzes
5           China      CHN   385   163       5     105
10          Spain      ESP   113    34      33      30
16       Bulgaria      BUL   212    51      46      77
18   South Africa      RSA    70    20      20      26
27 Czechoslovakia            143    49      33      45
31           Cuba      CUB   194    67      38      63
```

8. We use the `dplyr()` filter function to perform similar operations. In this example, we search and filter data where `Silver` medals are greater than 15. The code and the corresponding output are provided below:

```
> ## Using dplyr::filter()
> #filter( ) Function using 'dplyr'
> copying_to_anothervariable<-filter(sample_data, Silvers > 15)
> head(copying_to_anothervariable)
     Country NOC.CODE Total Golds Silvers Bronzes
1   Pakistan      PAK    10     3      25       4
2   Countr B        A    NA     1      51      NA
3    Ireland      IRL    23     8      39       8
4      Spain      ESP   113    34      33      30
5    Georgia      GEO    18     5      16      11
6 Azerbaijan      AZE    16     4      25       9
```

17.2.1 Handling NAs and NULL Values in the Data Set

In this section, we demonstrate functions to check NAs. The function is `is.na()`. `Is.na()` lists all the NAs present in the dataframe. There are a total of 14 NAs present in this data set. The code and the corresponding output are provided below:

```
> ##Check is data has any NAs
> sum(is.na(sample_data))
[1] 14
```

17.2.2 Apply() Functions in R

We will check the proportion of NAs in each column using the `apply()` function. Although the `for()` and `while()` loops are useful programming tools, curly brackets and structuring functions can sometimes be cumbersome, especially when dealing with large data sets. R has some cool functions that implement loops in a compact form to make data analysis simpler and more effective. R supports the following functions, which we'll look at from a data analysis perspective:

`apply()`: Evaluates a function to a section of an array and returns the results in an array

`lapply()`: Loops over a list and evaluates on each element or applies the function to each element

`sapply()`: A user-friendly application of `lapply()` that returns a vector, matrix, or array

`tapply()`: Usually used over a subset of a data set

These functions are used to manipulate and slice data from matrices, arrays, lists, or dataframes. These functions traverse an entire data set, either by row or by column, and avoid loop constructs. For example, these functions can be called to do the following:

- Calculate the mean, sum, or any other manipulation on a row or a column
- Transform or perform subsetting

The apply() function is simpler in form, and its code can be very few lines (actually, one line) while helping to perform effective operations. The other, more complex forms are lapply(), sapply(), vapply(), mapply(), rapply(), and tapply(). Using these functions depends on the structure of the data and the format of the output you need to help your analysis.

The following example demonstrates the proportion of NAs in each column and the row number of the presence of NA. In both cases, we are using the apply() function. The first apply() function checks the NAs in each column and provides the proportion of NAs. In the second case, it is providing the row numbers that have NAs. The code and the corresponding output are provided below:

```
> apply(sample_data, 2, function(col)sum(is.na(col))/length(col))
   Country   NOC.CODE      Total      Golds    Silvers    Bronzes
0.00000000 0.00000000 0.04761905 0.03174603 0.00000000 0.03174603
> #Using apply() function to identifying
> # the rows with NAs
> rownames(sample_data)[apply(sample_data, 2, anyNA)]
 [1] "3"   "4"   "6"   "9"   "10"  "12"  "15"  "16"  "18"  "21"
[11] "22"  "24"  "27"  "28"  "30"  "33"  "34"  "36"  "39"  "40"
[21] "42"  "45"  "46"  "48"  "51"  "52"  "54"  "57"  "58"  "60"
[31] "63"  "64"  "66"  "69"  "70"  "72"  "75"  "76"  "78"  "81"
[41] "82"  "84"  "87"  "88"  "90"  "93"  "94"  "96"  "99"  "100"
[51] "102" "105" "106" "108" "111" "112" "114" "117" "118" "120"
[61] "123" "124" "126"
```

Na.omit() removes all the records that have NAs in the data set. The code and the corresponding output are provided below:

```
> sample_data %>% na.omit()
                                      Country NOC.CODE     Total
1                                     Bermuda      BER    1.0000
2                                    Pakistan      PAK   10.0000
```

```
3                                    Barbados      BAR     1.0000
4                                      Latvia      LAT    17.0000
5                                       China      CHN   385.0000
6                                    Countr B        A   100.7833
```

17.2.3 lapply()

The `lapply()` function outputs the results as a list. `lapply()` can be applied to a list, dataframe, or vector. The output is always a list with the same number of elements as the object passed.

The following example demonstrates the `lapply()` function. We use `lapply()` to find out the mean of the total, gold, and bronze medals. The code and the corresponding output are provided below:

```
> lapply(select(medal_df, -Country,-NOC.CODE,-Silvers),mean, na.rm = TRUE)
$Total
[1] 102.4701

$Golds
[1] 33.08824

$Bronzes
[1] 35.06618
```

17.2.4 sapply()

The difference between `sapply()` and `lapply()` is the output result. The result of `sapply()` is a vector, whereas the result for `lapply()` is a list. You can use the appropriate functions depending on the kind of data analysis you are doing and the result format you need. The following example demonstrates the use of `sapply()` for the same Summer Olympics Medals data set example through the code and the corresponding output:

```
> ##SAPPLY()
> sapply(medal_df['Golds'],mean, na.rm = TRUE)
   Golds
33.08824
> sapply(select(medal_df, -Country,-NOC.CODE,-Silvers),mean, na.rm = TRUE)
    Total     Golds   Bronzes
102.47015  33.08824  35.06618
```

17.3 Removing Duplicate Records in the Data Set

The next command is to remove all the duplicates from the data set. You can check duplicate records either on one single column or on the entire data set. Since the data set has no duplicate records, it just returns the same number of records. The code and the corresponding output are provided below:

```
> ##Remove duplicates only on one column
> X2 = distinct(sample_data, Country)
> count(X2)
    n
1 126
> ##Remove duplicate rows using distinct()
> X1 = distinct(sample_data)
> count(X1)
    n
1 126
> #duplicated() - remove duplicates
> head(sample_data[!duplicated(sample_data$Golds), ])
   Country NOC.CODE    Total Golds Silvers   Bronzes
1  Bermuda      BER   1.0000     0       1   1.00000
2 Pakistan      PAK  10.0000     3      25   4.00000
4   Latvia      LAT  17.0000     2       3   4.00000
```

```
5     China       CHN 385.0000    163         5 105.00000
6 Countr B          A 100.7833      1        51  34.81148
7  Ireland        IRL  23.0000      8        39   8.00000
```

17.4 split()

The `split()` function divides the data set into groups defined by the argument. The `split()` function splits data into groups based on factor levels chosen as an argument. This function uses two arguments: (x, y). x is the dataframe, and y is the level (split). We will divide the data set into three "bins" we have created. The `split()` command divides the data set into three groups as shown through the code and corresponding output provided here:

```
> #Split() dataset into 5 groups based on "Total" medals
> bins <- c("Bin 1", "Bin 3", "Bin 2")
> c2 <- split(medal_df, f=factor(bins))
> head(c2)
$`Bin 1`
                                       Country NOC.CODE Total
1                                          USA      USA  2297
4                                Great Britain      GBR   714
7                                       Sweden      SWE   475
...
$`Bin 2`
                                     Country NOC.CODE Total
3   Germany (includes W but not E.Germany)      GER   851
6                                      Italy      ITA   521
9                                  Australia      AUS   432
12                                     Japan      JPN   360
```

The unsplit() function reverses the split() results. The code and the corresponding output are provided below:

```
> ##unsplit() example
> c3 <-unsplit(c2, list("Bin 1", "Bin 2", "Bin 3"))
> head(c3)
         Country NOC.CODE Total Golds Silvers Bronzes
1            USA      USA  2297   930     728     639
4  Great Britain      GBR   714   207     255     252
7         Sweden      SWE   475   142     160     173
10  East Germany            409   153     129     127
13        Russia      RUS   317   108      97     112
16        Poland      POL   261    62      80     119
```

17.5 Writing Your Own Functions in R

Just as in any other programming language, you can write your own functions in R. Functions are written if a sequence of tasks needs to be repeated multiple times in the program. Functions are also written if the code needs to be shared with others and the general public. A function reduces programming complexity by creating an abstraction of the code; only a set of parameters needs to be specified. Functions in R can be treated as an object. In R, functions can be passed as an argument to other functions, such as apply() or sapply().

Functions in R are stored as an object, just like data types, and defined as function(). Function objects are data types with a class object defined as a function. The following example shows how to define a simple function and pass an argument to the function.

In the following example, myfunc_count() prints only those that have total gold medals of greater than 25 (a number). The function my_function() takes two arguments. The first argument is the dataframe, and the second argument is the number. Once the function is initiated, you call the function by passing two arguments as shown. As you can see, we have passed the medal dataframe and 25. The output is to print all the data whose gold medals (column in the data set) are greater than 25. The code and the corresponding output are provided below:

```
> myfunc_count<- function(df, number)
+ {
+   head(df[df$Golds > number, ])
+
+ }
> myfunc_count(medal_df, 25)
                               Country NOC.CODE Total Golds Silvers
1                                  USA      USA  2297   930     728
2                         Soviet Union           1010   395     319
3 Germany (includes W but not E.Germany)    GER   851   247     284
4                        Great Britain      GBR   714   207     255
5                               France      FRA   638   192     212
6                                Italy      ITA   521   190     157
  Bronzes
1     639
2     296
3     320
4     252
5     234
6     174
```

17.6 Chapter Summary

In this chapter, we covered converting a data type, finding missing values and NAs, removing the rows or columns with NAs, generating a sample dataset, splitting the dataset, removing the duplicate records from the dataset, filtering the dataset based on the condition specified, etc.

We also covered various built-in looping `apply()` functions. We also covered writing your own functions.

We have only covered the basics of R required for the basic analytics problems that you will come across. This should be used only as a quick-reference guide. If you want to learn more complex functions and R programming, you may refer to an R book or the R documentation.

CHAPTER 18

Python Programming for Analytics

This chapter introduces Python. We discuss the fundamentals of Python that are required to perform business analytics, data mining, and machine learning. This chapter provides enough basics to start using Python programming for data analysis. We will explore the pandas and NumPy tools using Jupyter Notebook so you can perform basic data analytics to create models. We will be discussing the pandas DataFrame, NumPy arrays, data manipulation, data types, missing values, data slicing and dicing, as well as data visualization.

18.1 Introduction

Python has been around since the 1990s. It gained popularity as an interactive programming language and is easy to learn. It is known for its concise, modular, and simplistic approach without the need for any complicated code. Just like other programming languages such as C, C++, and Java, Python is a structured, procedural, functional, object-oriented language. It supports statements, control flows, arithmetic operations, scientific operations, and regular expressions. All these features have made Python a popular language for data analytics and machine learning. Several open-source libraries are available to support data analytics and machine learning including scikit-learn, TensorFlow, Keras, H2O, PyTorch, and many more. Scikit-learn is an open-source machine learning library that supports supervised and unsupervised learning. It also provides various tools for data wrangling, data preprocessing, model selection, model evaluation, and many other utilities. As you may recall, many of our regression and classification learning models were developed using scikit-learn libraries. Similarly, pandas and NumPy are the two most commonly used libraries for manipulating data and performing analysis on data.

U. R. Hodeghatta and U. Nayak, *Practical Business Analytics Using R and Python*,
https://doi.org/10.1007/978-1-4842-8754-5_18

Jupyter Notebook is the original web interface for writing Python code for data science. The notebook also can be shared with others as a web document. It offers a simple, web-based, interactive experience. JupyterLab is the latest development that provides an interactive development environment for coding. It provides a flexible interface to configure and arrange workflows in data science, scientific computing, and development of machine learning models.

This chapter assumes you are familiar with the basic programming of Python. We will not cover Python programming. We focus only on the pandas and NumPy libraries and how to use them for data analysis. All our work is done using Jupyter Notebook. We want users to write and practice the code rather than copying from the text. The full notebook is available on GitHub for download.

18.2 pandas for Data Analytics

pandas is an open-source Python library package providing fast, flexible tools for easily accessing and processing data. It supports a number of operations to read data from various sources including CSV, JSON, XML, and web data. There are two primary data structures in pandas: Series and DataFrame. The pandas Series is a one-dimensional ndarray with axis labels, and it can store heterogeneous data types. The pandas DataFrame is a two-dimensional data structure and can store multiple data types. The pandas DataFrame is similar to a table and is easy to manipulate; it is used by the majority of use cases in finance, statistics, and scientific functions that the R programming language can provide. pandas is built on top of NumPy and is intended to integrate well within a scientific computing environment with many other third-party libraries.

It supports real-world data analysis with the following functionalities:

- Handling missing data (`NULL` or `NaN`)
- Inputting missing values with mean, median, or your own functions
- Iteration over the data set (called DataFrame in Pandas) for the purpose of data analytics
- Feature engineering, merging features, deleting columns, separating features, etc.

- Statistical analysis of data and specific data features/columns
- Database operations such as slicing, indexing, subsetting, merging, joining data sets, and groupby operations
- Time-series operations such as date range generation, frequency conversion, moving window statistics, date shifting, and lagging
- Easy to convert differently indexed data in other Python and NumPy data structures into DataFrame objects
- Powerful plots and graphs for statistical analysis

The following examples demonstrate the techniques using pandas for data analytics that we have mentioned. We encourage you to type each line of the code and practice. If you need the full Jupyter Notebook, then you can download it from GitHub.

The first step is to import the pandas library and create a pandas DataFrame. sys and os are Python system libraries.

Here is the input code:

```
import pandas as pd
import sys
import os
```

There will not be any output shown. These packages will be imported into the memory and will be available for further use.

Each column in a pandas DataFrame is a Series. The pandas DataFrame is similar to a spreadsheet, database, or table. Basically, you have rows and columns.

Here is the input code to create a DataFrame and print out first few records of the data:

```
income = pd.Series([2000,4000,6000,1000])

income.head()
```

Here is the output along with the input shown above:

```
In [4]:  income = pd.Series([2000,4000,6000,1000])

In [5]:  income.head()

Out[5]: 0    2000
        1    4000
        2    6000
        3    1000
        dtype: int64
```

Now that we have created a one-dimensional dataframe, we can perform analysis on this simple dataframe. The same functions can be used for multidimensional data, which we demonstrate next.

Here is the input:

```
income.describe()

income.max()

income.min()

income.sum()

income.mean()
```

Here is the output along with the input code provided above:

```
In [6]:  income.describe()

Out[6]: count       4.000000
        mean     3250.000000
        std      2217.355783
        min      1000.000000
        25%      1750.000000
        50%      3000.000000
        75%      4500.000000
        max      6000.000000
        dtype: float64

In [7]:  income.max()

Out[7]: 6000

In [8]:  income.min()

Out[8]: 1000

In [9]:  income.sum()

Out[9]: 13000

In [10]:  income.mean()

Out[10]: 3250.0
```

In our next example, we will read four CSV files, `stores.csv`, `products.csv`, `salesperson.csv`, and `totalsales.csv`, which contain information about products, salespeople, sales, and units sold. Each file has multidimensional data. We read these data set files as a pandas DataFrame and demonstrate performing the data analysis. We store each data file as a dataframe. For example, we store `stores.csv` in a `stores` dataframe, `products.csv` in a `products` dataframe, and so on.

Here is the input:

```
datadir = 'E:/Umesh-MAY2022/Personal-May2022/BA2ndEdition/2ndEdition/Book
Chapters/Chapter-18 Python For Analytics/dataset'
os.chdir(datadir)
os.getcwd()

'E:\\Umesh-MAY2022\\Personal-May2022\\BA2ndEdition\\2ndEdition\\Book Chapters\
\Chapter 18 - Python For Analytics\\dataset'

stores = pd.read_csv("stores.csv")
products=pd.read_csv("products.csv")
agents=pd.read_csv("salesperson.csv")
sales=pd.read_csv("totalsales.csv")
```

Here is the output along with the input code shown above:

```
In [11]:  datadir = 'E:/Umesh-MAY2022/Personal-May2022/BA2ndEdition/2ndEdition/Book Chi
          os.chdir(datadir)
          os.getcwd()

Out[11]: 'E:\\Umesh-MAY2022\\Personal-May2022\\BA2ndEdition\\2ndEdition\\Book Chapte
         rs\\Chapter 18 - Python For Analytics\\dataset'

In [12]:  stores = pd.read_csv("stores.csv")
          products=pd.read_csv("products.csv")
          agents=pd.read_csv("salesperson.csv")
          sales=pd.read_csv("totalsales.csv")
```

Reading a dataframe using the `head()` or `tail()` function is shown next. Similar to R, Python has a function to get a glimpse of the data using the `head()` or `tail()` function. The function `head()` provides the top five rows (default), and `tail()` provides the bottom five rows of the data. You can also specify more rows to be displayed by specifying the number within the parentheses. For example, `head(10)` displays 10 rows.

Here is the input:

```
sales.head(10)

stores

products

agents
```

Here is the output along with the input code shown above:

```
In [22]:  sales.head(10)
Out[22]:
```

	Date	Store #	Salesperson	Product	Units sold
0	7/16/2011	2301243.0	John Davis	AR400	6.0
1	7/18/2011	2301345.0	Super Shevan	NK350	1.0
2	7/18/2011	2301504.0	Jacob Caravan	NK350	1.0
3	7/18/2011	2301748.0	John Davis	TB340	3.0
4	7/18/2011	2309823.0	Sarah Doom	NK350	6.0
5	7/19/2011	2307430.0	Jennifer Mohami	HP351	2.0
6	7/19/2011	2307430.0	Jennifer Mohami	HP351	1.0
7	7/19/2011	2307430.0	Super Shevan	RK200	4.0
8	7/19/2011	2307430.0	Jacob Caravan	HP351	2.0
9	7/19/2011	2307430.0	Jennifer Mohami	TB340	2.0

If you have limited data, we can print the whole dataframe, as shown here:

```
In [13]:  stores
Out[13]:
```

	Store#	Adress	city	state	ZIP	Phone
0	2301243	10100 Normandale Blvd,	Bloomington	MN	55347	9525554242
1	2301345	10660 Hampshire Ave	S, Bloomington,	MN	55438	952 334 2838
2	2301504	6150 Egan Dr	Savage,	MN	55378	612 366 2626
3	2301748	12009 Flying Cloud Drv	Chahansen	MN	55324	651 738 8377
4	2309823	13240 Singletree Lane	Eden Prairie	MN	55344	952 399 9383
5	2307430	23400 Gill Lane	Edina	MN	55432	6512342233

```
In [19]: products
```

Out[19]:

	Product	Product Name	Unit Price
0	AR400	Arli - Pro 2.4 Camera	629.99
1	NK350	Nikkon DSLR	450
2	TB340	Toshiba 50Inch TV	$299.00
3	HP351	HP 15 - inch Laptop	$349.00
4	RK200	Roku Streaming/Stick	39.99
5	AZ102	Amazon Firestick	69.99

```
In [20]: agents
```

Out[20]:

	First Name	Last Name	Commission
0	John	Davis	8%
1	Super	Shevan	6%
2	Jacob	Caravan	8%
3	James	Jumperman	8%
4	Sarah	Doomer	10%
5	Jennifer	Mohami	10%

dtypes provides the data types of the variables in the data set, and info provides more information about the data set, including null and non-null counts as shown, which is useful for further analysis. shape provides the size of the data such as the number of rows × the number of columns. In this example, the sales dataframe has 396 rows and 5 columns of data.

Here is the input:

```
sales.dtypes
sales.info()  #more details about the dataset
sales.shape
```

Here is the output along with the input code shown above:

```
In [24]:   sales.dtypes

  Out[24]: Date            object
           Store #        float64
           Salesperson     object
           Product         object
           Units sold     float64
           dtype: object
```

```
#  The data types in this "sales" DataFrame are integers (int64), floats
(float64) and strings (object).
```

```
In [25]:   sales.info()  #more details about the dataset

           <class 'pandas.core.frame.DataFrame'>
           RangeIndex: 396 entries, 0 to 395
           Data columns (total 5 columns):
            #   Column       Non-Null Count  Dtype
           ---  ------       --------------  -----
            0   Date         396 non-null    object
            1   Store #      395 non-null    float64
            2   Salesperson  396 non-null    object
            3   Product      396 non-null    object
            4   Units sold   385 non-null    float64
           dtypes: float64(2), object(3)
           memory usage: 15.6+ KB
```

```
In [27]:   sales.shape

  Out[27]: (396, 5)
```

18.2.1 Data Slicing Using pandas

In this section, we focus on data slicing. The `sales` dataframe is multidimensional. We can slice the dataframe to access certain columns and features. `tail(10)` provides the bottom 10 rows of the data as shown. The `head()` function provides a glimpse of the top five rows of the dataframe. There are multiple ways to slice the data in pandas. Two ways are shown next.

Here is the input:

```
sales['Product'].tail(10)

sales.Product.head()
```

Here is the output along with the input code shown above:

```
In [30]:  sales['Product'].tail(10)

Out[30]: 386     NK350
         387     AR400
         388     RK200
         389     TB340
         390     TB340
         391     NK350
         392     HP351
         393     HP351
         394     RK200
         395     HP351
         Name: Product, dtype: object
```

```
In [31]:  sales.Product.head()

Out[31]: 0    AR400
         1    NK350
         2    NK350
         3    TB340
         4    NK350
         Name: Product, dtype: object
```

If we want to access multiple columns, we can slice the dataframe as shown next. In this example, we are slicing the dataframe to access three columns out of five (Date, Product, and Salesperson) from the sales dataframe.

Here is the input:

```
sales[['Date','Product','Salesperson']].head()
```

Here is the output along with the input code shown above:

```
In [33]: sales[['Date','Product','Salesperson']].head()
Out[33]:
```

	Date	Product	Salesperson
0	7/16/2011	AR400	John Davis
1	7/18/2011	NK350	Super Shevan
2	7/18/2011	NK350	Jacob Caravan
3	7/18/2011	TB340	John Davis
4	7/18/2011	NK350	Sarah Doom

In the following example, we slice the dataframe to display only specific rows in the dataframe.

Here is the input:

```
sales[146:152]  ## Slicing - display only 146 to 151 records
sales['Product'][160:170] # slicing rows 160-169 of product column only
```

Here is the output along with the input code shown above:

```
In [34]: sales[146:152]  ## Slicing - display only 146 to 151 records
Out[34]:
```

	Date	Store #	Salesperson	Product	Units sold
146	8/24/2011	2301345.0	James Bedlock	AR400	2.0
147	8/27/2011	2301504.0	Sarah Doom	RK200	3.0
148	8/28/2011	NaN	James Bedlock	NK350	NaN
149	8/28/2011	2301243.0	Super Shevan	RK200	3.0
150	8/28/2011	2301345.0	Jacob Caravan	TB340	3.0
151	8/29/2011	2301504.0	Sarah Doom	NK350	2.0

```
In [37]:  sales['Product'][160:170]

Out[37]: 160    TB340
         161    TB340
         162    HP351
         163    HP351
         164    HP351
         165    RK200
         166    TB340
         167    HP351
         168    NK350
         169    AR400
         Name: Product, dtype: object
```

The next set of functions demonstrates data analysis on the pandas DataFrame. For example, search for the sales data of Units sold greater than 4 units.

Here is the input:

```
sales[sales['Units sold'] >4.0] #Units Sold greater than 4
```

Here is the output along with the input code shown above:

```
In [20]:  sales[sales['Units sold'] >4.0] #Units Sold greater than 4

Out[20]:
```

	Date	Store #	Salesperson	Product	Units sold
0	7/16/2011	2301243.0	John Davis	AR400	6.0
4	7/18/2011	2309823.0	Sarah Doom	NK350	6.0
10	7/20/2011	2301243.0	Jennifer Mohami	TB340	5.0
26	7/24/2011	2301243.0	Jacob Caravan	NK350	6.0
45	7/27/2011	2307430.0	James Bedlock	RK200	5.0
79	8/8/2011	2307430.0	John Davis	AR400	6.0
80	8/9/2011	2301243.0	Sarah Doom	NK350	6.0
99	8/13/2011	2301748.0	Jacob Caravan	AR400	5.0
103	8/15/2011	2301243.0	Jennifer Mohami	RK200	6.0

Similarly, in the following example we view sales data of product AR400 sold greater than or equal to 5. This data slicing and dicing is similar to extracting data from a database using SQL.

Here is the input:

```
sales[sales['Units sold'].ge(5.0) & sales['Product'].isin(['AR400'])]
```

Here is the output along with the input code shown above:

```
##Select product AR400 sold greater than 5 units
https://pandas.pydata.org/docs/reference/api/pandas.DataFrame.ge.html
```

```
In [48]: sales[sales['Units sold'].ge(5.0) & sales['Product'].isin(['AR400'])]
```

Out[48]:

	Date	Store #	Salesperson	Product	Units sold
0	7/16/2011	2301243.0	John Davis	AR400	6.0
79	8/8/2011	2307430.0	John Davis	AR400	6.0
99	8/13/2011	2301748.0	Jacob Caravan	AR400	5.0
153	8/29/2011	2309823.0	Jennifer Mohami	AR400	6.0
213	9/7/2011	2309823.0	James Bedlock	AR400	6.0
251	9/14/2011	2301504.0	Jacob Caravan	AR400	5.0
302	10/2/2011	2307430.0	James Bedlock	AR400	6.0
329	10/5/2011	2301748.0	James Bedlock	AR400	5.0
357	10/15/2011	2301345.0	Jacob Caravan	AR400	5.0
372	10/16/2011	2301748.0	Jacob Caravan	AR400	6.0

`df.loc` and `df.iloc` also provide index-based data slicing by position. `df.loc()` accesses a group of rows and columns by label or a Boolean array, whereas `df.iloc()` uses integer location-based indexing for selection by position.

Here is the input:

```
sales.iloc[[10,12]]
sales.iloc[[10]]
sales.iloc[:3]
```

Here is the output along with the input code shown above:

```
In [50]: sales.iloc[[10,12]]
Out[50]:
```

	Date	Store #	Salesperson	Product	Units sold
10	7/20/2011	2301243.0	Jennifer Mohami	TB340	5.0
12	7/20/2011	2301504.0	James Bedlock	NK350	3.0

```
In [51]: sales.iloc[[10]]
Out[51]:
```

	Date	Store #	Salesperson	Product	Units sold
10	7/20/2011	2301243.0	Jennifer Mohami	TB340	5.0

```
In [52]: sales.iloc[:3]
Out[52]:
```

	Date	Store #	Salesperson	Product	Units sold
0	7/16/2011	2301243.0	John Davis	AR400	6.0
1	7/18/2011	2301345.0	Super Shevan	NK350	1.0
2	7/18/2011	2301504.0	Jacob Caravan	NK350	1.0

In the case of `df.loc()`, you can use a row index to slice the data. It can be a number or label. In our case, rows are numbered, so we can use numbers. In the first example, we try to get the first record, and in the second example, we try to get the first and fifth records. In the third example we fetch the second and the third records.

Here is the input:

```
sales.loc[1]

sales.loc[[1,5]]

sales.loc[2:3]
```

Here is the output along with the input code shown above:

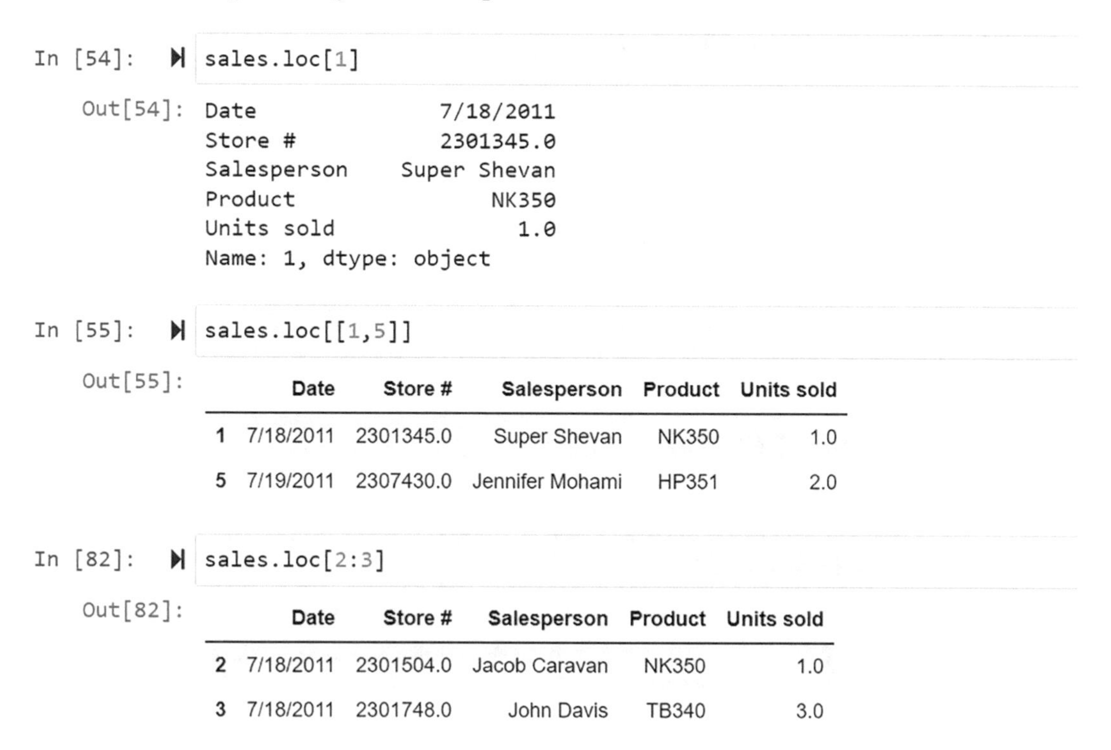

```
In [54]:  sales.loc[1]

Out[54]: Date                7/18/2011
         Store #             2301345.0
         Salesperson      Super Shevan
         Product                 NK350
         Units sold                1.0
         Name: 1, dtype: object

In [55]:  sales.loc[[1,5]]

Out[55]:
```

	Date	Store #	Salesperson	Product	Units sold
1	7/18/2011	2301345.0	Super Shevan	NK350	1.0
5	7/19/2011	2307430.0	Jennifer Mohami	HP351	2.0

```
In [82]:  sales.loc[2:3]

Out[82]:
```

	Date	Store #	Salesperson	Product	Units sold
2	7/18/2011	2301504.0	Jacob Caravan	NK350	1.0
3	7/18/2011	2301748.0	John Davis	TB340	3.0

18.2.2 Statistical Data Analysis Using pandas

This section demonstrates how to conduct statistical analysis on the pandas DataFrame. There are several functions supported by pandas, including functions to calculate the maximum value, minimum value, mean, etc. `value_counts()` counts the categorical values present in the data. Similarly, `unique` shows the unique values present in the data.

Here is the input:

```
sales['Units sold'].max()

sales['Units sold'].mean()

sales['Units sold'].std()

sales['Units sold'].describe()

sales.Product.unique()

sales['Product'].value_counts()

sales['Store #'].value_counts()
```

Here is the output along with the input code shown above:

```
In [21]:    sales['Units sold'].max()

   Out[21]: 6.0

In [61]:    sales['Units sold'].mean()

   Out[61]: 2.3324675324675326

In [22]:    sales['Units sold'].std()

   Out[22]: 1.5064480815849024

In [23]:    sales['Units sold'].describe()

   Out[23]: count    385.000000
            mean       2.332468
            std        1.506448
            min        1.000000
            25%        1.000000
            50%        2.000000
            75%        3.000000
            max        6.000000
            Name: Units sold, dtype: float64
```

```
In [42]:  sales['Store #'].value_counts() ## Number of Store CLass

Out[42]: 2301748.0    80
         2309823.0    76
         2307430.0    75
         2301345.0    69
         2301504.0    66
         2301243.0    29
         Name: Store #, dtype: int64
```

```
In [59]:  sales['Product'].value_counts() ##Number of Product class

Out[59]: HP351    93
         AR400    80
         TB340    77
         NK350    75
         RK200    71
         Name: Product, dtype: int64
```

```
In [60]:  sales.Product.unique()

Out[60]: array(['AR400', 'NK350', 'TB340', 'HP351', 'RK200'], dtype=object)
```

18.2.3 Pandas Database Functions

pandas supports many SQL database type of data slicing and accessing data, including `groupby()`, merging, join, etc. You should refer to the pandas documentation on how to use those functions. In the first example below, we are aggregating the Units sold per each individual salesperson. In the second example, we are aggregating (summing up) the units sold based on date.

Here is the input:

```
sales.groupby("Salesperson")['Units sold'].sum()

func = {'Units sold' :'sum', 'Date':'first'}
sales.groupby(['Date'], as_index=False).agg(func)
```

Here is the output along with the input code shown above:

```
In [22]:  sales.groupby("Salesperson")['Units sold'].sum()

Out[22]: Salesperson
         Jacob Caravan       177.0
         James Bedlock       129.0
         Jennifer Mohami     153.0
         John Davis          139.0
         Sarah Doom          155.0
         Super Shevan        145.0
         Name: Units sold, dtype: float64

In [65]:  func = {'Units sold' :'sum', 'Date':'first'}
          sales.groupby(['Date'], as_index=False).agg(func)

Out[65]:
```

	Units sold	Date
0	8.0	10/1/2011
1	7.0	10/10/2011
2	2.0	10/11/2011
3	10.0	10/12/2011
4	21.0	10/15/2011

18.2.4 Data Preprocessing Using pandas

In this section, we briefly touch upon identifying missing values, substituting missing values with some other values, imputing missing values, etc. We may not be able to demonstrate all the possible options; this is just a reference to understanding the concepts and how to apply them using Pandas. `isna()` shows whether a dataframe has any NAs or NULL values. There are several ways to use the `isna()` function. The following function also shows how many NAs are in dataframe.

Here is the input:

```
sales.isna().sum()

pd.isnull(sales).sum()

sales['Store #'].isna().sum()
```

Here is the output along with the input code shown above:

```
In [44]:  sales.isna().sum()

Out[44]: Date            0
         Store #         1
         Salesperson     0
         Product         0
         Units sold     11
         dtype: int64

In [45]:  pd.isnull(sales).sum() ## NULL values

Out[45]: Date            0
         Store #         1
         Salesperson     0
         Product         0
         Units sold     11
         dtype: int64

In [68]:  sales['Store #'].isna().sum()

Out[68]: 1
```

The previous `isna()` did not give the location of the NAs. To find out the location of the NAs, meaning specific rows and columns, you have to slice the data further, as shown in the next set of functions. The following function provides the location of the NAs in the dataframe.

Here is the input:

```
sales[sales['Store #'].isna()]

sales[sales['Units sold'].isna()]
```

Here is the output along with the input code shown above:

```
In [48]: sales[sales['Store #'].isna()]
```

Out[48]:

	Date	Store #	Salesperson	Product	Units sold
148	8/28/2011	NaN	James Bedlock	NK350	NaN

```
In [49]: sales[sales['Units sold'].isna()]
```

Out[49]:

	Date	Store #	Salesperson	Product	Units sold
13	7/20/2011	2301748.0	John Davis	RK200	NaN
36	7/24/2011	2301504.0	Sarah Doom	AR400	NaN
60	8/1/2011	2301243.0	Sarah Doom	TB340	NaN
89	8/10/2011	2301504.0	Super Shevan	AR400	NaN
125	8/17/2011	2307430.0	Sarah Doom	NK350	NaN
148	8/28/2011	NaN	James Bedlock	NK350	NaN
179	8/31/2011	2301345.0	James Bedlock	AR400	NaN
210	9/6/2011	2309823.0	Jacob Caravan	TB340	NaN
245	9/13/2011	2309823.0	Jacob Caravan	NK350	NaN
344	10/9/2011	2301504.0	Sarah Doom	AR400	NaN
368	10/16/2011	2301748.0	Super Shevan	NK350	NaN

Once you identify the NAs and NULL values in the data, it is up to you how you want to substitute these NAs. We have shown a simple example here, but you can extend it to any technique of assigning values to such locations with NAs. In the following example, we have just used the number 3.0 to fill the NAs. However, you can substitute the NAs with mean or mode or median values. You can use the same `fillna()` function to perform inputting mean or mode or any other value.

Here is the input:

```
sales['Units sold'] = sales['Units sold'].fillna(3.0)

sales['Units sold'].iloc[36] ##Check for others
```

Here is the output along with the input code shown above:

```
In [70]:  sales['Units sold'] = sales['Units sold'].fillna(3.0)

In [71]:  sales['Units sold'].iloc[36] ##Check for others
 Out[71]: 3.0
```

18.2.5 Handling Data Types

Sometimes while reading a data set from a file, pandas treats the date, categorical and numerical variables as objects. It becomes necessary to convert inappropriate data types to appropriate data types. The following example demonstrates one such example of converting into an `integer` variable. We use the `astype()` function to perform this operation.

Here is the input:

```
sales['Units sold'] = sales['Units sold'].astype(int)
##Read about as type() function

sales.dtypes
```

Here is the output along with the input code shown above:

```
In [77]:  sales['Units sold'] = sales['Units sold'].astype(int) ##Read about astype() function

In [79]:  sales.dtypes
 Out[79]: Date            object
          Store #        float64
          Salesperson     object
          Product         object
          Units sold       int32
          dtype: object
```

18.2.6 Handling Dates Variables

The pandas DataFrame treats Date variable as a regular object. It is necessary to convert the Date variable to the appropriate data type to manipulate the dates. The following example demonstrates how to convert dates to a proper data type and break down the date into months, year, and day for further analysis.

pandas reads the date as a numerical object. The first step is to convert to the datetime data type and then split the date into day, month, year and time, if it has one; then we can split it further to hours, minutes, and seconds.

Here is the input:

```
sales.dtypes

sales['Date'] = pd.to_datetime(sales['Date'],errors='coerce')

sales.dtypes

### divide DATE into day, month and year columns
sales['Day'] = sales['Date'].dt.day
sales['Month'] = sales['Date'].dt.month
sales['Year'] = sales['Date'].dt.year

sales.head()
```

Here is the output along with the input code shown above:

```
In [81]:  sales.dtypes

Out[81]: Date            object
         Store #        float64
         Salesperson     object
         Product         object
         Units sold       int32
         dtype: object

In [82]:  sales['Date'] = pd.to_datetime(sales['Date'], errors='coerce')

In [61]:  sales['Date'] =  pd.to_datetime(sales['Date'],errors='coerce')    #Date and T

In [83]:  sales.dtypes

Out[83]: Date          datetime64[ns]
         Store #              float64
         Salesperson           object
         Product               object
         Units sold             int32
         dtype: object
```

```
In [88]:  ### divide DATE into day, month and year columns
           sales['Day'] = sales['Date'].dt.day
           sales['Month'] = sales['Date'].dt.month
           sales['Year'] = sales['Date'].dt.year

In [89]:  sales.head()

Out[89]:
```

	Date	Store #	Salesperson	Product	Units sold	Day	Month	Year
0	2011-07-16	2301243.0	John Davis	AR400	6	16	7	2011
1	2011-07-18	2301345.0	Super Shevan	NK350	1	18	7	2011
2	2011-07-18	2301504.0	Jacob Caravan	NK350	1	18	7	2011
3	2011-07-18	2301748.0	John Davis	TB340	3	18	7	2011
4	2011-07-18	2309823.0	Sarah Doom	NK350	6	18	7	2011

18.2.7 Feature Engineering

The pandas DataFrame supports splitting columns, merging columns, etc. The following section demonstrates how two columns can be combined. In this example, the first name and last name columns are combined. There are couple ways we can do this. We have shown both types. The agg() function aggregates using one or more operations over the specified axis.

Here is the input:

```
agents['Full-Name'] = agents[['First Name', 'Last Name']].agg('-'.join, axis=1)
##combine First name and Last name into one column FULL NAME

#Since both are strings, I can simply do this:
agents['Fullname'] = agents['First Name'] + " " + agents['Last Name']

#agents['Fullname'].str.split(" ")

agents['newfirstname'] = agents['Fullname'].str.split(" ")

agents
```

Here is the output along with the input code shown above:

```
In [41]:  ##You can use AGG() function

          agents['Full-Name'] = agents[['First Name', 'Last Name']].agg('-'.join, axis=1)
```

```
In [28]:  ##combine First name and Last name into one column FULL NAME

In [34]:  agents['Fullname'] = agents['First Name'] + " " + agents['Last Name']

In [35]:  #Since both are strings, I can simply do this:
           agents['Fullname'] = agents['First Name'] + " " + agents['Last Name']

In [36]:  #agents['Fullname'].str.split(" ")

In [39]:  agents['newfirstname'] = agents['Fullname'].str.split(" ")

In [42]:  agents
Out[42]:
```

	First Name	Last Name	Commission	Fullname	newfirstname	Full-Name
0	John	Davis	8%	John Davis	[John, Davis]	John-Davis
1	Super	Shevan	6%	Super Shevan	[Super, Shevan]	Super-Shevan
2	Jacob	Caravan	8%	Jacob Caravan	[Jacob, Caravan]	Jacob-Caravan
3	James	Jumperman	8%	James Jumperman	[James, Jumperman]	James-Jumperman
4	Sarah	Doomer	10%	Sarah Doomer	[Sarah, Doomer]	Sarah-Doomer
5	Jennifer	Mohami	10%	Jennifer Mohami	[Jennifer, Mohami]	Jennifer-Mohami

18.2.8 Data Preprocessing Using the apply() Function

pandas supports the `apply()` function, which is similar to the `apply()` function in R. `apply()` is a powerful function to manipulate the entire dataframe without using `loops`. In the following example, we will demonstrate a simple preprocessing of text strings with the help of `apply()`. We will display `products` data before and after the preprocessing of text. `pre_proc()` is a simple Python function that is self-explanatory. It just uses the Python `replace()` function.

Here is the input:

```
def pre_proc(x):
    x = x.replace('$',"")
    x = x.replace('[&]',"")
    x = x.strip()
    x = x.replace('[/]',"")
    x = x.lower()
    return(x)

products

products['Unit Price'] = products['Unit Price'].apply(pre_proc)

products
```

Here is the output along with the input code shown above:

```
In [101]:  def pre_proc(x):
               x = x.replace('$',"")
               x = x.replace('[&]',"")
               x = x.strip()
               x = x.replace('[/]',"")
               x = x.lower()
               return(x)

In [99]:   products
```

Out[99]:

	Product	Product Name	Unit Price
0	AR400	Arli - Pro 2.4 Camera	629.99
1	NK350	Nikkon DSLR	450
2	TB340	Toshiba 50Inch TV	$299.00
3	HP351	HP 15 - inch Laptop	$349.00
4	RK200	Roku Streaming/Stick	39.99
5	AZ102	Amazon Firestick	69.99

```
In [111]:  ## Some values in Unit price column has $ sign and some do not. To keep the values consistent
           ## we will remove $ sign from the values.
           ## We will call pre_proc() function

In [102]:  products['Unit Price'] = products['Unit Price'].apply(pre_proc)

In [118]:  products
```

Out[118]:

	Product	Product Name	Unit Price
0	AR400	arli - pro 2.4 camera	629.99
1	NK350	nikkon dslr	450
2	TB340	toshiba 50inch tv	299.00
3	HP351	hp 15 - inch laptop	349.00
4	RK200	roku streaming/stick	39.99
5	AZ102	amazon firestick	69.99

18.2.9 Plots Using pandas

In conjunction with the `matplotlib()` libraries, pandas supports many visualization plots necessary for the data analytics. Here we demonstrate a few plots with examples using pandas. We use the same sales dataset used in earlier examples. We just plot a box plot and bar chart for the Units Sold feature.

Here is the input:

```
sales['Units sold'].plot.box()

sales["Salesperson'].value_counts().plot.bar()
```

Here is the output along with the input code shown above (a box plot and a bar chart):

```
In [105]:  sales['Units sold'].plot.box()
   Out[105]: <AxesSubplot:>
```

Units sold

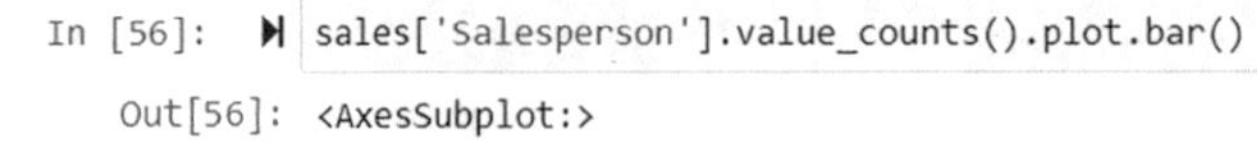

In [56]: sales['Salesperson'].value_counts().plot.bar()
Out[56]: <AxesSubplot:>

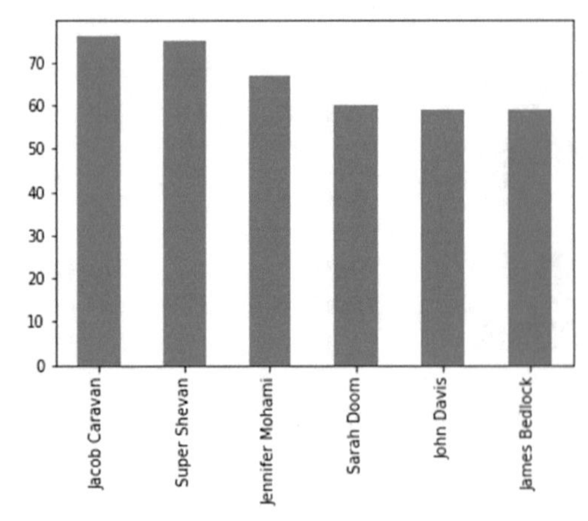

18.3 NumPy for Data Analytics

NumPy is another package used in data analytics and machine learning models. It is a Python data structure of a multidimensional array object. It supports a number of functions for easy data manipulation on the NumPy data object. It includes many scientific and mathematical operations, including shape manipulation, basic linear algebra, matrix operations, statistical operations, random number generations, and tensor operations.

This section will cover some of the basics of NumPy arrays that may be useful for any data analytics operations. For further information and usages, you can refer to the NumPy documentation and user guide provided in the reference (`https://numpy.org/`).

A NumPy array is represented as an ndarray object that encapsulates n-dimensional arrays of homogeneous data types. A NumPy ndarray is created with a fixed data size. If the size of the ndarray is changed, then a new array is created, and the original one is deleted. Using NumPy arrays, advanced mathematical and other types of operations on large numbers of data can be performed more efficiently and with fewer lines of code than Python. A growing number of machine learning and deep learning packages have been built using NumPy arrays, including TensorFlow, Keras, PyTorch, etc.

The first step is to import the NumPy library functions to the development environment using the import function and create a simple ndarray of shape 3 by 5 as shown next. In the first example, we generate 15 numbers and reshape to a 3 rows by 5 columns matrix. There are a set of functions supported by NumPy to find out the dimension of the array, data type, etc.

Here is the input:

```
import numpy as np

a = np.arange(15).reshape(3, 5)
print(a)

a.shape
a.ndim
a.dtype.type
a.size
type(a)
```

Here is the output along with the input code shown above:

```
In [1]:  import numpy as np

In [3]:  a = np.arange(15).reshape(3, 5)
         print(a)

         [[ 0  1  2  3  4]
          [ 5  6  7  8  9]
          [10 11 12 13 14]]

In [4]:  a.shape

Out[4]:  (3, 5)

In [9]:  a.ndim

Out[9]:  2

In [5]:  a.dtype.type

Out[5]:  numpy.int32

In [6]:  a.size

Out[6]:  15

In [16]:  type(a)

Out[16]:  numpy.ndarray
```

Following is another way to create the NumPy array:

Here is the input:

```
c = np.array([(1.2,3.4,5.2), (7,8,10)])
c.dtype
```

Here is the output along with the input code shown above:

```
There are several ways to create nparray

In [12]:  c = np.array([(1.2,3.4,5.2), (7,8,10)])

In [13]:  c

Out[13]:  array([[ 1.2,  3.4,  5.2],
                 [ 7. ,  8. , 10. ]])

In [14]:  c.dtype

Out[14]:  dtype('float64')
```

18.3.1 Creating NumPy Arrays with Zeros and Ones

This is useful for many scientific operations such as calculating eigen values, principal component analysis, linear programming, etc. In this example, we create a NumPy array of zeros and another empty array with nothing in it, just null values. `np.empty()` returns a new array of given shape and type, without initializing entries.

Similarly, you can create a NumPy array with ones.

Here is the input:

```
d = np.zeros((3,4))
e=np.ones((3,4))
f = np.empty((2,4))
d
e
f
```

Here is the output along with the input code shown above:

The function zeros creates an array full of zeros

```
In [44]:  d = np.zeros((3,4))

In [45]:  d

Out[45]: array([[0., 0., 0., 0.],
                [0., 0., 0., 0.],
                [0., 0., 0., 0.]])
```

The function ones - creates an array full of ones

```
In [15]:  e=np.ones((3,4))

In [16]:  e

Out[16]: array([[1., 1., 1., 1.],
                [1., 1., 1., 1.],
                [1., 1., 1., 1.]])
```

The function empty

creates an array whose initial content is random and depends on the state of the memory

```
In [17]:  f = np.empty((2,4))

In [18]:  f

Out[18]: array([[0.00000000e+000, 0.00000000e+000, 0.00000000e+000,
                 0.00000000e+000],
                [0.00000000e+000, 4.16991405e-321, 7.56599806e-307,
                 8.90104238e-307]])
```

Many times, arrays need reshaping for vector, matrix, and algebraic operations. We can perform such operations using the `reshape()` function of NumPy. The `np.reshape()` function gives a new shape to an array without changing its data. The following example demonstrates the different sizes of the matrix using the `np.reshape()` function.

Here is the input:

```
g = np.arange(12)
print(g.reshape(3,4))
g.reshape(4,3)
g.reshape(6,2)
g.reshape(12,1)
```

Here is the output along with the input code shown above:

```
In [19]:   g = np.arange(12)

In [20]:   g

 Out[20]:  array([ 0,  1,  2,  3,  4,  5,  6,  7,  8,  9, 10, 11])

In [21]:   print(g.reshape(3,4))

           [[ 0  1  2  3]
            [ 4  5  6  7]
            [ 8  9 10 11]]

In [23]:   g.reshape(4,3)

 Out[23]:  array([[ 0,  1,  2],
                  [ 3,  4,  5],
                  [ 6,  7,  8],
                  [ 9, 10, 11]])

In [24]:   g.reshape(6,2)

 Out[24]:  array([[ 0,  1],
                  [ 2,  3],
                  [ 4,  5],
                  [ 6,  7],
                  [ 8,  9],
                  [10, 11]])

In [25]:   g.reshape(12,1)

 Out[25]:  array([[ 0],
                  [ 1],
                  [ 2],
                  [ 3],
                  [ 4],
                  [ 5],
                  [ 6],
                  [ 7],
                  [ 8],
                  [ 9],
                  [10],
                  [11]])
```

Some of the matrix operations on NumPy arrays are demonstrated in the following example. The examples are self-explanatory. Operations include addition, multiplication, and subtractions on two or three different NumPy arrays.

Here is the input:

```
b = np.array([[10,11,12],[14,15,16]])
b
b.shape
c = np.array([(1.2,3.4,5.2), (7,8,10)])
c
c.shape
b+c
b-c
c-b
b*c
```

Here is the output along with the above lines of code (Please Note: Order of code lines here are a little different from the above):

```
In [30]:  b

  Out[30]: array([[10, 11, 12],
                  [14, 15, 16]])

In [31]:  c

  Out[31]: array([[ 1.2,  3.4,  5.2],
                  [ 7. ,  8. , 10. ]])

In [27]:  b.shape

  Out[27]: (2, 3)

In [29]:  c.shape

  Out[29]: (2, 3)

In [32]:  b+c

  Out[32]: array([[11.2, 14.4, 17.2],
                  [21. , 23. , 26. ]])

In [33]:  b-c

  Out[33]: array([[8.8, 7.6, 6.8],
                  [7. , 7. , 6. ]])

In [34]:  c-b

  Out[34]: array([[-8.8, -7.6, -6.8],
                  [-7. , -7. , -6. ]])

In [35]:  b*c ##elementwise multiplication

  Out[35]: array([[ 12. ,  37.4,  62.4],
                  [ 98. , 120. , 160. ]])
```

You can also perform mathematical and statistical analysis on the NumPy arrays. This includes finding the square root, finding the mean, finding exponential, etc. The following example demonstrates some of the mathematical operations on NumPy data. In this example we will add two NumPy arrays, find the square root, and find exponential, as well as calculate the square root of NumPy data.

Here is the input:

```
np.add(b,c)

np.square(c)

np.sqrt(c)

np.exp(c)

np.round_(c)
```

Here is the output along with the code input provided above (Please Note: order of the code shown here differs a little from the above):

```
In [36]:    np.add(b,c)

   Out[36]: array([[11.2, 14.4, 17.2],
                   [21. , 23. , 26. ]])

In [100]:   np.sqrt(c)

  Out[100]: array([[1.09544512, 1.84390889, 2.28035085],
                   [2.64575131, 2.82842712, 3.16227766]])

In [37]:    np.square(c)

   Out[37]: array([[  1.44,  11.56,  27.04],
                   [ 49.  ,  64.  , 100.  ]])

In [38]:    np.round_(c)

   Out[38]: array([[ 1.,  3.,  5.],
                   [ 7.,  8., 10.]])

In [39]:    np.exp(c)

   Out[39]: array([[3.32011692e+00, 2.99641000e+01, 1.81272242e+02],
                   [1.09663316e+03, 2.98095799e+03, 2.20264658e+04]])

In [40]:    b.shape

   Out[40]: (2, 3)

In [41]:    c.shape

   Out[41]: (2, 3)
```

Note These operations must match the fundamental rules of matrix operations. If the shape of the array is not according to the rules of operations, you will get an error.

18.3.2 Random Number Generation and Statistical Analysis

Sometimes we may need to generate random numbers, take samples for analysis and learning purposes. We can generate random numbers using the `np.random()` function. We can also use the `min()`, `max()`, `mean()`, and `standard_deviation()` functions on the array. In the following example, we will create two NumPy arrays, one of size (5,2) and the other of size (3,4), and perform the basic operations `min()`, `max()`, and `standard_deviation()`.

Here is the input:

```
h = np.random.random_sample((5,2))
l = np.random.random_sample((3,4))
h.min()
h.max()
```

Here is the output along with the above input code lines:

```
In [3]:  h = np.random.random_sample((5,2))

In [4]:  l = np.random.random_sample((3,4))

In [5]:  h.shape

Out[5]: (5, 2)

In [6]:  h

Out[6]: array([[0.31817782, 0.66307187],
               [0.52541306, 0.52163656],
               [0.93431217, 0.80106959],
               [0.8476229 , 0.37359513],
               [0.3648866 , 0.0298522 ]])

In [7]:  h.min()

Out[7]: 0.02985220080647555

In [8]:  h.max()

Out[8]: 0.9343121658332714
```

The standard deviation is computed along the axis. The default is to compute the standard deviation of the flattened array (axis 0 is a column and 1 is a row).

Here is the input:

```
h.std(axis=0)

h.std(axis=1)
```

Here is the output along with the input code shown above:

```
In [13]:  h.std(axis=0)

Out[13]: array([0.25033122, 0.26543037])

In [14]:  h.std(axis=1)

Out[14]: array([0.17244703, 0.00188825, 0.06662129, 0.23701389, 0.1675172 ])
```

18.3.3 Indexing, Slicing, and Iterating

Just like with pandas, we can slice and dice the NumPy array. This is required for data processing and data manipulation, particularly while creating machine learning models. In this example, we will demonstrate how to slice the data and change certain values (similar to inputting missing values). The code is self-explanatory. If you want to access the first element, use h[0][0]; similarly, to access the entire third row, use h[3:1].

Here is the input:

```
h[2]
h[4]
h[4][0]
h[0][0]
h[0:1]
h[0:1] = 10
```

You can also change a particular value of an array element by assigning values as shown.

Indexing, Slicing and Iterating

```
In [65]: h

Out[65]: array([[0.7613513 , 0.59405937],
                [0.00882519, 0.19933056],
                [0.78336269, 0.92969025],
                [0.95003044, 0.87588577],
                [0.03372778, 0.09245768]])

In [66]: h[2]

Out[66]: array([0.78336269, 0.92969025])

In [67]: h[4]

Out[67]: array([0.03372778, 0.09245768])

In [68]: h[4][0]

Out[68]: 0.033727775884743516

In [69]: h[0][0]

Out[69]: 0.7613513045624339

In [113]: ##Print 0.493106 - X41

In [70]: h[0:1]

Out[70]: array([[0.7613513 , 0.59405937]])

In [71]: h[0:1] = 10

In [72]: h

Out[72]: array([[1.00000000e+01, 1.00000000e+01],
                [8.82518844e-03, 1.99330557e-01],
                [7.83362692e-01, 9.29690250e-01],
                [9.50030444e-01, 8.75885766e-01],
                [3.37277759e-02, 9.24576754e-02]])
```

18.3.4 Stacking Two Arrays

The NumPy `hstack()` and `vstack()` functions are similar to the merging function in pandas. In this case, two arrays are stacked together as shown in this section.

`np.vstack()` stacks arrays in sequence vertically (row wise). This is equivalent to concatenation along the first axis after 1-D arrays of shape(N,) have been reshaped to (1,N). This is more useful for pixel-data height (first axis), width (second axis), and r/g/b channels (third axis). The functions `concatenate`, `stack`, and `block` provide more general stacking and concatenation operations. (Reference: `www.numpy.org`)

`np.hstack()` stacks arrays in sequence horizontally (column wise). This is equivalent to concatenation along the second axis, except for 1-D arrays where it concatenates along the first axis. This is more useful in the case of pixel data (computer vision, image data) with height data on the first axis, width (second axis), and r/g/b channels (third axis). (Reference: `www.numpy.org`)

Here is the input:

```
l = np.vstack((b,c))
m = np.hstack((b,c))
l
m
```

Here is the output:

Stacking together two different array

```
In [171]:  b.shape
  Out[171]: (2, 3)

In [172]:  c.shape
  Out[172]: (2, 3)

In [176]:  b
  Out[176]: array([[ 6,  7,  8],
                   [ 9, 10, 11]])

In [177]:  c
  Out[177]: array([[ 1.2,  3.4,  5.2],
                   [ 7. ,  8. , 10. ]])

In [174]:  l = np.vstack((b,c))

In [178]:  l
  Out[178]: array([[ 6. ,  7. ,  8. ],
                   [ 9. , 10. , 11. ],
                   [ 1.2,  3.4,  5.2],
                   [ 7. ,  8. , 10. ]])

In [180]:  m = np.hstack((b,c))

In [181]:  m
  Out[181]: array([[ 6. ,  7. ,  8. ,  1.2,  3.4,  5.2],
                   [ 9. , 10. , 11. ,  7. ,  8. , 10. ]])
```

18.4 Chapter Summary

In this chapter we discussed the fundamentals of Python for performing business analytics, data mining, and machine learning using pandas and NumPy.

We explored data manipulation, data types, missing values, data slicing and dicing, as well as data visualization.

We also discussed the `apply()` looping functions, statistical analysis on dataframes, and NumPy arrays.

We have only covered the parts of pandas and NumPy required for basic analytics problems that arise. Therefore, this chapter should be used only as a quick-reference guide. If you want to learn more complex functions and Python programming, you may refer to a Python book or the Python, pandas, or NumPy documentation.

References

1. BAESENS, BART. (2014). Analytics in a Big Data World, The Essential Guide to Data Science and Its Applications. Wiley India Pvt. Ltd.
2. MAYER-SCHONBERGER, VIKTOR & CUKIER KENNETH. (2013). Big Data, A Revolution That Will Transform How We Live, Work and Think. John Murray (Publishers), Great Britain
3. LINDSTROM, MARTIN. (2016). Small Data - The Tiny Clues That Uncover Huge Trends. Hodder & Stoughton, Great Britain
4. FREEDMAN, DAVID; PISANI, ROBERT & PURVES, ROGER. (2013). Statistics. Viva Books Private Limited, New Delhi
5. LEVINE, DAVID.M. (2011). Statistics for SIX SIGMA Green Belts. Dorling Kindersley (India) Pvt. Ltd., Noida, India
6. DONNELLY, JR. ROBERT.A. (2007). The Complete Idiot's Guide to Statistics, 2/e. Penguin Group (USA) Inc., New York 10014, USA
7. TEETOR, PAUL. (2014). R Cookbook. Shroff Publishers and Distributors Pvt. Ltd., Navi Mumbai
8. WITTEN, IAN.H.; FRANK, EIBE & HALL, MARK.A. (2014). Data Mining, 3/e - Practical Machine Learning Tools and Techniques. Morgan Kaufmann Publishers, Burlington, MA 01803, USA
9. HARRINGTON, PETER. (2015). Machine Learning in Action. Dreamtech Press, New Delhi
10. ZUMEL, NINA & MOUNT, JOHN. (2014). Practical Data Science with R. Dreamtech Press, New Delhi
11. KABACOFF, ROBERT.I. (2015). R In Action - Data analysis and graphics with R. Dreamtech Press, New Delhi

U. R. Hodeghatta and U. Nayak, *Practical Business Analytics Using R and Python*,
https://doi.org/10.1007/978-1-4842-8754-5

12. USUELLI, MICHELE. (2014). R Machine Learning Essentials. Packt Publishing

13. BALI, RAGHAV & SARKAR, DIPANJAN. (2016). R Machine Learning By Example. Packt Publishing

14. DAVID, CHIU & YU-WEI. (2015). Machine Learning with R Cookbook. Packt Publishing

15. LANTZ, BRETT. (2015). Machine Learning with R, 2/e. Packt Publishing

16. Data Mining - Concepts and Techniques By Jiawei Han, Micheline Kamber and Jian Pei, 3e, Morgan Kaufmann

17. S. Agarwal, R. Agrawal, P. M. Deshpande, A. Gupta, J. F. Naughton, R. Ramakrishnan, and S. Sarawagi. On the computation of multidimensional aggregates. VLDB'96

18. D. Agrawal, A. E. Abbadi, A. Singh, and T. Yurek. Efficient view maintenance in data warehouses. SIGMOD'97

19. R. Agrawal, A. Gupta, and S. Sarawagi. Modeling multidimensional databases. ICDE'97

20. S. Chaudhuri and U. Dayal. An overview of data warehousing and OLAP technology. ACM SIGMOD Record, 26:65–74, 1997

21. E. F. Codd, S. B. Codd, and C. T. Salley. Beyond decision support. Computer World, 27, July 1993

22. J. Gray, et al. Data cube: A relational aggregation operator generalizing group-by, cross-tab and sub-totals. Data Mining and Knowledge Discovery, 1:29-54, 1997

23. Swift, Ronald S. (2001) Accelerating Customer Relationships Using CRM and Relationship Technologies, Prentice Hall

24. Berry, M. J. A., Linoff, G. S. (2004) Data Mining Techniques. Wiley Publishing

25. Ertek, G. Visual Data Mining with Pareto Squares for Customer Relationship Management (CRM) (working paper, Sabancı University, Istanbul, Turkey)

26. Ertek, G., Demiriz, A. A framework for visualizing association mining results (accepted for LNCS)

27. Kumar, V., Reinartz, W. J. (2006) Customer Relationship Management, A Databased Approach. John Wiley & Sons Inc.

28. Spence, R. (2001) Information Visualization. ACM Press

29. Dyche, Jill, The CRM Guide to Customer Relationship Management, Addison-Wesley, Boston, 2002

30. Gordon, Ian. "Best Practices: Customer Relationship Management" Ivey Business Journal Online, 2002, pp. 1–6

31. Data Mining for Business Intelligence: Concepts, Techniques, and Applications in Microsoft Office Excel with XLMiner [Hardcover] By Galit Shmueli (Author), Nitin R. Patel (Author), Peter C. Bruce (Author)

32. A. Gupta and I. S. Mumick. Materialized Views: Techniques, Implementations, and Applications. MIT Press, 1999

33. J. Han. Towards on-line analytical mining in large databases. ACM SIGMOD Record, 27:97–107, 1998

34. V. Harinarayan, A. Rajaraman, and J. D. Ullman. Implementing data cubes efficiently. SIGMOD'96

35. C. Imhoff, N. Galemmo, and J. G. Geiger. Mastering Data Warehouse Design: Relational and Dimensional Techniques. John Wiley, 2003

36. W. H. Inmon. Building the Data Warehouse. John Wiley, 1996

37. R. Kimball and M. Ross. The Data Warehouse Toolkit: The Complete Guide to Dimensional Modeling. 2 ed. John Wiley, 2002

38. P. O'Neil and D. Quass. Improved query performance with variant indexes. SIGMOD'97

39. A. Shoshani. OLAP and statistical databases: Similarities and differences. PODS'00

40. S. Sarawagi and M. Stonebraker. Efficient organization of large multidimensional arrays. ICDE'94

41. E. Thomsen. OLAP Solutions: Building Multidimensional Information Systems. John Wiley, 1997

42. P. Valduriez. Join indices. ACM Trans. Database Systems, 12:218–246, 1987

43. J. Widom. Research problems in data warehousing. CIKM'95

44. Building Data Mining Applications for CRM, Alex Berson, Stephen Smith and Kurt Thearling (McGraw Hill, 2000)

45. Building Data Mining Applications for CRM, Alex Berson, Stephen Smith, Kurt Thearling (McGraw Hill, 2000)

46. Introduction to Data Mining, Pang-Ning, Michael Steinbach, Vipin Kumar, 2006 Pearson Addison-Wesley

47. Data Mining: Concepts and Techniques, Jiawei Han and Micheline Kamber, 2000, Morgan Kaufmann Publishers

48. Data Mining In Excel, Galit Shmueli Nitin R. Patel Peter C. Bruce, 2005

49. Principles of Data Mining by David Hand, Heikki Mannila, and Padhraic Smyth ISBN: 026208290x The MIT Press © 2001 (546 pages)

50. Creswell, J. W. (2013). Research design: Qualitative, quantitative, and mixed methods approaches. Sage Publications, Incorporated

51. Advance Data Mining Techniques, Olson, D.L, Delen, D, 2008, Springer

52. Phyu, Nu Thair, "Survey of Classification Techniques in Data Mining," Proceedings of the International MultiConference of Engineers and Computer Scientists 2009 Vol I IMECS 2009, March 18–20, 2009, Hong Kong

53. Myatt, J. Glenn, "Making Sense of Data – A practical Guide to Exploratory Data Analysis and Data Mining," 2007, WILEY-INTERSCIENCE A JOHN WILEY & SONS, INC., PUBLICATION

54. Fawcett, Tom, "An Introduction to ROC analysis," Pattern Recognition Letters 27 (2006) 861–874

55. Sayad, Saeed. "An Introduction to Data Mining," Self-Help Publishers (January 5, 2011)

56. Delmater, Rhonda, and Monte Hancock. "Data mining explained." (2001)

57. Alper, Theodore M. "A classification of all order-preserving homeomorphism groups of the reals that satisfy finite uniqueness." Journal of mathematical psychology 31.2 (1987): 135–154

58. Narens, Louis. "Abstract measurement theory." (1985)

59. Luce, R. Duncan, and John W. Tukey. "Simultaneous conjoint measurement: A new type of fundamental measurement." Journal of mathematical psychology 1.1 (1964): 1–27

60. Provost, Foster J., Tom Fawcett, and Ron Kohavi. "The case against accuracy estimation for comparing induction algorithms." ICML. Vol. 98. 1998

61. Hanley, James A., and Barbara J. McNeil. "The meaning and use of the area under a receiver operating characteristic (ROC) curve." Radiology 143.1 (1982): 29–36

62. Ducker, Sophie Charlotte, W. T. Williams, and G. N. Lance. "Numerical classification of the Pacific forms of Chlorodesmis (Chlorophyta)." Australian Journal of Botany 13.3 (1965): 489-499

63. Kaufman, Leonard, and Peter J. Rousseeuw. "Partitioning around medoids (program pam)." Finding groups in data: an introduction to cluster analysis (1990): 68–125

64. Minsky, M., & Papert, S. (1969). Perceptrons. M.I.T. Press

65. Activation Functions in Artificial Neural Networks: A Systematic Overview Johannes Lederer Department of Mathematics Ruhr-University Bochum, Germany johannes.lederer@rub.de January 26, 2021

66. ACTIVATION FUNCTIONS IN NEURAL NETWORKS, Siddharth Sharma, Simone Sharma, International Journal of Engineering Applied Sciences and Technology, 2020 Vol. 4, Issue 12, ISSN No. 2455-2143, Pages 310-316 Published Online April 2020 in IJEAST (http://www.ijeast.com)

67. Kingma, Diederik, and Jimmy Ba. "Adam: A method for stochastic optimization." arXiv preprint arXiv:1412.6980 (2014)

68. Quinlan, J. R. (1986). Induction of decision trees. Machine learning, 1(1), 81–106

69. Gower J. C. A general coefficient of similarity and some of its properties // Biometrics, 1971, 27, 857–872

70. Podani, J. Extending Gower's general coefficient of similarity to ordinal characters // Taxon, 1999, 48, 331–340

71. Forgy, E. W. (1965). Cluster analysis of multivariate data: efficiency vs interpretability of classifications.Biometrics, 21, 768–769

72. Hartigan, J. A. and Wong, M. A. (1979). Algorithm AS 136: A K-means clustering algorithm. Applied Statistics, 28, 100–108. doi: 10.2307/2346830

73. Lloyd, S. P. (1957, 1982). Least squares quantization in PCM. Technical Note, Bell Laboratories. Published in 1982 in IEEE Transactions on Information Theory, 28, 128–137

74. MacQueen, J. (1967). Some methods for classification and analysis of multivariate observations. InProceedings of the Fifth Berkeley Symposium on Mathematical Statistics and Probability, eds L. M. Le Cam & J. Neyman, 1, pp. 281–297. Berkeley, CA: University of California Press

75. Becker, R. A., Chambers, J. M. and Wilks, A. R. (1988). The New S Language. Wadsworth & Brooks/Cole. (S version.)

76. Everitt, B. (1974). Cluster Analysis. London: Heinemann Educ. Books

77. Hartigan, J.A. (1975). Clustering Algorithms. New York: Wiley

78. Sneath, P. H. A. and R. R. Sokal (1973). Numerical Taxonomy. San Francisco: Freeman

79. Anderberg, M. R. (1973). Cluster Analysis for Applications. Academic Press: New York

80. Gordon, A. D. (1999). Classification. Second Edition. London: Chapman and Hall / CRC

81. Murtagh, F. (1985). "Multidimensional Clustering Algorithms," in COMPSTAT Lectures 4. Wuerzburg: Physica-Verlag (for algorithmic details of algorithms used)

82. McQuitty, L.L. (1966). Similarity Analysis by Reciprocal Pairs for Discrete and Continuous Data.Educational and Psychological Measurement, 26, 825–831. doi: 10.1177/001316446602600402

83. Legendre, P. and L. Legendre (2012). Numerical Ecology, 3rd English ed. Amsterdam: Elsevier Science BV

84. Murtagh, Fionn and Legendre, Pierre (2014). Ward's hierarchical agglomerative clustering method: which algorithms implement Ward's criterion? Journal of Classification, 31, 274–295. doi:10.1007/s00357-014-9161-z

85. Tomas Mikolov, Kai Chen, Greg Corrado, and Jeffrey Dean. Efficient estimation of word representationsin vector space. ICLR Workshop, 2013

86. Miller, George A. "WordNet: A Lexical Database for English." Communications of the ACM 38.11 (1995): 39–41

87. The Stanford Natural Language Processing Group. Stanford TokensRegex, (software). Last accessed June 15, 2020

88. Jurafsky, Dan and James H. Martin. Speech and Language Processing, Third Edition (Draft), 2018

89. Mikolov, Tomas, et al. "Distributed representations of words and phrases and their compositionality." Advances in neural information processing systems. [2013]

90. [Online] https://cran.r-project.org/

91. [Online] www.r-project.org/

92. https://www.mordorintelligence.com/industry-reports/chatbot-market)

93. WordNet | A Lexical Database for English (princeton.edu)

94. https://wordnet.princeton.edu/

95. Google AI Blog: All Our N-gram are Belong to You (googleblog.com)

96. SRILM - The SRI. Language Modeling Toolkit - STAR Laboratory: SRI. Language Modeling Toolkit

97. https://cran.r-project.org/web/packages/rpart/rpart.pdf

98. https://www.rdocumentation.org/packages/rpart/versions/4.1.16

99. API Reference — scikit-learn 1.1.1 documentation; https://scikit-learn.org/stable/modules/classes.html#module-sklearn.preprocessing

100. https://cran.r-project.org/web/packages/rpart/rpart.pdf.

101. https://www.rdocumentation.org/packages/rpart/versions/4.1.16

102. API Reference — scikit-learn 1.1.1 documentation; https://scikit-learn.org/stable/modules/classes.html#module-sklearn.preprocessing

103. OLAP council. MDAPI specification version 2.0. In http://www.olapcouncil.org/research/apily.htm, 1998

104. Microsoft. OLEDB for OLAP programmer's reference version 1.0. In http://www.microsoft.com/data/oledb/olap, 1998

105. Kurt Thearling. Data Mining. http://www.thearling.com, kurt@thearling.com

106. http://scikit-learn.org/stable/modules/generated/sklearn.tree.DecisionTreeClassifier.html

107. http://paginas.fe.up.pt/~ec/files_1011/week%2008%20-%20Decision%20Trees.pdf

108. http://www.quora.com/Machine-Learning/Are-gini-index-entropy-or-classification-error-measures-causing-any-difference-on-Decision-Tree-classification

109. http://www.quora.com/Machine-Learning/Are-gini-index-entropy-or-classification-error-measures-causing-any-difference-on-Decision-Tree-classification

110. https://rapid-i.com/rapidforum/index.php?topic=3060.0

111. http://stats.stackexchange.com/questions/19639/which-is-a-better-cost-function-for-a-random-forest-tree-gini-index-or-entropy

112. COMPUTERWORLD FROM IDG. (2016). 8 big trends in big data analysis. [Online] Available from: http://www.computerworld.com/article/2690856/big-data/8-big-trends-in-big-data-analytics.html

113. WELLESLEY INFORMATION SERVICES, MA 02026, USA. (2016). Big Data Analytics Predictions for 2016. Available from: http://data-informed.com/big-data-analytics-predictions-2016/

114. COMPUTERWORLD FROM IDG. (2016). 11 Market Trends in Advanced Analytics. [Online] Available from: http://www.computerworld.com/article/2489750/it-management/11-market-trends-in-advanced-analytics.html#tk.drr_mlt

115. WELLESLEY INFORMATION SERVICES, MA 02026, USA. (2016). 5 Big Trends to Watch in 2016. [Online] Available from: http://data-informed.com/5-big-data-trends-watch-2016/

116. ZHANG, NANCY.R. Ridge Regression, LARS, Logistic Regression. [Online] Available from: http://statweb.stanford.edu/~nzhang/203_web/lecture12_2010.pdf

117. QIAN, JUNYANG & HASTIE, TRAVOR. (2014). Glmnet Vignette. [Online] Available from: http://web.stanford.edu/~hastie/glmnet/glmnet_alpha.html

Dataset CITATION

Justifying recommendations using distantly-labeled reviews and fine-grained aspects, Jianmo Ni, Jiacheng Li, Julian McAuley, Empirical Methods in Natural Language Processing (EMNLP), 2019

Index

A

U. R. Hodeghatta and U. Nayak, *Practical Business Analytics Using R and Python*,
https://doi.org/10.1007/978-1-4842-8754-5

B

C

D

E

F

I

J

K

L

M

N

O

P

Q

R

S

T

U

V

W, X, Y

Z

Printed in the United States
by Baker & Taylor Publisher Services